CURRENT DIRECTIONS IN INSULIN-LIKE GROWTH FACTOR RESEARCH

ADVANCES IN EXPERIMENTAL MEDICINE AND BIOLOGY

CURRENT DIRECTIONS IN INSULIN-LIKE GROWTH FACTOR RESEARCH

Edited by

Derek LeRoith

National Institutes of Health
Bethesda, Maryland

and

Mohan K. Raizada

University of Florida
Gainesville, Florida

SPRINGER SCIENCE+BUSINESS MEDIA, LLC

Library of Congress Cataloging-in-Publication Data

Current directions in insulin-like growth factor research / edited by
Derek LeRoith and Mohan K. Raizada.
    p.   cm. -- (Advances in experimental medicine and biology ; v.
343)
    "Proceedings of the Fourth International Symposium on Insulin,
IGFs, and Their Receptors, held April 20-23, 1993, in Woods Hole,
Massachusetts"--Copr. p.
    Includes bibliographical references and index.
    ISBN 978-1-4613-6301-9   ISBN 978-1-4615-2988-0 (eBook)
    DOI 10.1007/978-1-4615-2988-0
    1. Somatomedin--Congresses. 2. Insulin-like growth factor-binding
proteins--Congresses.   I. LeRoith, Derek, 1945-   . II. Raizada,
Mohan K. III. International Symposium on Insulin, IGFs, and their
Receptors (4th : 1993 : Woods Hole, Mass.) IV. Series.
QP552.S65C87 1993
612'.015756--dc20   93-46649
                                                            CIP

Proceedings of the Fourth International Symposium on Insulin, IGFs, and Their Receptors, held April
20–23, 1993, in Woods Hole, Massachusetts

ISBN 978-1-4613-6301-9

© 1993 Springer Science+Business Media New York
Originally published by Plenum Press, New York in 1993
Softcover reprint of the hardcover 1st edition 1993

PREFACE

The study of the insulin-like growth factor (IGF) family has become an exciting area of investigation. Initially, this family consisted of ligands (insulin, IGF-I and IGF-II) and receptors (the insulin receptor, the type I or IGF-I receptor and the type II or IGF-II/M-6-P receptor). Subsequently, it was discovered that six specific binding proteins (IGFBPs 1-6) play a major role in the actions of this growth factor family. In addition, there are now more potential receptors when one considers the possible roles of the insulin-receptor related receptor (IRR) and hybrid receptor dimers composed of insulin and IGF-I receptor (half-receptors).

Another important aspect of this area of research is the realization that the IGFs are not only essential for normal growth and development but, in addition play an important role in the normal specialized function(s) of all tissues of the body, including the nervous system, skeleton, reproductive system, kidney, and the immune system, to name but a few.

The development of recombinant human IGF-I for clinical testing has been a major breakthrough for investigators. Potential uses include wound healing, reversal of catabolic states, diabetes, bone remodeling, recovery from acute renal failure and many others. Further investigations in this area will determine both its use and its potential hazards.

These proceedings follow the very successful meeting of the IVth International Symposium on Insulin and Insulin-like Growth Factors held in Woods Hole, Massachusetts, in April, 1993. The meeting was held under the auspices of the American Diabetes Association and supported by educational grants from Ciba-Geigy, Genentech, Kabi Pharmaceuticals, Eli Lilly, and the Upjohn Company. This compilation of reviews should give the reader insights into the current research in the field as well as future directions.

Derek Le Roith, M.D., Ph.D.
Mohan K. Raizada, Ph.D.

CONTENTS

STRUCTURE, EXPRESSION, AND REGULATION OF THE IGF-I GENE

Martin L. Adamo,[1] Stefan Neuenschwander,[2] Derek LeRoith,[2]
and Charles T. Roberts, Jr.[2]

[1]Department of Biochemistry
University of Texas Health Science Center
San Antonio, TX 78284

[2]Diabetes Branch
Building 10, Room 8S-239, NIDDK
National Institutes of Health
Bethesda, MD 20892

INTRODUCTION

The insulin-like growth factors, IGF-I and IGF-II, are important regulators of growth, differentiation and maintenance of differentiated function (Daughaday and Rotwein, 1989), primarily through activation of the IGF-I receptor, a transmembrane tyrosine kinase (Ullrich et al., 1986). In addition to these typical growth factor-like effects, the IGFs, by virtue of their homology to insulin, and the similarity of the insulin and IGF-I receptors, can also elicit insulin-like metabolic effects. The latter effects are presumably the result of "heterologous" activation of the insulin receptor by the IGFs. As described elsewhere in this volume, the bioavailability and action of the IGFs are modulated by a family of IGF-binding proteins.

The liver is a major source of the IGFs, and, in adult rodents, hepatic biosynthesis is sufficient to account for the majority of circulating IGF-I (Schwander et al., 1983). There is a great deal of evidence, however, that the IGFs are produced by many extrahepatic tissues as well at various stages of development, as assayed by levels of either protein or mRNA. Given their widespread distribution, both temporally and spatially, and their multiple functions, it could be predicted that the biosynthesis of these important regulatory peptides would be carefully controlled, potentially at multiple levels. Below, we will review the structure of the IGF-I gene and the relationship of this structure to the regulation of IGF-I gene expression. An accompanying chapter will deal with similar issues with respect to the IGF-II gene.

Current Directions in Insulin-Like Growth Factor Research,
Edited by D. LeRoith and M.K. Raizada, Plenum Press, New York, 1994

1

IGF-I GENE STRUCTURE

To date, the structure of the IGF-I genes have been determined for a number of species, including man (Rotwein et al., 1986), rat (Shimatsu and Rotwein, 1987), sheep (Dickson et al., 1991), chicken (Kajimoto and Rotwein, 1991), and salmon (Kavsan et al., 1993), and cDNAs have been isolated from these species and others, including mouse (Bell et al., 1986), pig (Müller and Brem, 1990), cow (Fotsis et al., 1990), guinea pig (Bell et al., 1990), and *Xenopus* (Shuldiner et al., 1990; Kajimoto and Rotwein, 1990). Additionally, a cDNA encoding an IGF-like molecule has been characterized from hagfish (Nagamatsu et al., 1991). As shown in Figure 1, the "minimal" IGF-I gene (i.e., those components common to all IGF-I genes so far analyzed) consists of four exons. The first exon encodes the 5'-untranslated region (UTR) and, potentially, the amino terminus of the signal peptide. The next exon encodes the remainder of the signal peptide and the amino-terminal portion of the B domain of the mature IGF-I peptide. The third exon encodes the remainder of the B domain, the C, A, and D domains (which complete the mature peptide) and the first part of the carboxy-terminal E peptide moiety of the IGF-I prohormone. The last exon encodes the remainder of the E peptide, which is generally cleaved prior to secretion of the mature hormone, and the 3'-UTR.

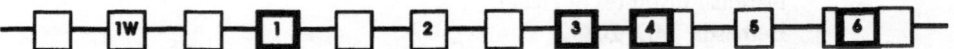

Figure 1. Schematic representation of IGF-I gene structure. Exons found in all IGF-I genes are in bold. Additional boxes represent the relative positions of potential extra exons and exonic sequences present in various IGF-I genes as described in the text. Sizes of exons and introns are not to scale. Exon nomenclature is that of mammalian IGF-I genes (Holthuizen et al., 1991).

It is instructive to analyze the variations of IGF-I gene structure seen in different species based upon the available genomic and cDNA clones characterized so far. One of the first changes evident is an apparent increase in the size of introns. Whereas the characterized chum salmon IGF-I gene (Kavsan et al., 1993) is less than 20 kilobases (kb) in length, the chicken IGF-I gene is ~50 kb in length (Kajimoto and Rotwein, 1991), and the rat and human genes are from 80 to 100 kb in length (Rotwein et al., 1986; Shimatsu and Rotwein, 1987).

Another complication of IGF-I gene structure appears to be the use of multiple polyadenylation sites in the 3'-UTR encoded by the last exon, resulting in IGF-I mRNAs which differ sufficiently in size to be resolved as discrete molecular weight species on Northern blots (Lund et al., 1989). I.e., salmon IGF-I mRNAs appear to be a single species, whereas *Xenopus*, chicken, and mammalian IGF-I mRNAs (at least in liver and in some extrahepatic tissues) occur as multiple species which differ primarily due to the length of the 3'-UTR. As described below, the use of more 3' polyadenylation sites results in the inclusion of potentially destabilizing AUUUA-like sequence elements in the human and rat higher-molecular-weight IGF-I mRNAs.

A third aspect of IGF-I gene structure involves the diversification of mechanisms to produce multiple E peptide sequences. In the chicken and sheep IGF-I genes, the E peptide is encoded by the 3' end of the exon which also encodes the carboxy terminus of the mature peptide and the 5' end of the last exon, which also encodes the 3'-UTR. In the salmon and mammalian IGF genes, however, there are a number of types of alternate splicing which produce IGF-I mRNAs which encode multiple E peptides and, therefore, multiple forms of the IGF-I prohormone. The salmon genes, in particular, are an extreme example of this type of diversity, in that they contain an additional 36-base pair (bp) "miniexon" found between the third and fourth "canonical" exons shown in Figure 1, as well as alternate splice donor and acceptor sites whose usage produces extended versions of the third (+81 bases (b)) and fourth canonical (+36 b) exons, respectively (Cao et al., 1989; Wallis and Devlin, 1993; Kavsan et al., 1993). Alternate splicing of the mini-exon and the use of multiple acceptor and donor splice sites could theoretically produce as many as eight different mRNAs, all encoding different E peptide sequences. A number of the possible sequence combinations have, in fact, been found in salmon IGF-I cDNAs. This analysis of the situation in salmon, as in *Xenopus*, is complicated by the presence of duplicate, non-allelic IGF-I genes.

In mouse and rat, divergent E peptide sequences arise from the alternate splicing of a 52-b miniexon (exon 5) sequence whose inclusion in the mRNA adds new amino acids and, because of its size, changes the translational reading frame of the E-peptide coding region of the last exon (Bell et al., 1986; Lowe et al., 1988). In the human gene, the splice donor site at the 3' end of this miniexon has apparently been mutated sufficiently so that it is no longer functional, resulting in a longer version of exon 5 which encodes an extended carboxy terminus of the E peptide and a unique 3'-UTR (Rotwein et al., 1986). The preceding exon is, therefore, spliced in a mutually exclusive manner to this exon or to the last exon. Interestingly, the sheep IGF-I gene has not yet been found to contain the equivalent of exon 5, nor have any cDNAs been characterized to date which would predict the existence of any other type of alternate splicing in this region.

A final aspect of IGF-I gene structure is the presence in the chicken and mammalian IGF-I genes of multiple leader exons. The simplest example of this to date is represented by the rat and human genes which both contain an additional leader exon (exon 2) a short distance (~1-2 kilobases (kb)) downstream of the first exon. The existence of this exon was first predicted on the basis of cDNA sequences (Roberts et al., 1987; Tobin et al., 1990) and subsequently corroborated by genomic sequence information (Bucci et al., 1989). This exon encodes a unique 5'-UTR and, due to the presence of an AUG translation initiation codon that is in-frame with an AUG in the following exon as well as with the remainder of the preproIGF-I open reading frame (ORF), may also encode an amino-terminally extended version of the signal peptide. Thus, the presence of in-frame AUG codons in both leader exons and in the exon encoding the beginning of the mature hormone predicts the existence of as many as three signal peptides.

The presence of this extra leader exon has been demonstrated in the sheep gene directly by analysis of both cDNA (Wong et al., 1989) and genomic (Dickson et al., 1991) clones and inferred for the mouse IGF-I gene on the basis of cDNA sequences (Bell et al., 1986). In both the ovine and murine genes, this exon 2 sequence contains an upstream AUG codon in the same reading frame as the proIGF-I ORF, so that the potential for both divergent signal peptide and 5'-UTR sequences exists in these species also. The latter two genes contain, in addition to leader exons 1 and 2, an additional leader exon whose sequence is spliced to the exon encoding the amino terminus of the mature IGF-I peptide (exon 3) in fully processed IGF-I mRNAs. This additional exon in the sheep gene (termed 1W) has been mapped to ~1 kb upstream of the canonical first exon (exon 1) and could potentially encode a 31-amino acid signal peptide.

In the case of the mouse gene, cDNAs in which sequences corresponding to exon 2 and to another putative leader exon are spliced to the exon encoding the first part of the mature peptide (exon 3) have been isolated (Bell et al., 1986). To date, no exon 1-derived cDNAs have been described, but the results of solution hybridization/RNase protection assays suggest the presence of a major exon 1-derived IGF-I mRNA species in mouse liver (C.T.R.; unpublished observations).

The existence of multiple leader exons in the chicken IGF-I gene is suggested by the characterization of genomic (Kajimoto and Rotwein, 1991) and cDNA (Kajimoto and Rotwein, 1989; Fawcett and Bulfield, 1990) clones. The canonical first exon sequence (exon 1) has been found in cDNA clones and mapped in genomic clones. Two other potential leader exons are suggested by the sequences of cDNAs isolated from chicken liver. One sequence is spliced to the second canonical exon (exon 3 in mammalian genes) and the second appears to be spliced between exon 1 and exon 3 sequences in one cDNA clone. The exon encoding this sequence would, thus, have to lie between exons 1 and 3. Neither sequence, however, corresponds to the exon 2 sequence characterized in the human, rat, mouse, and ovine genes.

A rigorous assessment of the importance of these putative additional leader sequences inferred to exist in the mouse and chicken genes will require the eventual characterization of genomic clones corresponding to these sequences.

To date, no evidence has been adduced for the presence of leader exons other than exon 1 in the *Xenopus* and salmon genes, and exon 2-like sequences specifically do not occur in the salmon gene (Kavsan et al., 1993). The existence of multiple leader exons in these IGF-I genes can not be ruled out at this point, however.

Based upon the preceding description it is clear that, while the basic structure of the IGF-I gene is relatively simple, every species exhibits some degree of additional complexity, with the mammalian genes being perhaps the most complicated. In the next two sections we will focus on the rat gene as a paradigm for mammalian IGF-I genes, and discuss its expression and regulation and their relationship to basic gene structure. Most, if not all, of the conclusions drawn from studies of the rat genes appear to be relevant to the human gene, these being the most extensively analyzed genes to date, and undoubtedly to other mammalian IGF-I genes as well.

EXPRESSION OF THE IGF-I GENE

Transcription of the rat IGF-I gene is initiated from distinct start sites in the two leader exons, exon 1 and exon 2. Transcription initiation in exon 1 is disperse, occurring from two major and two minor sites spread over several hundred bp. Transcription initiation in exon 2, on the other hand, occurs principally from a single cluster of sites (Adamo et al., 1991a). This situation appears to be the result of a lack of core promoter elements such as TATA and CAAT box motifs upstream of the exon 1 transcription start site and the presence of TATA and CAAT-like elements at appropriate positions upstream of the cluster of start sites in exon 2. The human (Jansen et al., 1991) and ovine (Pell et al., 1993) exhibit a qualitatively similar pattern of transcription initiation in these exons. An additional feature of the rat exon 1 sequence is a 186-b segment flanked by splice junctions which is not present in some rat IGF-I cDNAs (Shimatsu and Rotwein, 1988). The existence of this fully spliced IGF-I mRNA species was subsequently confirmed by solution hybridization/RNase protection assays (Foyt et al., 1991). Interestingly, this splicing event apparently does not occur in the human IGF-I gene, despite the apparent conservation of the pertinent splice junctions (J.S. Sussenbach, personal communication).

The available evidence suggests that transcription of both exon 1 and exon 2 is controlled by distinct promoter regions. Specifically, DNA sequences flanking exon 1 (Hall et al., 1992; Lowe and Teasdale, 1992; Adamo et al., 1993) and exon 2 (Adamo et al., 1993) of the rat IGF-I gene exhibit promoter activity in transient expression assays. Similar results have been obtained with sequences flanking exon 1 (Kim et al., 1991; Jansen et al., 1992) and exon 2 (Jansen et al., 1992) of the human IGF-I gene and exon 1 of the chicken gene (Kajimoto and Rotwein, 1991).

The presence of two leader exons and multiple transcription start sites and alternate splicing in the first exon results in the production of IGF-I mRNAs containing a variety of 5'-UTRs and (due to the presence of in-frame AUG translation initiation codons) signal peptide coding regions. Figure 2 illustrates the 5' ends of the principal mRNAs transcribed from the rat gene (the most upstream transcription start sites in exons 1 and 2 serve to define the 5' ends of these exons, but are not used to any appreciable extent (Adamo et al., 1991a, b).

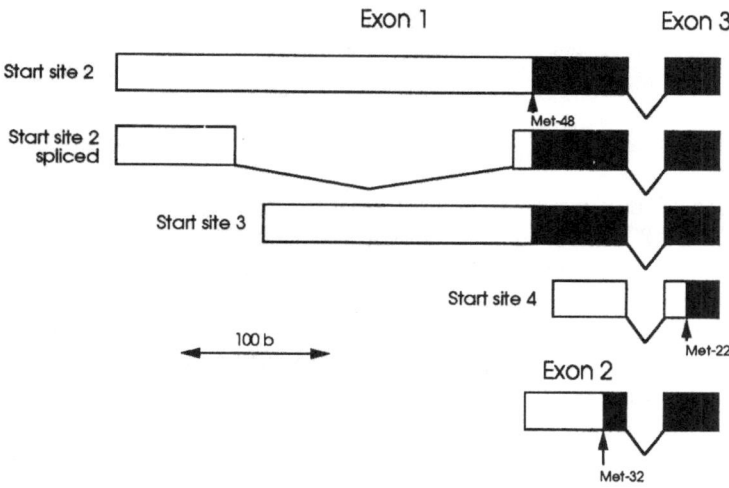

Figure 2. Schematic representation of major splicing variants in the 5' region of the rat IGF-I gene. 5'-UTR sequences are open boxes and closed boxes represent potential signal peptide coding regions initiated at the Met codons shown.

Rather extensive studies (Hoyt et al., 1988; Adamo et al., 1989; Shemer et al., 1992) have demonstrated that, in the rat, exon 1-derived transcripts predominate in every tissue expressing the IGF-I gene with exon 2-derived transcripts occurring at relatively high levels in liver, at low levels in some extrahepatic tissues such as kidney, testes, lung and stomach and being absent in many tissues (Figure 3). Less extensive, but qualitatively similar results have been obtained with human tissues (Jansen et al., 1991; Hernandez et al., 1992).

With respect to the various exon 1-derived mRNAs themselves, liver again appears to be the most diverse with respect to transcriptional complexity in this region, expressing all of the exon 1-derived IGF-I mRNA species (Adamo et al., 1991a, b), with full-length transcripts initiated at start site 2 and those initiated at start site 3 (see Figure 2) being the most abundant. While some extrahepatic tissues such as lung and kidney exhibit some of the heterogeneity in exon 1 transcription characteristic of liver, most exon 1 transcripts (and essentially all IGF-I transcription in most extrahepatic tissues) is initiated at start site 3 (Hall et al., 1992; Shemer et al., 1992).

5

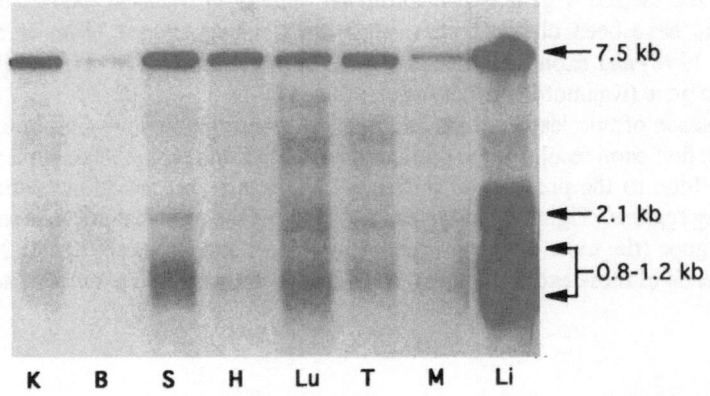

Figure 3. Relative proportions of exon 1 and exon-2 derived transcripts in various rat tissues. Liv., liver; K., kidney; St., stomach; L., lung; T., testes; M., muscle; Br., brain; H., heart.

As described in a preceding section, alternate splicing of exon 5 in the rat gene produces IGF-I mRNA species which encode two different E peptide sequences. IGF-I mRNAs which lack the exon 5 sequence and encode the type A E peptide sequence are generally ≥10 times more abundant than exon 5-containing mRNAs (which encode the type B E peptide sequence in rat tissue (Lowe et al., 1988). Limited observations suggest that this also holds true for the alternatively-spliced exon 5 variants of the human IGF-I gene (Hernandez et al., 1992).

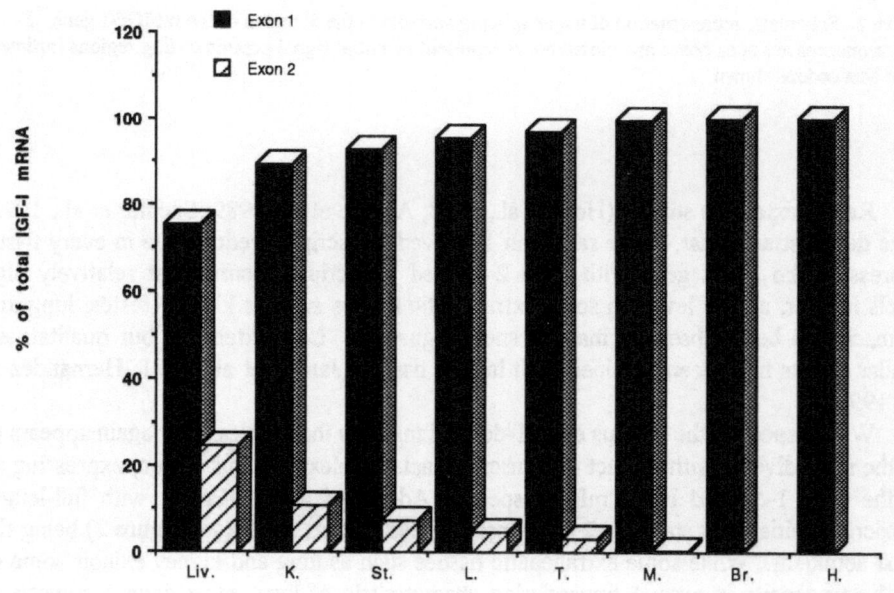

Figure 4. Northern blot analysis of total RNA from rat tissues with a rat IGF-I cDNA probe. K, kidney; B., brain; S., stomach; H., heart; Lu., lung; T., testes; M., muscle; Li., liver.

A final aspect of IGF-I gene expression to be considered is the use of alternative polyadenylation sites in exon 6. Use of polyadenylation sites that are proximal to the end of the preproIGF-I ORF generate the smaller-molecular-weight IGF-I mRNA species seen on Northern blots (~1 and 2 kb), whereas the use of a much more distal polyadenylation site produces the high-molecular-weight IGF-I mRNA species of ~7.5 kb. RNase H digestion studies have established that these differences in the length of the 3'-UTR are, in fact, responsible for most of the differences in size of IGF-I mRNAs resolved on Northern blots (Lund et al., 1989). The high-molecular-weight species represents ~30% of total IGF-I mRNA in liver, but constitutes the majority (and in many tissues all) of the IGF-I mRNA present. This is illustrated in Figure 4 for rat tissues.

REGULATION OF IGF-I GENE EXPRESSION

The complex architecture of the IGF-I gene, particularly in mammals, suggests that the expression of this gene may be regulated at many levels. IGF-I mRNA levels are regulated by GH status, nutritional state, by specific hormones such as insulin, estradiol and glucocorticoids, and by as yet uncharacterized tissue and developmental stage-specific factors (Daughaday and Rotwein, 1989; Adamo et al., 1991c). Of these various stimuli, GH (Mathews et al., 1986; Doglio et al., 1987; Tollet et al., 19990; Johnson et al., 1991; Bichell et al., 1992), insulin, (Pao et al., 1992), estradiol (Ernst and Rodan, 1991) and developmental factors (Kikuchi et al., 1992) have been shown to regulate mammalian IGF-I gene expression at the transcriptional level. Recently, AP-1-mediated transcriptional regulation of the chicken IGF-I gene by TPA has been described (Kajimoto et al., 1993).

If the expression of the mammalian IGF-I genes is, in fact, regulated by distinct promoter regions flanking exons 1 and 2, the available evidence suggests that these promoters may contain both common and distinct regulatory elements. If one assumes that changes in RNA levels reflect, to some extent, changes in transcriptional activity, the observation that exon 1 and exon 2-derived mRNAs are affected coordinately by nutritional deprivation and experimental diabetes (Adamo et al., 1991b) suggests that the pertinent regulatory mechanisms are shared by both promoters. On the other hand, the differential expression of exon 1 and exon 2 transcripts as a function of tissue (Adamo et al., 1989; see Figure 3), development (Adamo et al., 1988, 1991b) and growth hormone status (Lowe et al., 1987; Saunders et al., 1991; Foyt et al., 1992; Pell et al., 1993) supports the concept of promoter-specific regulatory elements.

As described above, the existence of alternative leader exons and multiple transcription start sites in exon 1 result in a collection of IGF-I mRNAs with different 5'-UTRs and signal peptide-coding sequences (Figure 2). The differential association of these different mRNAs with polysomes suggest that they may be differentially translated (Foyt et al., 1991, 1992). Recent studies (Adamo et al., submitted) have identified a particular sequence motif in the 5'-UTR (Figure 5) which may be responsible for the differential translatability of exon 1-derived mRNAs. This sequence, which consists of an AUGUGA translation initiation-termination codon pair, may influence translation of the preproIGF-I ORF by requiring reinitiation of protein synthesis by scanning ribosomal subunits. This latter process may itself be controlled by regulation of eukaryotic initiation factor-2 (eIF-2) activity through phosphorylation (Hershey, 1989; Samuel, 1993).

Another post-transcriptional aspect of IGF-I gene regulation concerns the different 3'-UTRs that result from the use of alternate polyadenylation sites in exon 6. *In vivo* and *in vitro* studies (Hepler et al., 1990; Foyt et al., 1991; Thissen and Underwood, 1992) have suggested that he high-molecular-weight IGF-I mRNA species may be less stable than the low-molecular weight species. This may be correlated with the presence of potential A/U-rich destabilizing sequences in the exon 6-encoded 3'-UTR downstream of the proximal polyadenylation sites in the human (Steenbergh et al., 1991) and rat (Hoyt et al., 1992) genes (Figure 5). Thus, differential 3' end processing and polyadenylation site selection may influence mRNA stability and, therefore, IGF-I biosynthesis.

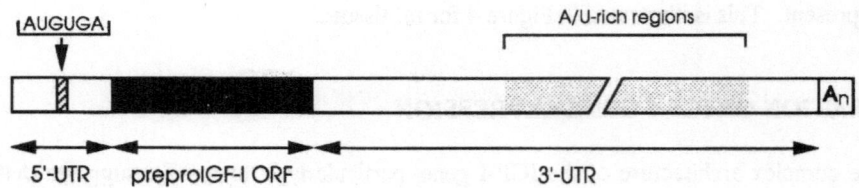

Figure 5. Schematic representation of the predominant IGF-I mRNA species in extrahepatic tissues.

Taken together, the results of the various solution hybridization and Northern blot analyses described in preceding sections suggest that a major hepatic IGF-I mRNA and the predominant extrahepatic IGF-I mRNA is an exon 1-derived transcript initiated at start site 3 which also contains an extended 3'-UTR (Figure 5). Given the relative levels of this particular IGF-I mRNA species, it is possible that the regulation of the translatability and stability of this transcript could be a major determinant of overall IGF-I biosynthesis.

Finally, there are two possible post-translational elements of IGF-I gene expression that may regulate IGF-I biosynthesis. The various IGF-I mRNA species predict the existence of as many as three signal peptides in mammalian IGF-I precursors, and these forms can be generated *in vitro* (Rotwein et al., 1987; Adamo et al., submitted). Should alternative forms of preproIGF-I be produced *in vivo*, they may be subject to differential processing, transport or degradation. Similarly, the two versions of the E peptide resulting from alternative splicing of exon 5 may also be subject to differential processing. This may be influenced by the potential glycosylation of one form of the E peptide, as has been demonstrated *in vitro* (Bach et al., 1990).

CONCLUSION

The structure of the IGF-I gene is complex and the specific aspects of this structure vary from species to species. The mammalian genes in particular are sufficiently complicated so as to accommodate multiple levels of regulation. While "classical" regulation at the level of transcription initiation controlled by various promoter elements undoubtedly contributes significantly to IGF-I gene expression and IGF-I biosynthesis, the role of other types of regulation is becoming more apparent. In particular, post-transcriptional control of mRNA translatability and stability may be as important as transcriptional control in many instances. Although less-characterized, post-translational aspects of IGF-I gene expression may also influence IGF-I biosynthesis. A complete understanding of the regulation of the IGF-I gene must consider all these levels of control, which are made possible by the architecture of the IGF-I gene itself.

REFERENCES

Adamo, M.L., Lowe, W.L., Jr., LeRoith, D., and Roberts, C.T., Jr., 1989, Insulin-like growth factor I messenger ribonucleic acids with alternative 5'-untranslated regions are differentially expressed during development of the rat, *Endocrinology*, 124:2737-2744.

Adamo, M.L., Ben-Hur, H., LeRoith, D., and Roberts, Jr., C.T., 1991a, Transcription initiation in the two leader exons of the rat IGF-I gene occurs from disperse versus localized sites, *Biochem. Biophys. Res. Commun.*, 176:887-893.

Adamo, M.L., Ben-Hur, H., Roberts, Jr., C.T., and LeRoith, D., 1991b, Regulation of start site usage in the two leader exons of the rat insulin-like growth factor I gene by development, fasting and diabetes, *Mol. Endocrinol.* 5:1677-1686.

Adamo, M.L., Bach, M.A., Roberts, C.T. Jr., and LeRoith ,D., 1991c, Regulation of insulin, IGF-I, and IGF-II gene expression. In: LeRoith D (Ed) *Insulin-like Growth Factors: Molecular and Cellular Aspects*. CRC Press, Boca Raton, pp 271-303.

Adamo, M.L., Lanau, F., Neuenschwander, S., Werner, H., LeRoith, D., and Roberts, Jr., C.T., 1993, Distinct promoters in the rat insulin-like growth factor-I (IGF-I) gene are active in CHO cells, *Endocrinology*, 132:935-937.

Bach., M.A, Roberts, Jr., C.T., Smith, E.P., and LeRoith, D., 1990, Alternative splicing produces messenger RNAs encoding insulin-like growth factor-I prohormones that are differentially glycosylated in vitro, *Mol. Endocrinol.*, 4:899-904.

Bell, G.I., Stempien, M.M., Fong, N.M., and Rall, L.B., 1986, Sequences of liver cDNAs encoding two different mouse insulin-like growth factor I precursors, *Nucleic Acids Res.*, 14:7873-7882.

Bell, G.I., Stempien, M.M., Fong, N.M., and Seino, S., 1990, Sequence of a cDNA encoding guinea pig IGF-I, *Nucleic Acids Res.*, 18:4275.

Bichell, D.P., Kikuchi, K., and Rotwein, P., 1992, Growth hormone rapidly activates insulin-like growth factor-I gene transcription *in vivo*, *Mol. Endocrinol.*, 6:1899-1908.

Bucci, C., Mallucci, P., Roberts, C.T. Jr., Frunzio, R., and Bruni, C.B., 1989, Nucleotide sequence of a genomic fragment of the rat IGF-I gene spanning an alternate 5' noncoding exon, *Nuc. Acids Res.*, 9:3596.

Cao, Q-P., Duguay, S.J., Plisetskaya, E., Steiner, D.F., and Chan, S.J., 1989, Nucleotide sequence and growth hormone-regulated expression of salmon insulin-like growth factor I mRNA, *Mol. Endocrinol.*, 3:2005-2010.

Dickson, M.C., Saunders, J.C., and Gilmour, R.S., 1991, The ovine insulin-like growth factor-I gene: characterization, expression and identification of a putative promoter, *J. Mol. Endocrinol.*, 6:17-31.

Doglio, A., Dani, C., Fredrikson, G., Grimaldi, P., and Ailhaud, G., 1987, Acute regulation of insulin-like growth factor-I gene expression by growth hormone during adipose cell differentiation, *EMBO J.*, 6:4011-4016.

Ernst, M., and Rodan, G.A., 1991, Estradiol regulation of insulin-like growth factor-I expression in osteoblastic cells: evidence for transcriptional control, *Mol. Endocrinol.*, 5:1081-1089.

Fawcett, D.H., and Bulfield, G., 1990, Molecular cloning, sequence analysis and expression of putative chicken insulin-like growth factor-I cDNAs, *J. Mol. Endocrinol.*, 4:201-211.

Fotsis, T., Murphy, C., and Gannon, F., 1989, Nucleotide sequence of the bovine insulin-like growth factor I (IGF-I) and its IGF-IA precursor, *Nucleic Acids Res.*, 3:676.

Foyt, H.L., LeRoith, D., and Roberts, Jr., C.T., 1991, Differential association of insulin-like growth factor I mRNA variants with polysomes in vivo, *J. Biol. Chem.*, 266:7300-7305.

Foyt, H.L., Lanau, F., Woloschak, M., LeRoith, D., and Roberts, Jr., C.T., 1992, Effect of growth hormone on levels of differentially processed insulin-like growth factor I mRNAs in total and polysomal mRNA populations, *Mol. Endocrinol.*, 6:1881-1888.

Hall, I.K., Kajimoto, Y., Bichell, D., Kim., S-J., James, P.L., Counts, D., Nixon, L.J., Tobin, G., and Rotwein, P., 1992, Functional analysis of the rat insulin-like growth factor I gene and identification of an IGF-I gene promoter, *DNA Cell Biol.*, 11:301-313.

Hepler, J.E., Van Wyk, J.J., and Lund, P.K., 1990, Different half-lives of insulin-like growth factor mRNAs that differ in length of 3'-untranslated sequence. *Endocrinology*, 127:1550-1552.

Hernandez, E.R., Hurwitz, A., Pellicer, A., Adashi, E.Y. LeRoith, D., and Roberts, Jr., C.T., (1992), Expression of the insulin-like growth factor gene family in the human ovary, *J. Clin. Endocrinol. and Metab.*, 74:419-425.

Hershey, J.W.B., 1989, Protein phosphorylation controls translation rates, *J. Biol. Chem.*, 264:20823-20826.

Holthuizen, E., LeRoith, D., Lund, P.K., Roberts, Jr., C.T., Rotwein, P., Spencer, E.M., and Sussenbach, J.S., 1991, Revised nomenclature for the insulin-like growth factor genes and transcripts, *in*: "Modern concepts of insulin-like growth factors," E.M. Spencer, ed., Elsevier, New York.

Hoyt, E.C., Hepler, J.E., Van Wyk, J.J., and Lund, P.K., 1992, Structural characterization of exon 6 of the rat IGF-I gene, *DNA Cell Biol.*, 11:433-442.

Jansen, E., Steenbergh, P.H., LeRoith, D., Roberts, Jr., C.T., and Sussenbach, J.S., 1991, Identification of multiple transcription start sites in the human insulin-like growth factor-I gene, *Mol. Cell. Endocrinol.*, 78:115-125.

Jansen, E., Steenbergh, P.H., van Schaik, F.M.A., and Sussenbach, J.S., 1992, The human IGF-I gene contains two cell type-specifically regulated promoters, *Biochem. Biophys. Res. Commun.*, 187:1219-1226.

Johnson, T.R., Rudin, S.D., Blossey, B.K., Ilan, J., and Ilan, J., 1991, Newly synthesized RNA: simultaneous measurement in intact cells of transcription rates and RNA stability of insulin-like growth factor I, actin, and albumin in growth hormone-stimulated hepatocytes, *Proc. Natl. Acad. Sci. USA*, 88:5287-5291.

Kajimoto, Y., and Rotwein, P., 1989, Structure and expression of a chicken insulin-like growth factor I precursor, *Mol. Endocrinol.*, 3:1907-1913.

Kajimoto, Y., and Rotwein, P., 1990, Evolution of insulin-like growth factor I (IGF-I): structure and expression of an IGF-I precursor from *Xenopus laevis*, *Mol. Endocrinol.*, 4:217-226.

Kajimoto, Y., and Rotwein, P., 1991, Structure of the chicken insulin-like growth factor I gene reveals conserved promoter elements, *J. Biol. Chem.*, 266:9724-9731.

Kajimoto, Y., Kawamori, R., Umayahara, Y., Iwama, N., Imano, E., Morishima, T., Yamasaki, Y., and Kamada, T., 1993, An AP-1 enhancer mediates TPA-induced transcriptional activation of the chicken insulin-like growth factor I gene, *Biochem. Biophys. Res. Commun.*, 190:767-773.

Kavsan, V.M., Koval, A.P., Grebenjuk, V.A., Chan, S.J., Steiner, D.F., Roberts, Jr., C.T., and LeRoith, D., 1993, Structure of the chum salmon insulin-like growth factor-I gene, *DNA Cell Biol.*, in press.

Kikuchi, K., Bichell, D.P., and Rotwein, P., 1992, Chromatin changes accompany the developmental activation of insulin-like growth factor I gene transcription, *J. Biol. Chem.*, 267:21505-21511.

Kim, S-W., Lajara, R., and Rotwein, P., 1991, Structure and function of a human insulin-like growth factor I gene promoter, *Mol. Endocrinol.*, 5:1964-1972.

Lowe, Jr., W.L., Roberts, Jr., C.T., Lasky, S.K., and LeRoith, D., 1987, Differential expression of alternative 5'-untranslated regions in mRNAs encoding rat insulin-like growth factor-I, *Proc. Natl. Acad. Sci. USA*, 84:8946-8950.

Lowe, Jr., W.L., Lasky, S.R., LeRoith, D., and Roberts, Jr., C.T., 1988, Distribution and regulation of rat insulin-like growth factor I mRNAs encoding alternative carboxy terminal E-peptides: evidence for differential processing and regulation in liver, *Mol. Endocrinol.*, 2:528-525.

Lowe, Jr., W.L., and Teasdale, R.M., 1992, Characterization of a rat insulin-like growth factor I gene promoter, *Biochem. Biophys. Res. Commun.*, 189:972-978.

Lund, P.K., Hoyt, E., and Van Wyk, J., 1989, The size heterogeneity of rat insulin-like growth factor I mRNAs is due primarily to differences in the length of 3'-untranslated sequence, *Mol. Endocrinol.*, 3:2054-2061.

Mathews, L.S., Norstedt, G., and Palmiter, R.D., 1986, Regulation of insulin-like growth factor I gene expression by growth hormone. *Proc. Natl. Acad. Sci. USA*, 83:9343-9347.

Müller, M., and Brem, G., 1990, Nucleotide sequence of porcine insulin-like growth factor I: 5' untranslated region, exons 1 and 2 and mRNA, *Nucleic Acids Res.*, 18:364.

Nagamatsu, S., Chan, S.J., Falkmer, S., and Steiner, D.F., 1991, Evolution of the insulin gene superfamily: sequences of a preproinsulin-like growth factor cDNA from the Atlantic hagfish, *J. Biol. Chem.*, 266:2397-2402.

Pao, C-I., Farmer, P.K., Begovic, S., Goldstein, S., Wu, G.-J., and Phillips, L.S., 1992, Expression of hepatic insulin-like growth factor-I and insulin-like growth factor-binding protein-I genes is transcriptionally regulated in streptozotocin-diabetic rats, *Mol. Endocrinol.*, 6:969-977.

Pell, J.M., Saunders, J.C., and Gilmour, R.S., 1993, Differential regulation of transcription initiation from insulin-like growth factor-I (IGF-I) leader exons and of tissue IGF-I expression in response to changed growth hormone and nutritional status in sheep, *Endocrinology* 132:1797-1807.

Rotwein, P.S., Pollock, K., Didier, D., and Krivi, G., 1986, Organization and sequence of the human insulin-like growth factor I gene, *J. Biol. Chem.*, 261:4828-4832.

Rotwein, P.S., Folz, R.J., and Gordon, J.I., 1987, Biosynthesis of human insulin-like growth factor I (IGF-I), *J. Biol. Chem.*, 262:11807-11812.

Samuel, C.E., 1993, The eIF-2α protein kinases, regulators of translation in eukaryotes from yeasts to humans, *J. Biol. Chem.*, 268:7603-7607.

Saunders, J.C., Dickson, M.C., Pell, J.M., and Gilmour, R.S., 1991, Expression of a growth hormone-responsive exon of the ovine insulin-like growth factor-I gene, *J. Mol. Endocrinol.*, 7:233-240.

Schwander, J., Hauri, C., Zapf, J., and Froesch, E., 1983, Synthesis and secretion of insulin-like growth factor and its binding proteins by the perfused liver: dependence on growth hormone status, *Endocrinology*, 113:297-305.

Shemer, J., Adamo, M.L., Roberts, Jr., C.T., and LeRoith, D., (1992), Tissue-specific transcription start site usage in the leader exons of the rat IGF-I gene: evidence for differential regulation in the developing kidney, *Endocrinology*, 131:2793-2799.

Shimatsu, A., and Rotwein, P., 1987, Mosaic evolution of the insulin-like growth factors, *J. Biol. Chem.*, 262:7894-7900.

Shimatsu, A., and Rotwein, P.S., 1987, Sequence of two rat insulin-like growth factor I mRNAs differing within the 5'-untranslated region, *Nucleic Acids Res.*, 15:7196.

Shuldiner, A.R., Nirula, A., Scott, L.A., and Roth, J., 1990, Evidence that *Xenopus Laevis* contains two different nonallelic insulin-like growth factor-I genes, *Biochem. Biophys. Res. Commun.*, 166:223-230.

Steenbergh, P.H., Koonen-Reemst, A.M.C.B., Cleutiens, C.B.J.M., and Sussenbach, J.S., 1991, Complete nucleotide sequence of the high molecular weight human IGF-I mRNA, *Biochem. Biophys. Res. Commun.*, 175:507-514.

Thissen, J-P., and Underwood, L.E., 1992, Translational status of the insulin-like growth factor-I mRNAs in liver of protein-restricted rats, *J. Endocrinol.*, 132:141-147.

Tobin, G., Yee, D., Brunner, N., and Rotwein, P., 1990, A novel human insulin-like growth factor I messenger RNA is expressed in normal and tumor cells, *Mol. Endocrinol.*, 4:1914-1920.

Tollet, P., Enberg, B., and Mode, A., 1990, Growth hormone (GH) regulation of cytochrome P-450 11C12, insulin-like growth factor-I (IGF-I) and GH receptor messenger RNA expression in primary rat hepatocytes: a hormonal interplay with insulin, IGF-I thyroid hormone, *Mol. Endocrinol.*, 4:1934-1942.

Ullrich, A., Gray, A., Tam, A.W., Yang-Feng, T., Tsubokawa, M., Collins, C., Hanzel, W., LeBon, T., Kathuria, S., Chen, E., Jacobs, S., Francke, U., Ramachandran, J., and Fujita-Yamaguchi., Y., 1986, Insulin-like growth factor I receptor primary structure: comparison with insulin receptor suggests structural determinants that define functional specificity, *EMBO J.*, 5:2503-2512.

Wallis, A.E., and Devlin, R.H., 1993, Duplicate insulin-like growth factor-I genes in salmon display alternative splicing pathways, *Mol. Endocrinol.*, 7:409-422.

DIFFERENTIAL REGULATION OF IGF-I LEADER EXON TRANSCRIPTION

Jennifer M. Pell and R. Stewart Gilmour

Growth Regulation Group
AFRC Babraham Institute
Cambridge, CB2 4AT
UK

INTRODUCTION

Growth hormone (GH) and nutritional status are major determinants of growth rate and final body size. IGF-I is involved in mediating the anabolic effects of these stimuli, its circulating concentrations being sensitive to both GH (Zapf et al., 1981) and nutritional (Prewitt et al., 1982) status. Therefore, in addition to the ubiquitous autocrine/paracrine synthesis of IGF-I by tissues (D'Ercole et al., 1984), circulating IGF-I may exert an important stimulation for growth control. The liver has been identified as the primary source of circulating IGF-I since it has several-fold greater IGF-I mRNA abundance compared with other tissues (Murphy et al., 1987) and its synthesis of IGF-I can account for IGF-I turnover in the circulation (Schwander et al., 1983). The increased total hepatic expression of IGF-I in response to GH has been shown to be due to changes in the rate of gene transcription (Mathews et al., 1986).

The mammalian IGF-I gene displays considerable basic structural homology and is described in detail in this volume (see chapter by Roberts et al.). One of the more interesting properties of the gene in terms of regulation of IGF-I gene expression, is the existence of two 5' exons (termed exons 1 and 2) which are differentially spliced to exon 3 producing class 1 and 2 transcripts, respectively (Fig. 1; Lowe et al., 1987) and encode specific leader sequences of prepro IGF-I. As in the rat, ovine class 2 transcripts are liver specific whereas class 1 transcripts have been found in all sheep tissues examined so far (Dickson et al., 1991; Saunders et al., 1991). In contrast with other species, evidence for a third leader sequence has been found in sheep (Wong et al., 1989; Dickson et al., 1991) but we have found no evidence for its use in regulating IGF-I gene expression (Pell et al., 1993).

Exon 1 and 2 usage has been examined in fasted and diabetic rats; co-ordinate decreases in expression were observed which were reversed on refeeding or insulin therapy (Adamo et al., 1991). This is perhaps not surprising since both of these are severe catabolic insults which may override mechanisms of normal physiological regulation. Even though both class 1 and 2 transcripts increase in GH-treated hypophysectomised rats, class 2 transcripts show a greater

Current Directions in Insulin-Like Growth Factor Research,
Edited by D. LeRoith and M.K. Raizada, Plenum Press, New York, 1994

13

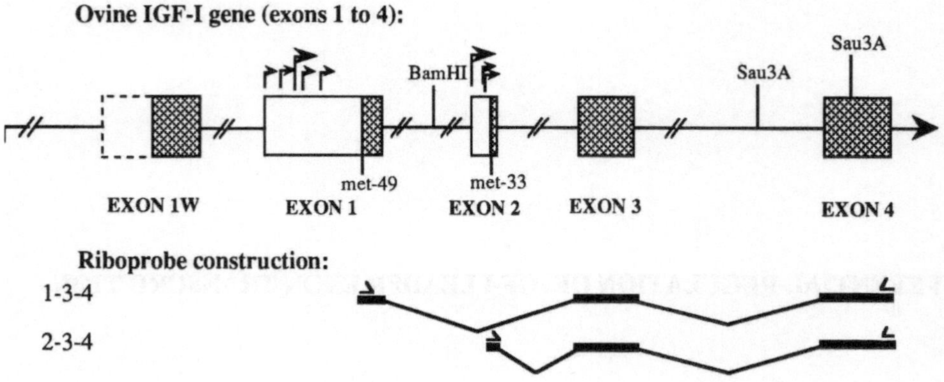

Ovine IGF-I gene (exons 1 to 4):

Riboprobe construction:

1-3-4

2-3-4

Figure 1. Upstream end of the ovine IGF-I gene. The three putative leader exons (1W, 1 and 2) and the common coding exons (3 and 4) are illustrated; exons 5 (not yet demonstrated in sheep transcripts) and 6 are not shown. Exons only are drawn to scale; coding regions are shaded. Arrows above exons show known transcription initiation sites as determined by RNase protection. AUG initiation codons are shown beneath exons 1 and 2. Selection of PCR sites for template construction and antisense riboprobe synthesis for probes 1-3-4 and 2-3-4 are indicated by one-way arrows.

magnitude of response (Lowe et al., 1987). In addition, a differential regulation of class 1 and 2 transcripts is observed during normal development with class 2 transcripts increasing at the time of onset of GH-dependent linear growth (Adamo et al., 1989; Adamo et al., 1991). When all these studies are considered together, it appears that there is evidence for greater GH sensitivity of exon 2 than exon 1 and that during normal physiological development, exons 1 and 2 exhibiting differential regulation. We therefore decided to investigate the regulation of class 1 and 2 transcripts using normal animals in which growth rate was manipulated in a moderate manner within normal physiological limits observed *in vivo*. We report here the effects of changed GH and nutritional status on IGF-I gene expression in two experiments. In the first, growth rate was stimulated by either improved energy or protein supply in the diet or by chronic GH treatment. In the second, growth rate was moderately reduced by acute immunisation against growth hormone-releasing hormone (GRF). Riboprobes were developed so that leader exon-specific or total IGF-I gene expression could be determined in RNase protection assays and are described in Pell et al., 1993, which also reports part of the first study. In the final sections of this chapter, we speculate on the factors which regulate IGF-I gene expression and the physiological relevance of the alternate leader exons.

IGF-I LEADER EXON TRANSCRIPTION INITIATION DURING GROWTH STIMULATION BY IMPROVED NUTRITION OR EXOGENOUS GH

Forty eight lambs were allocated to one of two protein diets containing either a moderate (12%, termed L) or high (20%, termed high) crude protein content. These were offered at either a restricted energy intake (30 g/kg liveweight, termed R) or at a high energy intake (*ad libitum*, approximately 50 g/kg liveweight, termed A). Within each diet, lambs were treated with either vehicle (termed -) or GH (0.1 mg kg.day, termed +). All treatments commenced at 9 weeks of age and continued for 10 weeks.

Whole body and tissue growth

Table 1 shows whole body, liver and vastus lateralis muscle weights as well as mean GH, insulin and IGF-I concentrations in blood sampled from the jugular vein. It is clear that in this experimental design, the major determinant of growth rate was energy intake, the 1.67-fold difference in intake accounting for as much as a 15 kg difference in final liveweight. GH treatment exerted statistically significant increases in all growth indices measured, although in quantitative terms these were not as great as those stimulated by energy supplementation. Whereas liver was the only tissue measured to respond to increased dietary protein and energy and GH, skeletal muscle weight was significantly increased by energy and GH.

Circulating hormone concentrations and hepatic IGF-I and GH receptor mRNA

Fig. 2 shows total hepatic IGF-I gene expression (left panel) and GH receptor mRNA (right panel) determined using RNase protection assays and a riboprobe encoding an intracellular portion of the receptor (specific region in legend). The abundance of both mRNAs responded similarly to changed GH and nutritional status, being increased by GH, the high protein diet and a high energy intake. Similar changes in abundance were also obtained for the GH receptor using a riboprobe derived from the extracellular region of the gene. There was a greater abundance of GH receptor mRNA compared with IGF-I mRNA in the liver, although the relative intensity units presented in Fig. 2 should not be compared directly because they were independently quantified.

Circulating GH, insulin and IGF-I concentrations are presented in Table 1. IGF-I concentrations exhibited similar changes to those observed for hepatic IGF-I mRNA and GH receptor. GH (data presented are a mean of six samples taken every 3 h from 7 am) and insulin concentrations did not respond to dietary protein and could not be readily correlated with the hepatic mRNA levels. GH concentrations were, of course, greater in the lambs administered exogenous GH and decreased significantly in the energy-restricted animals, confirming many

Table 1. Initial and final liveweights, liver and vastus lateralis (VL) muscle weights and circulating GH, insulin (Ins) and IGF-I concentrations in lambs of different nutritional and GH status. Differences between treatment means were assessed by analysis of variance and the probabilities of differences between treatment main effects of energy, protein and GH are summarised.

Energy:	Restricted				Ad libitum				S.E.D.	Main effects		
Protein:	Low		High		Low		High		(n=6)	Energy	Protein	GH
GH:	-	+	-	+	-	+	-	+				
Initial livewt (kg)	20.9	20.7	21.1	20.6	22.1	21.1	20.9	22.2	1.7	NS	NS	NS
Final livewt(kg)	28.5	30.1	28.0	29.7	43.3	45.3	44.0	47.0	1.6	<.001	NS	0.017
Liver wt(g)	467	475	618	611	996	1104	1108	1401	77	<.001	<.001	0.013
VL muscle wt(g)	89.1	118.4	85.8	94.1	127.1	162.7	144.7	150.7	11.2	<.001	NS	0.001
GH concn (ng/ml)	3.40	5.02	2.79	5.22	1.99	4.73	2.07	4.76	0.68	0.04	NS	<.001
Ins concn (μg/ml)	10.0	20.5	9.30	19.2	19.5	43.1	17.9	43.3	5.32	<.001	NS	<.001
IGF-I concn (ng/ml)	227	572	435	528	509	766	711	876	91	<.001	0.026	<.001

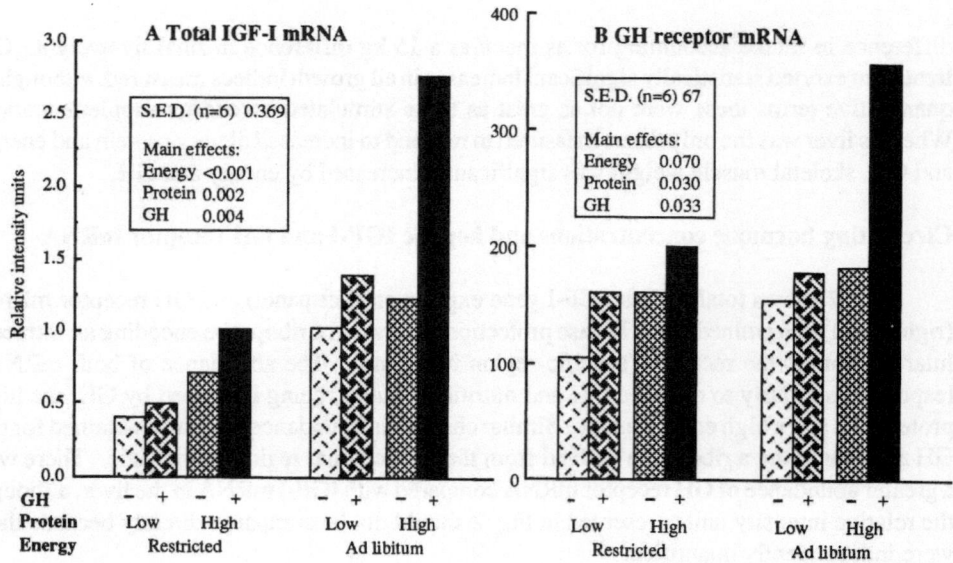

Figure 2. Effects of GH status and dietary energy and protein on A, total hepatic IGF-I mRNA and B, GH receptor mRNA. The GH receptor DNA template was produced by PCR using ovine liver cDNA to generate a 141 bp fragment of the intracellular domain of the receptor extending from nucleotides 1051-1191 of the published sheep GH receptor sequence (Adams et al., 1990)

previous studies in man and ruminants. Insulin concentrations were clearly related to energy intake and also displayed the characteristic "diabetogenic" response to GH.

Hepatic IGF-I transcription initiation from class 1 and 2 transcripts

Fig. 3 (left panel) shows an example of an autoradiogram of RNase protection assays for hepatic IGF-I mRNA using riboprobe templates derived from ovine IGF-I exons 1-3-4 or 2-3-4. Two protected bands can be seen for each assay corresponding to the full-length exon-specific protected class 1 or class 2 hybrid (1-3-4 or 2-3-4) or to the common coding exons 3-4. Fig. 4 shows quantification of these bands, expressed as a percent increase of the mean relative intensity for the low protein, restricted energy, saline lambs i.e. as a change compared with basal levels. The same scale has been used for both panels and it is immediately clear that, even though class 1 transcripts do respond to changed dietary energy and protein, these changes are only very modest when compared with the percent change in class 2 transcripts. Similarly, little change in exon 1 transcripts could be detected in GH-treated lambs although the dose of GH administered was only very modest, as the changes in mean circulating GH concentrations demonstrate (Table 1). Exon 2 transcripts displayed a very different sensitivity to GH, mean abundance increasing at all levels of nutrition and displaying significant synergy in the high protein and energy lambs.

IGF-I gene expression in skeletal muscle

Fig. 3 (right panel) illustrates IGF-I gene expression in the vastus lateralis muscle using riboprobes 1-3-4 and 2-3-4 together with a liver sample for comparison. As expected, only class 1 transcripts could be detected, all the expression determined using the exon 2 riboprobe being in the common coding 3-4 band. IGF-I mRNA abundance was much lower in skeletal muscle than in liver and was not responsive to the moderate changes in nutritional and GH status imposed in this study (the extremes of treatment groups are presented).

IGF-I TRANSCRIPTION INITIATION AFTER SUPPRESSION OF GROWTH BY IMMUNISATION AGAINST GRF

Fourteen 6-week-old lambs were immunised using a synthetic peptide corresponding to sequence 1-29 of ovine GRF; seven further lambs were "sham-immunised" to serve as controls. All animals were boosted at 4 week intervals over the next 18 weeks. The GRF-immunised lambs all produced antibodies against GRF as determined by standard ELISA techniques with titres ranging from 1:2,000 to 1:10,000. After 18 weeks, six of the anti-GRF group were treated for one week with a high dose of GH (0.25 mg/kg.day). All lambs were slaughtered at 19 weeks. Data from the study were analysed using t-tests, comparing the responses of control versus anti-GRF lambs and GH-treated versus anti-GRF lambs.

Whole body weight gain and tissue weights

Fig. 5 shows the whole body weights of the lambs from 4 weeks of age. All lambs showed a typical post-weaning check at 9 weeks but displayed no differences between treatment group at this point. Over the next few weeks, the GRF-immunised lambs exhibited a weight gain which was 20% less than that for the controls (P=0.001). Liver, vastus lateralis muscle and omental fat weights were recorded at slaughter (Table 2). GRF-immunised lambs tended to have more omental fat and less muscle then the control animals and these effects were counteracted by GH treatment, which also induced hepatic hypertrophy.

Liver and muscle IGF-I gene expression

As illustrated in Fig. 7 (left panel), total hepatic IGF-I gene expression was significantly decreased by about two-fold (P=0.003) in GRF-immunised animals and this was more than

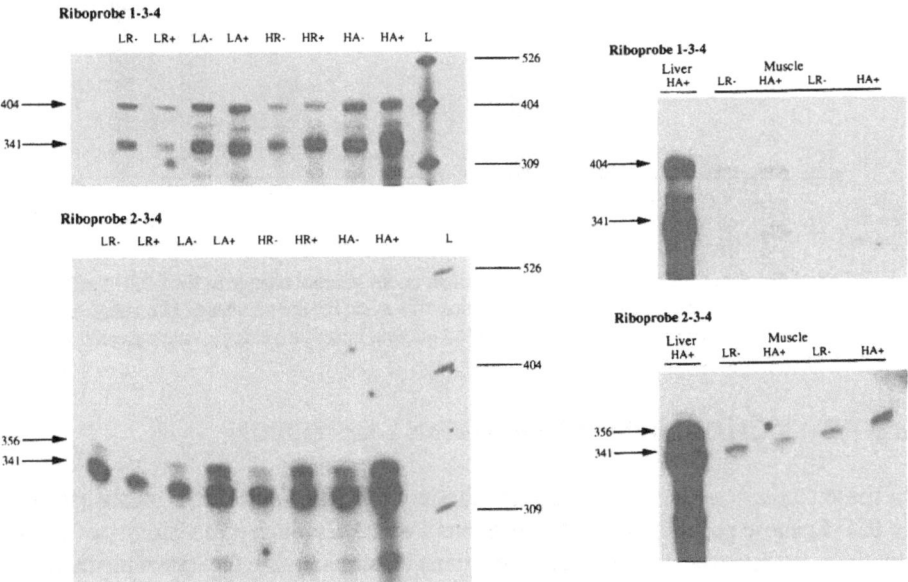

Figure 3. Effects of GH status and dietary energy and protein on sheep IGF-I mRNA in liver and skeletal muscle determined by solution hybrization/RNase protection (100μg total RNA). Left panel, liver; right panel, liver, (lane labelled liver) and vastus lateralis muscle (lanes labelled muscle). Tissue was derived from lambs fed a low (L) or high (H) protein diet at a restricted (R) or ad libitum (A) energy intake and treated with either saline (-) or GH (+). Each lane represents and individual animal. Lanes labelled L indicate the size and position of markers (32P-terminally labelled HpaII digest of pBR322).

compensated by the subsequent nine-fold increase induced by only one week of the high dose of GH (P<0.001). Circulating IGF-I concentrations were correlated with hepatic IGF-I mRNA, being: control, 158 ± 20; anti-GRF, 59 ± 7; GH, 288 ± 17 ng/ml; means ± S.E.M.; difference between all groups :P<0.001. Examples of RNase protection assays using IGF-I riboprobes 1-3-4 and 2-3-4 are shown in Fig. 6 for the liver (left panel) and percent changes from control lambs for individual class 1 and 2 transcripts are presented in Fig. 7 (right panel). The anti-GRF lambs had significantly reduced IGF-I expression derived from exon 2 (P=0.001). There was also a negative change in the class 1 transcripts although, as observed previously, the magnitude of the response was much greater for class 2 than 1 transcripts (P=0.013) and, in absolute terms, only the class 2 transcripts displayed a significant decrease in abundance when compared with the controls. GH induced a significant increase in both class 1 and 2 transcripts but once again, the change in class 2 abundance was significantly greater than that for class 1 (P<0.001).

Muscle IGF-I gene expression is illustrated in Fig. 6 (right panel). Again, essentially only class 1 transcripts were detected and these did not display any consistent trends between the treatment groups, further demonstrating the apparent insensitivity of skeletal muscle IGF-I mRNA to moderate changes in GH status.

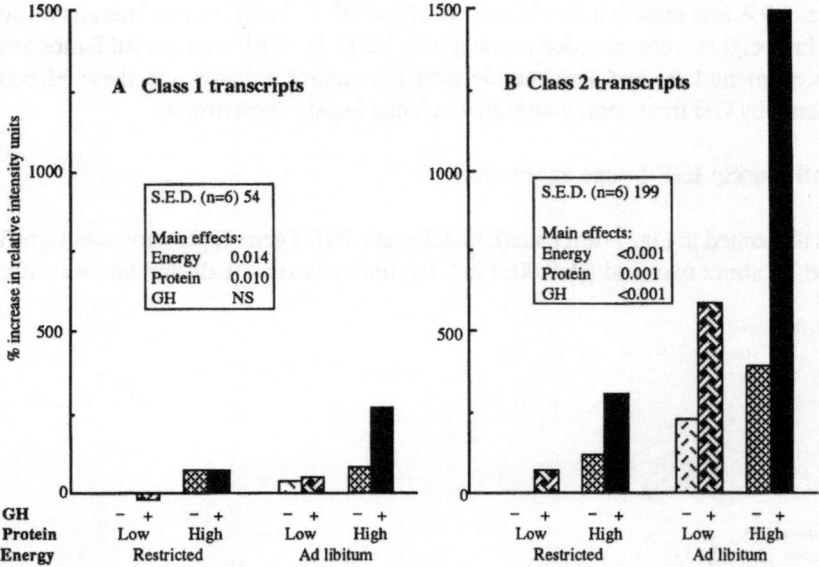

Figure 4. Effect of GH status and dietary energy and protein on the percent change in liver IGF-I mRNA on A, Class 1 transcripts and B, Class 2 transcripts. Correction was made for the number of U residues in each transcript. Percent increases were calculated relative to low-protein, restricted-energy, saline-treated lambs which therefore have a value of zero.

THE REGULATION OF HEPATIC IGF-I GENE EXPRESSION

In these studies, we have demonstrated that both GH and nutritional status can both regulate IGF-I gene expression derived from exons 1 and 2, although he sensitivity of exon 2 is greater than that for exon 1. A key objective must be to determine the mechanisms which account for these changes and these can be considered at at least two levels: 1) extracellular messengers which must stimulate 2) intracellular signals and factors which interact with the IGF-I gene itself.

The actions of GH are likely to be direct and via the GH receptor. The importance of the GH receptor itself is illustrated by the similar responses of GH receptor and total IGF-I mRNA. It is well established that chronic treatment with exogenous GH will up-regulate GH binding to hepatic membranes (e.g. Pell et al., 1990) and therefore the amount of GH receptor protein or rate of recycling. However, the precise intracellular signalling pathways from the receptor to the IGF-I gene itself are unclear at present. A variety of signalling mechanisms involving the regulation of cAMP levels, the production of diacylglycerol in the absence of phosphoinositide lipid turnover and the presence of GH receptor-associated tyrosine kinase activity have all been separately proposed (for reveiw see Kelly et al., 1991). However, none to date has been demonstrated unequivocally to act as the principal second messenger which mediates the actions of GH. An alternative approach to the signalling question, namely the identification of "GH response elements" within the IGF-I gene promoters, has not been informative. Promoter activity has been demonstrated for the gene sequence 5' to exon 1 when linked to a reporter gene and transfected into cultured cells (Adamo et al., 1993) but deletion studies do not suggest a simple colinear relationship between promoter potential and DNA sequence (Hall et al., 1992). Similar experiments with the exon 2 promoter failed to detect significant promoter activity; however, as with previous studies cultured cells of non-hepatic origin were used and hence the absence of an appropriate cellular milieu may have contributed to the lack of activity (Hall et al., 1992). The picture is further complicated by the finding that multiple DNase hypersensitive sites (and hence potential gene regulatory foci) are distributed across the entire span of the rat IGF-I gene. The only site which is truly GH dependent, as judged from GH administration to hypophysectomized rats, is located in the intronic DNA between exons 2 and 3 rather than in either of the putative promoter regions (Bichell et al., 1992). These findings suggest that the genomic mechanisms which regulate IGF-I transcription may be considerably more complex than previously imagined.

Throughout this chapter, we have used the phrase "nutritional status" which is a diffuse way of summarising the responses to nutrient supply in terms of the quality and quantity of substrate absorbed and the endocrine responses to these. Changes in total energy intake were manipulated simply by the amount of feed consumed and since protein is an essential component of muscle growth, dietary protein content was also varied. Circulating insulin concentrations were sensitive to energy intake and therefore insulin could be considered as a possible key regulator of hepatic IGF-I gene expression. Indeed, it has been demonstrated that insulin can stimulate IGF-I gene expression in vitro (Boni-Schnetzler et al., 1991). An insulin response element which is 90% homologous to the consensus TAGTCAAACA has been

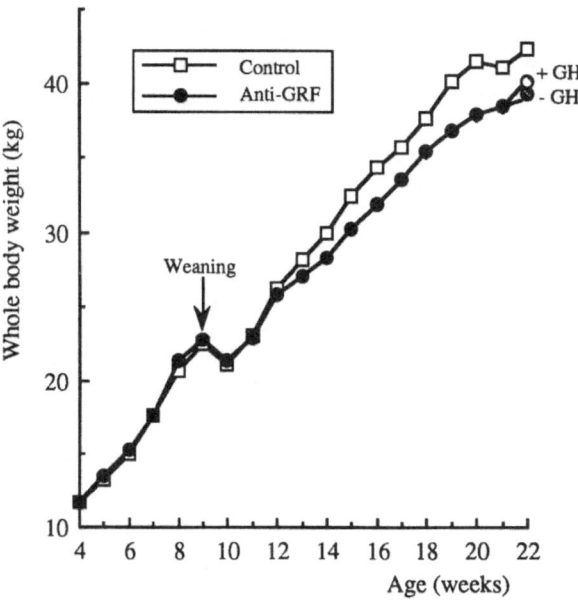

Figure 5. Whole body daily gain of control (n=7) and GRF-immunised lambs (n=14); six of the GRF-immunized lambs were treated with GH (0.25 mg/kg.day) during week 21. Average daily gains of the control and GRF-immunized lambs between weeks 10 and 21 were 276±11 and 230±6 g/day (means ± S.E.M.) and were significantly different (P=0.001).

identified in the ovine IGF-I gene -810 bp from the initiation codon of exon 2 (Dickson et al., 1991). Interestingly, circulating insulin concentrations apparently did not respond to changed dietary protein but clearly, some component of this protein did reach the liver as liver weight and circulating IGF-I concentrations increased. It is therefore possible that dietary protein may be able to directly stimulate IGF-I gene expression, perhaps via some specific amino acid, or that it stimulates some other indirect effector.

PHYSIOLOGICAL RELEVANCE OF THE IGF-I GENE 5' LEADER EXONS

The nutritional and hormonal treatments in these studies all induced changes in growth rate and circulating hormone concentrations which were within normal physiological limits and therefore several important conclusions can be made concerning growth regulation *in vivo*.

Skeletal muscle IGF-I gene expression apparently did not respond to GH or changed nutrition but muscle mass, and presumably hypertrophy, did change. If these are mediated via IGF-I then endocrine and not locally-produced IGF-I must mediate this anabolism, assuming no change in IGF-I bioactivity. Therefore the role of endocrine IGF-I in normal growth regulation should not be underestimated.

Generally, animals with lower growth rates exhibited a predominance of hepatic class 1 transcripts, whereas when increased growth rate was stimulated, class 2 transcripts predominated. This implies that the abundance of exon 2-derived mRNA will be more closely correlated with "anabolic potential" than will total hepatic IGF-I gene expression. In addition, the increased circulating IGF-I concentrations which occurred during increased growth are likely to be derived preferentially from class 2 transcripts.

It is intriguing that there is a switch in exon usage from exon 1 to exon 2 when maximum growth rate occurs resulting in a concomitant four-fold increases in circulating plasma IGF-I levels from apparently adequate basal levels. Why should this not occur through a single (exon 1) promoter? One explanation might be the need to separately regulate a "tonic" IGF-I synthesis required for maintaining basic metabolic processes from an anabolic IGF-I synthesis necessary to drive growth. Another explanation is based on the finding that in a few non-hepatic sources (e.g. brain, colostrum), a truncated IGF-I lacking the first three N-terminal amino acids is formed (Francis et al., 1988). This is more biologically active than its full-length counterpart due to reduced affinity for IGF binding proteins. Francis et al. (1988) suggest that truncated IGF-I is formed following post translational modification of the mature IGF-I peptide, rather than at the gene level. The proposal has been made that tissue (i.e. local autocrine/paracrine) IGF-I is derived from class 1 transcripts and that circulating (endocrine) IGF-I is derived from

Table 2. Initial and final liveweights and wights of liver, omental fat and vastus lateralis muscle of lambs immunised against GRF and treated with GH. Data are means ± S.E.M.; superscripts represent means which differ significantly: a, b P<0.08 and c, d P<0.05.

	Control	Anti-GRF	GH
Initial wt (kg)	11.60 ± 0.58	12.76 ± 0.86	11.00 ± 0.66
Final wt (kg)	42.20 ± 0.87[b]	38.56 ± 1.56[a]	41.4 ± 1.20[ab]
Liver wt (g)	819 ± 53[c]	855 ± 55[c]	1022 ± 17[d]
Omental fat (g)	766 ± 47	790 ± 53	676 ± 44
Vastus lateralis (g)	279 ± 17[ab]	249 ± 14[a]	283 ± 7[b]

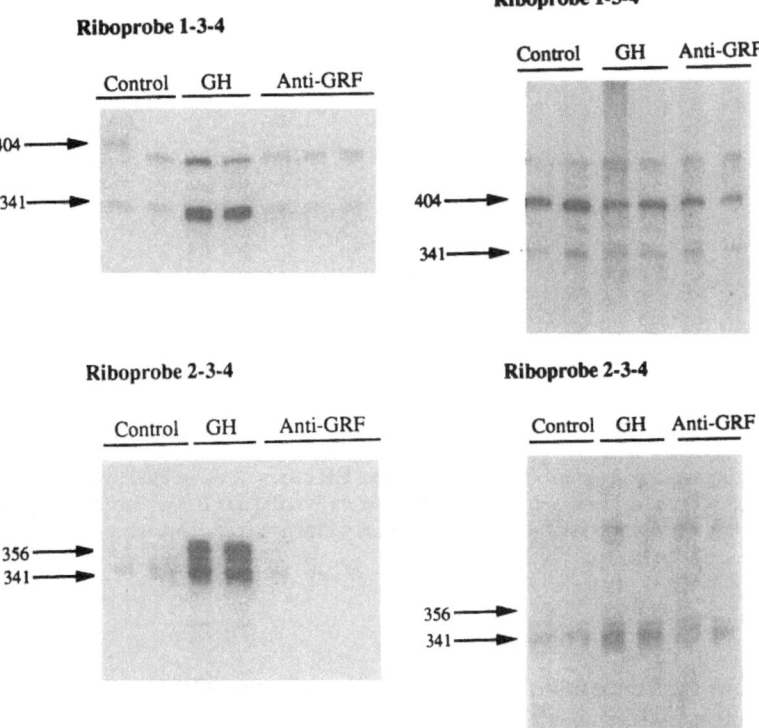

Figure 6. Effects of immunisation against GRF and subsequent GH treatment on sheep IGF-I mRNA in liver (left panel) and skeletal muscle (right panel) determined by solution hybridization/RNase protection (50 μg total RNA).

class 2 transcripts (LeRoith and Roberts, 1991; Pell et al., 1993). The significance of the alternate exons 1 and 2 is that they encode signal peptide extension peptides which might determine whether subsequent processing produces truncated or full-length forms of IGF-I (Roberts and LeRoith, 1991). In arguing against this interesting idea, it should be pointed out that truncated IGF-I has not been shown to be the form in all non-hepatic sites of IGF-I synthesis; also, the liver is a major producer of IGF-I from class 1 transcripts in certain conditions but only full-length peptide is found in the circulation. However, it may still be that the choice of exon may predetermine or direct IGF-I towards specific intracellular pathways through a different mechanism and the exon 1 and exon 2 leader peptides may mediate this. Their intracellular localisation of has not yet been determined but may help to elucidate the functions of the respective IGF-I peptides.

The differential regulation of mRNA arising from exons 1 and 2 may have an explanation in the subsequent biological activity of the transcripts. Roberts et al. (see chapter in this book) have suggested that the different start sites within exon 1 and between exons 1 and 2 will confer differences in subsequent translatability over and above those expected from just the lengths of sequence. If transcripts from exon 2 are readily translated, then class 2 transcripts will induce more rapid changes in IGF-I synthesis and it is then important to regulate class 2 abundance closely. Exons 5 and 6 (encoding alternate Eb and Ea terminal peptide sequence) and the 3' untranslated region of IGF-I have also been implicated in the regulation of IGF-I gene expression, particularly in terms of mRNA stability (Hepler et al., 1990) but there is little clear evidence for a link between exon 1 and 2 and the final choice of 3' exon usage (Hoyt et al., 1988) even though some correlation between class 2 and Eb abundance has been demonstrated in GH-treated hypophysectomised rat liver (LeRoith et al., 1990).

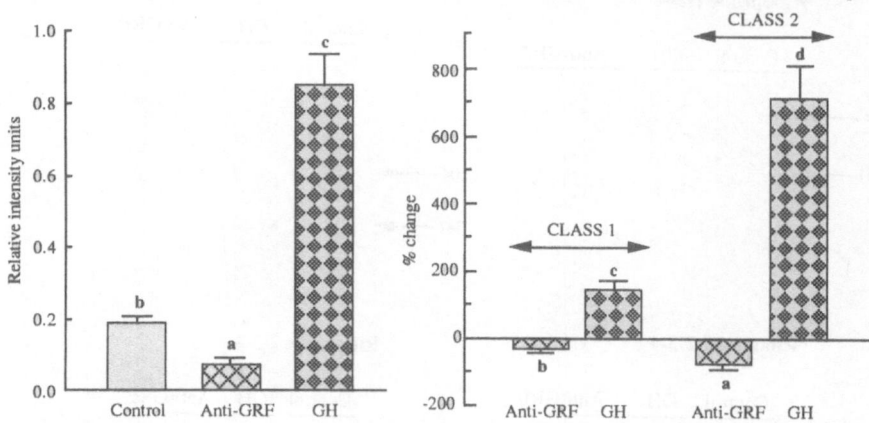

A Total IGF-I mRNA   B % change in class 1 and class 2 transcripts

Figure 7. Effects of immunisation against GRF and subsequent GH treatment on A, total liver IGF-I mRNA (data are means ± S.E.M.; a, b, c: means were statistically different P<0.01) and B, the percent change in hepatic class 1 and class 2 transcripts relative to mean levels in control lambs (data are means ± S.E.M.; statistical significance: a,b P=0.013; c, d P<0.001).

SUMMARY

We have shown that there is differential expression of the two IGF-I leader exons during physiological changes in GH and nutritional status in liver, with exon 2 being more responsive than exon 1. It may be that this exon is responsible for endocrine IGF-I synthesised in response to specific anabolic stimuli inducing subsequent whole body and skeletal muscle growth.

ACKNOWLEDGEMENTS

We wish to thank P.C. Bates, T.A. Calder and J.C. Saunders for their invaluable contributions to the work presented here.

REFERENCES

Adams, T.E., Baker, L., Fiddes, R.J. and Brandon, M.R., 1990, The sheep growth hormone receptor: molecular cloning and ontogeny of mRNA expression in the liver, *Molec. Cell Endocrinol.* 73:135.

Adamo, M.L., Ben-Hur, H., Roberts, C.T. Jr., LeRoith, D., 1991, Regulation of start site usage in the leader exons of the rat insulin-like growth factor-I gene by development, fasting, and diabetes, *Mol. Endocrinol.* 5:1677.

Adamo, M.L., Lowe, W.L. Jr., LeRoith, D., Roberts, C.T. Jr., 1989, Insulin-like growth factor I messenger ribonucleic acids with alternative 5'-untranslated regions are differentially expressed during development in the rat, *Endocrinology* 124:2737.

Adamo, M.L., Lanau, F., Neuenschwander, S., LeRoith, D. and Roberts, C.T., 1993, Distinct promoters in the rat insulin-like growth factor-I (IGF-I) gene are active in CHO cells, *Endocrinology* 132:935.

Bichell, D.P., Kikuki, K. and Rotwein, P., 1992, Growth hormone rapidly activates insulin-like growth factor-I gene transcription in vitro, *Molec. Endocrinol.* 11:1.

Boni-Schnetzler, M., Schmidt, C., Meier, P.J., Froesch, E.R., 1991, Insulin regulates insulin-like growth factor I mRNA in rat hepatocytes, *Am. J. Physiol.* 260:E846.

D'Ercole, A.J., Stiles, A.D., Underwood, L.E., 1984, Tissue concentrations of somatomedin C: further evidence for multiple sites of synthesis and paracrine or autocrine mechanisms of action, *Proc. Natl. Acad. Sci. USA* 81:935.

Dickson, M.C., Saunders, J.C., Gilmour, R.S., 1991, The ovine insulin-like growth factor-I gene: characterization, expression and indentification of a putative promoter, *J. Mol. Endocrinol.* 6:17.

Francis, G.L., Upton, F.M., Ballard, F.J., McNiel, K.A. and Wallace, J.C., 1988, Insulin-like growth factors 1 and 2 in bovine colostrum, *Biochem. J.* 251:95.

Hall, L.J., Kajimoto, Y., Bichell, D., Kim, S-W., James, P.L., Counts, D., Nixon, L.J., Tobin, G. and Rotwein, P., 1992, Functional analysis of rat insulin-like growth factor-I gene and identification of an IGF-I gene promoter, *DNA and Cell Biol.* 11:301.

Hepler, J.E., Van Wyk, J.J. and Lund, P.K., 1990, Different half lives of insulin-like growth factor-I mRNAs that differ in length of 3'-untranslated regions, *Endocrinology* 127:1550.

Kelly, P.A., Djane, J., Postel-Vinay, M-C. and Edery, M., 1991, The prolactin/growth hormone receptor family, *Endocrine Rev.* 12:235.

Lowe, W.L. Jr., Roberts, C.T. Jr., Lasky, S.R., LeRoith, D., 1987, Differential expression of alternative 5' untranslated regions in mRNAs encoding rat insulin-like growth factor I, *Proc. Natl. Acad. Sci. USA* 84:8946.

LeRoith, D., Adamo, M. and Roberts, C.T. Jr., 1990, Regulation of insulin-like growth factor-I gene expression, *in*: "Growth Factors: From Genes to Clinical Application", V.R. Sara, K.Hall and H. Low, eds., Raven Press, New York.

LeRoith, D. and Roberts, C.T. Jr., 1991, Insulin-like growth factor-I (IGF-I): a molecular basis for endocrine versus local action? *Mol. Cell. Endocrinol.* 77:C57.

Mathews, L.S., Norstedt, G. and Palmiter, R.D., 1986, Regulation of insulin-like growth factor I gene expression by growth hormone, *Proc. Soc. Nat. Acad. Sci. USA* 83:9343.

Murphy, L.J., Bell, G.I. and Friesen, H.G., 1987, Tissue distribution of insulin-like growth factor I and II messenger ribonucleic acid in the adult rat, *Endocrinology* 120:1279.

Pell, J.M., Elcock, C., Harding, R.L., Morrell, D.J., Simmonds, A.D. and Wallis, M., 1990, Growth, body composition, hormonal and metabolic status in lambs treated long-term with growth hormone, *Brit. J. Nutr.* 63, 431.

Pell, J.M., Saunders, J.C. and Gilmour, R.S., 1993, Differential expression of transcription initiation from insulin-like growth factor-I (IGF-I) leader exons and of tissue IGF-I expression in response to changed growth hormone and nutritional status in sheep, *Endocrinology* 132:1797.

Prewitt, T.E., D'Ercole, A.J., Switzer, B.R., Van Wyk, J.J. 1982, Relationship of serum immunoreactive somatomedin-C to dietary protein and energy in growing rats, *J. Nutr.* 112:144.

Saunders, J.C., Dickson, M., Pell, J.M., Gilmour, R.S., 1991, Expression of a growth hormone-responsive exon of the ovine insulin-like growth factor-I gene, *J. Mol. Endocrinol.* 7:233.

Schwander, J., Hauri, C., Zapf, J., Froesch, E.R., 1983, Synthesis and secretion of insulin-like growth factor and its binding protein by the perfused liver:dependence on growth hormone status, *Endocrinology* 113:297.

Wong, E.A., Ohlsen, S.M., Godfredson, J.A., Dean, D.M., Wheaton, J.E. 1989, Cloning of ovine insulin-like growth factor-I cDNAs: heterogeneity in the mRNA population, *DNA* 8:649.

Zapf, J., Froesch, E.R., Humbel, R.E. 1981, The insulin-like growth factors (IGF) of human serum - chemical and biological characterization and aspects of their possible physiological role, *Curr. Top. Cell. Regul.* 19:257.

INSULIN AND IGF-I ANALOGS: NOVEL APPROACHES TO IMPROVED INSULIN PHARMACOKINETICS

Lawrence J. Slieker[1], Gerald S. Brooke[1], Ronald E. Chance[1], Li Fan[1],
James A. Hoffmann[1], Daniel C. Howey[2], Harlan B. Long[1], John Mayer[1],
James E. Shields[1], Karen L. Sundell[1] and Richard D. DiMarchi[1]

1 Diabetes Research Division
2 Clinical Pharmacology Administration
Lilly Research Laboratories
Eli Lilly and Co.
Lilly Corporate Center
Indianapolis, IN 46285

INTRODUCTION

Current insulin formulations do not mimic the normal glucose-induced release of insulin by the pancreas in a physiological manner.[1] One limitation is the delayed absorption of hexameric insulin from the subcutaneous site of injection, such that soluble insulins (currently the most rapid acting formulations) are too slow and have too long a duration of action.[2] Another limitation is that longer acting insulin formulations, such as human ultralente, exhibit too short a duration of action, show a pronounced peak in activity and are suspensions, resulting in variability in administration.[3] The use of recombinant DNA technology and peptide chemistry have allowed the generation of insulin analogs with a wide variety of amino acid substitutions, which in turn have been useful in mapping regions of the insulin nucleus that are associated with Zn^{2+} binding, dimer formation and insulin receptor interaction. This report will review the physical, biological and clinical characterization of several insulin analogs that have been designed to improve absorption characteristics and pharmacodynamics. Because of the structural homology between insulin and insulin-like growth factor-I (IGF-I), we have investigated specific IGF-like modifications in the insulin sequence to determine if these will transfer to pharmacokinetic differences in insulin absorption and clearance.

The rate at which insulin is absorbed from the site of injection is limited by the rate of dissociation of hexamers to dimers and monomers.[3] That reduced self-association of an insulin analog would translate to improved absorption kinetics has been demonstrated.[4,5,6] Because of the reduced tendency of IGF-I to self-associate, it was postulated that inversion of the 28-29 position of the insulin B chain to that of the homologous region of IGF-I (see Fig. 1) would alter association state. The rationale for this was that the homology between the C-terminal 12 residues of insulin (which represents the dimer interface region) and the corresponding region of IGF-I is quite high, except for the obvious inversion of sequence at B^{28}-B^{29}. Indeed, $Lys^{B28}Pro^{B29}$ insulin self associates to a much lesser degree than does insulin, and it has been demonstrated clinically to be more rapid acting than regular insulin.[6] Because of the concern for the possibility that modifications in this region will alter the growth promoting and/or metabolic activity of insulin, $Lys^{B28}Pro^{B29}$ insulin and several other analogs in structurally related series were characterized in terms of insulin and

Current Directions in Insulin-Like Growth Factor Research,
Edited by D. LeRoith and M.K. Raizada, Plenum Press, New York, 1994

25

IGF-I receptor affinity, and potential to stimulate the growth of human mammary epithelial cells.

IGF-I and -II circulate bound to a family of binding proteins (IGFBPs), of which 6 members have been cloned and sequenced. These proteins have differing affinities for IGF-I or -II, but are all characterized by not binding insulin.[7] Since IGF-I analogs with reduced affinity for IGFBPs have been shown to be cleared more rapidly than native IGF-I,[8] IGFBP affinity was engineered into insulin to determine if this would result in an analog with decreased clearance and delayed time action, but characterized by solubility at physiological pH. Although modifications of both the insulin and IGF-I nucleus were investigated, the initial approach was to exploit previous identification of the 3-4, 15-16 and 49-51 regions of IGF-I, which have been determined previously to mediate a large portion of the selectivity of IGF-I for the IGFBPs.[9]

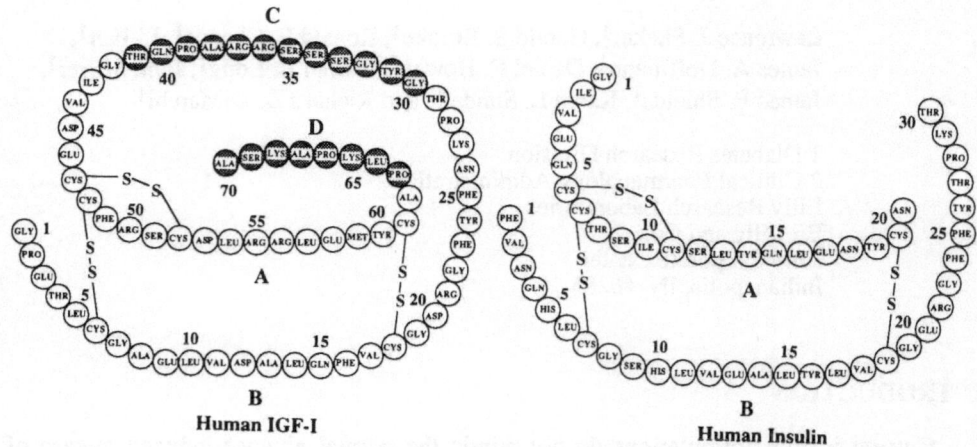

Figure 1. Structures of human IGF-I and insulin

MATERIALS AND METHODS

The insulin analogs were made either by recombinant methods, by trypsin-catalyzed semi-synthesis from des-octapeptide (B$^{23-30}$)insulin (or des-octapeptide Asp^{B10} insulin) and the corresponding octapeptide modified in the B$^{28-29}$ position, or by chain combination of modified A and B chains. IGF-I analogs were prepared by solid phase peptide synthesis and chain combination.

Insulin and IGF-I competitive binding assays were performed with membranes prepared from full term human placenta isolated essentially as described by Gruppuso.[10] Approximately 30 µg of protein were incubated with 50,000 cpm of ^{125}I[A14]-insulin or ^{125}I-IGF-I and increasing doses of cold competing ligand in 0.5 ml of 100 mM Hepes, pH 7.8, 120 mM NaCl, 5 mM KCl, 1.2 mM MgSO$_4$, 8 mM glucose and 0.25% BSA. After 18 hrs at 4 °C, membranes were collected on glass fiber filters. IGFBP binding assays were performed using the bovine gamma globulin/polyethylene glycol procedure of Clemmons et al.[9] for assays involving IGF-I analogs, and the charcoal adsorption assay for modified insulin analogs.[11] Acid stable serum IGF binding protein[11] was provided by Dr. Michele Smith (Virology Division, Lilly Research Laboratories) and purified IGFBP1[12] was provided by Dr. Ronald Bowsher (Drug Disposition and Bioanalytical Research Division, Lilly Research Laboratories). Competitive binding data were fit by non-linear regression to a 4 parameter model using JMP (SAS Institute, Inc.) for determination of EC$_{50}$ values.

Mitogenicity was assessed by measuring insulin analog stimulated growth of human mammary epithelial cells (HMEC) in culture. HMEC were obtained from Clonetics (San Diego, CA, USA) at passage 7 and were expanded and frozen at passage 8. A fresh ampule was brought up for each experiment so that cells were not passaged beyond passage 9. Cells were maintained in MCDB170 medium containing bovine insulin (5 µg/ml), recombinant human EGF (10 ng/ml), hydrocortisone (0.5 µg/ml), bovine pituitary extract (50 µg/ml) and gentamycin/ amphotericin B. For a growth experiment, cells were plated in 96-well trays at a density of 4000 cells per well in the above medium modified as follows: 0.1% bovine serum albumin was added and 5 µg/ml bovine insulin was substituted by a graded dose of human insulin or analog from 0 to 1000 nM final concentration. Trays were incubated for 72 hours and the cells were counted by Coulter counter (Coulter Electronics, Hialeah, FL). Typically, the maximal growth response was between 3 - and 4-fold stimulation over basal, and it did not differ between analogs. Dose response data were fit by non-linear regression as described above.

Models of the $Lys^{B28}Pro^{B29}$ insulin structure were generated from extensive calculations on a Cray-2 supercomputer. The program INSIGHT was used to produce the images, and calculations were conducted using DISCOVER (INSIGHT and DISCOVER from Biosym Technologies, Inc.).

RESULTS AND DISCUSSION

Fast-Acting Analogs

Approximately 35 insulin analogs in the $Xaa^{B28}Pro^{B29}$ insulin series and structurally related series were prepared. Previous reports have demonstrated the monomeric nature of these analogs. Specifically, substitution of the proline at B^{28} with acidic, basic or neutral aliphatic residues while maintaining Lys at B^{29} reduces the dimer association constant up to 200 fold compared to insulin.[13] Variation of B^{28} in conjunction with substitution of Pro for Lys at B^{29} further reduces aggregation from 300-fold ($Lys^{B28}Pro^{B29}$) to 3000-fold ($Asp^{B28}Pro^{B29}$) relative to human insulin, as well as essentially eliminates Zn^{2+} - induced higher association as determined by ultracentrifugation.[13] Substitution of His^{B10} with Asp, either by itself or in conjunction with substitutions in the $B^{28}B^{29}$ position, also decreases self-association as well as dramatically increases physical and chemical stability.[14] A 600 psec molecular dynamics trajectory performed on a Cray-2 supercomputer has predicted the solution conformation of $Lys^{B28}Pro^{B29}$ insulin in a drop of water to deviate from the crystal structure of insulin by subtle conformational changes in the extended beta strand of the C-terminal 8 residues of the B chain (Fig. 2). This new conformation is likely to be incompatible with dimer formation, and would explain the weakened self-association characteristic of this insulin analog in solution. [15]

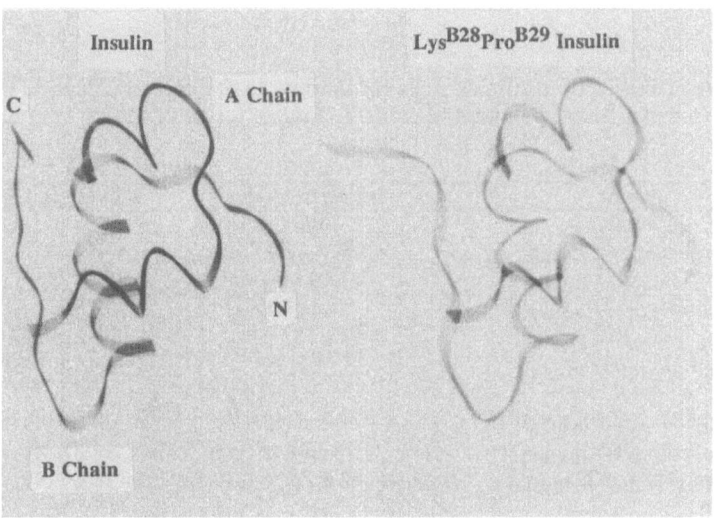

Figure 2. Ribbon structures of insulin (based on crystal structure) and $Lys^{B28}Pro^{B29}$ insulin (molecular dynamics calculation of solution conformation).

Lys^{B28}Pro^{B29} insulin has been demonstrated to have a more rapid onset and shorter duration of action than regular insulin after subcutaneous injection in clinical studies employing normal individuals. Using the euglycemic clamp, peak levels of insulin and Lys^{B28}Pro^{B29} insulin were observed at 2.3 and 0.6 hrs, respectively, and peak glucose infusion rates were observed at 2.2 and 1.0 hrs, respectively.[16] This has also been confirmed in preliminary studies employing both insulin-dependent and noninsulin-dependent diabetic subjects, which demonstrated that equal or better glucose control could be obtained with Lys^{B28}Pro^{B29} insulin being given much closer to mealtime than regular insulin (Diabetes, in press).

Because of the slight increased homology of Lys^{B28}Pro^{B29} insulin to IGF-I, the binding affinity of the Xaa^{B28}Pro^{B29} insulin analog series for both the human insulin and IGF-I receptors was compared. Substitutions in the B^{28} position, with the exception of Phe, Trp, Ile, Leu and Gly, had relatively little effect on insulin receptor affinity.[17] However, the IGF-I receptor affinity was more sensitive to charge differences, as well as steric bulk, in this region. Specifically, basic residues (Lys, Arg and Orn) increased relative IGF-I receptor affinity approximately 1.5 to 2 fold, while acidic residues decreased it two fold, relative to insulin (Table 1). This effect of charge alteration in the C terminal region of the B chain was particularly evident in analogs having an increased number of basic residues, such as Lys^{B28}Lys^{B29} and Arg^{B31}Arg^{B32} insulin (interestingly, the latter analog is a natural intermediate in the enzymatic processing of proinsulin). These effects were also additive to the enhancement of IGF-I receptor affinity induced through the Asp^{B10} substitution.

These data indicate that the C terminal end of the B chain of insulin (and presumably IGF-I) as well as basic C region extensions of the B chain are important in mediating IGF-I receptor selectivity, but are relatively unimportant for insulin receptor affinity. These results are consistent with those reported for an insulin/IGF-I hybrid containing the C region of IGF-I plus the C-terminal 8 residues of the B region (residues 22-41) added to a scaffold of des-octapeptide insulin.[18] This molecule had appreciable increased affinity for the IGF-I receptor relative to insulin, and only moderate reduction of affinity for the insulin receptor, demonstrating that the selectivity sites for each receptor are at least partially distinct.

Mitogenicity assays employing normal human mammary epithelial cells have confirmed that analogs with particularly high affinity for the IGF-I receptor (such as Asp^{B10} insulin and Arg^{B31}Arg^{B32} insulin) are more potent than insulin at stimulating cell growth, and that their relative potency is proportional to their enhanced IGF-I receptor

Table 1. Relative affinities of insulin analogs for binding to the human placental insulin and IGF-I receptors (mean ± SEM, n in parentheses). Affinities for each analog are relative to insulin = 100%

Analog	Insulin Receptor	IGF-I Receptor
Insulin	100	100
Lys^{B28}Pro^{B29} Insulin	83 ± 6 (8)	157 ± 15 (7)
Arg^{B28}Pro^{B29} Insulin	91 ± 11 (4)	189 ± 14 (4)
Orn^{B28}Pro^{B29} Insulin	88 ± 8 (4)	159 ± 24 (4)
Lys^{B28}Lys^{B29} Insulin	94 ± 18 (2)	228 ± 29 (2)
Arg^{B31}Arg^{B32} Insulin	153 ± 21 (5)	1020 ± 185 (6)
Glu^{B28}Pro^{B29} Insulin	87 ± 6 (4)	38 ± 3 (4)
Asp^{B28}Pro^{B29} Insulin	78 ± 9 (4)	52 ± 8 (4)
CysSO$_3$^{B28}Pro^{B29} Insulin	88 ± 12 (5)	36 ± 7 (4)
Asp^{B10} Insulin	198 ± 17 (7)	302 ± 38 (9)
Asp^{B10}Lys^{B28}Pro^{B29} Insulin	232 ± 10 (5)	453 ± 42 (4)
Asp^{B10}Asp^{B28}Pro^{B29} Insulin	224 ± 15 (3)	138 ± 8 (3)

affinity. $Lys^{B28}Pro^{B29}$ insulin, which has only a marginal greater affinity for the IGF-I receptor, is equipotent to insulin in stimulating cell growth (Fig. 3). These data confirm previous reports of enhanced mitogenicity of Asp^{B10} insulin, relative to insulin, in human[19] and rat[20] aortic smooth muscle cells, and are of particular interest since Asp^{B10} insulin was removed from clinical trials because of an association with breast tumors observed in 12-month toxicity studies in rats.[21]

Long-Acting Analogs

One approach to prolonging insulin's duration of action would be to induce insulin binding to endogenous factors, thereby decreasing its rate of clearance. High titers of insulin antibodies have been suggested to contribute to the longer time action of some formulations of animal source vs. human insulins[22]. IGF-I and -II circulate bound to members of a family of IGF binding proteins (IGFBPs). The physiological role of these proteins is complex, and may involve targeting IGFs to particular locations, potentiating IGF action or transporting IGFs across capillary surfaces[23]. IGFBPs are characterized by generally having greater affinity for IGF-II over IGF-I, and having no measurable affinity for insulin. IGF-I analogs that do not bind IGFBPs have shorter half lives in vivo, suggesting that binding of IGFs to BPs might increase clearance times,[8] and differential rates for clearance of IGF-I have been observed in human subjects where the differences were ascribed to the association with endogenous IGF-binding proteins.[24] Because residues 3, 4, 15, 16 and 49-51 of IGF-I have been reported to be important in mediating affinity to varying degrees for all of the IGFBPs,[7,9] these residues were substituted into the homologous regions of insulin. The following insulin analogs were prepared by solid phase peptide synthesis and chain combination: $Phe^{A8}Arg^{A9}Ser^{A10}$ insulin, $Glu^{B4}Gln^{B16}Phe^{B17}$ insulin and $Phe^{A8}Arg^{A9}Ser^{A10}Glu^{B4}Gln^{B16}Phe^{B17}$ insulin. Table 2 shows the affinity of these analogs for the insulin and IGF-I receptors, and IGFBP.

$Phe^{A8}Arg^{A9}Ser^{A10}Glu^{B4}Gln^{B16}Phe^{B17}$ insulin demonstrated low, but measurable affinity for both the acid stable serum binding proteins (mostly IGFBP3) and IGFBP1 (EC_{50} approx 1-2 μM), but at the expense of a dramatic reduction in insulin receptor affinity. The decrease in insulin receptor binding induced by both the A and B chain substitutions are additive, but the magnitude of the effect of the B chain modification alone confirms the importance of the B^{16-17} region in insulin receptor recognition[25]. Interestingly, a moderate, but quantitatively smaller reduction in IGF-I receptor affinity was also observed with these analogs, suggesting that the B^{16-17} region plays a less important, but measurable, role in mediating insulin binding to the IGF-I receptor. This is quantitatively somewhat different from the demonstration of Bayne et al. that substitution of $Gln^{15}Phe^{16}$ of IGF-I with the homologous $Tyr^{B16}Leu^{B17}$ sequence of insulin had no effect on affinity for the IGF-I receptor.[25] In fact, substituting the entire first 17 residues of IGF-I with the corresponding region of the insulin B chain only reduced type I receptor affinity 2 fold.[25]

Neither of the single chain modified analogs showed measurable affinity for either binding protein preparation, implying that both B chain and A chain regions act synergistically to impart binding activity. It is interesting that no difference between

Table 2. Receptor and IGFBP affinities for insulin/IGF-I hybrids. IGFBP represents the acid stable binding protein from human serum. Purified IGFBP1 gave similar results. Affinities of each analog are relative to insulin or IGF-I = 100% (± SEM, n in parentheses).

Analog	Ins Rec	IGF-I Rec	IGFBP
Insulin	100	0.7	- - -
IGF-I	nd	100	100
$Phe^{A8}Arg^{A9}Ser^{A10}$ Insulin	32 ± 7 (2)	0.27 ± 0.01 (2)	- - -
$Glu^{B4}Gln^{B16}Phe^{B17}$ Insulin	4.6 ± 0.71 (2)	0.13 ± 0.01 (2)	- - -
$Phe^{A8}Arg^{A9}Ser^{A10}Glu^{B4}Gln^{B16}Phe^{B17}$ Insulin	1.8 ± 0.78 (2)	0.28 ± 0.05 (2)	0.01

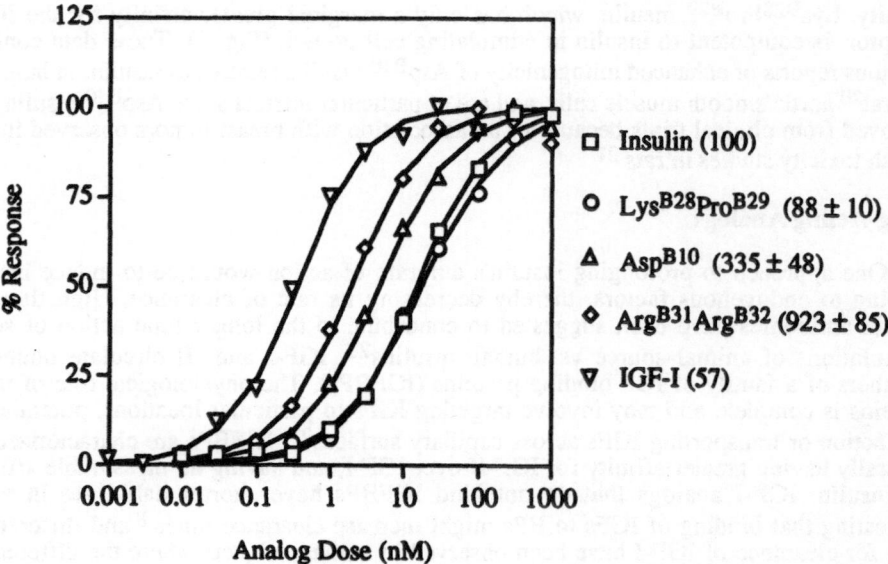

Figure 3. Stimulation of growth of normal human mammary epithelial cells by insulin analogs and IGF-I. Individual points represent the average of 4 separate experiments, while the solid lines indicate the pooled data fit by non-linear regression. Numbers in parentheses in legend indicate potency relative to insulin = 100 %).

IGFBP1 or the mixed serum IGFBPs was observed in this respect, particularly since previous reports have suggested that the B chain mutations in IGF-I have a much greater effect than do the A chain mutations on reducing affinity for serum IGFBP, while for IGFBP1, the opposite appears to be true.[26]

Because modification of the insulin nucleus generated analogs with low affinity for both the insulin receptor and the IGFBPs, another approach was taken by starting with the IGF-I nucleus and increasing the affinity for the insulin receptor. The following two-chain IGF-I analogs were prepared by solid-phase peptide synthesis and chain combination: B:A, BC:A, B:CA and B:AD, where A, B, C and D refer to domains of IGF-I and the colon indicates a break in the peptide primary sequence. Table 3 shows the affinity of these analogs for the insulin and IGF-I receptors, and IGFBP.

B:A IGF-I has high affinity for the serum IGFBPs, moderate affinity for the insulin receptor and significantly reduced IGF-I receptor affinity. This latter point confirms the importance of the C domain of IGF-I for receptor selectivity.[27] The D region has no effect on IGF-I receptor affinity (B:AD vs. B:A), as reported previously,[25] but it does reduce affinity for both the insulin receptor and IGFBP. Maintaining the C region but generating a

Table 3. Relative receptor and IGFBP binding affinity of IGF-I analogs. IGFBP refers to the acid stable binding protein from human serum. Affinities of each analog are relative to insulin or IGF-I = 100% (± SEM, n in parentheses).

Analog	IGF-I Rec	Ins Rec	IGFBP
IGF-I	100	1.2	100
Insulin	0.37	100	- - -
B:A	7.7 ± 0.54 (3)	17.8 ± 3.7 (2)	151 ± 15 (3)
BC:A	127 ± 45 (2)	9.7 ± 2.5 (2)	226 ± 66 (4)
B:CA	2.7 ± 0.13 (3)	1.4 ± 0.4 (2)	154 ± 37 (4)
B:AD	7.8 ± 1.2 (2)	7.2 ± 0.01 (2)	69 ± 17 (3)

two-chain molecule by clipping between residues 41-42 (BC:A) had no effect on IGF receptor affinity, but clipping on the other side of the C domain at residues 29-30 dramatically reduced affinity, suggesting that the absolute presence of the C domain is insufficient if the conformation is not restricted. Interestingly, removal or derestriction of the C domain appears to enhance affinity for the serum IGFBPs.

The unexpectedly high affinity of B:A IGF-I for the insulin receptor combined with high affinity for IGFBPs suggests that this analog might be useful as a foundation for the development of long-acting basal insulin agonists.

References

1. M. Berger, H. J. Cüppers, H. Hegner, V. Jörgens and P. Berchtold, Absorption kinetics and biologic effects of subcutaneously injected insulin preparations, *Diabetes Care* 5:77 (1982).

2. J. A. Galloway, C. T. Spradlin, R. L. Nelson, S. M. Wentworth, J. A. Davidson and J. L. Swarner, Factors influencing the absorption, serum insulin concentration, and blood glucose responses after injections of regular insulin and various insulin mixtures, *Diabetes Care* 4:366 (1981).

3. P. Hildebrandt , P. Sejrsen , S. L. Nielsen, K. Birch and L. Sestoft, Diffusion and polymerization determines the insulin absorption in diabetic patients, *Scand J Clin Lab Invest* 45:685 (1985).

4. J. Brange, U. Ribel, J. F. Hansen, G. Dodson, M. T. Hensen, S. Havelund, S. G. Melburg, F. Norris, K. Norris, L. Snel, A. R. Sørensen and H. O. Voigt, Monomeric insulins obtained by protein engineering and their medical implications, *Nature* (Lond) 333:679 (1988).

5. S. Kang, F. M. Creagh, J. R. Peters, J. Brange, A. Vølund and D. R. Owens, Comparison of subcutaneous soluble human insulin and insulin analogues (Asp^{B9}, Glu^{B27}; Asp^{B10}; Asp^{B28}) on meal-related plasma glucose excursions in type I diabetic subjects, *Diabetes Care* 14:571 (1991).

6. R. D. DiMarchi, J. P. Mayer, L. Fan, D. N. Brems, B. H. Frank, L. K. Green, J. A. Hoffmann, D. C. Howey, H. B. Long, W. N. Shaw , J. E. Shields, L. J. Slieker , K. S. E. Su, K. L. Sundell and R. E. Chance, Synthesis of a fast-acting insulin analog based upon structural homology with insulin-like growth factor-I, in: Peptides: Chemistry and Biology. Proceedings of the Twelfth American Peptide Symposium, J. A. Smith and J. E. Rivier, eds., 26-28, ESCOM, Leiden (1992).

7. Y. Oh, H. L. Müller, D-Y. Lee, P. J. Fielder and R. G. Rosenfeld, Characterization of the affinities of insulin-like growth factor (IGF)-binding proteins 1-4 for IGF-I, IGF-II, IGF-I/insulin hybrid, and IGF-I analogs, *Endocrinology* 132:1337 (1993).

8. M. A. Cascieri, R. Saperstein, N. Hayes, B. G. Green, G. G. Chicchi, J. Applebaum and M. L. Bayne, Serum half-life and biological activity of mutants of human insulin-like growth factor I which do not bind to serum binding proteins, *Endocrinology* 123:373 (1988).

9. D. R. Clemmons, D. L. Dehoff, W. H. Busby, M. L. Bayne and M. A. Cascieri, Competition for binding to insulin-like growth factor (IGF) binding protein-2,3,4 and 5 by the IGFs and IGF analogs, *Endocrinology* 131:890 (1992).

10. P. A. Gruppuso, B. H. Frank and R. Schwartz, Binding of proinsulin and proinsulin conversion intermediates to human placental insulin-like growth factor-I receptors, *J Clin Endocrinol Metab* 67:194 (1988).

11. M. L. Bayne, M. A. Cascieri, B. Kelder, J. Applebaum, G. G. Chicchi, J. A. Shapiro, F. Pasleau and J. J. Kopchick, Expression of a synthetic gene encoding human insulin-like growth factor I in cultured mouse fibroblasts, *Proc Natl Acad Sci USA* 85:2638 (1987).

12. W. H. Busby, D. G. Clapper and D. R. Clemmons, Purification of a 31-kDa IGF binding protein from human amniotic fluid, *J Biol Chem* 263:14302 (1988).

13. D. N. Brems, L. A. Alter, M. J. Beckage, R. E. Chance, R. D. DiMarchi, L. K. Green, H. B. Long, A. H. Pekar, J. E. Shields and B. H. Frank, Altering the association properties of insulin by amino acid replacement, *Protein Engineering* 5:527 (1992).

14. D. N. Brems, P. L. Brown, C. Bryant, R. E. Chance, L. K. Green, H. B. Long, A. A. Miller, R. Millican, J. E. Shields and B. H. Frank, Improved insulin stability through amino acid substitution, *Protein Engineering* 5:519 (1992).

15. M. A. Weiss, Q-X. Hua, C. S. Lynch, B. H. Frank and S. E. Shoelson, Heteronuclear 2D NMR studies of an engineered insulin monomer: Assignment and characterization of the receptor-binding surface by selective ^{2}H and ^{13}C labeling with application to protein design, *Biochemistry* 30:7373 (1991).

16. D. C. Howey, S. A. Hooper and R. R. Bowsher, [Lys(B28), Pro(B29)]-Human insulin: An equipotent analog of insulin with rapid onset and short duration of action, *Diabetes* 40(Supp 1): 423A (1991).

17. L. J. Slieker and K. L. Sundell, Modifications in the 28-29 position of the insulin B-chain alter binding to the IGF-I receptor with minimal effect on insulin receptor binding, *Diabetes* 40(Supp 1): 168A (1991).

18. J. F. Cara, R. G. Mirmira, S. H. Nakagawa and H. S. Tager, An insulin-like growth factor I/insulin hybrid exhibiting high potency for interaction with the type I insulin-like growth factor and insulin receptor of placental plasma membranes, *J Biol Chem* 265:17820 (1990).

19. H. Wolpert, L. J. Slieker, K. Sundell and G. King, Identification of an insulin analog with enhanced growth effect in aortic smooth muscle cells, *Diabetes* 39(Supp 1):140A (1990).

20. K. E. Bornfeldt, R. A. Gidlöf, A. Wasteson, M. Lake, A. Skottner and H. J. Arnqvist, Binding and biological effects of insulin, insulin analogues and insulin-like growth factors in rat aortic smooth muscle cells. Comparison of maximal growth promoting activities, *Diabetologia* 34:307 (1991).

21. K. Drejer, The bioactivity of insulin analogues from in vitro receptor binding to in vivo glucose uptake, *Diabetes/Metabolism Reviews* 8:259 (1992).

22. S. Gray, P. Cowan, U. Di Mario, R. A. Elton, B. F. Clarke and L. P. J. Duncan, Influence of insulin antibodies on pharmacokinetics and bioavailability of recombinant human insulin and highly purified beef insulins in insulin-dependent diabetics, *Br Med J* 290:1687 (1985).

23. R. S. Bar, D. R. Clemmons, M. Boes, W. H. Busby, B. A. Booth, B. L. Dake and A. Sandra. *Endocrinology* 127:1078 (1990).

24. H-P. Guler, J. Zapf and E. R. Froesch, Short-term metabolic effects of recombinant human insulin-like growth factor I in healthy adults, *N Engl J Med* 317:137 (1987).

25. M. L. Bayne, J. Applebaum, G. G. Chichi, N. S. Hayes, B. G. Green and M. A. Cascieri, Structural analogs of human insulin-like growth factor I with reduced affinity for serum binding proteins and the type II insulin-like growth factor receptor, *J Biol Chem* 263:6233 (1988).

26. D. R. Clemmons, M. A. Cascieri, C. Camacho-Hubner, R. H. McCusker and M. L. Bayne, Discrete alterations of the insulin-like growth factor I molecule which alter its affinity for insulin-like growth factor-binding proteins result in changes in bioactivity, *J Biol Chem* 265:12210 (1990).

27. M. L. Bayne, J. Applebaum, D. Underwood, G. G. Chicchi, B. G. Green, N. S. Hayes and M. A. Cascieri, The C region of human insulin-like growth factor (IGF) I is required for high affinity binding to the type I IGF receptor, *J Biol Chem* 264:11004 (1988).

ANALYSIS OF THE INTERACTION OF IGF-I ANALOGS
WITH THE IGF-I RECEPTOR AND IGF BINDING PROTEINS

Margaret A. Cascieri and Marvin L. Bayne

Department of Molecular Pharmacology and
Biochemistry
Merck Research Laboratories
Rahway, NJ 07065

The characterization of the various species of IGF binding proteins (IGFBPs) by molecular cloning has revealed a large family of highly homologous soluble proteins with high affinity for both IGF-I and IGF-II. The type 1 IGF receptor and the IGFBPs appear to form a complex regulatory system for transducing and modulating the activity of IGF-I and IGF-II. The determination of the roles of these proteins in IGF action has been facilitated by the use of IGF-I analogs that selectively bind to either the receptor or to IGFBPs.

The 21 amino acid A-region and the 29 amino acid B-region of IGF-I are highly homologous to the A-chain and B-chain of insulin. These two regions are linked by a 12 amino acid C-region that is not present in insulin, and an 8 amino acid D-region extends from the carboxyl terminus of the A-region. Since insulin has poor affinity for the IGF-I receptor and does not bind to IGFBPs, analogs were synthesized in which selected regions of IGF-I were replaced with the homologous regions of insulin, or in which regions of IGF-I that are not present in insulin were deleted. These analogs were produced using site-directed mutagenesis of a synthetic gene encoding IGF I (1), were expressed in Saccharomyces cerevisiae and were purified to homogeneity from the conditioned media (2).

Current Directions in Insulin-Like Growth Factor Research,
Edited by D. LeRoith and M.K. Raizada, Plenum Press, New York, 1994

33

RECEPTOR BINDING DOMAIN

Blundell and his colleagues have proposed a model for the tertiary structure of IGF-I based on its extensive homology with insulin (3,4). Determination of the solution structure of IGF-I by NMR has confirmed the validity of this model (5). The structure predicts that certain insulin receptor binding domains are conserved between insulin and IGF-I and may account for the cross-reactivity of IGF-I for the insulin receptor. Several groups have shown that the aromatic residues at positions 24, 25 and 26 in the B-chain of insulin are important in maintaining affinity of insulin for the insulin receptor (6,7).

In IGF-I, these residues are conserved as Phe^{23}, Tyr^{24}, and Phe^{25}. Substitution of Tyr^{24} with serine or leucine results in a dramatic loss in affinity for the IGF-I receptor, while binding to the IGFBPs in acid-treated human serum is maintained (8). Removal of the eight amino acid D-region of IGF-I has little effect on affinity for the IGF-I receptor (9), however, [Leu^{24}, 1-62] IGF-I has 100-fold reduced affinity (8).

Analogs in which much of the C-region has been replaced with a 4-residue glycine bridge ([1-27, Gly^4, 38-70] IGF-I and [1-27, Gly^4, 38-62] IGF-I) also have 30 to 40-fold reduced affinity for the IGF-I receptor while maintaining normal affinity for acid-treated human serum IGFBPs (9). Replacement of the tyrosine at position 31 in this region with alanine results in a 6-fold loss in affinity for the receptor (10). The analog containing the double mutation, [Leu^{24}, Ala^{31}] IGF-I has 200-fold reduced affinity. These data suggest that residues in the carboxyl terminal extended domain of the B-region and residues in the C-region proximal to this are involved in receptor binding, and that the tyrosines at residues 24 and 31 are involved in specific interactions with the receptor (Figure 1).

ANALOGS WITH REDUCED AFFINITY FOR THE TYPE 2 IGF RECEPTOR

The type 2 IGF receptor has high affinity for IGF-II, 100-fold lower affinity for IGF-I, and no measurable affinity for insulin (11). Analogs of IGF-I in which the A-region was replaced with the homologous residues in the A-chain of insulin (A-chain mutant) or in which residues 49-51 were so substituted ([Thr^{49}, Ser^{50}, Ile^{51}] IGF-I) have >10-fold reduced affinity for this receptor, while maintaining normal affinity for the IGF-I receptor and for the IGFBPs in acid-treated human serum (12). Analogous substitutions have been made in IGF-II by several groups to confirm that this region

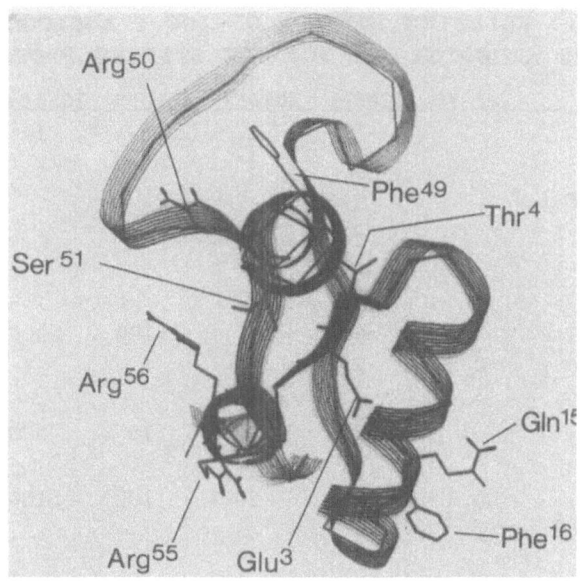

Figure 1. Structure of human IGF-I highlighting tyosines 24 and 31 in the receptor binding domain.

is involved in maintaining high affinity for the type 2 IGF receptor (13-15). These analogs may prove useful in establishing the role of this receptor in IGF action (13-15).

ANALOGS WITH REDUCED AFFINITY FOR IGF BINDING PROTEINS

We originally showed that analogs in which selected residues within the amino terminal helical domain of the B-region were substituted with the homologous residues in insulin have reduced affinity for the IGFBPs in acid-treated human serum and for the IGFBPs in conditioned media from BALB/C 3T3 cells while maintaining normal affinity for the IGF-I receptor (16,17). More recently we have collaborated with Dr. Robert Baxter and Dr. David Clemmons to determine the affinity of these analogs for purified human IGFBP3 (18) and purified IGFBP1, 2, 3, 4, and 5, respectively (19,20). A summary of the data from these experiments is presented in Table 1.

Substitution of the first 16 residues of IGF-I with the first 17 amino acids of the B-chain of insulin (B-chain mutant) results in a dramatic loss in affinity for all of the species of binding protein tested (18-20). In contrast, when residues 15 and 16 of IGF-I were changed from Gln-Phe to Tyr-Leu, the affinity for IGFBP1, 2 and 3 is unaltered, but the affinity for IGFBP4 and IGFBP5 is reduced 10-fold and 100-fold, respectively (20). The analog in which Glu^3-Thr^4 of IGF-I is replaced with Gln-Ala has 10-fold reduced affinity for IGFBP3 and more dramatically reduced affinity for the

TABLE 1. RELATIVE POTENCY OF IGF-I ANALOGS FOR THE IGF RECEPTOR AND FOR IGF BINDING PROTEINS

ANALOG	IGF-IR	IGFBP1	IGFBP2	IGFBP3	IGFBP4	IGFBP5
IGF-I	1	1	1	1	1	1
$[1\text{-}27, G^4, 38\text{-}70]$IGF-I	30	-	-	0.5	-	-
$[SER^{24}]$IGF-I	17	-	-	2	-	-
$[ALA^{31}]$IGF-I	6	0.5	-	-	-	-
B-CHAIN MUTANT[1]	2	>40	>20	>100	>200	>100
$[TYR^{15}, LEU^{16}]$ IGF-I	0.8	2	2	1	13	100
$[GLN^3, ALA^4]$ IGF-I	1	>20	>20	10	200	>100
$[GLN^3, ALA^4, TYR^{15}, LEU^{16}]$IGF-I	0.9	20	20	100	>200	>100
A-CHAIN MUTANT[1]	0.65	>40	>20	6	13	7
$[THR^{49}, SER^{50}, ILE^{51}]$IGF-I	1.4	>40	>20	1	60	>100

Assays were performed using the IGF-I receptor from human placental membranes, purified human IGFBP1, purified bovine IGFBP2, purified bovine or human IGFBP3, purified human IGFBP4 and purified human IGFBP5.

[1]B-CHAIN MUTANT=$[Phe^{-1}, Val^1, Asn^2, Gln^3, His^4, Ser^8, His^9, Glu^{12}, Tyr^{15}, Leu^{16}]$ IGF-I   A-CHAIN MUTANT= $[Ile^{41}, Glu^{45}, Gln^{46}, Thr^{49}, Ser^{50}, Ile^{51}, Ser^{53}, Tyr^{55}, Gln^{56}]$ IGF-I

other species of binding proteins (20). $[Gln^3, Ala^4, Tyr^{15}, Leu^{16}]$ IGF-I has 20-fold reduced affinity for IGFBP1 and 2, and >100-fold reduced affinity for IGFBP-3, 4, and 5. In sum, the data suggest that the amino terminal and helical domains of the B-region play an important role in the binding of IGF-I to IGFBPs.

The A-chain mutant and $[Thr^{49}, Ser^{50}, Ile^{51}]$ IGF-I have 6-fold lower affinity and normal affinity for IGFBP3, respectively (18, 20). In contrast, these analogs have dramatically reduced affinity for IGFBP1 and 2 (20). The A-chain mutant has 13-fold and 7-fold reduced affinity for IGFBP4 and 5 at their pH optima of 6, although the affinity is reduced more dramatically at physiological pH (20). $[Thr^{49}, Ser^{50}, Ile^{51}]$ IGF I has poor affinity for IGFBP4 and 5.

In summary, the specificity of the binding of IGF-I analogs to IGFBP1 and IGFBP2 is identical, suggesting that the binding sites for these two proteins are quite conserved. In contrast, IGFBP3 has normal affinity for analogs with mutations in residues 49-51 of the A-region. These residues are located at the end of a helical domain and are in close proximity to the amino terminus of the B-region (Figure 2). The data suggest that the amino terminal residues of IGF-I, but not the residues in the A-region, are involved in binding to IGFBP3. The specificity of analog binding to IGFBP4 and 5 is similar and differs from that of the binding to IGFBP1, 2, and 3 in that [Tyr^{15}, Leu^{16}] IGF-I has reduced affinity. These data suggest that the binding pockets of IGFBP4 and 5 are larger than those for IGFBP1, 2, and 3.

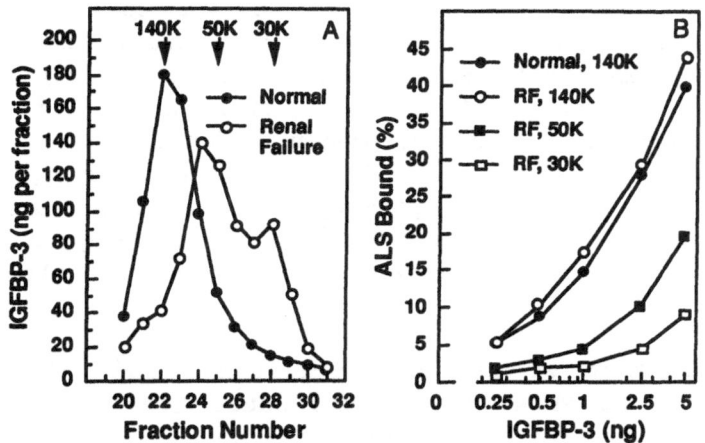

Figure 2. *Structure of human IGF-I highlighting the positions of the helical domains in the A- and B-regions, and the residues involved in IGFBP binding.*

SPECIFICITY OF THE FORMATION OF A TERNARY COMPLEX BETWEEN IGF-I, IGFBP3 AND THE ACID-LABILE α-SUBUNIT

In contrast to the other species of IGFBPs, IGFBP3 circulates as a ternary complex of IGF-I and an acid-labile protein of 85kDa (21). Ternary complex formation requires prior formation of a binary complex between IGF-I and IGFBP3. In collaboration with Dr. Robert Baxter, the structural requirements for ternary complex formation were examined by performing saturation binding isotherms using ^{125}I-α-subunit as the ligand (18). Using this paradigm, the association constants and binding site concentrations for the IGF-IGFBP3 complex using several analogs of IGF-I were determined. In general, analogs with reduced affinity for IGFBP3 also had reduced binding site concentration for ^{125}I-α-subunit consistant with a reduced potential for binary complex formation (18). However, despite having normal affinity for IGFBP3, analogs with mutations in residue 24 of the B-region and analogs in which the D-region was deleted, formed binary complexes with a 50 to 80% decrease in the association constant for ^{125}I-α-subunit (18). These data suggest that regions of IGF-I that are not directly involved in binding to IGFBP3 can alter the potential for IGFBP3 binding to the α-subunit.

CHARACTERIZATION OF THE BIOLOGICAL ACTIVITY OF SELECTIVE IGF-I ANALOGS

The IGF-I analogs with selective affinity for receptors or binding proteins have been utilized to probe the roles of these proteins in IGF-I action. The B-chain mutant and [Gln^3, Ala^4, Tyr^{15}, Leu^{16}] IGF-I, analogs that have reduced affinity for IGFBPs, are 10-fold more potent than IGF-I in stimulating DNA synthesis in BALB C/3T3 cells (17) and are 2 to 4-fold more potent in stimulating ^{14}C-glucose incorporation into muscle glycogen *in vivo* (22). These data suggest that binding of IGF-I to some binding proteins impairs their biological potency. These two analogs also have a reduced serum half-life in rats, suggesting that binding to serum IGFBPs attenuates the rate of clearance of IGF-I from serum (22).

Tollefson et al. have utilized [Gln^3, Ala^4, Tyr^{15}, Leu^{16}] IGF-I to show that IGFBPs cause IGF-I resistance in the fibroblasts from a patient with short stature (23). The fibroblasts from this patient required 3-fold higher levels of IGF-I to stimulate DNA synthesis and amino acid uptake than did fibroblasts from normal subjects and from other short children. However, the fibroblasts responded normally to [Gln^3, Ala^4, Tyr^{15}, Leu^{16}] IGF-I. They further showed that these fibroblasts have a 10-fold increase in the cell surface expression of a 32 kDa IGFBP (23).

In contrast to these studies, Clemmons et al. showed that inclusion of IGFBP1 potentiated the stimulation of DNA synthesis by IGF-I and [Leu^{24}, 1-62] IGF-I, but not by the B-chain mutant in porcine aortic smooth muscle cells (19). These data suggest that binding of an analog to this binding protein can increase its biological potency.

SUMMARY

Distinct domains of IGF-I are important for maintaining high affinity for the IGF-I receptor and for the various species of IGFBPs. The analogs that selectively bind the receptor have proven useful in determining the relative importance of IGFBPs in the regulation of the biological activity of IGF-I. Analogs with poor affinity for the receptor have also been useful in order to demonstrate that a given activity of IGF-I is mediated by the type 1 IGF receptor. These studies confirm that the role of these various proteins in IGF-I action is complex, and may be cell or tissue-type specific.

REFERENCES

1. Bayne, M.L., Cascieri, M.A., Kelder, B., Applebaum, J., Chicchi, G., Shapiro, J.A., Pasleau, F., and Kopchick, J.J. (1987) Proc. Natl. Acad. Sci. USA 84:2638-2642.
2. Bayne, M.L., Applebaum, J., Chicchi, G.G., Hayes, N.S., Green, B.G., and Cascieri, M.A. (1988) Gene 66:235-244.
3. Blundell, T.L., Bedarker, S., Rinderknecht, E., and Humbel, R.E. (1978) Proc. Natl. Acad. Sci., USA 75:180-184.
4. Blundell, T.L., Bedarker, S., and Humbel, R.E. (1983) Fed. Proc., Fed. Am. Soc. Exp. Biol. 42:2592-2597.
5. Cooke, R.M., Harvey, T.S., Campbell, I.D. (1991) Biochemistry 30:5484-5491.
6. Tager, H., Thomas, N., Assoian, R., Rubenstein, A., Solkow, M., Olefsky, J., and Kaiser, E.T. (1980) Proc. Natl. Acad. Sci., USA 77:3181-3185.
7. Kobayashi, M., Ohgaku, S., Iwasaki, M., Maegawa, H., Shigeta, Y., and Inouye, K. (1982) Biochem. J. 206:597-603.
8. Cascieri, M.A., Chicchi, G.G., Applebaum, J., Hayes, N.S., Green, B.G., and Bayne, M.L. (1988) Biochemistry 27:3229-3223.
9. Bayne, M.L., Applebaum, J., Underwood, D., Chicchi, G.G., Green, B.G., Hayes, N.S., and Cascieri, M.A. (1988) J. Biol. Chem. 264:11004-11008.
10. Bayne, M.L., Applebaum, J., Chicchi, G.G., Miller, R.E., and Cascieri, M.A. (1990) J. Biol. Chem. 26:15648-15652.
11. Morgan, D.O., Edman, J.C., Standring, D.N., Fried, V.A., Smith, M.C., Roth, R.A., and Rutter, W.J. (1987) Nature 329:301-307.
12. Cascieri, M.A., Chicchi, G.G., Applebaum, J., Green, B.G., Hayes, N.S., and Bayne, M.L. (1989) J. Biol. Chem. 264:2199-2202.
13. Beukers, M.W., Oh, Y., Shang, H., Ling, N., and Rosenfeld, R.G. (1991) Endocrinology 128:1201-1203.
14. Sakano, K., Enjoh, T., Numata, F., Fujiwara, H., Marumoto, Y., Higashihashi, N., Sato, Y., Perdue, J.F., and Fujita-Yamaguchi, Y. (1991) J. Biol. Chem. 266:20626-20635.

15. Burgisser, D.M., Roth, B.V., Giger, R., Luthi, C., Weigl, S., Zarn, J., and Humbel, R.E. (1991) J. Biol. Chem. 266:1029-1033.
16. Bayne, M.L., Applebaum, J., Chicchi, G.G., Hayes, N.S., Green, B.G., and Cascieri, M.A. (1988) J. Biol. Chem. 263:6233-6239.
17. Cascieri, M.A., Hayes, N.S., and Bayne, M.L. (1989) J. Cell. Physiol. 139:181-188.
18. Baxter, R.C., Bayne, M.L., and Cascieri, M.A. (1992) J. Biol. Chem. 267:60-65.
19. Clemmons, D.R., Cascieri, M.A., Camacho-Hubner, C., McCusker, R.H., and Bayne, M.L. (1990) J. Biol. Chem. 265:12210-12216.
20. Clemmons, D.R., Dehoff, M.L., Busby, W.H., Bayne, M.L. and Cascieri, M.A. (1992) Endocrinology 131:890-895.
21. Baxter, R.C., Martin, J.C., and Beniac, V.A. (1989) J. Biol. Chem. 264:11843-11848.
22. Cascieri, M.A., Saperstein, R., Hayes, N.S., Green, B.G., Chicchi, G.G., Applebaum, J., and Bayne, M.L. (1988) Endocrinology 123:373-381.
23. Tollefson, S.E., Heath-Monnig, E., Cascieri, M.A., Bayne, M.L., and Daughaday, W.H. (1991) J. Clin. Inv. 87:1241-1250.

SYNTHESIS AND CHARACTERIZATION OF IGF-II ANALOGS: APPLICATIONS IN THE EVALUATION OF IGF RECEPTOR FUNCTION AND IGF-INDEPENDENT ACTIONS OF IGFBPS

Youngman Oh,[1] Hermann L. Müller,[1] Heping Zhang,[2] Nicholas Ling,[2] and Ron G. Rosenfeld[1]

[1]Dept. of Pediatrics, Div. of Endocrinology
 Stanford University School of Medicine
 Stanford, CA 94305
[2]Dept. of Molecular Endocrinology
 The Whittier Institute for Diabetes and Endocrinology
 La Jolla, CA 92037

INTRODUCTION

The elucidation of the mechanisms involved in the biological actions of the IGFs has been hampered by the complexity of the IGF system: three peptide ligands (IGF-I, IGF-II and insulin),[1] three receptors (type 1 and type 2 IGF receptors and insulin receptors)[2] and six distinct, but structurally related binding proteins (IGFBP-1 to IGFBP-6).[3] The physiological interactions among these components of the IGF system are not completely understood, although they have all been cloned and sequenced.[4]

In particular, the biological role of IGF-II has remained puzzling, in part because of its relatively high affinity for all three receptors, in addition to the IGFBPs.[5-8] Furthermore, the high affinity of IGFBPs for IGF-II and the presence of these IGFBPs (especially IGFBP-3 and IGFBP-5) on the cell surface,[9,10] have further complicated our efforts to understand the interaction between IGF peptides and their receptors.

Previous studies from our laboratory and others have led to the development and use of anti-receptor antibodies as probes of receptor function.[11-15] Although such investigations have resulted in considerable insight, studies with receptor antibodies are often limited by: 1) lack of specificity of many anti-receptor antibodies; 2) toxic effects of antibodies on cells studied; and 3) the ability of some antibodies to function as both agonists and antagonists.[12,13,16,17] We have, consequently, attempted to develop three different classes of IGF-II analogs: 1) IGF-II analogs with affinity for IGFBPs and the type 2, but not type 1, IGF receptor; 2) IGF-II analogs with affinity for receptors, but not for IGFBPs; and 3) IGF-II analogs with affinity for only the type 2 IGF receptor.

In this review, we characterize these IGF-II analogs in terms of their binding affinities for the six IGFBPs and three receptors. In addition, we demonstrate the use of IGF-II analogs as excellent probes of type 2 IGF receptor function and IGF-independent IGFBP action.

SYNTHESIS AND CHARACTERIZATION OF IGF-II ANALOGS

Synthesis of IGF-II Analogs

The elegant studies of Humbel et. al. on the tertiary structures of the IGFs[18] and the work of

Current Directions in Insulin-Like Growth Factor Research,
Edited by D. LeRoith and M.K. Raizada, Plenum Press, New York, 1994

41

Cascieri, Bayne and co-workers on IGF-I structure-function relationships[19-22] have demonstrated that the binding sites for IGF receptors and insulin receptor are different from those for IGFBPs. In addition, the data implied that the type 2 IGF receptor binding domain differs from those of the type 1 IGF and insulin receptors. Based on the structural homology among IGF peptides and insulin and the observations cited above, we have designed IGF-II analogs which have preferential binding affinity for the type 2 IGF receptor {[Leu^{27}]IGF-II and [Ser^{27}]IGF-II}, and/or have alterations in affinity for IGFBPs {[Gln^6,Ala^7,Tyr^{18},Leu^{19}]IGF-II and [Gln^6,Ala^7,Tyr^{18},Leu^{19},Leu^{27}]IGF-II}, as illustrated in Fig. 1.

Human [Leu^{27}]IGF-II, [Ser^{27}]IGF-II, [Gln^6,Ala^7,Tyr^{18},Leu^{19}]IGF-II and [Gln^6,Ala^7, Tyr^{18},Leu^{19},Leu^{27}]IGF-II were synthesized by a solid phase peptide synthesis procedure[23] using a Bio-Glu(Bzl)-PAM-resin on a model 990 peptide synthesizer (Beckman, Palo Alto, CA). Derivatized amino acids used in the synthesis were of the L-configuration and were purchased from Peninsula Laboratories, Inc. (Belmont, CA). The N$^\alpha$-amino functional group was protected exclusively with the t-butyloxycarbonyl group. The side-chain protecting groups were as follows: Bzl for Asp, Thr, Ser, Glu; p(Me)Bzl for Cys; 2,6-(Cl$_2$)Bzl for Tyr; 2-(Cl)Z for Lys and Tos for Arg. After the last residue was coupled on, the growing peptide was treated with the low-high hydrogen fluoride cleavage procedure to remove the peptide from the resin anchor and deprotect the side-chain functional groups.[24]

The crude peptide was extracted, cyclized by air oxidation and dialyzed against 1 M acetic acid. The recovered dialysate was lyophilized and the crude product gel filtered on a Sephadex G-50 fine column in 1 M acetic acid to remove the polymeric byproduct. The recovered

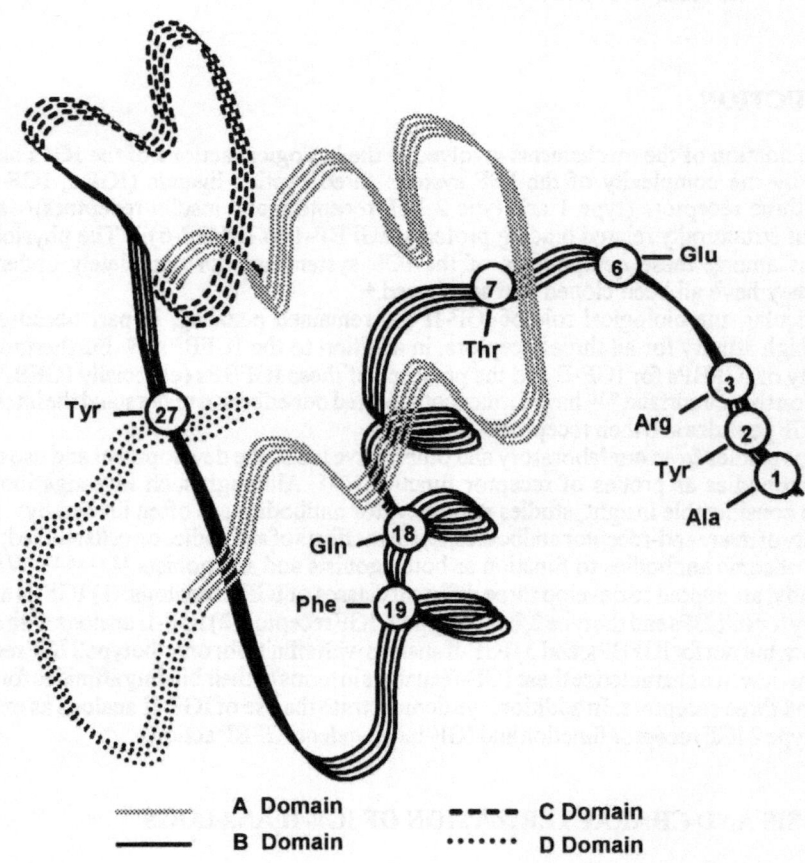

Figure 1. Schematic representation of the tertiary structure of IGF-II. The mutation portions of IGF-II analog are indicated by *numbers*. Modified from Blundell et al.[18]

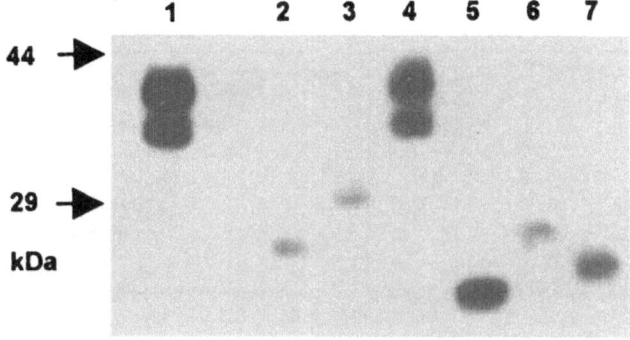

Figure 2. Western ligand blot of purified IGFBPs 1-6 and human serum. Samples were separated by 10% SDS-PAGE, electroblotted to nitrocellulose, incubated with [^{125}I]IGF-II, and autoradiographed. Lane 1, human serum; lane 2, hIGFBP-1; lane 3, hIGFBP-2; lane 4, rec-hIGFBP-3; lane 5, hIGFBP-4; lane 6, rec-hIGFBP-5; lane 7, rec-hIGFBP-6.

monomeric product was purified batchwise by three steps of HPLC, using the conditions as previously described.[125] IGF-II analogs were monitored by a type 2 IGF receptor binding assay, using rat placental membranes. The purified product has the correct amino acid composition and its sequence was verified by amino acid sequence analysis, using a model 470A gas-phase protein sequenator (Applied Biosystems, Inc., Foster City, CA).

Preparation of IGFBPs (IGFBP-1 to -6) and Receptors

HPLC-purified human (h) IGFBP-1 from human amniotic fluid was a generous gift from Dr. D. R. Powell (Baylor College of Medicine, Houston, TX). hIGFBP-2 and hIGFBP-4 were HPLC-purified from conditioned media (CM) of primary human prostatic cell cultures and prostatic epithelial cells, respectively, as previously described.[26] Recombinant (rec) hIGFBP-3[CHO], in a fully glycosylated form expressed in Chinese hamster ovary cells, was kindly provided by Celtrix (Santa Clara, CA). rec-hIGFBP-5 was purified from hIGFBP-5 expressed in yeast extracts, provided by Dr. M. C. Kiefer[27] and rec-hIGFBP-6 was purchased from Austral Biologicals (San Ramon, CA). All pure IGFBPs were confirmed by western ligand blot, as shown in Fig. 2. The sizes of each IGFBP are as follows: hIGFBP-1, 26-kDa; hIGFBP-2, 32-kDa; rec-hIGFBP-3[CHO], 39- and 41-kDa; hIGFBP-4, 24-kDa; rec-hIGFBP-5, 28-kDa; rec-hIGFBP-6, 25-kDa.

Crude microsomal human and rat placental membranes were prepared by differential centrifugation for insulin receptor, type 1 IGF receptor, and type 2 IGF receptor binding assays, respectively.[28]

CHARACTERIZATION OF IGF-II ANALOG AFFINITIES FOR IGFBPs AND RECEPTORS

Binding Affinity of IGF-II Analogs for IGFBPs

To characterize the binding affinity of IGF-II analogs for each IGFBP, charcoal assays were performed as previously described.[26] Briefly, the purified IGFBPs (4 ng hIGFBP-1; 1 ng hIGFBP-2, rec-hIGFBP-3[CHO] and hIGFBP-4; 0.1 ng rec-hIGFBP-5; 5 ng rec-hIGFBP-6) were incubated with [^{125}I]IGF-II in the presence or absence of various concentrations of unlabeled IGF-I, IGF-II and IGF-II analogs. After incubation, 0.5% activated charcoal in PBS containing 1% BSA with protamine sulfate (0.2 mg/ml) was applied, and samples were centrifuged to separate [IGF-IGFBP] complexes from unbound IGF-II. It is of note that protamine sulfate was not added to the IGFBP-1 and IGFBP-5 assays, because of its inhibitory effect on binding.

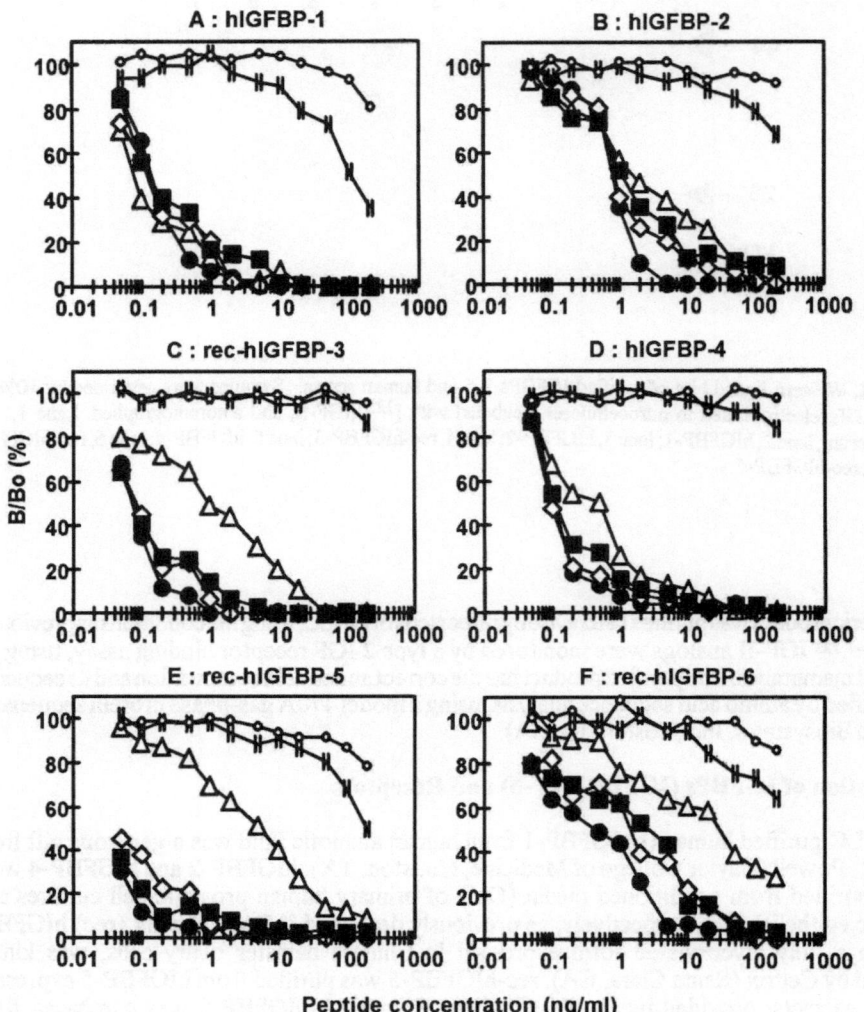

Figure 3. Competitive protein binding assay between [^{125}I]IGF-II and unlabeled IGFs and IGF-II analogs for binding to IGFBPs 1-6 (A-F). Increasing concentrations of IGF-I (△); IGF-II (●); [Leu^{27}]IGF-II (◇); [Ser^{27}]IGF-II (■); [Gln^6,Ala^7,Tyr^{18},Leu^{19}]IGF-II (||); and [Gln^6,Ala^7, Tyr^{18},Leu^{19},Leu^{27}]IGF-II (o) were added to all the binding assays. Results shown are means of duplicate determinations performed in two independent experiments for each IGFBP.

Binding curves of [^{125}I]IGF-II binding to individual IGFBPs in the presence of various concentrations of IGF-II analogs are shown in Fig. 3. In addition, the relative potency of each analog in competition studies is presented in Table 1 as the mean IC$_{50}$ of IGF-II divided by the mean IC$_{50}$ of the IGF-II analog. As can be seen in Fig. 3, IGF-I and IGF-II were approximately equipotent in their affinities for hIGFBP-1, hIGFBP-2 and hIGFBP-4, as assessed by their abilities to compete with [^{125}I]IGF-II. In contrast, different binding affinities for IGF-I and IGF-II were seen with rec-hIGFBP-3, rec-hIGFBP-5 and rec-hIGFBP-6 (Fig. 3 C, E and F). These IGFBPs showed a higher affinity for IGF-II than for IGF-I, i.e., more than 12-fold higher affinity for rec-hIGFBP-3, 250-fold for rec-hIGFBP-5 and 50-fold for rec-hIGFBP-6. The IC$_{50}$ values of IGF-II for hIGFBP-1, hIGFBP-2, rec-hIGFBP-3, hIGFBP-4, rec-hIGFBP-5 and rec-hIGFBP-6 were 0.15 ± 0.01, 0.9 ± 0.07, 0.07 ± 0.028, 0.1 ± 0.01, 0.02 ± 0.009 and 0.5 ± 0.14 ng/ml, respectively. Modifications of Glu^6, Thr^7, Gln^{18} and Phe^{19} in the B domain of IGF-II resulted in severe decreases in affinity for binding to all IGFBPs tested. [Gln^6,Ala^7,Tyr^{18},Leu^{19}]IGF-II and [Gln^6,Ala^7,Tyr^{18},Leu^{19},Leu^{27}]IGF-

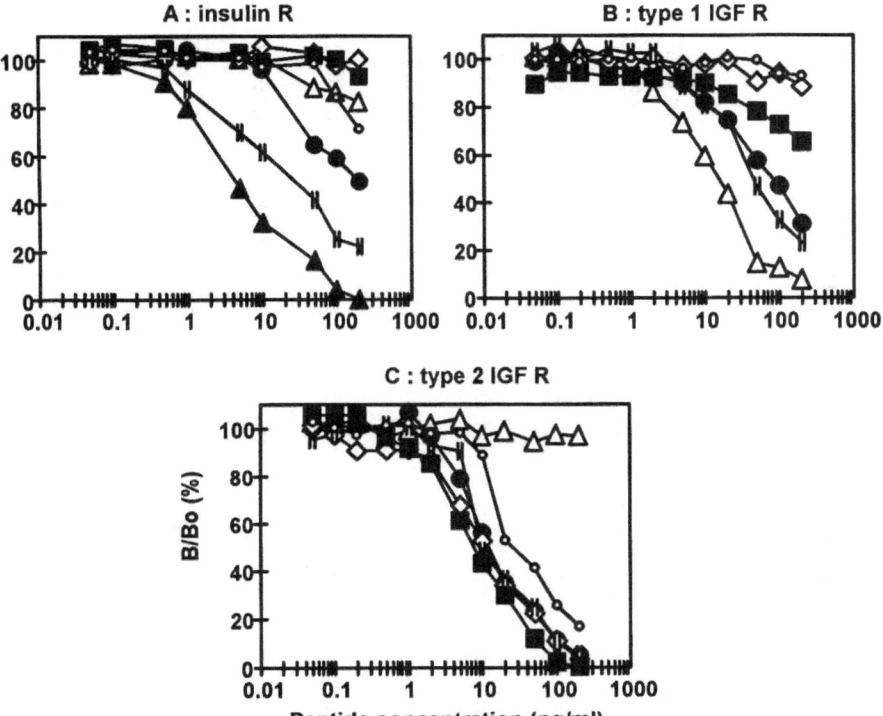

Figure 4. Inhibition of ligand binding to human placental insulin receptor (A, [^{125}I]insulin), human placental type 1 IGF receptor (B, [^{125}I]IGF-I), and rat placental type 2 IGF receptor (C, [^{125}I]IGF-II) by insulin (▲), IGF-I (△); IGF-II (●);[Leu^{27}]IGF-II (◇);[Ser^{27}]IGF-II (■); [Gln^6,Ala^7,Tyr^{18},Leu^{19}]IGF-II (◫); and [Gln^6,Ala^7, Tyr^{18},Leu^{19},Leu^{27}]IGF-II (o). Results shown are means of duplicate determinations performed in two independent experiments for each receptor.

II demonstrated more than 1000-fold decreases in affinity for all of IGFBPs ($p<0.005$), except hIGFBP-2 (more than 250-fold, $p<0.05$), as shown in Table 1. Substitutions for tyrosine at position 27 with either leucine or serine resulted in no significant changes in binding affinity for any of the IGFBPs.

These data indicate that modifications in the B domain of IGF-II can dramatically affect affinity for IGFBPs, especially positions 6, 7, 18 and 19, which are critical for binding to IGFBPs, as has been shown with IGF-I analogs.[26] Our study, furthermore, demonstrates that each analog has striking consistency in its relative affinity for IGFBP-1-6, indicating that IGF-binding sites are similar among the IGFBPs. It is of note that relatively high sequence similarity exists among the six IGFBPs, with preservation of at least 16 cysteines. It is, therefore, not surprising that structural alterations of IGF-II which affect its affinity for one IGFBP are likely to affect affinities for the other IGFBPs, as well.

Binding Affinity of IGF-II Analogs for Receptors

The affinities of IGF-II analogs for the type 1 and type 2 IGF receptors and insulin receptor were evaluated by receptor competitive binding studies. Briefly, crude microsomal human and rat placental membranes were prepared by differential centrifugation.[28] Fifty μg of membranes were incubated with [^{125}I]-ligands in the presence or absence of various concentrations of unlabeled IGF-I, IGF-II, insulin and IGF-II analogs. Samples were centrifuged at 12, 000 x g for 30 minutes at 4C. The resulting pellets were counted in a gamma counter.

As depicted in Fig. 4, the binding affinity of IGF-II for each receptor was type 2 IGF receptor > type 1 IGF receptor > insulin receptor. The IC$_{50}$ values of IGF-II for insulin receptor, type 1 and type 2 IGF receptors were 195 ± 7.1, 80 ± 42.4 and 11 ± 5.7 ng/ml, respectively (Table 1).

Table 1. Relative potency of IGF-II analogs for competing with [^{125}I]IGF-II binding to IGFBPs and receptors
Relative potency is the ratio of IC_{50} of IGF-II/IC_{50} of the analog.
Experiments have been performed twice in duplicate and data are presented as mean ±S.D.
** p < 0.005, * p < 0.05 versus IGF-II.

Peptide	hIGFBP-1	hIGFBP-2	rec-hIGFBP-3	hIGFBP-4	rec-hIGFBP-5	rec-hIGFBP-6	insulin receptor	type 1 IGF receptor	type 2 IGF receptor
IGF-II (IC_{50} in ng/ml)	1 (0.15±0.01)	1 (0.9±0.07)	1 (0.07±0.028)	1 (0.1±0.01)	1 (0.02±0.009)	1 (0.5±0.14)	1 (195±7.1)	1 (80±42.4)	1 (11±5.7)
IGF-I	1.7±0.23	0.5±0.22	0.08±0.035	0.2±0.04	*0.004±0.001	*0.02±0.007	*<0.1	6.0±4.52	<0.07
[L^{27}]IGF-II	1.2±0.01	1.0±0.21	1.3±0.54	1.1±0.07	0.3±0.02	0.3±0.09	*<0.1	<0.05	0.8±0.17
[S^{27}]IGF-II	1.2±0.01	0.8±0.04	0.9±0.35	0.9±0.12	0.9±0.83	0.2±0.12	*<0.1	<0.05	1.0±0.29
[Q^{6}A^{7}Y^{18}L^{19}]IGF-II	**<0.001	**<0.005	*<0.001	*<0.001	*<0.001	*<0.001	*7.4±0.34	1.5±0.21	1.0±0.57
[QAYL-L^{27}]IGF-II	**<0.001	**<0.005	*<0.001	*<0.001	*<0.001	*<0.001	*<0.1	<0.05	0.5±0.04
insulin							*46.1±5.52		

All four IGF-II analogs, involving either modifications of position 27 and/or modifications of the B domain, showed no significant change in affinity for the type 2 IGF receptor (Fig. 4C). In contrast, analogs with modifications of position 27, such as [Leu[27]]IGF-II, [Ser[27]]IGF-II and [Gln[6],Ala[7],Tyr[18],Leu[19],Leu[27]]IGF-II, had significantly decreased affinities for insulin receptor (more than 5-fold, $p < 0.05$) and type 1 IGF receptor (more than 20 fold) (Fig. 4A and B). It is of note that Bürgisser et al.[29] and Sakano et al.[30] have shown similar changes in receptor affinity with their [Leu[27]] or [Glu[27]]IGF-II preparations.

Interestingly, [Gln[6],Ala[7],Tyr[18],Leu[19]]IGF-II showed significantly increased binding affinity, especially for the insulin receptor (approximately 7-fold increase, $p < 0.05$), compared with IGF-II (Table 1). This surprising observation can be explained by the existence of IGFBP-3 in human placental membrane preparations. Indeed, IGFBP-3 was identified in human placenta by northern blot and by [^{125}I]IGF-I or -II affinity cross-linking of human placental membranes (data not shown), indicating that membrane-bound IGFBP-3 can interfere in IGF binding to receptors because of its high affinity for IGFs. In particular, in studies of insulin receptor binding, membrane-bound IGFBP-3 affects only IGF ligands, but neither [^{125}I]insulin or cold insulin, resulting in the appearance of significantly decreased affinity of IGFs for the insulin receptor. However, [Gln[6],Ala[7],Tyr[18],Leu[19]]IGF-II, which has significantly decreased affinity for IGFBPs, can overcome the interference by IGFBP-3 in competition studies with [^{125}I]insulin, resulting in an approximately 7-fold increased affinity compared with IGF-II. In this matter, data from [Gln[6],Ala[7],Tyr[18],Leu[19]]IGF-II represents the true IGF-II affinity for insulin receptor; thus, IGF-II affinity is only 6-fold less than insulin for the insulin receptor, not the two orders of magnitude difference commonly reported. It is of note that [Gln[3],Ala[4],Tyr[15],Leu[16]]IGF-I showed 14-fold higher affinity than IGF-I for human placental insulin receptors, although IGF-I, itself, has significantly low affinity for the insulin receptor.[19]

In summary, these studies clearly demonstrate that position 27 of the IGF-II molecule is critical for binding to insulin and type 1 IGF receptors, whereas positions 6, 7, 18, and 19 are critical for binding to IGFBPs, implying that binding sites reside differently in the IGF-II molecule for the type 2 IGF receptor, type 1 IGF or insulin receptors, or IGFBPs. Indeed, Sakano et al have previously demonstrated that modifications at positions 48, 49 and 50, or 54 and 55 resulted in significantly reduced binding affinity for the type 2 IGF receptor only, implying that binding sites for the type 2 IGF receptor reside in the A domain of the IGF-II molecule.[30]

Taken together, we have successfully synthesized IGF-II analogs with selected affinities for IGFBPs and receptors : 1) IGF-II analogs with affinity for IGFBPs and only for the type 2 IGF receptor; [Leu[27]]IGF-II and [Ser[27]]IGF-II; 2) IGF-II analogs with affinity for receptors only; [Gln[6],Ala[7],Tyr[18],Leu[19]]IGF-II; 3) IGF-II analogs with affinity for only the type 2 IGF receptor; [Gln[6],Ala[7],Tyr[18],Leu[19],Leu[27]]IGF-II.

USE OF IGF-II ANALOGS AS PROBES OF RECEPTOR AND IGFBP FUNCTION

As mentioned above, it has been exceedingly difficult to correlate IGF-II's interaction with the various receptors and its consequent biological actions, because of the affinity of IGF-II for all three receptors, as well as for the IGFBPs. To address this matter, our IGF-II analogs, which have selective affinity for the type 2 IGF receptor with/without affinity for IGFBPs, or which retain full affinity for all three receptors but significantly reduced affinity for all IGFBPs, can be readily employed to discriminate among insulin and IGF receptors and to determine which receptor mediates a specific biological action of IGF-II. In addition, these analogs can be excellent tools to elucidate the effect of IGFBPs on the interactions between IGFs and their receptors, and to identify the mechanism for IGF-independent action of IGFBPs.

Application for Receptor Studies

Non-type 2 IGF Receptor Mediated IGF-II Action-The stimulation of DNA synthesis and cell proliferative effects of IGF-II are thought to be mediated via the type 1 IGF receptor or the insulin receptor, depending on the cell types studied.[1,2] To further investigate the mechanism of those IGF-II actions, we used Balb/c 3T3 cells to examine IGF-II's effect on DNA synthesis. As shown in Fig. 5, Balb/c 3T3 cells can be stimulated to incorporate [^{3}H]thymidine with nanomolar concentrations of either IGF-I or IGF-II. However, when [Leu[27]]IGF-II, which is highly selective for the type 2 IGF receptor, was applied, [^{3}H]thymidine incorporation rate

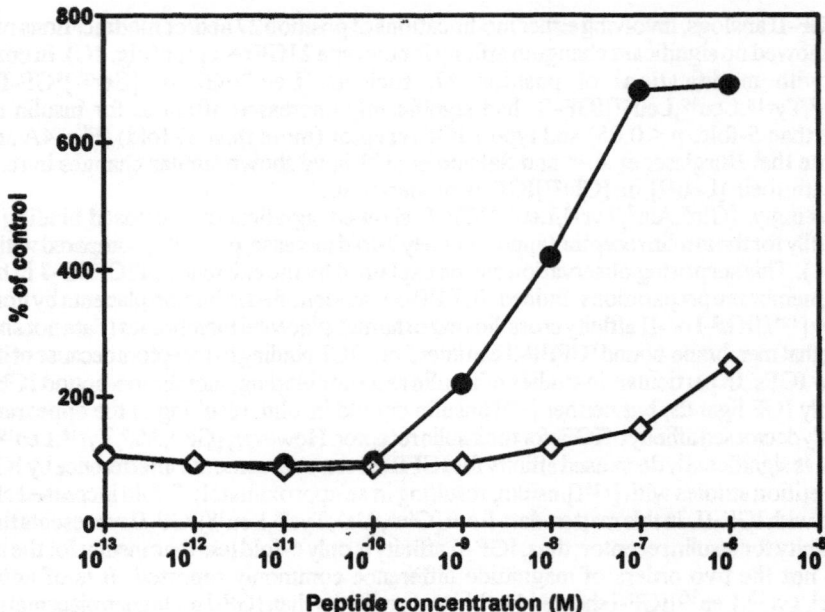

Figure 5. Stimulation of [³H]thymidine incorporation in Balb/c 3T3 cells by IGF-II (●) and [Leu²⁷]IGF-II
(◇). Data from Beukers et al.[25]

was markedly reduced (more than 100-fold). These data convincingly demonstrate that stimulation
of DNA synthesis in these cells by IGF-II is not mediated by the type 2 IGF receptor. On the contrary,
it is most likely that IGF-II stimulation of [³H]thymidine incorporation in these cells is through the type
1 IGF receptor. These conclusions are consistent with previous studies in other cell lines, employing
antibodies to the type 1 or type 2 receptors.[11,14,15]

Type 2 IGF Receptor Mediated IGF-II Action-The type 2 IGF receptor has proven to be
identical to the cation-independent mannose-6-phosphate receptor (CI-MPR).[6-8] The type 2
IGF-II/CI-MPR is a transmembrane monomeric molecule that binds the two classes of ligands
to different sites[8,31] and has no demonstrable tyrosine kinase activity. It has been reported that
IGF-II interaction with this receptor can modulate the binding of lysosomal enzymes,[31,32] and
Nishimoto et al.[33,34] have shown that this receptor is capable of interacting with a $G_{\alpha i2}$ GTP
binding protein and can couple to a calcium-permeable channel. Nevertheless, attempts at
defining the biological relevance of the binding of IGF-II to this receptor have been less
successful.

Recently, Helman et al.[35] have demonstrated that IGF-II functions as an autocrine growth
and motility factor in human rhabdomyosarcoma cell lines. The former effect was inhibited by
α-IR3, a blocking monoclonal antibody specific for the type 1 IGF receptor, whereas the latter
effect was not inhibited, suggesting type 2 receptor mediated action. To further investigate the
mechanism for IGF-II effect on cell motility, IGF-II analogs were applied to one line of human
rhabdomyosarcoma cells, RD cells.[36] As shown in Fig. 6, motility responses of RD cells can
be detected with either 10⁻⁸ M IGF-II or [Leu²⁷] IGF-II with equivalent potency. It is of note
that at the concentrations of [Leu²⁷]IGF-II employed in these experiments, there is minimal
binding of the analog to the type 1 IGF receptor.[36] In addition, 1 μg/ml α-IR-3, a blocking
antibody to the type 1 IGF receptor, did not inhibit the motility response elicited by either IGF-
II or [Leu²⁷]IGF-II (Fig. 6A). On the other hand, when RD cells were preincubated with the
purified IgG antibody fraction to the human type 2 IGF receptor, the purified human type 2
receptor Ab completely inhibited the [Leu²⁷]IGF-II-induced stimulation of RD cell motility,
whereas it had no effect on unstimulated motility (Fig. 6B).

These studies in RD cells clearly demonstrate that IGF-II can exert two different responses
mediated by distinct receptors: 1) a mitogenic response through the type 1 IGF receptor, and 2) a
motility response through the type 2 IGF receptor. The application of IGF-II analogs in these studies
thus provides further insight into type 1 and type 2 IGF receptor specific actions.

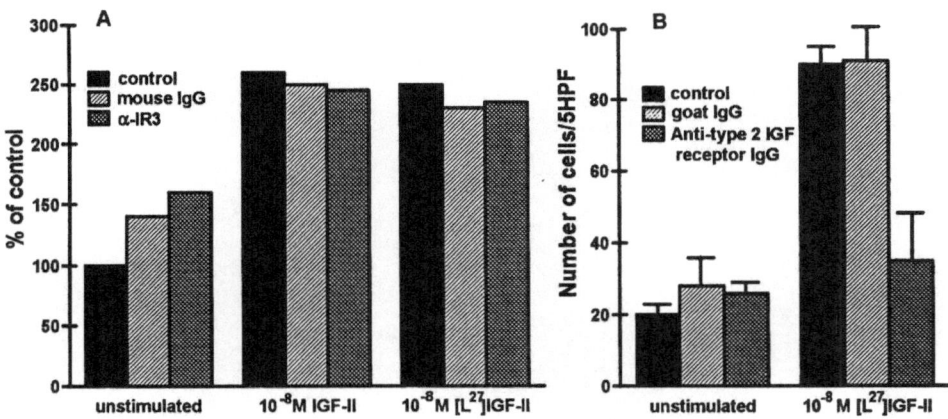

Figure 6. Motility response of rhabdomyosarcoma cells pretreated with anti-type 1 or anti-type 2 IGF receptor antibodies. RD cells were harvested, rested, and then incubated in serum free medium with α-IR3, monoclonal anti-type 1 IGF receptor Ab (**A**), or the purified anti-type 2 IGF receptor Ab (**B**). The incubation with α-IR3 did not inhibit IGF-II- nor [Leu27]IGF-II-induced motility (**A**), whereas purified IgG to type 2 IGF receptor totally abolished RD motility response to [Leu27]IGF-II (**B**). Data from Minniti et al.[36]

In summary, these studies exemplify the use of IGF-II analogs in receptor studies and prove that IGF-II analogs can be important tools in determining the biological roles of IGF-II and the type 1 and type 2 IGF receptors. Compared with the anti-receptor antibodies, IGF-II analogs have the additional advantage of not being influenced by cytotoxic or non-receptor mediated effects of anti-receptor antibodies.

Application for Studies of IGF-independent Actions of IGFBPs

Recently, our laboratory and others have observed the existence of IGFBP-3 on the cell surface.[9,37,38] In addition, we have demonstrated that cell surface-associated IGFBP-3 can be released into conditioned media by the addition of IGF peptide in Hs578T human breast cancer cells.[9,39] Further investigation has demonstrated that IGFBP-3 binding on the cell surface is a specific event and is facilitated by divalent cations, indicating involvement of specific receptor-mediated interaction.[40] For these studies, which are critical to our understanding of how IGFBPs modulate IGF-receptor interactions, IGF-II analogs can be of considerable discriminatory value.

[Leu27]IGF-II Effect on Unmasking IGF Receptors-As shown in Fig. 7A, [125I]IGF-I cross-linking to Hs578T microsomal crude membranes showed no detectable α-subunit of the type 1 IGF receptor (135-kDa) under reducing conditions. Instead, [125I]IGF-I binding to Hs578T membranes was associated with IGFBP-3, shown as a broad band of approximately 48-kDa, indicating that under normal conditions, IGF was preferentially bound to IGFBP-3 on Hs578T cell surface. This membrane-associated IGFBP-3 was displaced by cold IGF-I and IGF-II, but not by IGF-I/insulin hybrid or by [Gln3,Ala4,Tyr15,Leu16]IGF-I, which have decreased affinity for IGFBPs. However, coincubation of [Leu27]IGF-II, which competes for occupancy of the IGFBPs, but not for the type 1 IGF receptor, revealed a typical 135-kDa α-subunit and 270-kDa α-α dimer of the type 1 IGF receptor. Thus, because of the higher affinity of IGFBP-3 for IGFs, the type 1 IGF receptor is normally "masked", but can be unmasked by low concentrations of [Leu27]IGF-II. The analog preferentially binds to cell surface-associated IGFBPs, thereby permitting the type 1 IGF receptor to bind [125I]IGF-I.

Similarly, the type 2 IGF receptor was masked by the presence of membrane-bound IGFBP-3, as shown in Fig. 7B. Addition of low concentration of [Leu27]IGF-II displaces [125I]IGF-II from the IGFBPs, unmasking the type 2 IGF receptor. Addition of excess unlabeled IGF-I also reveals a 250-kDa type 2 IGF receptor, since IGF-I acts like the IGF-II analog, with preferential binding affinity for IGFBPs. The detection of membrane bound IGFBP-3 is not an artifact

49

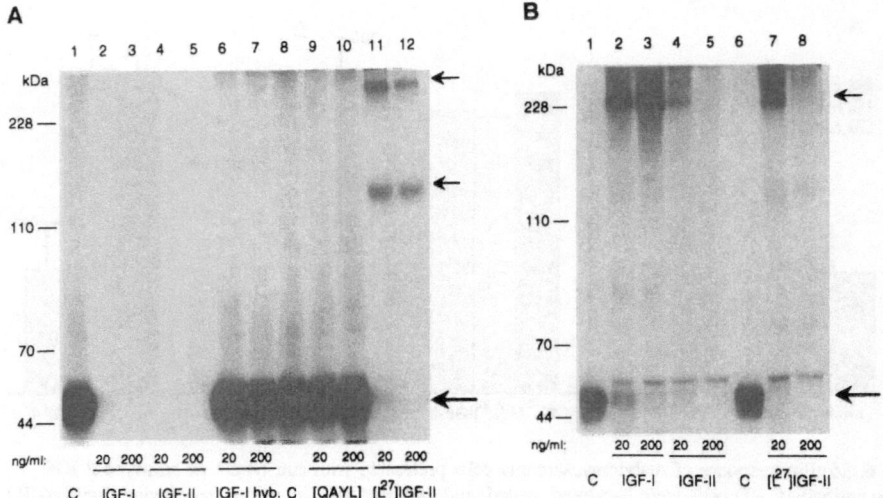

Figure 7. Autoradiogram of [^{125}I]IGF-I or -II cross-linked to crude microsomal membranes of Hs578T cells. Hs578T crude microsomal membrane were cross-linked with either [^{125}I]IGF-I (**A**) or [^{125}I]IGF-II (**B**) in the presence or absence of unlabeled IGF peptides and IGF analogs. 100 µg of membranes were either untreated (C, control) or preincubated with 20 or 200 ng/ml IGF-I, IGF-II, IGF-I/insulin hybrid, [Gln^3,Ala^4,Tyr^{15}, Leu^{16}]IGF-I and [Leu^{27}]IGF-II for 30 min at 23 C, before [^{125}I]IGF (150,000 cpm) binding and cross-linking. Solubilized complexes (reducing conditions) were separated by SDS-PAGE on a 6 % gel. Large arrows indicate IGFBP-3, whereas small arrows indicate α-subunit and α-α dimer of the type 1 IGF receptor (**A**), or the type 2 IGF receptor (**B**). Data from Oh et al.[40]

of microsomal crude membrane preparations, since IGFBP-3 is demonstrable on the cell surface by using monolayer cross-linking methods.[9] These studies demonstrate an example of the unique use of IGF analogs with high affinity for IGFBPs to identify "masked" IGF receptors.

Identification of IGF-independent Action of IGFBP-3–In most cell systems, IGFBPs inhibit IGF actions, presumably by competing with receptors for the IGF ligand. To investigate the biological significance of IGFBP-3 binding to the cell surface, IGF-II analogs were employed to evaluate whether the effects of IGFBP-3 result from blocking IGF actions by preventing IGF binding to receptors ("IGF-dependent" IGFBP-3 action), or result from unique, "IGF-independent" actions.

In the Hs578T breast cancer cell system, it has been observed that insulin, IGFs and EGF have no mitogenic effects on these cells,[41,42] presumably because cells possess the oncogenic *ras* protein, which was derived from a point mutation at position 12 of Hs578T *ras* protein.[43] This can be confirmed by IGF-II and IGF-II analog treatments, as shown in Fig. 8. Both native IGF-II and [Gln^6,Ala^7,Tyr^{18},Leu^{19},Leu^{27}]IGF-II, which does not bind to IGFBPs, have no effect on Hs578T cell monolayer growth in concentrations up to 100 nM. This indicates that the failure of IGF-II to stimulate Hs578T cells cannot be attributed to interference by endogenous IGFBPs, since both IGF-II and the IGF-II analog with reduced affinity for IGFBPs, were equivalently impotent.

However, exogenously added IGFBP-3$^{E.\ coli}$ still showed a significant inhibitory effect on monolayer growth of Hs578T cells (p<0.005). This inhibitory effect was dose-dependent, with 70% inhibition at a concentration of 100 nM, and was specific for IGFBP-3; no significant inhibitory effect was seen with addition of IGFBP-1 (data not shown). Furthermore, when 20 nM IGF-II was coincubated with various concentrations of IGFBP-3$^{E.\ coli}$, the IGFBP-3$^{E.\ coli}$ inhibitory effect on Hs578T cells was blocked, until IGFBP-3 concentrations reached 20 nM. In contrast, coincubation of with 20 nM [Gln^6,Ala^7,Tyr^{18},Leu^{19},Leu^{27}]IGF-II did not result in any attenuation of the IGFBP-3 inhibitory effect, even though this analog is equipotent with IGF-II for binding to the type 2 IGF receptor.

These data indicate that the IGFBP-3 inhibitory effect on cell growth does not result from blocking the mitogenic actions of IGF-II. Rather, the results suggest a cell surface-specific action of IGFBP-3, itself. The ability of IGF-II, but not the analog, to attenuate the inhibitory effect of IGFBP-3 correlates with the affinity of IGF-II for IGFBP-3, thereby preventing cell surface binding

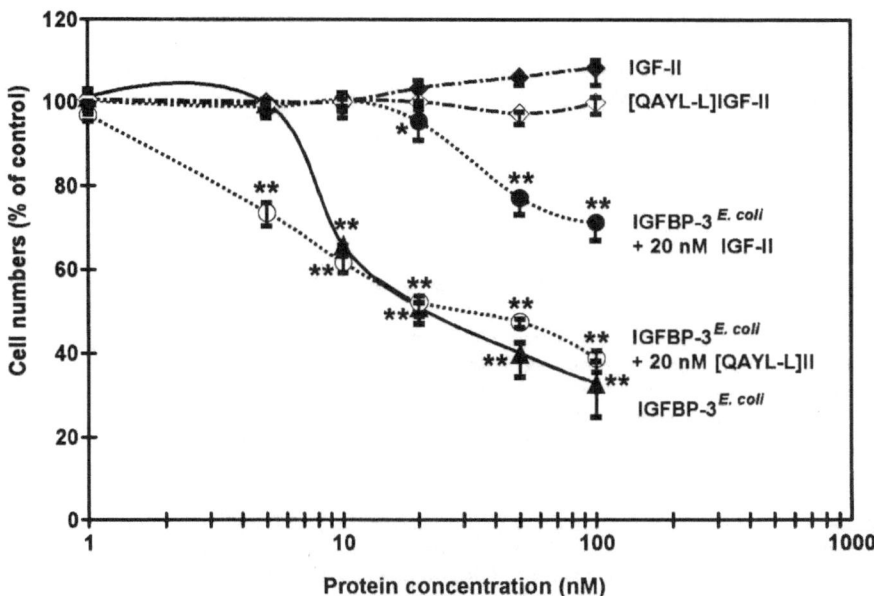

Figure 8. Effects of coincubation with IGFBP-3[E. Coli] and IGF-II or IGF-II analog on monolayer growth of Hs578T cells. Cells were treated with various concentrations of IGF-II (♦), [Gln[6],Ala[7],Tyr[18],Leu[19],Leu[27]] IGF-II (◇), IGFBP-3[E. Coli] alone (▲), or IGFBP-3 plus 20 nM IGF-II (●) or 20 nM [Gln[6],Ala[7],Tyr[18], Leu[19],Leu[27]]IGF-II (○). Statistical significance in comparison with control values is indicated by * (P < 0.05) or ** (P < 0.005). Data from Oh et al.[40]

of IGFBP-3. Analogs with reduced affinity for IGFBP-3 fail to dissociate the BP from the cell surface, and, therefore, fail to attenuate the inhibitory actions of IGFBP-3. Thus, analogs demonstrate that in this situation, rather than IGFBPs modulating IGF action, IGF peptides modulate IGFBP action.

In summary, IGF-II analogs, [Leu[27]]IGF-II, [Ser[27]]IGF-II, and [Gln[6],Ala[7], Tyr[18],Leu[19],Leu[27]]IGF-II, which have selective affinity for type 2 IGF receptor with/without affinity for IGFBPs, can be readily applied to identify type 2 receptor mediated IGF-II actions, and can be useful probes to discriminate among insulin and IGF receptors, to determine which receptor mediates a specific biological action of IGF-II. In addition, IGF-II analogs, [Gln[6],Ala[7],Tyr[18],Leu[19]]IGF-II and [Gln[6],Ala[7], Tyr[18],Leu[19],Leu[27]]IGF-II, which have significantly reduced affinity for all IGFBPs, can be excellent tools to elucidate the effect of IGFBPs on the interactions between IGFs and their receptors, and to identify the mechanism for IGF-independent action of IGFBPs, such as that demonstrated for Hs578T cells.

CONCLUSION

This review focuses on our recent findings concerning the synthesis, characterization and experimental applications of IGF-II analogs. Taken together with previous reports, these studies clearly demonstrate that the IGF-II molecule has different binding sites for receptors and IGFBPs. In particular, the type 2 IGF receptor binding site is different from binding sites for either the insulin or type 1 IGF receptors or IGFBPs. These observations led us to synthesize and characterize specific IGF-II analogs with selective affinities for IGFBPs and receptors: 1) IGF-II analogs with affinity for IGFBPs and only for the type 2 IGF receptor; [Leu[27]]IGF-II and [Ser[27]]IGF-II, 2) IGF-II analogs with affinity for receptors only; [Gln[6],Ala[7],Tyr[18],Leu[19]]IGF-II, 3) IGF-II analogs with affinity only for the type 2 IGF receptor; [Gln[6],Ala[7],Tyr[18],Leu[19],Leu[27]]IGF-II.

Although only a few illustrative examples for uses of IGF-II analogs are described in this review, we believe that IGF-II analogs will make it possible to further investigate type 2 IGF receptor specific biological actions. Furthermore, these analogs will be important tools for understanding the complexities of the IGF-IGFBP-IGF-receptor axis.

ACKNOWLEDGMENTS

Supported in part by NIH grant HD28703 (RGR) and Dr. Mildred Scheel Stipendium, Deutsche Krebshilfe e. V. (HLM).

REFERENCES

1. M.M. Rechler and S.P. Nissley, Insulin-like growth factors, in: "Handbook of Experimental Pharmacology", Vol. 95/1, "Peptide Growth Factors and Their Receptors I", M.B. Sporn and A.B. Roberts, eds., Springer-Verlag, Berlin (1990).

2. Y. Oh, H.L. Müller, E.K. Neely, G. Lamson, and R.G. Rosenfeld, New concepts in insulin-like growth factor receptor physiology, Growth Regulation (in press).

3. R.G. Rosenfeld, G. Lamson, H. Pham, Y. Oh, C. Conover, D.D. De Leon, S.M. Donovan, I. Ocrant, and L.C. Giudice, Insulin-like growth factor binding Proteins, Rec. Prog. Horm. Res. 46:99 (1990).

4. G. Steele-Perkins, J. Turner, J.C. Edman, J. Hari, S.B. Pierce, C. Stover, W.J. Rutter, and R.A. Roth, Expression and characterization of a functional human insulin-like growth factor I receptor, J. Biol. Chem. 263:11486 (1988).

5. S. Shimasaki and N. Ling, Identification and molecular characterization of insulin-like growth factor binding proteins (IGFBP-1,-2,-3,-4,-5, and -6), Prog. Growth Factor Res. 3:243 (1991).

6. D.O. Morgan, J.C. Edman, D.N. Standring, V.A. Fried, M.C. Smith, R.A. Roth, and W.J. Rutter, Insulin-like growth factor II receptor as a multifunctional binding protein, Nature 329:301 (1987).

7. R.G. MacDonald, S.R. Pfeffer, L. Coussens, M.A. Tepper, C.M. Brocklebank, J.E. Mole, J.K. Anderson, E. Chen, M.P. Czech, and A. Ullrich, A single receptor binds both insulin-like growth factor-II and mannose-6-phosphate, Science 239:1134 (1988).

8. P.Y. Tong, S.E. Tollefsen, and S. Kornfeld, The cation-independent mannose-6-phosphate receptor binds insulin-like growth factor II, J. Biol. Chem. 263:2585 (1988).

9. Y. Oh, H.L. Müller, H. Pham, G. Lamson, and R.G. Rosenfeld, Non-receptor mediated, post-transcriptional regulation of insulin-like growth factor binding protein-3 in Hs578T human breast cancer cells, Endocrinology 131:3123 (1992).

10. C. Camacho-Hubner, W.H.Jr. Busby, R.H. McCusker, G. Wright, and D.R. Clemmons, Identification of the forms of insulin-like growth factor-binding proteins produced by human fibroblasts and the mechanisms that regulate their secretion, J. Biol. Chem. 267:11949 (1992).

11. C.A. Conover, P. Misra, R.L. Hintz, and R.G. Rosenfeld, Effect of an anti-insulin-like growth factor I receptor antibody on insulin-like growth factor II stimulation of DNA synthesis in human fibroblasts, Biochem. Biophys. Res. Commun. 139:501 (1987).

12. J. Hari, S.B. Pierce, D.O. Morgan, V. Sara, M.C. Smith, and R.A. Roth, The receptor for insulin-like growth factor II mediates an insulin-like response, EMBO J. 6:3367 (1987).

13. I. Kojima, I. Nishimoto, T. Iiri, E. Ogata, and R.G. Rosenfeld, Evidence that type II insulin-like growth factor receptor is coupled to calcium gating system, Biochem. Biophys. Res. Commun. 154:9 (1988).

14. C. Mottola and M.P. Czech, The type II insulin-like growth factor receptor does not mediate increased DNA synthesis in H-35 hepatoma cells, J. Biol. Chem. 259:12705 (1984).

15. E.Y. Adashi, C.E. Resnick, and R.G. Rosenfeld, Insulin-like growth factor -I (IGF-I) hormonal action in cultured rat granulosa cells: Mediation via type I but not type II IGF receptors, Endocrinology 126:216 (1989).

16. G. Steele-Perkins and R.A. Roth, Monoclonal antibody alphaIR$_3$ inhibits the ability of insulin-like growth factor II to stimulate a signal from the type-I receptor without inhibiting its binding, Biochem. Biophys. Res. Commun. 171:1244 (1990).

17. G. Steele-Perkins and R.A. Roth, Insulin-mimetic anti-insulin receptor monoclonal antibodies stimulate receptor kinase activity in intact cells, J. Biol. Chem. 265:9458 (1990).

18. T.L. Blundell, S. Berdarka, E. Rinderknecht, and R.E. Humbel, Insulin-like growth factor: A model for tertiary structure accounting for immunoreactivity and receptor binding, Biochemistry 75:180 (1978).

19. M.A. Cascieri and M.L. Bayne, Identification of the domains of IGF-I which interact with the IGF receptors and binding proteins, in: "Molecular and Cellular Biology of Insulin-like Growth Factors and Their Receptor", D. ReRoith and M.K. Raizada, eds., Plenum Press, NY (1989).

20. M.L. Bayne, J. Applebaum, G.G. Chicchi, N.S. Hayes, B.G. Green, and M.A. Cascieri, Structural analogs of human insulin-like growth factor I with reduced affinity for serum binding proteins and the type 2 insulin-like growth factor receptor, J. Biol. Chem. 263:6233 (1988).

21. M.A. Cascieri, G.G. Chicchi, J. Applebaum, B.G. Green, N.S. Hayes, and M.L. Bayne, Structural analogs of human insulin-like growth factor (IGF) I with altered affinity for type 2 IGF receptors J. Biol. Chem. 264:2199 (1989).

22. M.L. Bayne, J. Applebaum, G.G. Chicchi, R.E. Miller, and M.A. Cascieri, The role of tyrosines 24, 31, and 60 in the high affinity binding of insulin-like growth factor-I to the type 1 insulin-like growth factor receptor, J. Biol. Chem. 265:15648 (1990).

23. N. Ling, F. Esch, P. Böhlen, P. Brazeau, W.B. Wehrenberg, and R. Guillemin, Isolation, primary structure and synthesis of human hypothalamic somatocrinin: Growth hormone-releasing factor, Proc. Nat. Acad. Sci. USA. 81:4302 (1984).

24. J.P. Tam, W.F. Heath, and R.B. Merrifield, S_N2 deprotection of synthetic peptides with a low concentration of HF in dimethylsulfide: Evidence and application in peptide synthesis, J. Am. Chem. Soc. 105:6442 (1983).

25. M.W. Beukers, Y. Oh, H. Zhang, N. Ling, and R.G. Rosenfeld, [Leu^{27}]insulin-like growth factor II is highly selective for the type-II IGF receptor in binding, cross-linking and thymidine incorporation experiments, Endocrinology 128:1201 (1991).

26. Y. Oh, H.L. Müller, D. Lee, P.J. Fielder, and R.G. Rosenfeld, Characterization of the affinities of insulin-like growth factor (IGF)-binding proteins 1-4 for IGF-I, IGF-II, IGF/insulin hybrid, and IGF-I analogs, Endocrinology 132:1337 (1993).

27. M.C. Kiefer, C.S. Schmid, M. Waldvogel, I.Schläpfer, E. Futo, F.R. Masiarz, K. Green, P.J. Barr, and J. Zapf, Characterization of recombinant human insulin-like growth factor binding proteins 4, 5, and 6 produced in Yeast, J. Biol. Chem. 267:12692 (1992).

28. Y. Oh, M.W. Beukers, H. Pham, P.A. Smanik, M.C. Smith, and R.G. Rosenfeld, Altered affinity of insulin-like growth factor II (IGF-II) for receptors and IGF-binding proteins, resulting from limited modifications of the IGF-II molecule, Biochem. J. 278:249 (1991).

29. D.M. Bürgisser, B.V. Rith, R. Giger, C. Lüthi, S. Weigl, J. Zarn, and R.E. Humbel, Mutants of human insulin-like growth factor II with altered affinities for the type 1 and type 2 insulin-like growth factor receptor, J. Biol. Chem. 266:1029 (1991).

30. K. Sakano, T. Enjoh, F. Numata, H. Fujiwara, Y. Marumoto, N. Higashihashi, Y. Sato, J.F. Perdue, and Y.F. Yamaguchi, The design, expression, and characterization of human insulin-like growth factor II (IGF-II) mutants specific for either the IGF-II/cation-independent mannose 6-phosphate receptor or IGF-I receptor, J. Biol. Chem. 266:20626 (1991).

31. W. Kiess, G.D. Blickenstaff, M.M. Sklar, C.L. Thomas, S.P. Nissley, and G.C. Sahagian, Biochemical evidence that type II insulin-like growth factor receptor is identical to the cation-independent mannose 6-phosphate receptor, J. Biol. Chem. 263:9339 (1988).

32. W. Kiess, C.L. Thomas, M.M. Sklar, and S.P. Nissley, β-galactosidase decreases the binding affinity of the insulin-like growth factor II/mannose 6-phosphate receptor for insulin-like growth factor II, Eur. J. Biochem. 190:71 (1990).

33. I. Nishimoto, Y. Hata, E. Ogata, and I, Kojima, Insulin-like growth factor II stimulates calcium influx in competent Balb/c 3T3 cells primed with epidermal growth factor: Characteristics of calcium influx and involvement of GTP-binding protein, J. Biol. Chem. 262:12120 (1987).

34. T. Okamoto, T. Katada, Y. Murayama, M. Ui, E. Ogata, and I. Nishimoto, A simple structure encodes G protein-activating function of the IGF-II/mannose 6-phosphate receptor, Cell 62:709 (1990).

35. O.M. El-Badry, C.P. Minniti, E.C. Kohn, P.J. Houghton, W.H. Daughaday, and L.J. Helman, Insulin-like growth factor II acts as an autocrine growth and motility factor in human rhabdomyosarcoma tumors, Cell Growth & Differ. 1:325 (1990).

36. C.P. Minniti, E.C. Kohn, J.H. Grubb, W.S. Sly, Y. Oh, H.L. Müller, R.G. Rosenfeld, and L.J. Helman, The insulin-like growth factor II (IGF-II)/mannose 6-phosphate receptor mediates IGF-II-induced motility in human rhabdomyosarcoma cells, J. Biol. Chem. 267:9000 (1992).

37. J. Delbé, C. Blat, G. Desauty, and L. Harel, Presence of IDF45 (mIGFBP-3) binding sites on chick embryo fibroblasts, Biochem. Biophys. Res. Commun. 179:495 (1991).

38. J.L. Martin, M. Ballesteros, and R.C. Baxter, Insulin-like growth factor-I (IGF-I) and transforming growth factor-β1 release IGF-binding protein-3 from human fibroblasts by different mechanisms, Endocrinology131:1703(1992).

39. Y. Oh, H.L. Müller, H. Pham, G. Lamson, and R.G. Rosenfeld, Insulin-like growth factor binding protein (IGFBP)-3 levels in conditioned media of Hs578T human breast cancer cells are post- transcriptionally regulated, Growth Regulation 3:84 (1993).

40. Y. Oh, H.L. Müller, G. Lamson, and R.G. Rosenfeld, Insulin-like growth factor (IGF)-independent action of IGF binding protein (BP)-3 in Hs578T human breast cancer cells: cell surface binding and growth inhibition, J. Biol. Chem. (in press).

41. D.D. DeLeon, D.M. Wilson, M. Powers, and R.G. Rosenfeld, Effects of insulin-like growth factors (IGFs) and IGF receptor antibodies on the proliferation of human breast cancer cells, Growth Factors 6:327 (1992).

42. B.W. Ennis, E.M. Valverius, S.E. Bates, M.E. Lippman, F. Bellot, R. Kris, J. Schlessinger, H. Masui, A. Goldenberg, J. Mendelsohn, and R.B. Dickson, Anti-epidermal growth factor receptor antibodies inhibit the autocrine-stimulated growth of MDA-468 human breast cancer cells, Mol. Endocrinol 3:1830 (1989).

43. M.H. Kraus, Y. Yuasa, and S.A. Aaronson, A position 12-activated H-ras oncogene in all Hs578T mammary carcinosarcoma cells but not normal mammary cells of the same patient, Biochemistry 81:5384 (1984).

TOWARDS IDENTIFICATION OF A BINDING SITE ON INSULIN-LIKE
GROWTH FACTOR-II FOR IGF-BINDING PROTEINS

Leon A Bach,[1] Susan Hsieh,[1] Katsu-ichi Sakano,[2]
Hiroyuki Fujiwara,[2] James F Perdue,[3] and Matthew M Rechler[1]

[1] Growth and Development Section
Molecular and Cellular Endocrinology Branch
National Institute of Diabetes and Digestive and Kidney Diseases
National Institutes of Health
Bethesda MD 20892
[2] Molecular Biology Research Laboratory
Daiichi Pharmaceutical Co, Ltd
16-13, Kitakasai 1-chome
Edogawa-ku, Tokyo 134, Japan
[3] Molecular Biology Laboratory
The Jerome H Holland Laboratory of the American Red Cross
Rockville MD 20855

INTRODUCTION

Insulin-like growth factor (IGF) -I and -II mediate mitogenesis, differentiation and insulin-like metabolic effects.[1] Their amino acid sequence is divided into 4 domains designated B-C-A-D beginning from the amino terminus. The B- and A-domains of IGF-I and IGF-II share substantial homology with each other and with the B- and A-chains of insulin.

The actions of IGFs are mediated through the IGF-I, IGF-II and insulin receptors. IGF action is also modulated by binding to a family of 6 specific IGF binding proteins (IGFBPs).[2] In most circumstances, the biological actions of IGFs are inhibited by binding to IGFBPs; however, IGF action may also be potentiated by interaction with IGFBPs.

Recently, in order to elucidate the structure/function relationships of IGF-II, a number of IGF-II mutants have been synthesized using recombinant DNA technology and characterized with respect to their binding to IGF-I, insulin and IGF-II/Mannose-6-Phosphate (Man 6-P) receptors.[3] These results are summarized in Table 1 and suggest that insulin and IGF-I receptors recognize structural determinants that are distinct from those of the IGF-II/Man 6-P receptor. The aim of the present study was to characterize the binding of these and other mutants of human IGF-II to purified human IGFBPs 1-6.

Current Directions in Insulin-Like Growth Factor Research,
Edited by D. LeRoith and M.K. Raizada, Plenum Press, New York, 1994

55

Table 1. Receptor binding specificities of IGF-II mutants.

	IGF-I receptor	Insulin receptor	IGF-II /Man-6-P receptor
[Ser[26]]IGF-II	↓	↓↓	=
[Leu[27]]IGF-II	↓↓	↓↓	=
[Leu[43]]IGF-II	↓↓	↓↓	=
[Thr[48],Ser[49],Ile[50]]IGF-II	=	=	↓↓
[Arg[54],Arg[55]]IGF-II	=	=	0

The indicated residues were substituted for Phe[26], Tyr[27], Val[43], Phe[48], Arg[49], Ser[50], Ala[54], and Leu[55] of native IGF-II respectively (Fig. 1). Affinities relative to IGF-II: =, decreased <3-fold; ↓, decreased ~5-fold; ↓↓, decreased >50-fold; 0, does not bind to the IGF-II/Man-6-P receptor. Results are summarized from [3].

MATERIALS AND METHODS

Purification of IGFBPs

IGFBP-2 and IGFBP-6 were purified from human cerebrospinal fluid by ammonium sulfate concentration, IGF-II affinity chromatography and reverse phase FPLC as previously described.[4,5] IGFBP-1 was purified from human amniotic fluid by acid gel filtration, IGF-II affinity chromatography and reverse phase FPLC.[5] Recombinant glycosylated IGFBP-3 was kindly provided by Drs Christopher Maack and Andreas Sommer (Celtrix, Santa Clara, CA). IGFBP-4 purified from media conditioned by human T98G glioblastoma cells and recombinant human IGFBP-5 expressed in Chinese hamster ovary cells were kindly provided by Dr David Clemmons (University of North Carolina, Chapel Hill).

IGF-II Mutants

[Ser[26]]IGF-II, [Leu[27]]IGF-II, and [Leu[43]]IGF-II were expressed and purified from *Bombyx mori* larvae as previously described.[3] Other mutants were expressed and purified from *E. Coli*.[3] Fig. 2 shows the tertiary structure of IGF-II with mutated residues highlighted.

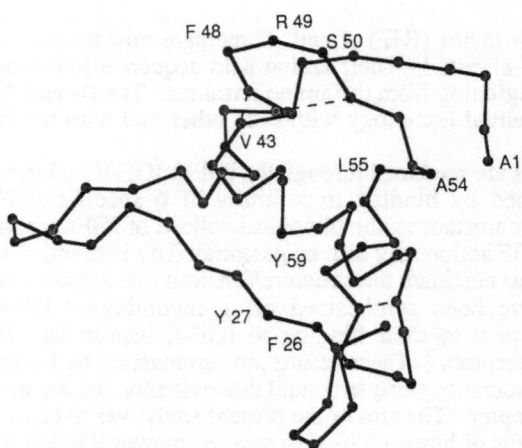

Figure 1. Tertiary structure of IGF-II. Residues are designated using the single letter code (A=Ala, F=Phe, L=Leu, R=Arg, S=Ser, V=Val, Y=Tyr). Mutated residues are labeled and denoted in black. Disulfide bonds are shown as broken lines. The picture was generated using MacImdad (Molecular Applications Group, Stanford, CA) based on the model described by Blundell et al. [6]

Competitive Binding

Individual IGFBPs were incubated with ^{125}I-IGF-II (12.5 pM) and 0.005-1.2 nM of unlabeled IGF-II, IGF-I or IGF-II mutant in 0.1 M NaPO$_4$/0.1% BSA/0.02% Na azide (4° C, 18 h, 0.4 ml final volume). Bound and free ligand were separated by addition of 0.5 ml ice-cold 5% charcoal/2% fatty acid-free bovine serum albumin, incubation on ice for 10 min, and centrifugation at 1300 g (4° C, 30 min). Bound radioactivity in supernatants was quantitated.[4,5] Results are shown as mean ± SD.

RESULTS

Binding of IGF-I and IGF-II to Purified Human IGFBPs 1-6

Of the IGFBPs, IGFBP-6 bound IGF-II with the highest affinity and had a 70-fold preferential binding affinity for IGF-II over IGF-I (Table 2). IGFBP-5 bound IGF-II with 7-fold higher affinity than IGF-I, whereas the other IGFBPs only had a slight preference for IGF-II.

Table 2. Binding of IGF-I and IGF-II to IGF-binding proteins.

	IGFBP-1	IGFBP-2	IGFBP-3	IGFBP-4	IGFBP-5	IGFBP-6
Ka IGF-II $(M^{-1} \times 10^{-10})$	K_1, 19 ± 8 K_2, 0.13 ± 0.02	1.7 ± 0.2	7.5 ± 1.0	1.1 ± 0.2	1.6 ± 0.5	39 ± 14
Relative affinity (% of IGF-II)						
IGF-I	64 ± 5	30 ± 1	40 ± 0	53 ± 21	14 ± 1	1.5 ± 0.4

Results are summarized from [5].

Binding of IGF-II Mutants to Purified Human IGFBPs 1-6

A-domain Mutants. Substitution of residues 48-50 with Thr[48], Ser[49], Ile[50] (the analogous residues in insulin) had the most profound effect on the binding of IGF-II to IGFBPs. IGFBPs -1, -4, -5 and -6 bound to this mutant with 1% or less of the affinity of IGF-II, whereas IGFBPs -2 and -3 bound the mutant with 10-15% of the affinity of IGF-II (Fig. 2, Table 3).

The other A-domain mutants studied also bound to IGFBPs with slightly decreased affinity. [Arg[54],Arg[55]]IGF-II, which does not bind to the IGF-II/Man 6-P receptor, bound IGFBPs 1-6 with 21-100% of the affinity of IGF-II (Fig. 2, Table 3). Substitution of Tyr[59], which has been implicated in maintaining the tertiary structure of IGF-II through interaction with Ile[42], with Ala decreased binding affinity 2.5-4-fold compared with IGF-II (Table 3).

B-domain Mutants. Both [Ser[26]]IGF-II and [Leu[27]]IGF-II bind to the IGF-I and insulin receptors with decreased affinity, with the latter mutation having the more pronounced effect. However, [Leu[27]]IGF-II bound to IGFBPs 1-6 with only slightly lower affinity than IGF-II, whereas substitution of Phe[26] with Ser reduced binding affinity for IGFBPs -1, -5 and -6 to less than 10% of that of IGF-II (Fig. 2, Table 3). Binding of [Ser[26]]IGF-II to the other IGFBPs was slightly less impaired. [Leu[26]]IGF-II bound to IGFBPs similarly to [Ser[26]]IGF-II, suggesting that the effect of substitution was due to loss of the aromatic side-chain of Phe (Table 3).

[Val[43]]IGF-II, which binds poorly to the IGF-I and insulin receptors, bound to IGFBPs 1-6 with unaltered affinity. Deletion of the first 5 amino-terminal residues of IGF-II also had no major effects on binding to IGFBPs.[5]

C- and D-domain Mutants. Substitution of residues 32, 35 and 36 with the amino acids from mouse IGF-II or deletion of the D-domain (residues 62-67) did not substantially affect binding to IGFBPs.[5]

Table 3. Binding of IGF-II mutants to IGF-binding proteins.

	IGFBP-1	IGFBP-2	IGFBP-3	IGFBP-4	IGFBP-5	IGFBP-6
Relative affinity (% of IGF-II)						
B domain mutants						
Ser[26]	1.3 ± 0.4	21 ± 9	19 ± 2	14 ± 0	6 ± 0	7 ± 3
Leu[26]	6 ± 1	21 ± 4	ND	52 ± 2	18 ± 6	6 ± 3
A domain mutants						
Ala[59]	25 ± 3	25 ± 6	ND	31 ± 5	32 ± 2	40 ± 8
Thr[48], Ser[49], Ile[50]	<1	15 ± 7	12 ± 2	<1	1	<1
Arg[54], Arg[55]	60 ± 0	100 ± 62	37 ± 5	25 ± 4	25	21 ± 2

ND, not done.
Results are summarized from [5].

DISCUSSION

The present study demonstrates that amino acids 48-50 of the A-domain and amino acid 26 of the B-domain of IGF-II are major determinants of binding of IGF-II to IGFBPs. These binding determinants more closely resemble the previously characterized binding determinants of IGF-II to the IGF-II/Man-6-P receptor than to the IGF-I or insulin receptors[3] (Table 1), but are distinct from both.

Interestingly, according to the three-dimensional structure of IGF-II proposed by Blundell et al.,[6] amino acids 48-50 are located on the surface of the molecule close to residues 6 and 7 (Fig. 3). These residues are the analogues of residues 3 and 4 of IGF-I which are important for binding to IGFBPs.[7] To a certain extent, the binding affinities of IGFs mutated at these sites for different IGFBPs change in parallel. For example, [Thr[48],Ser[49],Ile[50]]IGF-II and [Thr[49],Ser[50],Ile[51]]IGF-I[8] do not bind to IGFBP-1 but bind to IGFBP-3 with only moderately decreased affinity. Similarly, des(1-3)IGF-I[7] does not bind to IGFBP-1 and [Gln[3],Ala[4]]IGF-I[8] binds to IGFBP-1 with 50-fold decreased affinity compared with IGF-I; IGFBP-3 binds these mutants[7,8] and des(1-6)IGF-II[9] [the equivalent of des(1-3)IGF-I] with only 10-fold decreased affinity relative to the native ligand. The situation is less clear for binding to IGFBP-2 where different studies have shown different results.

Based on the three-dimensional structure of IGF-II, residues 26 and 48-50 are located on opposite sides of the molecule (Fig. 1) so it is difficult to reconcile a single binding site containing all of these residues. Some insight into this problem may be obtained by studies of Phe[B24], the analogous residue in the insulin molecule to Phe[26].[10,11] The aromatic side-chain of this residue projects inwards to the hydrophobic core of the molecule. Impairment of binding to the insulin receptor by substitution of this aromatic side-chain results from a generalized change in structure rather than direct disruption of the binding site, and an analogous situation could explain the effects of these substitutions in IGF-II on binding to the IGFBPs.

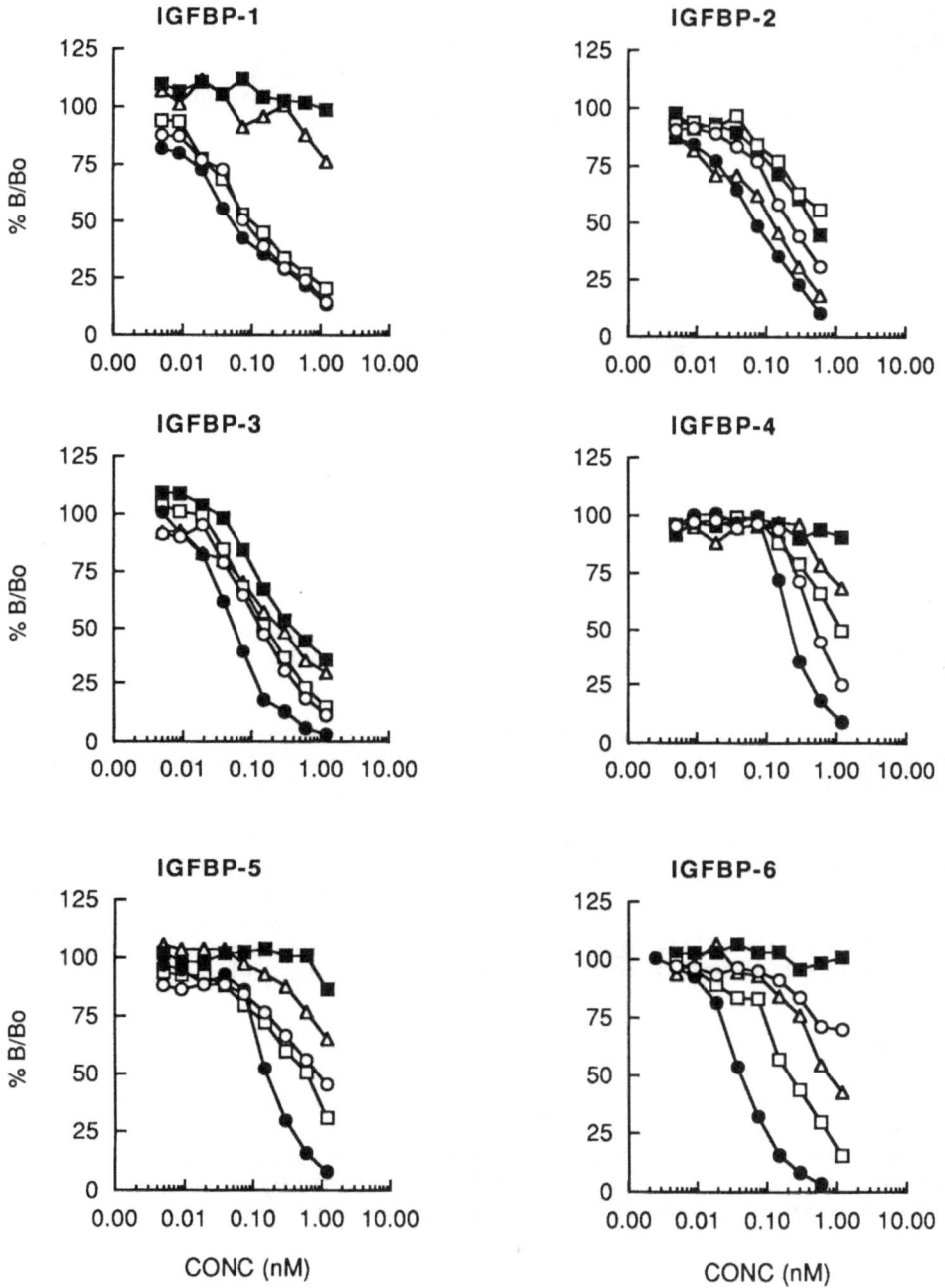

Figure 2. Binding of IGF-II mutants to IGFBPs 1-6. IGFBPs were incubated with ^{125}I-IGF-II and unlabeled IGF-II (filled circles), IGF-I (open circles), [Arg[54],Arg[55]]IGF-II (open squares), [Ser[26]]IGF-II (triangles), or [Thr[48],Ser[49],Ile[50]]IGF-II (filled squares). Results are expressed as the percentage of binding in the absence of unlabeled peptide (% B/Bo). (Redrawn from results previously reported in [5])

Another apparently paradoxical finding of the present study was that substitution of Tyr[27], which was more important than Phe[26] for binding to insulin and IGF-I receptors, did not substantially affect binding to the IGFBPs. In contrast to Phe[B24], the side-chain of Phe[B25] (the analogous residue to Tyr[27]) in the insulin molecule projects to the surface of the molecule and may be directly involved in the receptor binding site. Tyr[27] in IGF-II may therefore be directly involved in binding to insulin and IGF-I receptors. The IGF-I and insulin receptor binding domain on IGF-II, which includes amino acids 27 and 43, does not appear to be directly involved in binding to IGFBPs.

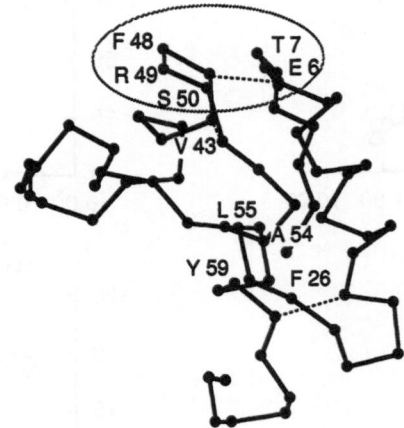

Figure 3. Tertiary structure of IGF-II demonstrating the proximity of amino acids 6-7 to residues 48-50. We propose that the circled region, which includes these residues, is the IGFBP binding domain.

The results of the present study taken together with those of previous studies suggest a binding site for IGFBPs on IGF-II involving residues 48-50 of the A-domain and residues 6-7 of the B-domain (Fig. 3). Substitution of residue 26 inhibits binding, most likely due to an effect on IGF-II conformation. Of the IGFBPs, IGFBP-1 and IGFBP-6 are most sensitive to changes in IGF-II structure despite the fact that IGFBP-1 binds IGF-I and IGF-II with equal affinity and IGFBP-6 has a marked preferential affinity for IGF-II over IGF-I. In contrast, IGFBP-2 and IGFBP-3 are best able to accomodate changes in IGF-II structure. Mutants of IGF-II with different binding specificities for IGF receptors and IGFBPs should prove useful tools for elaborating the role of the various components of the IGF system, and may lead to the design of agents with tissue-selective activity.

ACKNOWLEDGEMENTS

The authors would like to thank Dr Michael Weiss (Department of Medicine and Biological Chemistry and Molecular Pharmacology, Harvard Medical School) for his valuable comments. L.A.B. is a recipient of the J.J. Billings Travelling Fellowship of the Royal Australasian College of Physicians. S.H. was supported by the Juvenile Diabetes Foundation International Summer Student Program. This work was supported in part by National Institutes of Health Grant CA 47150 to J.F.P.

REFERENCES

1. M. M. Rechler and S. P. Nissley, Peptide Growth Factors and Their Receptors I. Insulin-like Growth Factors, in: "Handbook Exp. Pharm., Vol. 95," M. B. Sporn and A. B. Roberts, eds., Springer-Verlag, Berlin (1990), p 263.

2. M. M. Rechler, Insulin-like growth factor binding proteins, *Vitamins & Hormones* 47:1 (1993).

3. K. Sakano, T. Enjoh, F. Numata, H. Fujiwara, Y. Marumoto, N. Higashihashi, Y. Sato, J. F. Perdue, and Y. Fujita-Yamaguchi, The design, expression, and characterization of human insulin-like growth factor II (IGF-II) mutants specific for either the IGF-II/cation-independent mannose 6-phosphate receptor or IGF-I receptor, *J. Biol. Chem.* 266:20626 (1991).

4. L. A. Bach, N. R. Thotakura, and M. M. Rechler, Human insulin-like growth factor binding protein-6 is O-glycosylated, *Biochem. Biophys. Res. Commun.* 186:301 (1992).

5. L. A. Bach, S. Hsieh, K. Sakano, H. Fujiwara, J. F. Perdue, and M. M. Rechler, Binding of mutants of human insulin-like growth factor II (IGF-II) to insulin-like growth factor binding proteins 1-6, *J. Biol. Chem.* (1993). In press.

6. T. L. Blundell, S. Bedarkar, and R. E. Humbel, Tertiary structures, receptor binding, and antigenicity of insulinlike growth factors, *Federation Proc.* 42:2592 (1983).

7. B. Forbes, L. Szabo, R. C. Baxter, F. J. Ballard, and J. C. Wallace, Classification of the insulin-like growth factor binding proteins into three distinct categories according to their binding specificities, *Biochem. Biophys. Res. Commun.* 157:196 (1988).

8. D. R. Clemmons, M. L. DeHoff, W. H. Busby, M. L. Bayne, and M. A. Cascieri, Competition for binding to insulin-like growth factor (IGF) binding protein-2, 3, 4, and 5 by the IGFs and IGF analogs, *Endocrinology* 131:890 (1992).

9. C. Lüthi, B. V. Roth, and R. E. Humbel, Mutants of human insulin-like growth factor II (IGF II)-- Expression and characterization of truncated IGF II and of two naturally occurring variants, *Eur. J. Biochem.* 205:483 (1992).

10. S. E. Shoelson, Z. -X. Lu, L. Parlautan, C. S. Lynch, and M. A. Weiss, Mutations at the dimer, hexamer, and receptor-binding surfaces of insulin independently affect insulin-insulin and insulin-receptor interactions, *Biochemistry* 31:1757 (1992).

11. Q. X. Hua, S. E. Shoelson, K. Inouye, and M. A. Weiss, Paradoxical structure and function in a mutant human insulin associated with diabetes mellitus, *Proc. Natl. Acad. Sci. USA* 90:582 (1993).

TRANSCRIPTIONAL AND POST-TRANSCRIPTIONAL REGULATION OF THE HUMAN IGF-II GENE EXPRESSION

J.S. Sussenbach, R.J.T. Rodenburg, W. Scheper and P. Holthuizen

Laboratory for Physiological Chemistry
Utrecht University
Utrecht, The Netherlands

SUMMARY

The human insulin-like growth factor II (IGF-II) gene consists of nine exons and has four promoters (P1-4). The promoters exhibit a tissue-specific and developmental stage-dependent expression pattern. In fetal liver promoters P2-4 are expressed, but after birth these promoters are shut off and another promoter, P1, is activated.

We have investigated some properties of the human promoters P1 and P3 and identified a number of sequence elements, that are recognized by transcription factors. Promoter P1 is stimulated by the liver-enriched transcription factors C/EBP and LAP, whereas in the proximal region of P3 we have identified several elements that are recognized by transcription factors, including krox20/egr2 and krox24/egr1.

Besides transcriptional regulation of expression also regulation at the post-transcriptional level occurs. We have found that the IGF-II mRNAs are subjected to site-specific endonucleolytic cleavage yielding a labile 5' specific fragment and a stable polyadenylated 3' specific cleavage product of 1.8 kb. Two widely separated sequence elements within the last exon were identified that are able to interact and yield a double-stranded stem structure. It is likely that this structure is essential for post-transcriptional cleavage of IGF-II mRNAs.

INTRODUCTION

The human gene coding for IGF-II consists of nine exons 1-9, of which exons 7, 8 and the first part of exon 9 code for the IGF-II precursor. The 5' part of the gene consists of non-coding leader exons 1-6, of which exons 1, 4, 5 and 6 are preceded by a promoter (P1-P4)(De Pagter-Holthuizen et al., 1987, 1988; Holthuizen et al., 1990). Transcription from these promoters leads to the formation of IGF-II mRNAs with different untranslated leaders 1-4. The IGF-II promoters are subjected to developmental and tissue-specific

Current Directions in Insulin-Like Growth Factor Research,
Edited by D. LeRoith and M.K. Raizada, Plenum Press, New York, 1994

63

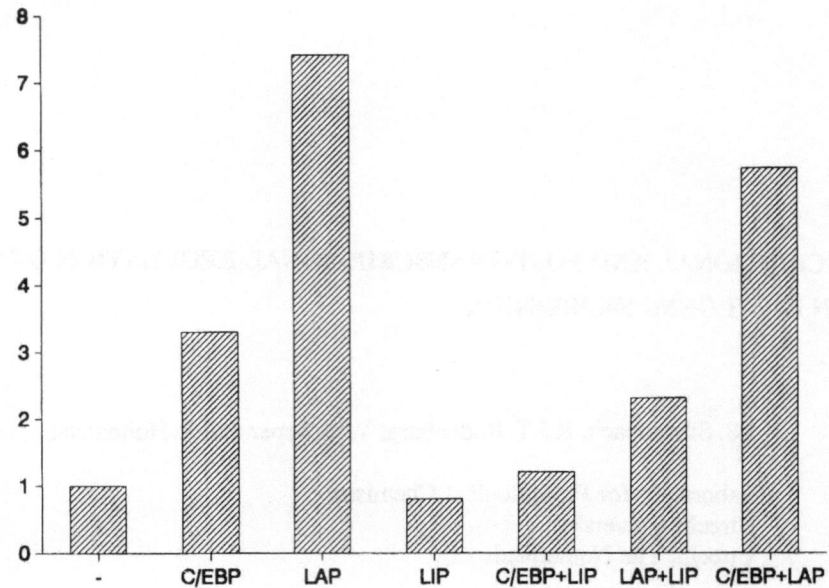

Figure 1. Transactivation of IGF-II promoter P1 by C/EBP and LAP in transient transfected Hep3B cells. A P1 region from positions -207 to + 52 linked to the luciferase gene was cotransfected with C/EBP, LAP and LIP expression plasmids. Cells were grown in 25 cm² flasks to 40% confluence and transfected with 2 μg of P1-plasmid, 1 μg of transcription factor expression plasmid and 0.5 μg of pRSV-LacZ, which served as an internal standard. Luciferase and β-galactosidase levels were measured 24 hrs after transfection.

regulation of expression. The best example of differential regulation of the IGF-II gene is found in the human liver. During fetal liver development promoters P2, P3 and P4 are active, of which P3 is the stronger promoter. After birth the activity of these promoters declines and promoter P1 becomes active in adult liver, yielding a 5.3 kb mRNA.

In addition to the IGF-II encoding mRNAs, a 1.8 kb RNA is expressed that is derived from the 3' untranslated part of the last exon and therefore does not code for IGF-II (De Pagter-Holthuizen et al., 1988). The 3' end of the 1.8 kb RNA is located at the major polyadenylation site of the IGF-II gene, whereas its 5' end has been mapped 1.65 kb upstream of this site. It was shown that the 1.8 kb RNA does not arise by transcriptional initiation within exon 9, but is formed through endonucleolytic cleavage of IGF-II mRNAs (Meinsma et al., 1991). Analysis of the sequence requirements for the cleavage reaction has led to the identification of two widely separated elements of approximately 300 nt long (Meinsma et al., 1992). It is likely that these elements interact and form a double-stranded structure of 70 nt long.

ACTIVATION OF PROMOTER P1

The expression of promoter P1 has only been observed in adult liver, suggesting that this organ contains a tissue-specific transcription factor involved in P1 activation. Previously, we have shown that the 42 kDa liver-enriched CAAT/Enhancer Binding Protein (C/EBP) (Johnson et al., 1987; Landschulz et al., 1988, 1989) might play a role in liver-specific expression of P1 (van Dijk et al., 1992). This factor is expressed predominantly in terminally differentiated hepatocytes, adipocytes and lung (Birkenmeier et al., 1989) and binds to sites in DNA that share the loose sequence homology $T^T/_GNNC/_TAAT/_G$ (Ryden and Beemon, 1989).

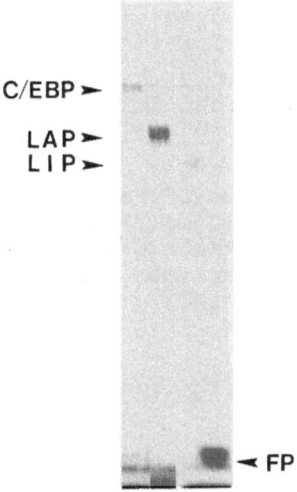

C/EBP ➤
LAP ➤
LIP ➤

◄ FP

Figure 2. Electrophoretic mobility shift assay (EMSA) showing binding of LAP, LIP and C/EBP to IGF-II promoter P1. 50 pg of a ^{32}P-labeled double-stranded oligonucleotide probe, covering P1 positions -112 to -89 was incubated with crude lysates of E. coli BL21(DE3) expressing C/EBP, LAP or LIP, respectively, and separated on a non-denaturing 6% acrylamide gel containing 0.01% NP40. FP = free probe.

When Hep3B cells were cotransfected with a construct bearing a fragment of promoter P1 (-889/+52) coupled to the luciferase gene, and the expression plasmid pMSV-C/EBPwt we observed an approximately 3- to 4-fold higher luciferase value than found when Hep3B cells were transfected with the same promoter P1 construct and pUC12 DNA (Fig. 1) (van Dijk et al., 1992). On the other hand, promoter P3 could not be activated by C/EBP. This indicates that the stimulatory effect of C/EBP is significant and specific for P1.

Identification of the C/EBP binding site in P1 revealed that this site is located at positions -99 to -91 where a C/EBP consensus sequence is found with one mismatch. The same site is theoretically also recognized by the 36 kDa transcription activating factor LAP (Liver-enriched Activator Protein) and a 22 kDa C-terminal nested fragment called LIP (Liver-enriched Inhibiting Protein) that originates by translation from an internal in frame AUG codon in the LAP open reading frame and has an inhibitory effect on LAP activity by forming inactive heterodimers (Descombes and Schibler, 1991). Cotransfections with a P1-luciferase reporter plasmid and C/EBP, LAP or LIP expression plasmids revealed that C/EBP stimulates promoter P1 with a factor of 3-4 and LAP with a factor of 7, whereas LIP has no effect (Fig. 1). When P1 constructs were cotransfected with combinations of C/EBP, LAP and LIP it was observed that LIP has an inhibitory effect on C/EBP and LAP stimulation. Furthermore, cotransfections of C/EBP and LAP show that the stimulating effects of LAP and C/EBP are not synergistic.

That these factors are indeed able to bind to the C/EBP consensus sequence between positions -99 and -91 was demonstrated when extracts of E.coli pLysS expressing C/EBP, LAP and LIP, respectively, were incubated with a double-stranded oligonucleotide corresponding to positions -112 to -89 in promoter P1 and subjected to electrophoretic mobility shift analysis (Fig. 2). These results indicate that in the adult liver C/EBP as well as LAP may be involved in the activation of promoter P1.

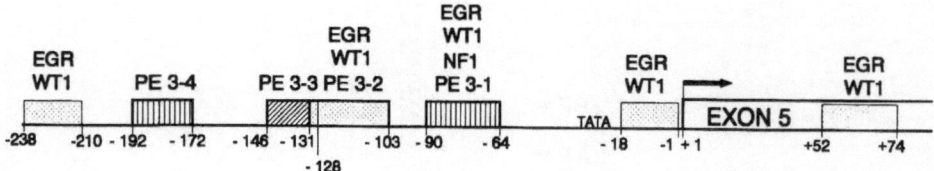

Figure 3. Schematic representation of the human IGF-II promoter P3 region. The domains that are protected by specific nuclear proteins PE3-1, PE3-2, PE3-3 and PE3-4 are indicated. In vitro transcription experiments show that PE3-1 and PE3-4 are involved in activation of transcription. In addition five putative binding sites for krox20/egr2 and krox24/egr1 (activators) are present. The P3 regions that can be protected by WT1 (repressor) were identified by Drummond et al.(1992).

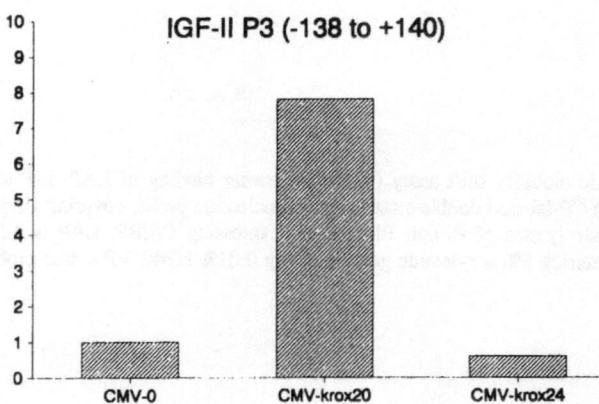

Figure 4. Cotransfections of IGF-II promoter P3-luciferase constructs (positions -138 to +140, relative to the initiation of transcription site) and expression vector only (CMV-0), CMV-krox20, or CMV-krox24, respectively. The luciferase activity of P3 -138 was arbitrarily set at 1. Hep3B cells were grown in 25 cm^2 flasks to 40% confluence. Each flask was transfected with 2 µg of promoter construct linked to the luciferase reporter gene and 2 µg of transcription factor expression plasmid. To correct for transfection efficiency 1 µg of pRSV-LacZ was included. All transfections were performed in triplicate and the standard deviation was within 15%. Cells were harvested 24 hrs after transfection and the luciferase activity was determined.

ACTIVATION OF PROMOTER P3

The human IGF-II promoter P3 is expressed in many fetal and non-hepatic adult tissues. DNase footprint analysis, electrophoretic mobility shift assays and in vitro transcription have revealed that in addition to a TATA box, the proximal region contains from position -288 to the cap site a number of elements that are recognized by nuclear proteins (PE3-1 to PE3-4, Fig. 3) (van Dijk et al., 1991). Only for the protein binding to PE3-3 no evidence could be obtained by in vitro transcription that it plays a role in promoter activation. PE3-1 and PE3-4 are bound by still unknown transcription factors (Fig. 3). PE3-2 (positions -131 to -103) is a very interesting promoter element. It contains the common recognition sequence for krox20/egr2 and krox24/egr1 (positions -131/-123) (Lemaire et al., 1988, 1990). Examination of the proximal promoter region revealed that here four extra putative binding sites for krox/egr are present (-229/-221, -84/-76, -15/-7 and +63/+71).

Interestingly, PE3-2 not only contains putative binding sites for krox/egr factors, but the same binding site (5'-CGCCCCCGC-3') can be recognized by the Wilms' tumor locus zinc finger protein WT1 (Rauscher III et al., 1990). Recently, it was shown by footprint

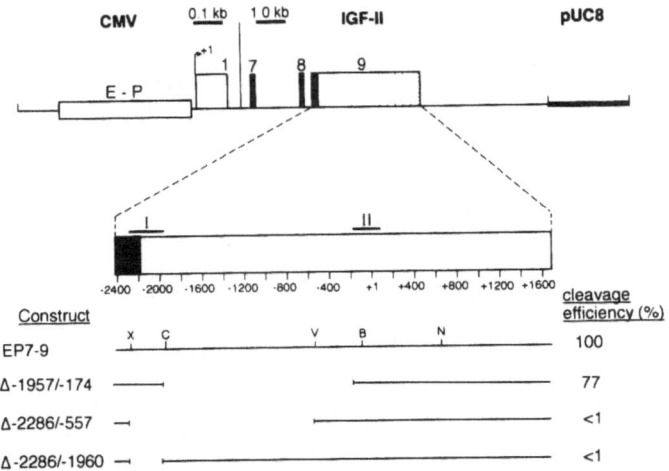

Figure 5. Structure of CMV-IGF-II minigenes. The enhancer-promoter and the first exon of the cytomegalovirus immediate early gene are linked to exons 7, 8 and 9 of the human IGF-II gene. The stippled region in the enlarged exon 9 represents the 1.8 kb RNA region. Positions (bp) are indicated relative to the cleavage site (+1). Elements I and II are indicated as bars. The amounts of 3' specific cleavage products were determined relative to the total amount of full length and 3' specific IGF-II RNAs. The cleavage efficiency obtained with construct EP7-9 was set at 100%.

analysis that WT1 can indeed bind to several IGF-II P3 regions (Drummond et al., 1992). Whereas krox24/egr1 and krox20/egr2 are transcriptional activators (Lemaire et al., 1990), WT1 suppresses transcription (Madden et al., 1991). Several Wilms' tumours have been found to contain elevated levels of IGF-II mRNA (Haselbacher et al., 1987; Paik et al., 1989), which might be an indication that WT1 is involved in IGF-II gene regulation, probably by interacting with the krox/egr and WT1 sites.

To investigate the role of the krox/egr binding elements in P3, Hep3B cells were cotransfected with a construct bearing a fragment of promoter P3 (-138/+140) coupled to the luciferase gene, and the expression plasmids pCMV-0, pCMV-krox20 and pCMV-krox24. We observed that expression of pCMV-krox20 causes approximately 7-8 fold higher luciferase values than found with pCMV-0 whereas expression of pCMV-krox24 has a weak inhibiting effect (Fig. 4). Electrophoretic mobility shift analysis using extracts from recombinant vaccinia virus infected cells producing egr1/krox24 and egr2/krox20 show that these factors can bind to several sites in the proximal promoter region of P3 (Fig. 3). Similar results were obtained by Drummond et al. (1992) for binding of WT1 to these sites. These data suggest that especially krox20 and probably also WT1 play a role in activation or repression of P3.

ENDONUCLEOLYTIC CLEAVAGE OF IGF-II mRNAs

Recently, we have described the properties of a 1.8 kb RNA derived from the 3' end of exon 9 of the IGF-II gene that arises by endonucleolytic cleavage of IGF-II mRNAs (Meinsma et al., 1991). To investigate the sequence requirements for the endonucleolytic cleavage of IGF-II mRNAs a nested set of deletions was created in a construct designated EP7-9 (Fig. 5). This construct contains the cytomegalovirus (CMV) enhancer-promoter, a genomic IGF-II fragment encompassing exons 7, 8, and 9 and about 5 kb downstream of the IGF-II gene. Transient expression yielded a full length 4.8 kb mRNA as well as the 3' specific cleavage product of 1.8 kb. After expression of the various deletion constructs in 293 cells the levels of IGF-II mRNAs and cleavage products

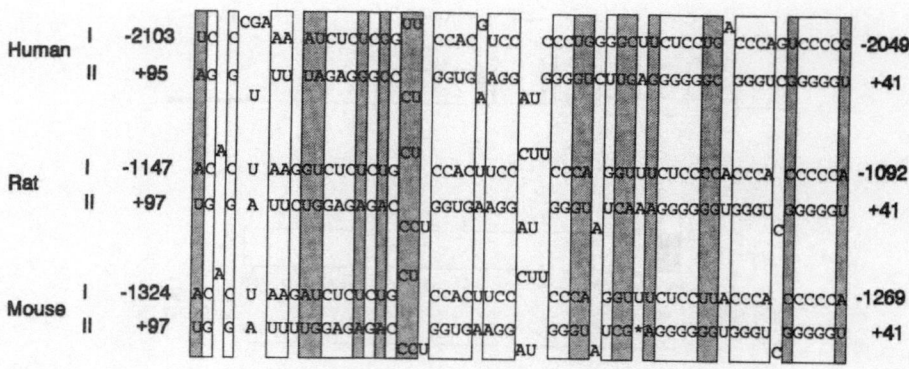

Figure 6. Computer-folding of elements I (black bar) and II (open bar) of human IGF-II mRNA according to Zuker and Stiegler (1981). For clarity the 3' and 5' sequences flanking the interaction are not shown. Within element II the conserved hairpins and G-rich region are represented by grey respectively dark grey bars. The cleavage site and the BglII restriction site (Meinsma et al. 1991) are indicated by arrows.

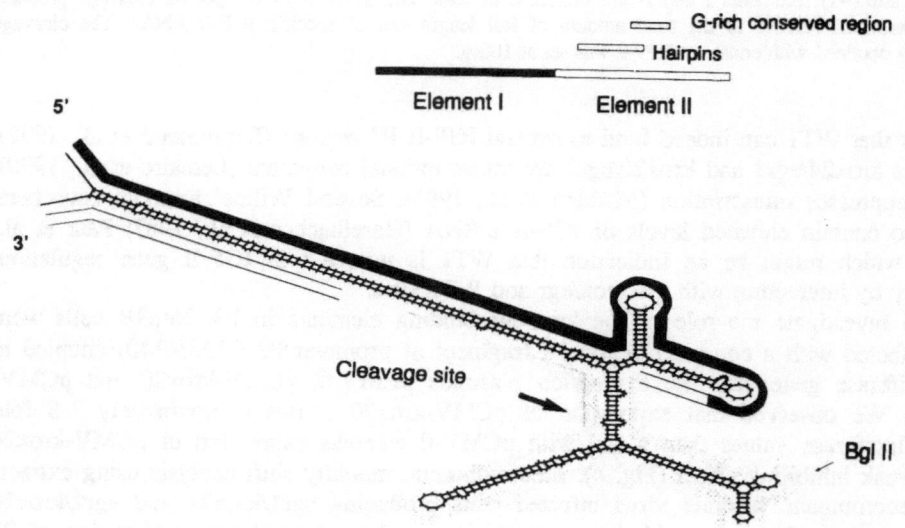

Figure 7. Alignment of the long range interaction between elements I and II of human, rat and mouse IGF-II mRNAs. Numbers indicate the positions in exon 9 (human) and exon 6 (mouse and rat). The structures are more extended both 5' and 3', but the interactions shown can be aligned and are always folded identical irrespective of 5' or 3' flanking sequences or sequences between the elements. Open boxes represent conservation of the primary structure in both elements as well as the secondary structure. Conservation of secondary but not primary structure is indicated by shaded boxes.

were determined by hybridization of Northern blots of total RNA with a 3' specific probe that detects both the 1.8 kb RNA and full length IGF-II RNAs. An element (II) essential for the cleavage reaction was identified from positions -174 to +151. Element II consists of two domains; one that can form two stem-loop structures from -139 to -3 which may constitute a recognition site for one or more proteins involved in the cleavage reaction, and a G-rich region directly downstream of the cleavage site (Meinsma et al., 1992).

Surprisingly, an additional essential element I was identified between positions -2286 and -1960, located about 2 kb upstream of element II, (Fig. 5). We found that both

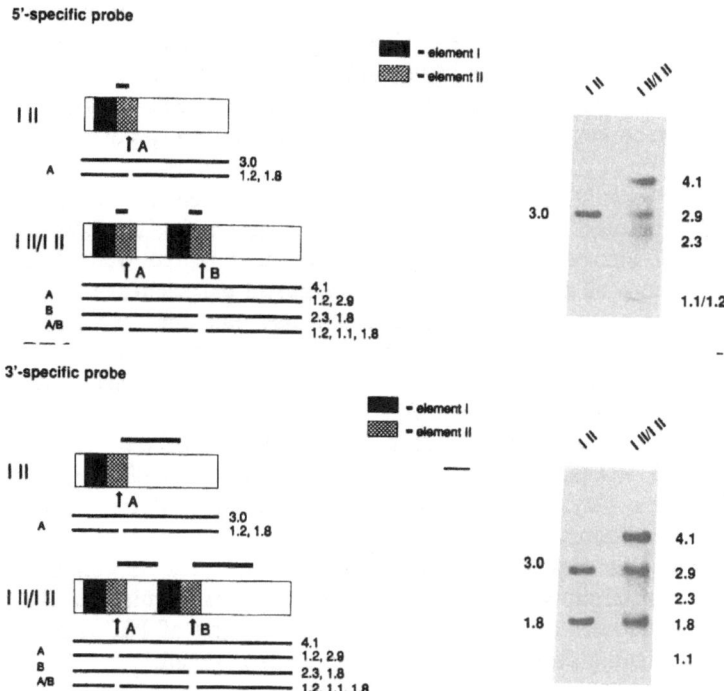

Figure 8. Schematic representation of human IGF-II exon 9 in the constructs I-II and I-II/I-II. These constructs contain a genomic fragment of exons 7-9 driven by the CMV promoter/enhancer. Elements I are represented by dark grey and elements II by grey boxes. Potential cleavage sites are indicated by arrows and called A (upstream site) or B (downstream site). The black bars drawn above the constructs show the positions of sequences that hybridize with the probes indicated that are specific for either sequences upstream (3' specific) or downstream of the cleavage site (5' specific). Thick lines below the constructs represent the products that result from cleavage at site A, site B or both sites. In the panels on the right hand side of the constructs Northern blots are shown of RNA isolated from 293 cells transiently transfected with the constructs and hybridized with the 5' and 3' specific probes, respectively.

sequence elements I and II have to be present simultaneously to obtain cleavage of IGF-II mRNA, (Meinsma et al., 1992).

An intriguing question is what the relation is between the widely separated elements I and II. Interestingly, computer-aided folding of elements I and II revealed the possibility that these elements form a stable secondary structure of considerable length (Fig. 6). It is very likely that this double-stranded RNA structure is indeed of physiological importance, specifically since strong conservation of the secondary structure in mouse, rat and human is observed (Fig. 7). Moreover, when comparing the sequences of the three species, base changes in element I are usually compensated for by another mutation in element II in order to maintain the double-stranded stem structure (Fig. 7). This further provides phylogenetic evidence that the stem structure of these mRNAs plays a major role in the cleavage reaction.

To further investigate the interaction between elements I and II in cleavage of IGF-II mRNAs in vivo, we constructed an expression plasmid based on EP7-9 (Fig. 5) containing two intact cleavage units in tandem (I-II/I-II) (Fig. 8). Transcripts derived from this construct can potentially be cleaved at two sites, indicated as A and B, resulting in the RNA products indicated in Fig. 8. Transfection of 293 cells with this construct revealed that all these RNA species could indeed be detected on a Northern blot by differential usage of probes specific for either sequences downstream (3' specific probe) or upstream

(5' specific probe) of the cleavage sites. This indicates that mRNAs containing two cleavage units can be processed at both sites, suggesting that when, after a first cleavage reaction, the resulting RNA product still contains a second intact cleavage unit this can be processed as well. The latter can be concluded from the presence of the 1.1 kb RNA that contains the sequence between the two sites and that is only produced if both sites are used.

In order to look at the role of the position of the elements within a cleavage unit we constructed a plasmid again containing two cleavage units, but now in the first unit 180 nucleotides of element I containing the interacting region were deleted (IΔII/I-II). Using similar methods as described above it was shown that cleavage could occur at both sites indicating that cleavage at site A from the upstream unit can occur by using the downstream element I from the second unit. This further shows that the relative position of element II towards element I is not important, but that the interaction itself is essential (Scheper et al., manuscript in preparation).

These data are in agreement with an important role of the stem structure in the cleavage of IGF-II RNAs. The physiological function of the cleavage reaction is still unknown; probably it plays a role in the rapid degradation of IGF-II mRNA levels under different growth conditions.

In conclusion, the expression of the human IGF-II gene is controlled at different levels including transcriptional as well as post-transcriptional regulation. The abundance of promoters and the process of post-transcriptional processing of IGF-II mRNAs make the IGF-II gene an intriguing gene and worthwhile to study in detail.

ACKNOWLEDGMENTS

The authors wish to thank Mrs. A.M.C.B. Koonen-Reemst and Mrs. W. Teertstra for expert technical assistance. This work was supported by a grant from the Netherlands Organization for Research (NWO).

REFERENCES

Birkenmeier, E.H., Gwynn, B., Howard, S., Jerry, J., Gordon, J.I., Landschulz, W.H. and McKnight S.L.,1989, Tissue-specific expression, developmental regulation, and genetic mapping of the gene encoding CCAAT/enhancer binding protein, *Genes Dev.* 3:1146.

Descombes, P. and Schibler, U., 1991, A liver-enriched transcriptional activator protein, LAP, and a transcriptional inhibitory protein, LIP, are translated from the same mRNA, *Cell*, 67:569.

De Pagter-Holthuizen, P., Jansen, M., Van Schaik, F.M.A., van der Kammen, R.A., Oosterwijk, C., Van den Brande, J.L. and Sussenbach, J.S., 1987, The human insulin-like growth factor II contains two development-specific promoters, *FEBS Lett.* 214:259.

De Pagter-Holthuizen, P., Jansen, M., van der Kammen, R.A., Van Schaik, F.M.A. and Sussenbach, J.S., 1988, Differential expression of the human insulin-like growth factor II gene. Characterization of the IGF-II mRNAs and an mRNA encoding a putative IGF-II-associated protein, *Biochim. Biophys. Acta* 950:282.

Drummond, I.A., Madden, S.L., Rohwer-Nutter, P., Bell, G.I., Sukhatme, V.P. and Rauscher III, F.J., 1992, Repression of the insulin-like growth factor II gene by the Wilms tumor suppressor WT1, *Science*, 257:674.

Haselbacher, G.K., Irminger, J.C., Zapf, J., Ziegler, W.H. and Humbel, R.E., 1987, Insulin-like growth factor II in human adrenal pheochromocytomas and Wilms tumours: expression at the mRNA and protein level, *Proc. Natl. Acad. Sci. USA*, 84:1104.

Holthuizen, P., van der Lee, F.M., Ikejiri, K., Yamamoto, M. and Sussenbach, J.S., 1990, Identification and initial characterization of a fourth leader exon and promoter of the human IGF-II gene, *Biochim. Biophys. Acta* 1087:341.

Johnson, P.F., Landschulz, W.H., Graves, B.J. and McKnight, S.L, 1987, Identification of a rat liver nuclear protein that binds to the enhancer core element of three animal viruses, *Genes Dev.*, 1:133.

Landschulz, W.H., Johnson, P.F., Adashi, E.Y., Graves, B.J. and McKnight, S.L., 1988, Isolation of a recombinant copy of the gene encoding C/EBP, *Genes Dev.* 2:786.

Landschulz, W.H., Johnson, P.F. and McKnight, S.L., 1989, The DNA binding domain of the rat liver nuclear factor C/EBP is bipartite, *Science* 243:1681.

Lemaire, P., Relevant, O., Bravo, R., and Charnay, P., 1988, Two mouse genes encoding potential transcription factors with identical DNA binding domains are activated by growth factors in cultured cells, *Proc. Natl. Acad. Sci. USA* 85:4691.

Lemaire, P., Vesque, C., Schnitt, J., Stunnenberg, H., Frank, R. and Charnay, P., 1990, The serum-inducible mouse gene Krox-24 encodes a sequence-specific transcriptional activator, *Mol. Cell. Biol.* 10:3456.

Madden, S.L., Cook, D.M., Morris, J.F., Gashler, A., Subhatme, V.P. and Rauscher-III, F.J., 1991, Transcriptional repression mediated by the WT1 Wilms tumor gene product, *Science* 253:1550.

Meinsma, D., Holthuizen, P., Van den Brande, J.L. and Sussenbach, J.S., 1991, Specific endonucleolytic cleavage of IGF-II mRNAs, *Biochem. Biophys. Res. Commun.* 179:1509.

Meinsma, D., Scheper, W., Holthuizen, P., Van den Brande, J.L. and Sussenbach, J.S., 1992, Site-specific cleavage of IGF-II mRNAs requires sequence elements from two distinct regions of the IGF-II gene, *Nucleic Acids Res.* 20:5003.

Paik, S., Rosen, N., Jung, W., You, J.M., Lippman, M.E., Perdue, J.F. and Yee, D., 1989, Expression of insulin-like growth factor-II mRNA in fetal kidney and Wilms' tumor. An in situ hybridization study, *Lab. Invest.* 61:522.

Rauscher-III, F.J., Morris, J.F., Tournay, O.E., Cook, D.M. and Curren, T., 1990, Binding of the Wilms' tumor locus zinc finger protein to the EGR-1 consensus sequence, *Science* 250:1259.

Ryden, T.A. and Beemon, K., 1989, Avian retroviral long terminal repeats bind CCAAT/Enhancer binding protein, *Mol. Cell. Biol.* 9:1155.

van Dijk, M.A., Van Schaik, F.M.A., Bootsma, H.J., Holthuizen, P. and Sussenbach, J.S., 1991, Initial characterization of the four promoters of the human IGF-II gene, *Mol. Cell. Endocrinol.* 81:81.

van Dijk, M.A., Rodenburg, R.J.T., Holthuizen, P. and Sussenbach, J.S., 1992, The liver-specific promoter of the human insulin-like growth factor II gene is activated by CCAAT/enhancer binding protein (C/EBP), *Nucleic Acids Res.* 20:3099.

Zuker, M. and Stiegler, P., 1981, Optimal computerfolding of large RNA sequences using thermodynamics and auxiliary information, *Nucleic Acids Res.* 18:3035.

SIGNIFICANT SPECIES DIFFERENCES IN LOCAL IGF-I AND -II GENE EXPRESSION

Carolyn A. Bondy, Edward Chin and Jian Zhou

Developmental Endocrinology Branch
National Institute of Child Health and Human Development, NIH
Bethesda, MD 20892

INTRODUCTION

Much thinking about the roles of IGF-I and -II is shaped by observations in the rat or mouse, although it is not clear to what extent these laboratory rodents reflect IGF system function or regulation in other species. For example, IGF-II is usually regarded primarily as a fetal growth factor because its postnatal expression is highly restricted in the rodent. However, in the human and other species, IGF-II levels are equal to or greater than IGF-I in the circulation and in various tissues of adults (reviewed ref. 1). In addition, current concepts about the importance of growth hormone induced IGF-I production in peripheral tissues are based on rodent studies and, to our knowledge, have not been supported by human data. In order to investigate the extent to which local patterns of IGF system expression are similar in the rat and human, we have compared IGF-I, IGF-II and IGF-I receptor gene expression in the kidney and ovary[2-5].

KIDNEY

Figure 1 compares IGF system gene expression in the rat and human kidney. In the rat, IGF-I mRNA is concentrated in the inner stripe of the outer medulla, where it is expressed by epithelial cells of the thick ascending limb[2]. IGF-II mRNA is not detected in the adult rat kidney, except in some perinephric blood vessels (Fig. 1B). In contrast, IGF-I mRNA is not detected in the human kidney, but IGF-II mRNA is abundant (Fig. 1 D&E). IGF-II mRNA is not localized in renal epithelium, but is expressed by the renal vasculature and medullary interstitial cells[3]. Thus, the difference in renal IGF expression

Current Directions in Insulin-Like Growth Factor Research,
Edited by D. LeRoith and M.K. Raizada, Plenum Press, New York, 1994

73

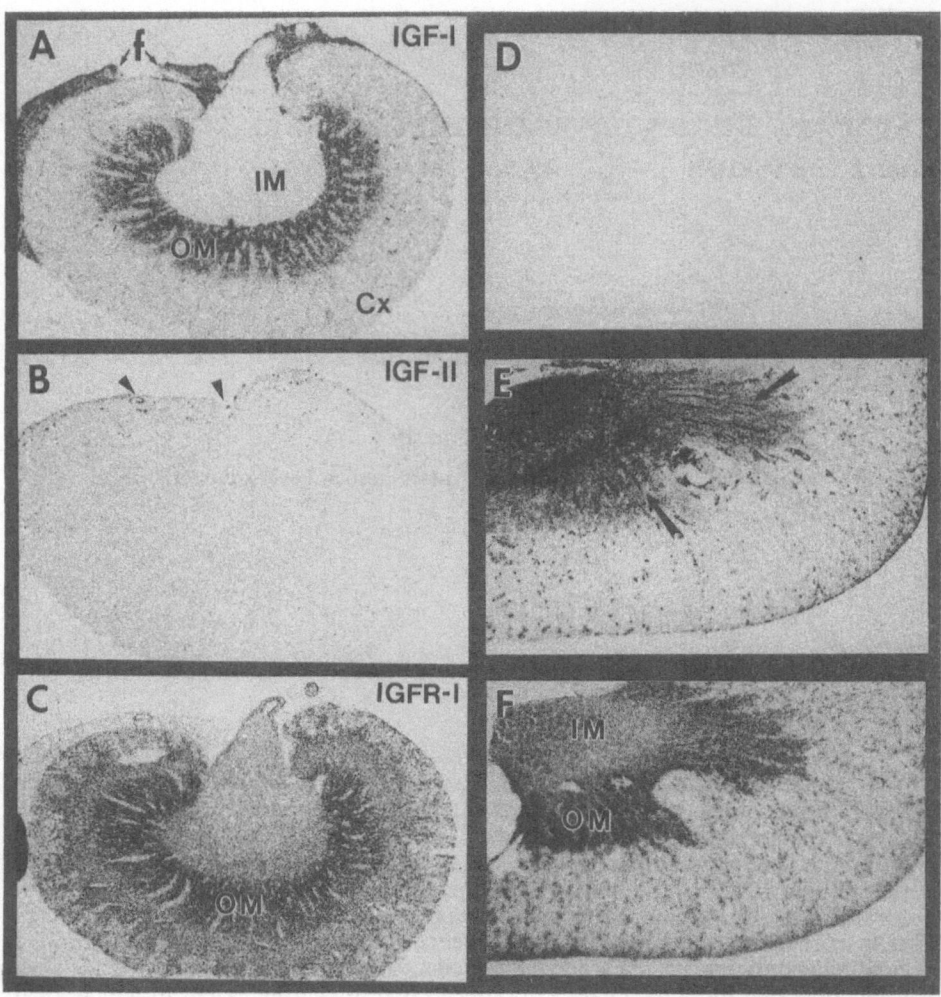

Fig. 1 IGF system gene expression in sequential sections from the rat (A-C) and human kidney (D-F).
Film autoradiographs shows sections hybridized to (rat or human, as appropriate) cRNA probes for IGF-I
(A&D), IGF-II (B&E) and the IGF-I receptor (C&F). IGF-I mRNA is concentrated in the outer medulla
(OM) of the rat kidney and is also present in the perinephric fat (f). IGF-II is detected only in perinephric
blood vessels (B, arrowheads) in the rat, but is abundant throughout the human medulla where it is
localized in interstitial cells and in vasa rectae (E, arrows). It is also localized in afferent arterioles, which
appear as punctate densities in the cortex. The pattern of IGF-I receptor mRNA distribution is very
similar in rat and human.

is not simply a substitution of IGF-II for IGF-I in the human, but also includes a different cellular distribution of IGF gene expression. Despite the different patterns of ligand gene expression, the renal distribution of IGF-I receptor gene expression is quite similar in the two species (Fig. 1 C&F).

The thick ascending limb (TAL), where IGF-I is focally expressed in the rat kidney, is distinguished by having the highest level of Na/K ATPase activity of any portion of the nephron, and by a unique dependence on glucose to fuel this pump activity, while the rest of the nephron preferentially uses other substrates. Given these special local metabolic characteristics and the observation that IGF-I is co-expressed in this segment with the facilitative glucose transporter GT4, we have suggested that IGF-I production by this segment might serve to augment TAL metabolic activity[6]. During high protein feeding, Na/K pump activity increases and local IGF-I mRNA levels also increase[7]. Renal adaption to sustained high protein feeding is associated with hypertrophy and hyperplasia of TAL epithelium, suggesting that chronically increased IGF-I production might stimulate local epithelial cell growth[7].

The situation of IGF-II in the human kidney is quite different from that of IGF-I in the rat kidney. It is expressed in the renal vasculature, but that feature is not specific to the kidney, for IGF-II gene expression is concentrated in blood vessels in many human tissues, including ovary[5], testis[8], brain and pituitary (our unpublished data). In the renal parenchyma, IGF-II mRNA is localized in medullary interstitial cells. It is possible that IGF-II produced by interstitial cells acts in a paracrine fashion on medullary epithelium to promote metabolic and morphological adaptation to alterations in protein and osmotic loads, as we have suggested for IGF-I in the rat kidney. IGF-I receptor mRNA is concentrated in the medullary tubular epithelium, supporting this possibility.

OVARY

Figure 2 compares IGF system gene expression in the rat and human ovary. IGF-I mRNA is highly abundant in granulosa cells of healthy-appearing but not in atretic follicles in the rat ovary[4, 9, 10], where IGF-II mRNA is detected only in blood vessels. In contrast, IGF-I mRNA is not detected in human granulosa cells, but IGF-II mRNA is found in granulosa cells of mature and atretic follicles in the human ovary[5, 11-13]. Little IGF-II mRNA is detected in granulosa cells of small growing follicles, however[5]. Interestingly, the pattern of IGF-I receptor gene expression is similar or identical in ovaries of both species. IGF-I receptor mRNA is abundant in granulosa cells of all follicular stages, from primordial through atretic. Furthermore, IGF-I mRNA is abundant in oocytes of both rat and human ovaries[4, 5]. Given the conservation of cellular patterns of IGF-I receptor expression across the species, it is possible that ligands derived from alternative sources serve the same basic functions, for example, early follicular growth might be stimulated by granulosa cell IGF-I production in the rat and by oocyte IGF-I or -II production in the human ovary.

SUMMARY

Comparison of renal and ovarian IGF system gene expression in the rat and human has shown that IGF-I mRNA is predominant in rat and IGF-II in human tissues, and that the two IGFs demonstrate different cellular patterns of expression in the homologous organs. IGF-I receptor gene expression, however, describes an apparently identical pattern in the tissues of both rat and human, suggesting that, although local sources and presumably regulation of IGF expression vary between the species, the ultimate functions served might be the same.

RAT   HUMAN

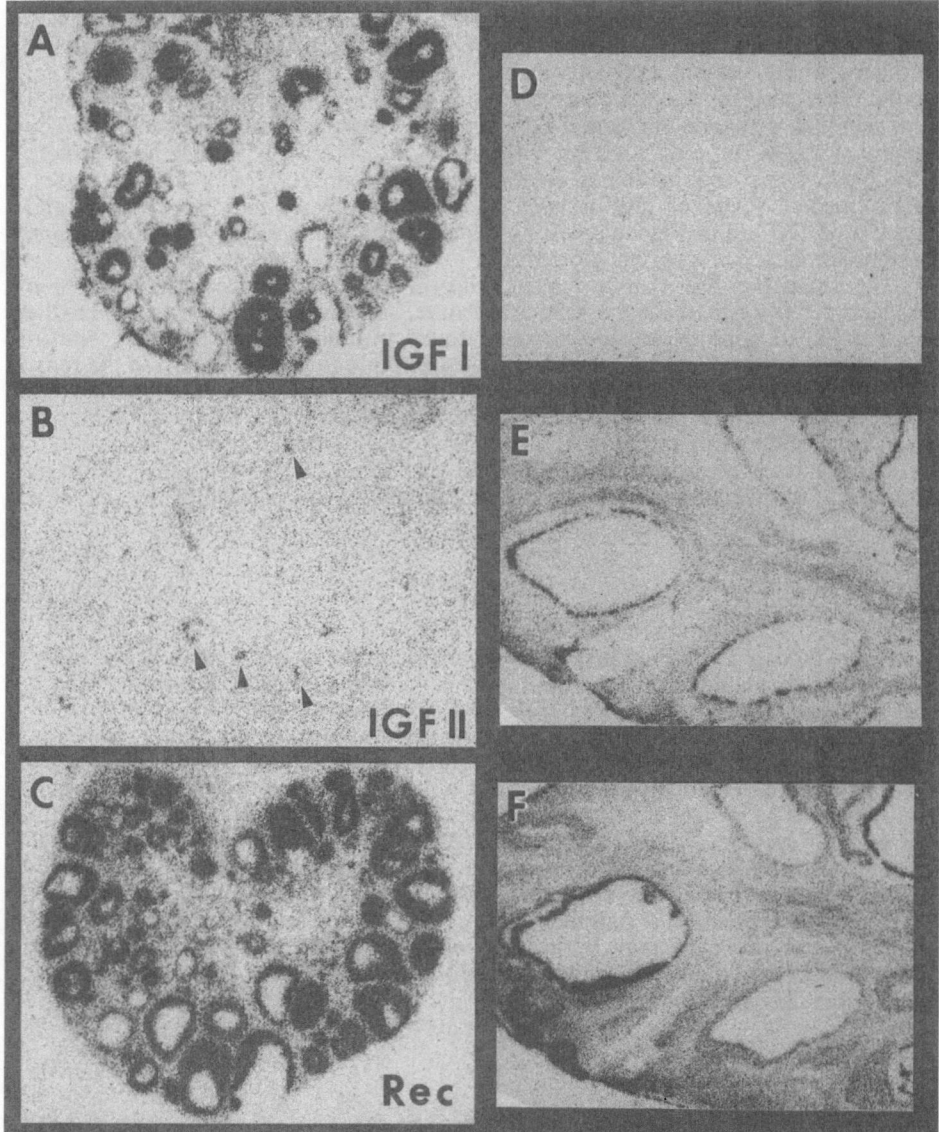

Fig. 2 IGF system gene expression in sequential sections from the rat (A-C) and human (D-F) ovary. IGF-I mRNA is abundant in many follicles of the rat ovary, IGF-II mRNA is localized in perifollicular blood vessels (arrowheads) and IGF-I receptor mRNA is abundant in all follicles. IGF-I mRNA is not detected in the human ovary; IGF-II mRNA is localized in the thin layer of granulosa cells lining the large atretic follicles, as is IGF-I receptor mRNA.

REFERENCES

1. W.H. Daughaday and P. Rotwein, Insulin-like growth factors I and II. Peptide messenger ribonucleic acid and gene structures, serum, and tissue concentration, *Endocrine Reviews* 10:68 (1989).
2. E. Chin, J. Zhou and C.A. Bondy, Anatomical relationships in the pattern of IGF-I, IGFBP-1 and IGF-I receptor gene expression in the rat kidney, *Endocrinology* 130:3237 (1992).

3. E. Chin and C.A. Bondy, IGF system gene expression in the human kidney, *J. Clin. Endocrinol. Metab.*75:962 (1992).

4. J. Zhou, E. Chin and C.A. Bondy, Cellular pattern of IGF-I and IGF-I receptor gene expression in the developing and mature follicle, *Endocrinol.*129:3281 (1991).

5. J. Zhou and C.A. Bondy, Anatomy of the Human Ovarian IGF System, *Biol. Reprod.* 48:467 (1993).

6. E. Chin, J. Zhou and C.A. Bondy, Anatomical and developmental patterns of facilitative glucose transporter gene expression in the rat kidney, *J. Clin .Invest* 91:810 (1993).

7. E. Chin and C.A. Bondy, Increased local IGF-I gene expression in protein-induced renal hypertrophy, *Prog. 75th Annual Meeting of Endo. Soc.* Abstract (1993).

8. J. Zhou and C.A. Bondy, The anatomy of the human testicular IGF system, *Prog 75th Annual Meeting of Endo Soc.* Abstract (1993).

9. E.R. Hernandez, C.T. Roberts, D. LeRoith and E.Y. Adashi, Rat ovarian IGF-I gene expression is granulosa cell-selective, *Endocrinology* 125:572 (1989).

10. J.E. Oliver, J.R. Aitman, J.F. Powell, C.A. Wilson and R.N. Clayton, IGF-I gene expression in the rat ovary is confined to the granulosa cells of developing follicles, *Endocrinology* 124:2671 (1989).

11. E.R. Hernandez, A. Hurwitz, A. Vera, A. Pellicer, E.Y. Adashi, D. LeRoith and C.T. Roberts, Expression of the genes encoding the IGFs and their receptors in the human ovary. *J. Clin. Endocrinol. Metab.* 74:419 (1992).

12. F. Geisthovel, I. Moretti-Rojas, R.H. Asch and F.J. Rojas, Expression of IGF-II but not IGF-I mRNA in human preovulatory granulosa cells, *Human Reprod.* 4:899 (1989).

13. A. El-Roeiy, X. Chen, V.J. Roberts and S.S.C. Yen, Localization of mRNAs encoding IGF-I, IGF-II and receptors of IGF-I, IGF-II and insulin in normal human ovaries, *Prog. 74th Annual meeting of Endo. Soc.* 237 (1992).

TRANSCRIPTIONAL REGULATION OF THE INSULIN RECEPTOR GENE PROMOTER

Catherine McKeon

Director, Metabolic Diseases and Gene Therapy Research Program
National Institute of Diabetes and Kidney and Digestive Diseases
National Institutes of Health, Bethesda, MD 20892

INTRODUCTION

The human insulin receptor is a membrane bound glycoprotein which is found on the surface of most cells. Like many growth factor receptors, the insulin receptor is a tyrosine kinase. The receptor is a heterotetramer composed of 2 alpha- and 2 beta-subunits. The alpha-subunit is located extracellularly where it binds insulin while the beta-subunit spans the membrane and contains the tyrosine kinase domain. The presence of this receptor is probably necessary for normal cell growth. Specialized cells such as myocytes and adipocytes have evolved mechanisms to elevate levels of this protein on their surface which allows these cells to be especially sensitive to the effects of insulin. In addition, the number of insulin receptor has been shown to be regulated by systemic factors such as hormones[1] and nutrients such as glucose.[2] Many investigators have cloned and analyzed the human insulin receptor promoter in an attempt to identify individual elements which are responsible for these effects. In this review, I will summarize the progress in this field and will discuss some of the conflicting data which has resulted.

THE STRUCTURE OF THE HUMAN INSULIN RECEPTOR GENE

The era of molecular biology of the insulin receptor began in 1985 when two independent groups reported the cloning and sequencing of the insulin receptor cDNA.[3,4] The structure of the insulin receptor gene was elucidated soon afterwards.[5] A single gene encodes both the alpha- and beta-subunit of the insulin receptor which is proteolytically cleaved. The gene spans over 150 kb of DNA and consists of 22 exons. The insulin receptor gene produces many different mRNA transcripts by using different start sites, different poly-A addition sites and alternate splicing to increase diversity.

Current Directions in Insulin-Like Growth Factor Research,
Edited by D. LeRoith and M.K. Raizada, Plenum Press, New York, 1994

79

When the insulin receptor mRNA is studied by Northern analysis, there are 5 different species of mRNA ranging in size from 5.2 Kb to 10 Kb. This size variation has been shown to result from the use of alternate poly-A addition sites at the 3' end of the molecule.[6] Several investigators have data suggesting that these different 3' nontranslated regions may be involved in translational control of insulin receptor levels.[2,6]

The two cDNAs which were originally isolated differed by the insertion of 36 nucleotides.[3,4] This difference has been shown to occur due to alternate splicing of exon 11. The proportion of mRNA including exon 11 or excluding exon 11 varies between tissues.[7,8] Recent studies have suggested that as a result of this alternate splicing the insulin receptor isoforms have different properties. The insulin receptor mRNA that includes the exon 11 sequence has been reported to be abnormally elevated in muscle cells of individuals with insulin resistance.[9]

HUMAN INSULIN RECEPTOR PROMOTER

Several groups have independently cloned and sequenced the 5' flanking region of the human insulin receptor gene.[5,6,10,11,12] Although initially the exon/intron junction was reported incorrectly,[10,11] once the second exon was sequenced it became clear that the first intron contains 100 nucleotides of coding region.[5] The first exon encodes the signal peptide and ends with the first G of the Valine codon for amino acid number 7 in the insulin receptor followed by a consensus splice donor site.[5,6,12] The sequences reported by each group are in general agreement and the differences have been tabulated previously.[12] The mouse insulin receptor promoter region has also been cloned and sequenced.[13] The sequence of the mouse promoter has stretches of homology with the human promoter which may mark regions of functional significance.

Inspection of the sequence reveals that the promoter region is GC-rich with these nucleotides comprising 82% of the first 575 nucleotides. There is no obvious TATA box or CCAAT box sequence upstream of the translation start site. These characteristics are typical for a class of so called "housekeeping" promoters. These are promoters which are active in many cells and is therefore an appropriate classification for the insulin receptor.

Like other housekeeping promoters, the insulin receptor gene does not have a discrete start site of transcription. Because of the secondary structure formed by the GC-rich promoter DNA, the methods used to determine the start sites of transcription are prone to artifacts. This probably accounts for the two initial reports that placed the start sites in locations which were not confirmed by functional studies.[5,10] The three subsequent publications are in general agreement that there are several start sites of transcription scattered over a 160 base pair region from -550 to -390.[6,11,12] This problem was best addressed by Tewari et. al.[6] who used both S1 nuclease digestion and primer extension to confirm the localization of the transcriptional start sites. Only the start sites located at -550, -464, -424 and -390 were observed by both methods.[6] Our data suggested that the most 5' start site was located at -537 and we have isolated a cDNA clone beginning at -535 which would correspond to transcription initiation at the most 5' start site proposed.[12] Therefore, the 5' untranslated region could vary by up to 160 bases depending on the choice of start sites. For this reason, the numbering of the insulin receptor flanking sequence used by most investigators uses the unique start site of translation as its origin +1.

Many investigators have performed functional studies of the human insulin receptor by fusing portions *i* the 5' flanking region to the chloramphenicol acetyl

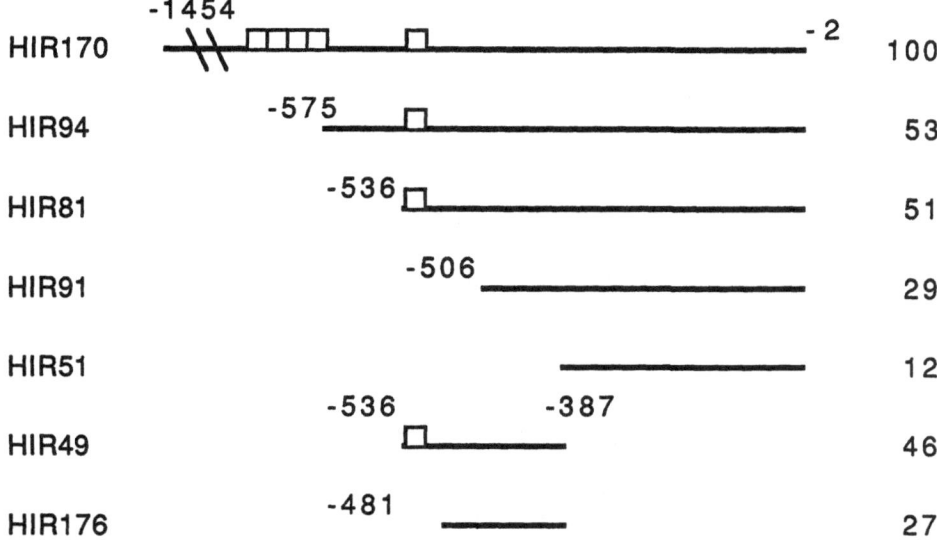

Figure 1. Deletion analysis of the human insulin receptor promoter. The promoter is represented schematically with Sp1 binding sites denoted by boxes. The deletion constructions are shown and the remaining elements are denoted. The activity of the full-length 1454 bp promoter is set equal to 100%. The percent activity of each deletion is compared to the full-length promoter HIR170 and is the mean of multiple determinations. Adapted from McKeon et al.[12]

transferase (CAT) reporter gene.[2,5,6,10,12,14,15] These construction are then transfected into tissue culture cells and the activity of the CAT enzyme is measured as an indication of the strength of the promoter. Although these experiments have been performed essentially the same by each group, there is disagreement in the resulting data. In Figure 1, are the results published by our groups to define the important functional domains of the promoter.[12] These results show that there are positive factors upstream of -575 and between -536 and -506. The data from all groups taken together confirms at least two regions which contain well defined positive regulators of transcription, a proximal enhancer region and the Sp1 enhancer cluster region.

The Proximal Enhancer Region

We first identified a positive element as a result of the deletion analysis shown in Figure 1. Deletions which included the region between -535 and -505 resulted in a consistent 2-fold decrease in the CAT activity.[12] To further study this region, we showed that when a fragment from -575 to -481 containing this region was placed in front of an enhancerless SV-40 promoter, it was able to enhance activity of the promoter 3-fold.[12] As shown in Table I, this activity was orientation independent since it occurred no matter which orientation the fragment was inserted. The presence of this enhancer has been confirmed by Cameron et al.[15] They showed that deletion of sequences between -575 to -489 resulted in a 3-fold decrease in activity and that this region could enhance another heterologous promoter, the thymidine kinase promoter.[15] Therefore, this region meets the criteria of an enhancer since it functions on a heterologous promoter and in either orientation.

Table 1. Relative Increase in CAT Activity by the Enhancer Sequence

	HepG2	3T3
pA10-CAT	1	1
pA10-CAT + 94 bp Normal orientation	3.0 ± 1.1	8.8 ± 1.1
pA10-CAT + 94 bp Reverse orientation	3.3 ± 0.5	3.1 ± 0.9

The values shown in the table are the means of three separate experiments where the activity of pA10-CAT was set equal to 1. They are followed by the SEM. In all three experiments, the constructions with the 94 bp inserts were more active than pA10-CAT. From McKeon et al.[12]

To further analyze the sequence of this enhancer we have synthesized a double stranded oligonucleotide with the sequence from -528 to -502. This sequence was cloned into the insulin receptor promoter to create a promoter with either 1, 2 or 3 copies of this sequence. In vectors with 2 copies, there was a 3- to 4-fold increase in CAT activity and with 3 copies, there was a 5- to 6-fold increase.[16] Therefore, this effect can be mediated by as little as 26 bp and is additive.

Three groups have studied this region to determine where proteins might bind.[14,15,16] We have used a 26 bp oligonucleotide with the sequence from -528 to -502 to look for proteins which bound to this enhancer. We have shown using an electrophoretic mobility shift assay (EMSA) that a protein could be identified in multiple tissue types which bound to this sequence.[16] DNA footprint studies in this region showed that in HepG2 cell extracts two factors bound to DNA between -550 and -530 and -520 and -500.[14,15] The factor which binds between -520 and -500 appears to be identical to the one in our study. Although the sequence in this region differs from the Sp1 consensus sequence, competition studies have shown that the identity of one protein binding to this region is Sp1. Protein binding is able to be competed by an oligonucleotide with a known Sp1 binding site.[15,16] In addition, purified Sp1 can bind to this oligonucleotide sequence with a reduced affinity. We have been able to further retard the DNA/protein binding complex by preincubation with an anti-Sp1 antibody confirming the participation of Sp1 in this complex.[17]

These studies show that the identity of at least one of the enhancer binding proteins is Sp1. The participation of this non-specific factor would help to explain the ubiquitous nature of the promoter defined by these studies. Sequence comparison between the mouse and the human promoters provides further evidence for a functional significance for this region. This element has 17 out of 18 base pair identity with the analogous region in the mouse insulin receptor promoter.[13,16] A second region that binds Sp1 (see below) has been shown to be important in the regulation of the human insulin receptor promoter. However unlike the proximal enhancer, there is no homologous region in the mouse promoter.

Sp1 Enhancer Cluster

All of the investigators studying the effect of upstream sequences on promoter activity agree that sequences between -1818 and -575 have a positive effect on promoter activity.[2,5,6,10,12,14,15] However, the magnitude of this effect differs from a low of 2-fold[5,12] to a high of 10-fold.[2,18] It is not readily apparent why different groups have obtained such different results from essentially the same experiment. A clue to the variable may have been discovered by Lee et al.[14] When they used the same vector in two different hepatocyte cell lines, they observed a 2-fold increase in the PLC cell line and a 8-fold increase in the HEPG2 cell line when the upstream region was added to the reporter vector.[14] Although all investigator are reportedly using the HepG2 cells, it is possible that subtle differences such as passage number and differentiation characteristics may account for the apparent discrepancy observed.

The most striking feature of this region is the presence of four tandem consensus Sp1 binding sites from -618 to -593. The effect of these binding sites on the activity of the human insulin receptor promoter has been studied by Araki et al. using a new CAT reporter vector.[18] This vector, pSVOOCAT has a lower background than the vector used by other investigators. In their studies, deletion of the region from -1452 to -575 resulted in a 9-fold decrease in activity. When only the sequence from -631 to -593 containing the 4 Sp1 sites was added back to the minimal promoter at -575, the activity increased 25-fold.[18] In addition, purified Sp1 has been shown to bind to these sequences. It would seem that these sites may be modulated by the surrounding sequences to account for the lesser effect of these Sp1 sites when embedded in the context of the larger fragment.

It is interesting to note that there is a sequence polymorphism found within one of these Sp1 binding sites in the promoter. The original sequence reported by Araki et al.[10] had a G at base pair -603 while subsequent sequences have found an A at this position.[5,6,] Sequencing alleles from unrelated individuals has confirmed the presences of both sequences (T. Kadowaki and S. I. Taylor, personal communication). Comparison of this change to other known Sp1 binding sites would suggest that the A at this position would reduce the affinity of Sp1 binding to this site. The implications of this polymorphism on the activation of the insulin receptor promoter has yet to be determined. However, the studies of the Sp1 binding sites reported by Araki et al.[18] use the sequence with the G in position -603 and this may explain why they observe a larger effect of this region than was observed by others.

A recent study has combined a deletion analysis and DNA footprinting of the region up to -1823.[15] Several new protein binding sites have been demonstrated by these techniques. Two of these sites have been postulated to bind negative regulatory factors.[15] The characterization of these new binding sites and their interaction with the ones already described will extend our appreciation of the complexity of this promoter.

Clearly, the regulation of the insulin receptor promoter will be intricate. The two promoter elements which have been described so far bind the ubiquitous factor, Sp1. The presence of these sites in the promoter would contribute to the expression of the insulin receptor in multiple cell types. Although the insulin receptor is transcribed in many cell types, it is transcribed at a greater rate in particular cells which are the target cells for insulin action and in response to particular hormones. Several of the mechanisms responsible for these phenomenon have been studied.

GLUCOCORTICOID REGULATION

One method to induce insulin receptors on the surface of cells is by the administration of glucocorticoids. Fantus et al.[1] initially showed that the number of insulin receptor increased on the surface of IM9 cells when physiologic doses of glucocorticoids were added to the culture medium. Dexamethasone was the most potent inducer however several members of this hormone family could produce similar effects.

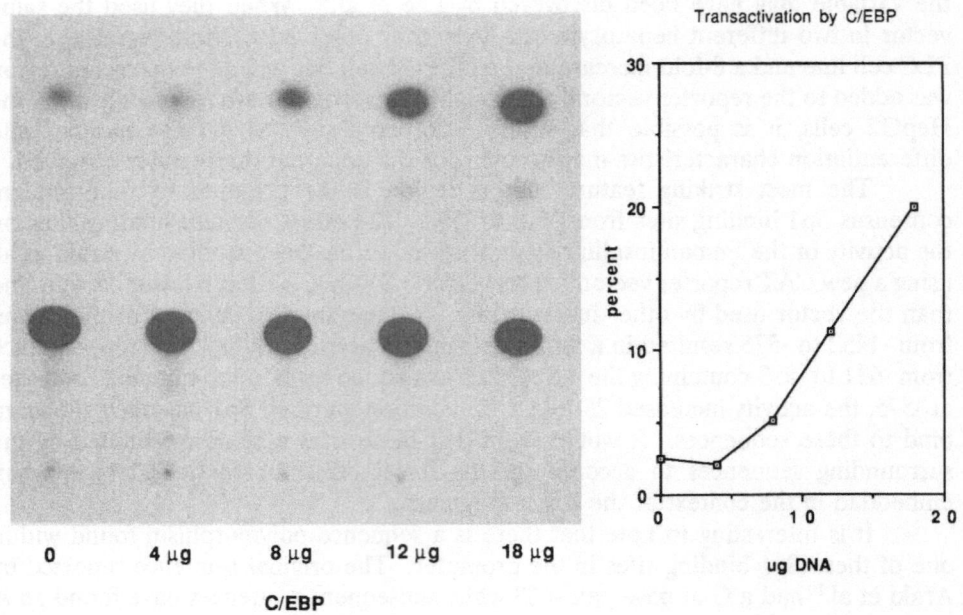

Figure 2. Insulin receptor mRNA levels increase after glucocorticoid treatment. RNA was isolated from IM9 cells following incubation with 1.4 μM hydrocortisone for 0, 2, 4, and 6 hours and quantitated by a nuclease protection assay. Lanes 0 -6 and B and D used 20 μg of RNA while lanes C and E used 40 μg of RNA. From Rouiller et al.[19]

The mechanism for insulin receptor increase has been shown to occur by transcriptional regulation. When mRNA was isolated from IM9 cells at various times After hydrocortisone treatment and quantitated by a nuclease protection assay, the insulin receptor mRNA increased over the 6 hour period while the ferritin control remained the same.[19] The insulin receptor mRNA increased about 3-fold in response to hydrocortisone (Figure 2).[19,20] Nuclear runoff studies which measure the rate of transcription initiation showed that this increase in mRNA resulted from an increased rate of mRNA transcription which could be seen as early as 1 hour after the addition of hormone.[19,21] When McDonald et al. used primer extension on mRNA isolated after glucocorticoid stimulation, they found that all of the mRNA start sites showed an coordinate increase in response to the hormone.[20]

Because these studies demonstrated that glucocorticoids had a direct effect on the human insulin receptor gene, several groups have searched for the glucocorticoid regulatory element in the insulin receptor promoter. There is no sequence present in the HIR promoter which corresponds to the GRE consensus sequence however there are several sequences with some homology to the published consensus sequence. Transfected insulin receptor promoter-CAT constructions containing as little as -692 base pairs have been shown to be induced by the addition of glucocorticoids.[14] Recently, Lee et al. have demonstrated by DNA footprinting that purified glucocorticoid receptor can bind to a putative site in the 5' flanking region of the insulin receptor from -357 to -345 which show homology to the GRE.[14] This sequence will need to be studied by mutational analysis to provide definitive evidence that it is the site of glucocorticoid binding in vivo.

MYOCYTE ENHANCER

Cell culture systems have been enormously useful in elucidating the mechanisms of differentiation. Several cell lines are available which develop morphologic characteristics of differentiation after growth under appropriate conditions. When these cells are grown in high serum and at low cell densities, they will grow as myoblasts. When confluent myoblasts are grown in low serum, the cell take on the morphologic characteristics of differentiated myocytes. Both mouse C2 cells and rat L6 cells will fuse to form primitive myotubes where as the mouse BH3 cell line is non-fusing. The insulin receptor is induced during differentiation of these muscle cells in culture. A 5- to 10-fold induction of insulin receptors was demonstrated in the mouse myoblast cell lines, BC3H-1 and C2.[22] This induction was accompanied by a 5- to 10-fold increase in steady state mRNA levels.[22] The rat muscle cell line, L6 has also been shown previously to have a 2-fold increase in insulin receptors during differentiation.[23] We have been able to show an induction of insulin receptor mRNA in the rat muscle cell line, L6 after the myoblasts fused to form primitive myotubes (H. Chen and C. McKeon, unpublished observations). One muscle-specific protein which could be a potential mediator of this effect has been shown to bind in the upstream region of the promoter.[24]

ADIPOCYTE ENHANCER

The induction of insulin receptors has also been studies in cell culture models of adipocyte differentiation. The mouse 3T3-L1 cell line can be induced to differentiate by the addition of isomethylbutylxanthine, dexamethasone and insulin.[25] The insulin receptor gene is induced about 10-fold in cultured 3T3-L1 cells during adipocyte differentiation. This increase is accompanied by an analogous increase in insulin receptor mRNA.[3,13] Previous attempts to identify an adipocyte regulatory region in the mouse promoter were not successful.[13] We believe that one sequence required for this induction in the human gene is located in the first intron of the insulin receptor gene. We have cloned segments of the insulin receptor gene into an expression vector consisting of the HIR promoter fused to the chloramphenicol acetyltransferase gene (CAT). By transfecting into 3T3-L1 cells, selecting stable transfectants and monitoring CAT activity during differentiation, several clones were identified in which the activity of CAT increased 6- to 8-fold. These clones had been transfected with a vector, HIRIN2, containing -535 of the human insulin receptor promoter and +15 to +1703 of the first exon and intron. Successive deletions of this segment have localized the region which contributed to the induction of the human insulin receptor during differentiation to a 577 bp fragment.[26]

This fragment contains several consensus binding sites for previously identified transcriptional regulators. There are 4 consensus binding sites for the transcription factor Sp1 and a consensus binding sites for the transcription factor C/EBP. Cotransfection of the vector containing the HIR promoter and intron (HIR2) with an expression vector for the factor C/EBP shows that C/EBP can transactivate this HIR vector 6- to 8- fold in a transient cotransfection assay.[27] In Figure 3, the amount of C/EBP expression vector was increased as indicated while the reporter construction was held constant. The CAT activity increased in a dose-dependent manner. Because C/EBP has been implicated in the induction of adipocyte-specific genes, these studies initially suggested to us that the C/EBP site might be important in the induction of the insulin receptor during adipocyte differentiation.[28] However,

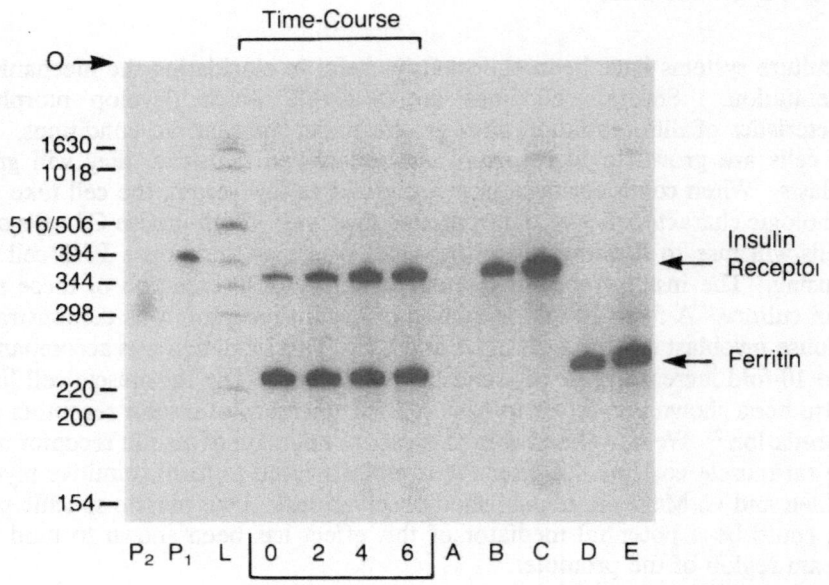

Figure 3. The C/EBP expression vector transactivates the insulin receptor promoter in a dose-dependent manner. 2 μg of HIRIN2 was transfected with increasing amounts of the C/EBP expression vector as indicated. A defective expression vector was added so that the DNA per transfection totaled 20 μg. CAT activity was determined after 48 hrs. The CAT assay is shown on the left and the percent conversion as determined by β-scanning is graphed on the right. From McKeon and Pham[27].

mutation of the consensus C/EBP site (+324 to +343) failed to abolish the induction by the C/EBP expression vector (McKeon, unpublished observations). Therefore, either C/EBP is binding to another sequence in the intron or the effects of the C/EBP expression vector are indirect. C/EBP may be acting by inducing another adipocyte factor which in turn induces the insulin receptor. These alternative hypothesises are being tested.

LOCALIZATION OF PROMOTER ELEMENTS

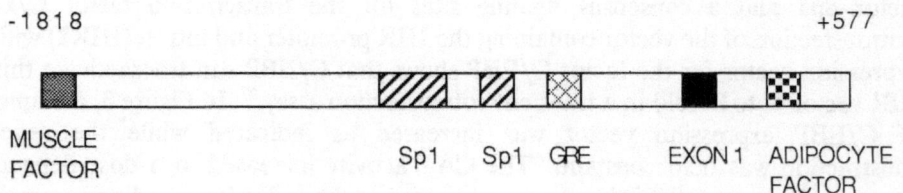

Figure 4. Relative position of the promoter elements in the human insulin receptor gene which have been discusssed.

SUMMARY

The insulin receptor is a highly regulated promoter. A schematic of several of the elements so far identified is shown in Figure 4. The gene has a basic "housekeeping" promoter which controls low level expression in all cells. This promoter seems to be regulated by the transcription factor, Sp1 at several locations upstream. There are in addition several potential Sp1 binding sites in the first intron. Specific enhancers are present to allow increased expression in certain cell types or in response to hormones. Several potential enhancers have been identified including a potential GRE binding site, muscle specific binding protein, and adipocyte binding protein. Clearly additional elements need to be identified in order to elucidate the complexed interactions which are required for appropriate regulation of the insulin receptor.

ACKNOWLEDGEMENTS

I wish to thank Simeon Taylor, Dominique Rouiller, Phillip Gorden, Domenico Accili, Gillian Walker, Tony Pham and Hui Chen with whom this work was done and to thank the JDFI for grants #188739 and #190916 which supported portions of this work.

REFERENCES

1. I.G. Fantus, G.A. Saviolakis, J.A. Hedo and P. Gorden, Mechanism of glucorticoid-induced increase in insulin receptors of cultured human lymphocytes, *J. Biol. Chem.* 257:8277 (1982).
2. J.R. Levy, G. Krystal, P. Glickman and F. Dastvan, Effects of media conditions, insulin, and dexamethasone on insulin-receptor mRNA and promoter activity in HepG2 cells, *Diabetes.* 40:58 (1991).
3. A. Ullrich, J.R. Bell, E.Y. Chen, R. Herrera, L.M. Petruzzelli, T.J. Dull, A. Gray, L. Coussens, Y.-C. Liao, M. Tsubokawa, A. Mason, P.H. Seeburg, C. Grunfeld, O.M. Rosen and J. Ramachandran, Human insulin receptor and its relationship to the tyrosine kinase family of oncogenes, *Nature.* 313:756 (1985).
4. Y. Ebina, L. Ellis, K. Jarnagin, M. Edery, L. Graf, E. Clauser, J. Ou, F. Masiarz, Y.W. Kan , I.D. Goldfine, R.A. Roth and W.J. Rutter, The human insulin receptor cDNA: the structural basis for hormone-activated transmembrane signalling, *Cell.* 40:747 (1985).
5. S. Seino, M. Seino, S. Nishi and G. Bell, Structure of the human insulin receptor gene and characterization of its promoter, *Proc. Natl. Acad. Sci. USA.* 86:114 (1989).
6. D.S. Tewari, D.M. Cook and R. Taub, Characterization of the promoter region and 3' end of the human insulin receptor gene, *J. Biol. Chem.* 264:16238 (1989).
7. D.E. Moller, A. Yokota, J.F. Caro and J.S. Flier, Tissue-specific expression of two alternatively spliced insulin receptor mRNAs in man. *Mol. Endocrinol.* 3:1263 (1989).
8. S. Seino and G.I. Bell, Alternative splicing of human insulin receptor messenger RNA. *Biochem. Biophys. Res. Comm.* 159:312 (1989).

9. L. Mosthaf, J. Eriksson, H. Haring, L. Groop, E. Widen and A. Ullrich, Insulin receptor isotype expression correlates with risk of non-insulin-dependent diabetes, *Proc. Natl. Acad. Sci. USA*. 90:2633 (1993).

10. E. Araki, F. Shimada, H. Uzawa, M. Mori and Y. Ebina, Characterization of the promoter region of the human insulin receptor gene, *J Biol Chem*. 262:16186 (1987).

11. P.W. Mamula, K. Wong, B.A. Maddux, A.R. McDonald and I.D. Goldfine, Sequence and analysis of promoter region of human insulin-receptor gene, *Diabetes*. 37:1241 (1988).

12. C. McKeon, V. Moncada, T. Pham, P. Salvatore, T. Kadowaki, D. Accili and S.I. Taylor, Structural and functional analysis of the insulin receptor promoter, *Mol. Endocrinol*. 4:647 (1990).

13. E. Sibley, T. Kastelic, T.J. Kelly and M.D. Lane, Characterization of the mouse insulin receptor gene promoter, *Proc. Natl. Acad. Sci. USA*. 86:9732 (1989).

14. J.K. Lee, J.W.O. Tam, M.J. Tsai and S.Y. Tsai, Identification of cis- and trans-acting factors regulating the expression of the human insulin receptor gene, *J. Biol. Chem*. 267:4638 (1992).

15. K.E. Cameron, J. Resnik and N.J.G. Webster, Transcriptional regulation of the human insulin receptor promoter, *J. Biol. Chem*. 267:17375 (1992).

16. H. Chen, G.E. Walker, S.I. Taylor and C. McKeon, Characterization of the proximal enhancer of the insulin receptor gene, *Endocrinology*. 130(Suppl):111 (1992).

17. H. Chen, G.E. Walker and C. McKeon, The proximal enhancer of the insulin receptor gene binds the transcription factor Sp1, *Endocrinology*. in press, (1993).

18. E. Araki, T. Murakami, T. Shirotani, F. Kanai, Y. Shinohara, F. Shimada, F. Mori, M. Shichiri and Y. Ebina, A cluster of four Sp1 binding sites required for efficient expression of the human insulin receptor gene, *J. Biol. Chem*. 266:3944 (1991).

19. D.G. Rouiller, C. McKeon, S.I. Taylor and P. Gorden, Hormonal regulation of insulin receptor gene expression: Hydrocortisone and insulin act by different mechanisms, *J. Biol. Chem*. 263:13185 (1988).

20. A.R. McDonald, A.S. Maddox, Y. Okabayashi, K.Y. Wong, D.M. Hawley, C.D. Logson and I.D. Goldfine, Regulation of insulin receptor mRNA levels by glucocorticoids, *Diabetes*. 36:779 (1987).

21. A.R. McDonald and I.D. Goldfine, Glucocorticoid regulation of insulin receptor gene transcription in IM-9 cultured lymphocytes, *J Clin Invest*. 81:499 (1988).

22. A. Brunetti, B.A. Maddox, K.Y. Wong and I.D. Goldfine, Muscle cell differentiation is associated with increased insulin receptor biosynthesis and messenger RNA levels, *J. Clin. Invest*. 83:192 (1989).

23. F. Beguinot, C.R. Kahn, A.C. Moses and R.J. Smith, The development of insulin receptors and responsiveness is an early marker of differentiation in the muscle cell line L6, *Endocinology*. 118:446 (1986).

24. A. Brunetti, K.Y. Wong, P.A. Goodman and I.D. Goldfine, Identification of nuclear binding proteins for the insulin receptor promoter, *Endocrinology*. 128(Suppl):222 (1991).

25. H. Green and M. Meath, An established pre-adipose cell line and its differentiation in culture, *Cell*. 3:127 (1974).

26. C. McKeon, T. Pham and G.E. Walker, A region in the intron of the insulin receptor gene is responsible for induction during adipocyte differentiation, *Endocrinology*. 128(suppl):231 (1991).

27. C. McKeon and T. Pham, Transactivation of the human insulin receptor gene by the CCAAT/Enhancer binding protein, *Biochem. Biophy. Res. Commun.* 174:721 (1991).

28. R.J. Christy, V.W. Yang, J. Ntambi M., D.E. Geiman, W.H. Landschulz, A.D. Friedman, Y. Nakabeppu, T.J. Kelly and M.D. Lane, Differentiation-induced gene expression in 3T3-L1 preadipocytes: CCAAT/enhancer binding protein interacts with and activates the promoters of two adipocyte-specific genes, *Genes Devel.* 3:1323 (1989).

THE REGULATION OF IGF-I RECEPTOR GENE EXPRESSION BY POSITIVE AND NEGATIVE ZINC-FINGER TRANSCRIPTION FACTORS

Haim Werner, Charles T. Roberts, Jr., and Derek LeRoith

Section on Molecular and Cellular Physiology, Diabetes Branch
National Institutes of Health, NIDDK
Building 10, Room 8S-239
Bethesda, Maryland 20892

INTRODUCTION

The insulin-like growth factor-I receptor (IGF-I-R) is a trans-membrane tyrosine kinase which mediates the trophic, metabolic, and differentiative effects of IGF-I and IGF-II[1,2]. Beginning at early organogenesis, the IGF-I-R gene is constitutively expressed by most body tissues, consistent with the putative role of IGF-I as a progression factor in the cell cycle[3]. Recent studies have shown that overexpression of the IGF-I-R in a fibroblast cell line abrogates all requirements for exogenous growth factors, suggesting that this receptor does, in fact, mediate a central mechanism in the cell cycle[4].

Developmental Regulation of the IGF-I-R Gene

Previous studies by us and others have shown that the expression of the IGF-I-R gene is developmentally regulated: high levels of IGF-I-R mRNA and binding are seen at embryonic and early postnatal stages, at which time the IGF-I-R is probably involved in differentiation. At later postnatal and adult stages, the levels of IGF-I-R mRNA and protein decrease to low levels, possibly due to down-regulation by high levels of circulating IGF-I[5]. Despite its down-regulation at adult stages, however, the IGF-I-R gene product is constitutively expressed by most body tissues. This suggests that the IGF-I-R gene can, to a certain extent, be considered to be a "housekeeping" gene.

Superimposed on this basal level of expression, the IGF-I-R gene has been shown to be highly regulated in a number of physiological and pathological states, including fasting, hormonal changes, diabetes, and neoplasia[6-10].

Current Directions in Insulin-Like Growth Factor Research,
Edited by D. LeRoith and M.K. Raizada, Plenum Press, New York, 1994

91

Cloning of the IGF-I-R Gene Promoter

To study the transcriptional control mechanisms that regulate IGF-I-R gene expression, we have cloned the promoter region of the rat IGF-I-R gene[11,12]. Analysis of this region indicates that the IGF-I-R gene promoter contains a number of interesting features that suggest it is a novel intermediate between promoters for highly-regulated genes and those for housekeeping genes.

The transcriptional start site of the IGF-I-R gene was determined by primer extension and RNase protection assays. Unlike the insulin receptor gene and many other housekeeping genes, in which transcription is initiated from several sites, the IGF-I-R gene has a unique start site. This site is contained within an "initiator" motif, and it defines a very long 5'-untranslated region of 940 bp in the rat gene and ~1 kb in the human gene[13,14]. This initiator element, which was previously described in other highly regulated genes such as the terminal deoxynucleotidyl transferase, can direct specific transcription initiation in the absence of a TATA element[15].

The proximal ~450 bp of 5'-flanking region is extremely GC-rich (~80%), and there are no obvious TATA or CAAT elements in the vicinity of the transcription start site. Several GC-boxes, which are putative binding sites for transcription factor Sp1, are found both in the 5'-flanking and 5'-untranslated regions (Fig. 1).

In addition, a perfectly conserved site and several related potential binding sites for transcription factors of the early growth response (EGR) family are found in this region. The involvement of these trans-activating factors in the regulation of the IGF-I-R gene promoter is discussed below.

The 5'-flanking region of the IGF-I-R gene also contains potential binding sites for at least three other known transcription factors: ETF, GCF, and AP-2. The participation of these elements in promoter activity has not yet been established, however.

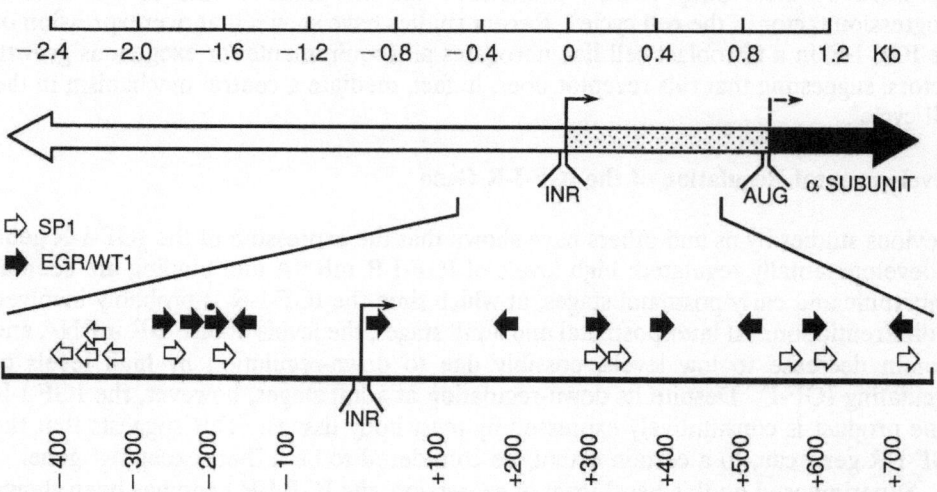

Figure 1. Schematic representation of the 5'-flanking and 5'-untranslated regions of the rat IGF-I-R gene. The coding region is black, the 5'-untranslated region is dotted, and the 5'-flanking region is open. The transcription initiation site is denoted by an arrow at the initiator (INR) motif. The dashed arrow denotes the translation start site. Open arrows are potential Sp1 binding sites. The black arrow with a dot is a consensus EGR/WT1 binding site, and the black arrows are EGR/WT1-related sites.

Mapping the IGF-I-R Gene Promoter Activity

To assay promoter activity exhibited by the 5'-flanking and 5'-untranslated regions of the IGF-I-R gene, different genomic fragments were fused to a firefly luciferase reporter gene in the promoterless expression vector pOLUC and used to transiently transfect buffalo rat liver (BRL3A) and Chinese hamster ovary (CHO) cells (Fig. 2). In both cell lines, no significant difference in promoter activity was seen between a fragment containing 2350 bp of 5'-flanking and 640 bp of 5'-untranslated regions [p(-2350/+640)LUC] and a fragment containing 416 bp of 5'-flanking and 232 bp of 5'-untranslated regions [p(-416/+232)LUC] (Nucleotide #1 corresponds to the transcription initiation site). To determine the contribution of the 5'-untranslated sequences contained in the -2350/+640 and -416/+232 fragments to overall promoter activity, we subcloned a fragment (-455 to +30) containing a minimal 5'-untranslated region sequence into pOLUC and measured its promoter activity after transfection into both BRL3A and CHO cells. Since the activity of this fragment was not significantly different from that of the -2350/+640 and -416/+232 fragments, all of the basal promoter activity exhibited in these cell lines is apparently contained in the proximal 5'-flanking region, and there is no significant contribution of 5'-untranslated sequences to transcriptional activity. The orientation of the initiator element in the sense orientation was crucial in order to generate high levels of expression in all of the above constructs. However, the initiator element, by itself, was devoid of any measurable promoter activity, as determined by experiments in which a synthetic oligonucleotide corresponding to the sequence from bp -27 to +21 was subcloned into pOLUC and transiently transfected into BRL3A and CHO cells.

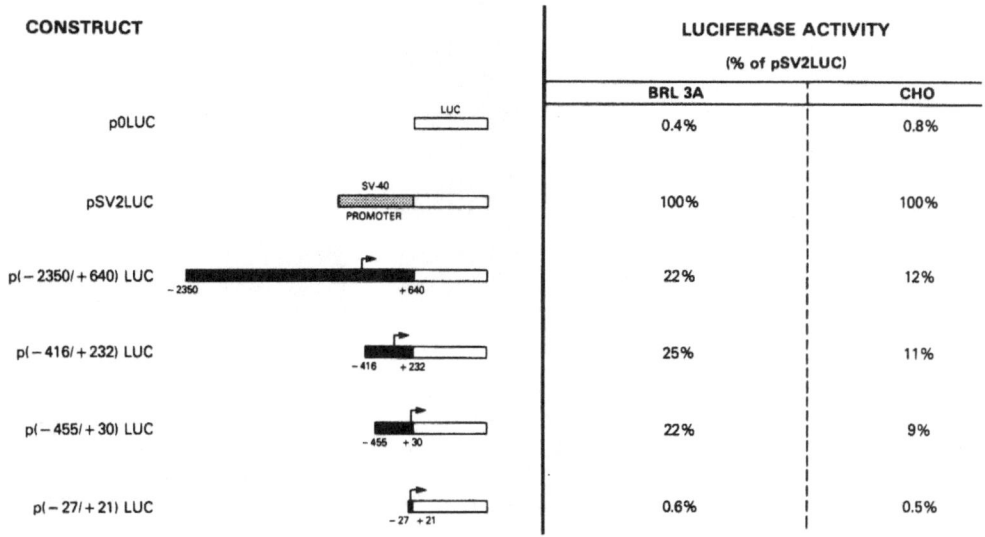

Figure 2. Promoter activity of 5'-flanking and 5'-untranslated fragments of the IGF-I-R gene. Fragments of the IGF-I-R gene containing different portions of the 5'-flanking and 5'-untranslated regions were subcloned in the promoterless pOLUC vector upstream of a firefly luciferase cDNA. Constructs were transiently transfected into BRL3A and CHO cells by electroporation, and the luciferase activity generated was measured with a luminometer. The values obtained were normalized by dividing them by the β-galactosidase activity generated by cotransfecting with a β-galactosidase expression vector. Results obtained are expressed as percentage of the luciferase activity obtained with a construct in which transcription of the luciferase cDNA is under the control of an SV40 enhancer/promoter (pSV2LUC).

In conclusion, the IGF-I-R promoter exhibits extremely high basal activity (10%-25% of the activity of an SV40 enhancer/promoter control), which may explain the widespread distribution of IGF-I-R mRNA and protein in most body tissues under basal conditions.

Trans-Activation of IGF-I-R Promoter by Sp1

As described above, most of the basal promoter activity of the IGF-I-R gene was mapped to the proximal ~450 bp of 5'-flanking region. Since this region contains six GC boxes, we investigated the capacity of Sp1 to trans-activate the IGF-I-R promoter. Sp1 is an ubiquitous nuclear transcription factor which binds GC-rich sequences by means of three zinc fingers, and activates transcription via glutamine-rich domains[16-17]. Coexpression experiments using the reporter plasmid p(-416/+232)LUC and an Sp1 expression vector driven by an alcohol dehydrogenase promoter (pPadh.Sp1-in) were performed in Schneider line 2 cells, a *Drosophila* cell line which lacks endogenous Sp1. The results of these studies showed that Sp1 trans-activated the IGF-I-R promoter by ~7- to 8-fold, whereas cotransfection of an Sp1 expression vector in which the Sp1 cDNA is translated out of frame had no effect (Fig. 3). These results suggest that the trans-activation effect of Sp1 was due to actively translated Sp1 protein.

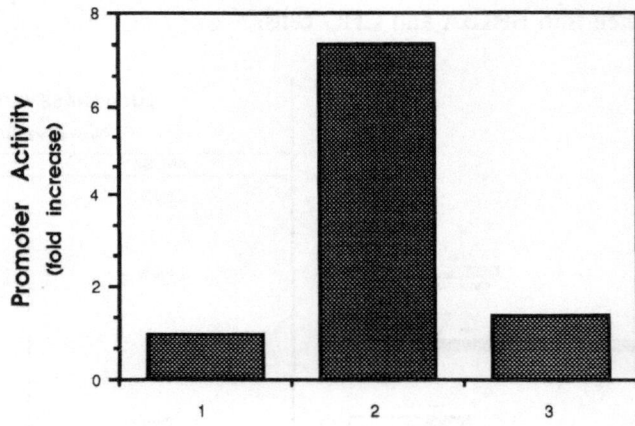

Figure 3. Trans-activation of the IGF-I-R promoter by Sp1. Five micrograms of the reporter plasmid p(-416/+232)LUC were cotransfected into *Drosophila* Schneider line 2 cells with 0 μg (1) or 15 μg (2) of the Sp1 expression vector pPadh.Sp1-in, or with 15 μg of the out-of-frame Sp1 expression vector pPadh.Sp1-out (3). Cells were transfected by the calcium phosphate method, as described (ref. 12). The values of luciferase activity were normalized for protein.

Sp1 Binds to Specific IGF-I-R Promoter Sequences

To study the protein-binding activity of the IGF-I-R promoter, gel-retardation experiments were performed using crude nuclear extracts from CHO cells and ^{32}P-labeled fragments of the 5'-flanking and 5'-untranslated regions. Double-stranded synthetic oligonucleotides corresponding to specific consensus binding sites for transcriptional regulatory factors were used as competitors in binding assays. The results of gel-retardation experiments suggested that proteins present in CHO nuclear extracts were able to specifically bind Sp1 binding sites. On the other hand, oligonucleotides directed against ETF and AP-2 binding sites were unable to prevent the formation of DNA-protein complexes, suggesting that these specific factors are not involved in binding to the IGF-I-R gene promoter, or are relatively deficient in CHO nuclear extracts.

Gel-retardation experiments were also performed using crude nuclear extracts from rat neonate brain and adult liver (Fig. 4). The rationale for using these sources of nuclear proteins was that neonate brain expresses very high levels of IGF-I-R mRNA, whereas receptor transcripts in adult liver are almost undetectable. The gel-retardation

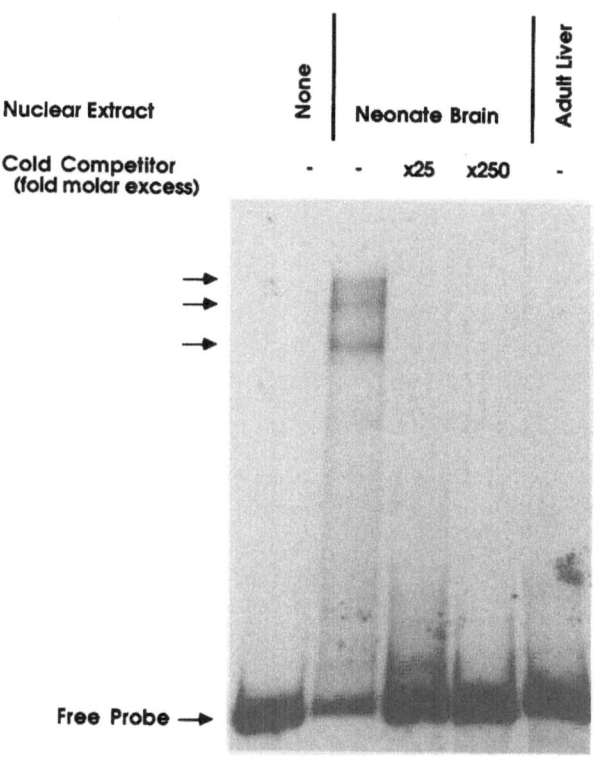

Figure 4. Gel-retardation analysis of the -416/-331 fragment of the IGF-I-R gene promoter. One microgram of crude nuclear extract from rat neonate brain or adult liver was incubated with ^{32}P-labeled -416/-331 fragment, in the absence or presence of 25- to 250-fold molar excess of unlabeled probe. Reaction products were run on a 4% polyacrylamide gel. The arrows indicate the position of the probe and of DNA-nuclear protein complexes.

pattern seen with the brain extract was very similar to that obtained with CHO nuclear extracts, suggesting that similar proteins in both brain and CHO cells are involved in binding to the IGF-I-R promoter. Liver nuclear extracts, at the concentrations assayed, did not generate any retarded bands.

Gel-retardation experiments were complemented by DNase I footprinting assays using purified Sp1 protein and brain and liver nuclear extracts (Fig. 5). This approach allowed us to map five GC boxes (out of a total of six) in the 5'-flanking region which were bound by Sp1. Consistent with the results of gel-retardation assays, some of these sites were also bound by protein from brain extracts but not by protein from liver extracts.

In conclusion, the capacity of Sp1 to trans-activate the IGF-I-R promoter is consistent with its ability to specifically bind to consensus sites present in this region. Furthermore, the difference between brain and liver extracts in their content of IGF-I-R promoter-binding proteins may, at least partially, explain the difference in IGF-I-R mRNA levels seen in these tissues.

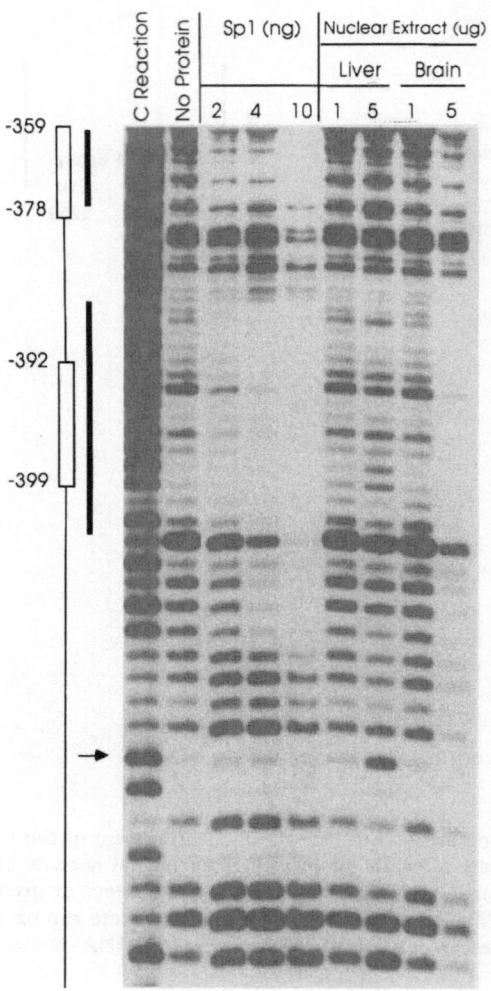

Figure 5. DNase I footprinting analysis of the IGF-I-R gene promoter. A ^{32}P-labeled probe extending from -416 to -135 was incubated with purified Sp1 protein or with crude nuclear extracts from adult liver or neonate brain, and subsequently digested with DNase I. The open box extending from -399 to -392 corresponds to the extreme left Sp1 site shown in Figure 1. The open box extending from -378 to -359 corresponds to the overlapping second and third Sp1 sites starting from the left side of Figure 1.

Are the Sp1 and IGF-I-R Genes Co-Regulated During Normal Development?

Although expressed ubiquitously, a recent study in the mouse has shown that there are substantial variations in expression of the Sp1 gene during development[18]. Sp1 mRNA levels were 100-fold higher in certain cell types than in others, and similar differences at the protein level were seen using immunocytochemistry. Fully differentiated cells with specialized functions generally exhibited the lowest levels of Sp1 mRNA and protein. In liver, for example, moderate amounts of Sp1 protein were seen in fetal hepatocytes, and much lower levels were seen in adult cells. This developmental pattern is similar to the pattern seen with IGF-I-R, i.e., high levels of IGF-I-R mRNA and binding were seen at embryonic and early postnatal stages whereas much lower levels were seen in the fully-differentiated adult hepatocyte[5]. In addition, gel-retardation assays revealed that the DNA-binding efficiency of Sp1 is significantly reduced in aged animals as compared to young ones[19].

In the early fetus (8.5 - 12.5 days), highest levels of Sp1 mRNA and protein were seen in neural tissues, consistent with our published observations that, at this early ontogenic stage, highest levels of IGF-I-R mRNA were found in the developing nervous system[20-21]. Low levels of Sp1 were seen in fully differentiated Purkinje cells.

A fairly good correlation between Sp1 and IGF-I-R distribution was also seen in the kidney, in which staining for Sp1 (and IGF-I-R) in the medulla was most pronounced in the collecting tubules, while in the cortex, highest levels were seen in the glomeruli[22].

This spatial and developmental correspondence between Sp1 and IGF-I-R is, however, not a universal phenomena. Thus, whereas weaning in the mice was associated with an increase in the levels of Sp1 protein in the stomach, which continued into adulthood, the levels of IGF-I-R mRNA in the adult rat stomach were ~5-fold lower than in the neonate animal.

In conclusion, although the role of transcription factor Sp1 in growth and development is far from clear, there is a good correlation in a number of instances in the localization and in the developmental regulation of the Sp1 and IGF-I-R genes. The fact that Sp1 and IGF-I-R are differentially regulated in certain circumstances is clear evidence that factors in addition to Sp1 are involved in the regulation of the IGF-I-R gene.

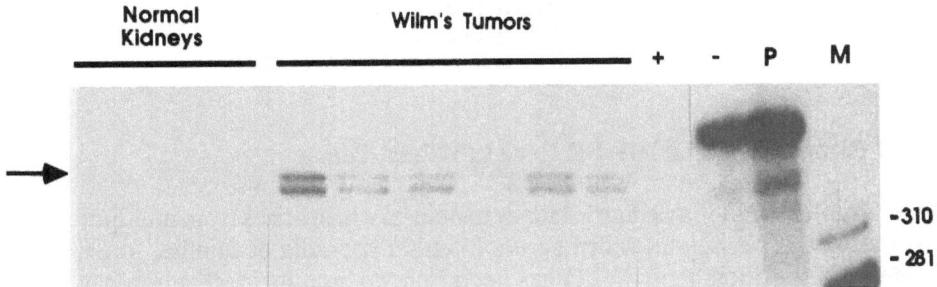

Figure 6. Expression of IGF-I-R gene in Wilms' tumor and normal adjacent kidney tissue. Levels of IGF-I-R mRNA were measured by solution hybridization/RNase protection assays using 10 μg of total RNA. +, probe alone with RNase; -, probe alone without RNase; P, native probe; M, molecular weight marker. For further details, see ref. 30.

De-Regulation of the IGF-I-R Gene in Neoplasia

As already mentioned, the expression of the IGF-I-R gene by most body tissues persists, albeit at low levels, until adulthood. Certain pathological conditions, including fasting, diabetes, and malignancies, are associated with dramatic increases in the levels of IGF-I-R mRNA and protein. Among these disease states, derangements in the expression of the IGF-I-R gene were especially well-characterized in various tumors, most notably those from breast and brain[9,10]. Many questions, however, remain unanswered. For instance, it is still unclear whether overexpression of the IGF-I-R gene precedes the transformation process, or that, on the contrary, high levels of IGF-I-R gene expression are a result of the transformed phenotype.

Artificial overexpression of a human IGF-I-R cDNA in NIH3T3 cells resulted in a ligand-dependent, highly transformed phenotype. In the presence of IGF-I (or high concentrations of insulin), the overexpressing cells formed aggregates in tissue culture dishes, colonies in soft agar, and tumors in nude mice. These results suggest that, when amplified, the IGF-I-R behaves like an oncogenic protein capable of promoting neoplastic growth *in vivo*[23].

Table 1. Expression of IGF-I-R mRNA in Wilms' tumor.

Sample	n	IGF-I-R mRNA[a]
Normal kidney	7	2.87 ± 0.70
Wilms' tumor[b]	25	16.77 ± 2.86[*]
With heterologous elements	8	23.28 ± 5.43
Without heterologous elements	10	10.83 ± 3.83[**]
Uncharacterized	7	17.81 ± 5.42

[a] Levels of IGF-I-R mRNA are expressed as arbitrary absorbance units.
[b] This group includes eight tumors with heterologous elements, ten tumors without heterologous elements and seven uncharacterized tumors.
[*] Significantly different from normal kidney tissue (p<0.02).
[**] Significantly different from WT with heterologous elements (p<0.01).

Increased Expression of the IGF-I-R Gene in Wilms' Tumor

Wilms' tumor (WT) is a pediatric kidney neoplasm which arises from multipotential cells of the metanephros, and which occurs in either sporadic or familial forms[24]. A candidate predisposition gene (WT1) has been isolated, and its inactivation has been postulated to be a key step in the etiology of WT[25,26].

IGFs are among the most potent mitogenic factors for kidney cells in culture[27]. Several lines of evidence suggested that IGFs may be involved in the etiology and/or progression of WT[28,29]: i) extracts of WT exhibit increased ^{125}I-IGF-I binding and tyrosine kinase activity as compared to normal kidney tissue; ii) an antibody to the human IGF-I-R (αIR-3) can inhibit ^{125}I-IGF-I binding and IGF-I-stimulated thymidine incorporation by WT cells in culture; and iii) intraperitoneal administration of αIR-3 to nude mice bearing WT heterotransplants can prevent tumor growth and results in partial tumor remission.

The expression of the IGF-I-R gene in WT was recently measured in a collection of 25 WT and 7 normal adjacent kidney samples by means of solution hybridization/ RNase protection assays[30](Fig. 6 and Table 1). The levels of IGF-I-R mRNA in the tumors were ~6-fold higher than in normal kidney tissue. Furthermore, the levels of IGF-I-R mRNA were ~2-fold higher in tumors containing heterologous stromal elements (striated muscle, cartilage, and bone) than in tumors without heterologous elements (i.e., blastema with epithelial differentiation). These findings suggest that the expression of the IGF-I-R gene is correlated with the degree of differentiation of the tumor.

In addition, when the levels of IGF-I-R mRNA in individual tumors were compared to the levels of WT1 mRNA, we found that there was a significant inverse correlation between these two parameters (R=-0.52, p<0.05) (Fig. 7). Specifically, tumors with low levels of WT1 mRNA (and with heterologous elements) had high levels of IGF-I-R mRNA. On the other hand, tumors with high levels of WT1 mRNA (and without heterologous elements) had lower levels of IGF-I-R mRNA.

Since levels of IGF-II mRNA are highly increased in WT[31,32], it is possible that locally produced IGF-II may interact with the high levels of IGF-I-R. Paracrine activation of this receptor may, in turn, elicit a mitogenic event which may be a key event in the etiology and/or progression of WT.

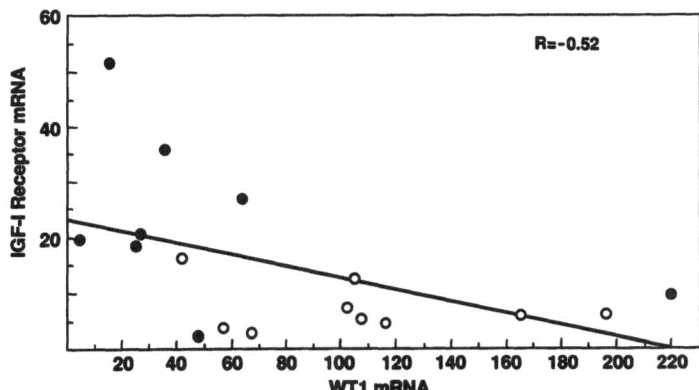

Figure 7. Multiple regression analysis of IGF-I-R mRNA and WT1 mRNA levels in WT samples. Black dots are WT samples with heterologous elements (striated muscle, cartilage and bone) and open dots are WT samples without heterologous elements. There is a significant inverse correlation between the levels of IGF-I-R mRNA and WT1 mRNA in individual tumors (R=-0.52).

Repression of IGF-I-R Promoter Activity by the WT1 Tumor Suppressor Gene

The WT predisposition gene, WT1, encodes a DNA-binding protein with a serine- and proline-rich NH_2-terminus and four Zn^{2+}-finger domains. The WT1 gene product is a member of the EGR family of transcriptional activators, whose common consensus binding site is the sequence GCGGGGGCG[33]. Unlike other members of this family, such as EGR1 and 2, WT1 represses the activity of promoters that contain this motif, including the promoter of the IGF-II gene[34,35].

As previously indicated, the 5'-flanking and 5'-untranslated regions of the IGF-I-R gene contain several putative EGR/WT1 binding sites[12](Fig. 1). Thus, it is possible that the IGF-I-R gene promoter is a target for the inhibitory action of the WT1 protein. Furthermore, it is conceivable that the increased levels of IGF-I-R mRNA in WT may result from loss of negative regulation by the WT1 gene product.

To investigate whether WT1 could indeed repress the activity of the IGF-I-R promoter *in vivo*, coexpression studies were performed using an IGF-I-R promoter-reporter construct [p(-2350/+640)LUC] and WT1 expression vectors (pCMVhWT and pCMVhWT-TTL) encoding full-length and truncated WT1 proteins, respectively. The results of these experiments indicated that the WT1 gene product inhibited IGF-I-R promoter activity in a dose-dependent manner (Fig. 8). On the other hand, a truncated WT1 expression vector lacking the Zn^{2+}-finger domain did not repress promoter activity[30].

In conclusion, the WT1 tumor suppressor gene product can repress the activity of the IGF-I-R promoter, presumably by interacting with potential WT1 binding sites in both the 5'-flanking and 5'-untranslated regions of this gene. Inactivation of the WT1 gene by underexpression, deletion, or mutation may result in an inactive protein which is unable to repress IGF-I-R promoter activity, leading to high levels of IGF-I-R gene expression and IGF-I binding. Current studies are intended to define the active WT1 sites in the IGF-I-R promoter, and the role of the IGF-I-R in the etiology of WT.

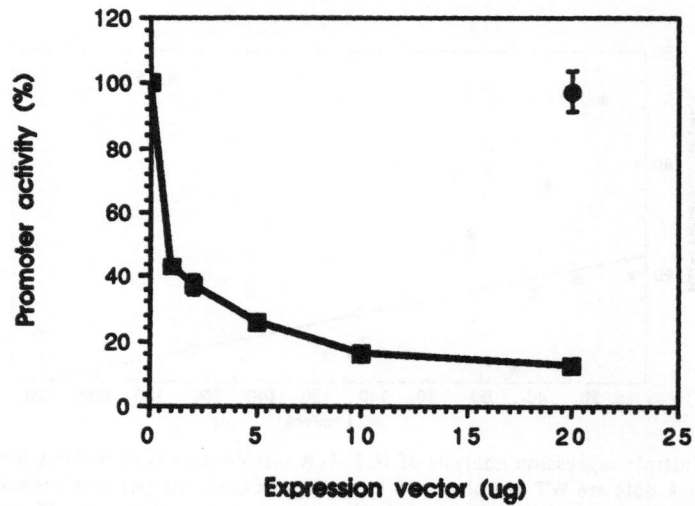

Figure 8. Repression of IGF-I-R promoter by WT1. One μg of the reporter plasmid p(-2350/+640)LUC was cotransfected into CHO cells with increasing amounts of the WT1 expression vector, pCMVhWT (■), or with 20 μg of a truncated WT1 expression vector, pCMVhWTTTL (●). The luciferase activity values were normalized for β-galactosidase activity.

SUMMARY

The IGF-I-R gene promoter is a TATA-less, CAAT-less, GC-rich promoter which contains potential binding sites for the Sp1 and WT1 zinc-finger transcription factors. We have shown that Sp1 positively activates the IGF-I-R promoter. Since both the Sp1 and IGF-I-R genes are widely expressed, it is possible that Sp1 is one of the main regulators of IGF-I-R gene expression. This is supported by the correlation between the distribution and developmental regulation of Sp1 and IGF-I-R gene expression, in that both genes appear to be co-regulated during normal development.

In a model of human neoplasm, WT, we have demonstrated increased expression of the IGF-I-R gene, which may result from loss of repression of the IGF-I-R promoter by another Zn^{2+}-finger protein, the WT1 tumor suppressor gene product. Future studies will define whether other disease states in which the IGF-I-R gene is highly expressed are also associated with loss of negative regulation of the IGF-I-R promoter by WT1 or other tumor suppressor gene products.

Acknowledgments

The authors wish to thank Dr. R. Tjian for providing Sp1 protein and expression vectors; Drs. V.P. Sukhatme, F.J. Rausher III, and I.A. Drummond for WT1 expression vectors; and Drs. G.G. Re, D.A. Sens, and A.J. Garvin for WT specimens.

REFERENCES

1. D. LeRoith, M. Adamo, H. Werner, and C.T. Roberts, Jr, Insulin-like growth factors, their binding proteins and receptors as growth regulators in normal physiology and pathological states, *Trends in Endocrinol. Metab.* 2:134-139 (1991).
2. H. Werner, M. Woloschak, B. Stannard, Z. Shen-Orr, C.T. Roberts, Jr., and D. LeRoith, The insulin-like growth factor I receptor: molecular biology, heterogeneity and regulation, *in:* "Insulin-Like Growth Factors: Molecular and Cellular Aspects," D. LeRoith, ed., CRC Press, Boca Raton (1991).
3. E. Di Cicco-Bloom and I.B. Black, Insulin growth factors regulate the mitotic cycle in cultured rat sympathetic neuroblasts, *Proc. Natl. Acad. Sci. USA.* 85:4066- (1988).
4. Z. Pietrzkowski, R. Lammers, G. Carpenter, A.M. Soderquist, M. Limardo, P.D. Phillips, A. Ullrich, and R. Baserga, Constitutive expression of insulin-like growth factor 1 and insulin-like growth factor 1 receptor abrogates all requirements for exogenous growth factors, *Cell Growth and Differentiation*, 3:199-205 (1992).
5. H. Werner, M. Woloschak, M. Adamo, Z. Shen-Orr, C.T. Roberts, Jr., and D. LeRoith, Developmental regulation of the rat insulin-like growth factor I receptor gene, *Proc. Natl. Acad. Sci. USA.* 86:7451-7455 (1989).
6. W.L. Lowe, Jr., M. Adamo, C.T. Roberts, Jr., and D. LeRoith, Regulation by fasting of rat insulin-like growth factor I and its receptor: effects on gene expression and binding, *J. Clin. Invest.* 84:619-626 (1989).
7. A. Ota, Z. Shen-Orr, C.T. Roberts, Jr., and D. LeRoith, TPA-induced neurite formation in a neuroblastoma cell line (SH-SY5Y) is associated with increased IGF-I receptor mRNA and binding, *Mol. Brain Res.* 6:69-76 (1989).
8. H. Werner, Z. Shen-Orr, B. Stannard, B. Burguera, C.T. Roberts, Jr., and D. LeRoith, Experimental diabetes increases insulin-like growth factor I and II receptor concentration and gene expression in kidney, *Diabetes* 39:1490-1497 (1990).
9. K.J. Cullen, D. Yee, W.L. Sly, J. Perdue, B. Hampton, M.E. Lippman, N. Rosen, Insulin-like growth factor receptor expression and function in human breast cancer, *Cancer Res.* 50:48-53 (1990).
10. R.P. Glick, R. Gettleman, K. Patel, R. Lakshman, J.C.M Tsibris, Insulin and insulin-like growth factor I in brain tumors: binding and *in vitro* effects, *Neurosurgery* 24:791-797 (1989).
11. H. Werner, B. Stannard, M.A. Bach, D. LeRoith, and C.T. Roberts, Jr., Cloning and characterization of the proximal promoter region of the rat insulin-like growth factor I (IGF-I) receptor gene, *Biochem. Biophys. Res. Commun.* 169:1021-1027 (1990).
12. H. Werner, M.A. Bach, B. Stannard, C.T. Roberts, Jr., and D. LeRoith, Structural and functional analysis of the insulin-like growth factor I receptor gene promoter, *Mol. Endocrinol.* 6:1545-1558 (1992).

13. D.W. Cooke, L.A. Bankert, C.T. Roberts, Jr., D. LeRoith, and S.J. Casella, Analysis of the human type I insulin-like growth factor receptor promoter region, *Biochem. Biophys. Res. Commun.* 177:1113-1120 (1991).

14. P.W. Mamula and I.D. Goldfine, Cloning and characterization of the human insulin-like growth factor-I receptor gene 5'-flanking region, *DNA and Cell Biology* 11:43-50 (1992).

15. S.T. Smale and D. Baltimore, The "initiator" as a transcription control element, *Cell* 57:103-113 (1989).

16. M.R. Briggs, J.T. Kadonaga, S.P. Bell, and R. Tjian, Purification and biochemical characterization of the promoter-specific transcription factor Sp1, *Science*, 234:47-52 (1986).

17. A.J. Courey and R. Tjian, Analysis of Sp1 *in vivo* reveals multiple transcriptional domains, including a novel glutamine-rich activation motif, *Cell* 55:887-898 (1988).

18. J.D. Saffer, S.P. Jackson, and M.B. Annarella, Developmental expression of Sp1 in the mouse, *Mol. Cell Biol.* 11:2189-2199 (1991).

19. R. Ammendola, M. Mesuraca, T. Russo, and F. Cimino, Sp1 DNA binding efficiency is highly reduced in nuclear extracts from aged rat tissues, *J. Biol. Chem.* 267:17944-17948 (1992).

20. C.A. Bondy, H. Werner, C.T. Roberts, Jr., and D. LeRoith, Cellular pattern of insulin-like growth factor-I (IGF-I) and type I IGF receptor gene expression in early organogenesis: comparison with IGF-II gene expression, *Mol. Endocrinol.* 4:1386-1398 (1990).

21. C.A. Bondy, H. Werner, C.T. Roberts, Jr., and D. LeRoith, Cellular pattern of type-I insulin-like growth factor receptor gene expression during maturation of the rat brain: comparison with insulin-like growth factors-I and -II, *Neuroscience* 46:909-923 (1991).

22. E. Chin, J. Zhou, and C.A. Bondy, Anatomical relationships in the patterns of insulin-like growth factor (IGF)-I, IGF binding protein -1, and IGF-I receptor gene expression in the rat kidney, *Endocrinology* 130:3237-3245 (1992).

23. M. Kaleko, W.J. Rutter, and A.D. Miller, Overexpression of the human insulin-like growth factor I receptor promotes ligand-dependent neoplastic transformation, *Mol. Cell Biol.* 10:464-473 (1990).

24. A.J. Altman and A.D. Schwartz. "Malignant Diseases of Infancy, Childhood and Adolescence," W.B. Saunders, Philadelphia (1978).

25. K.M. Call, T. Glaser, C.Y. Ito, A.J. Buckler, J. Pelletier, D.A. Haber, E.A. Rose, A. Kral, H. Yeger, W.H. Lewis, C. Jones, and D.E. Housman, Isolation and characterization of a zinc finger polypeptide gene at the human chromosome 11 Wilms' tumor locus, *Cell* 60:509-520 (1990).

26. E.A. Rose, T. Glaser, C. Jones, C.L. Smith, W.H. Lewis, K.M. Call, M. Minden, E. Champagne, W. Lewis, L. Bonetta, H. Yeger, and D.E. Housman, Complete physical map of the WAGR region of 11p13 localizes a candidate Wilms' tumor gene, *Cell* 60:495-508 (1990).

27. F.G. Conti, L.J. Striker, M.A. Lesniak, K. Mackay, J. Roth, and G.E. Striker, Studies on binding and mitogenic effects of insulin and insulin-like growth factor I in glomerular mesangial cells, *Endocrinology* 122:2788-2795 (1988).

28. T. Gansler, K.D. Allen, C.F. Burant, T. Inabnett, A. Scott, M.G. Buse, D.A. Sens, and A.J. Garvin, Detection of type I insulinlike growth factor (IGF) receptors in Wilms' tumors, *Am. J. Pathol.* 130:431-435 (1988).

29. T. Gansler, R. Furlanetto, T. Stokes-Gramling, K.A. Robinson, N. Blocker, M.G. Buse, D.A. Sens, and A.J. Garvin, Antibody to type I insulinlike growth factor receptor inhibits growth of Wilms' tumor in culture and in athymic mice, *Am. J. Pathol.* 135:961-966 (1989).

30. H. Werner, G.G. Re, I.A. Drummond, V.P. Sukhatme, F.J. Rauscher, III, D.A. Sens, A.J. Garvin, D. LeRoith, and C.T. Roberts, Jr., Increased expression of the insulin-like growth factor I receptor (IGF-I-R) gene in Wilms' tumor is correlated with modulation of IGF-I-R promoter activity by the WT1 Wilms' tumor gene product, Submitted (1993).

31. A.E. Reeve, M.R. Eccles, R.J. Wilins, G.I. Bell, and L.J. Millow, Expression of insulin-like growth factor II transcripts in Wilms' tumor, *Nature* 317:258-260 (1985).

32. G.K. Haselbacher, J.C. Irminger, J. Zapf, W.H. Ziegler, and R.E. Humbel, Insulin-like growth factor II in human adrenal pheochromocytomas and Wilms' tumors: expression at the mRNA and protein level, *Proc. Natl. Acad. Sci. USA* 84:1104-1106 (1987).

33. F.J. Rauscher, III, J.F. Morris, O.E. Tournay, D.M. Cook, and T. Curran, Binding of the Wilms' tumor locus zinc finger protein to the EGR-1 consensus sequence, *Science* 250:1259-1261 (1990).

34. S.L. Madden, D.M. Cook, J.F. Morris, A. Gashler, V.P. Sukhatme, and F.J. Rauscher, III, Transcriptional repression mediated by the WT1 Wilms tumor gene product, *Science* 253:1550-1553 (1991).

35. I.A. Drummond, S.L. Madden, P. Rohwer-Nutter, G.I. Bell, V.P. Sukhatme, and F.J. Rauscher, III, Repression of the insulin-like growth factor II gene by the Wilms tumor suppressor WT1, *Science* 257:674-678 (1992).

CELL CYCLE CONTROL BY THE IGF-1 RECEPTOR AND ITS LIGANDS

Renato Baserga*, Pierluigi Porcu, Michele Rubini and Christian Sell

Jefferson Cancer Institute
Thomas Jefferson University
Room 624A, Bluemle Life Sciences Building
233 S. 10th Street
Philadelphia, PA 19107

* corresponding author

ABBREVIATIONS

IGF-1 or IGF-II, insulin-like growth factor 1 or II; SV40, simian virus 40; PDGF, platelet derived growth factor; EGF, epidermal growth factor

INTRODUCTION

The interaction of the IGF-1 receptor with its ligands (IGF-1, IGF-II and insulin at supraphysiological concentrations) plays a major role in normal development and in the control of both normal and abnormal growth (for reviews, see Werner et al., 1991 and Lowe, 1991). The importance of the IGF-1 receptor in development is especially supported by the seminal experiments of Efstratiadis and co-workers (DeChiara et al., 1990, and personal communication). These investigators have shown that targeted disruption of the IGF-II gene results in progeny, which, at birth, has a body weight that is 70% the body weight of wild type litter mates. When both the IGF-II and the IGF-1 receptor genes are disrupted by homologous recombination, the homozygous mutant embryos at birth have a body weight that is only 30% the weight of wild type litter mates. Thus, it can be stated that the activation of IGF-1 receptor by its ligands (IGF-1 or IGF-II) accounts for 70% of embryonal murine growth.

Most cell types in culture have an absolute requirement for IGF-1 for growth (see review by Goldring and Goldring, 1991), and these cell types include fibroblasts, epithelial cells, smooth muscle cells, T lymphocytes, chondrocytes, osteoblasts as well as the stem cells of the bone marrow (Huang and Terstappen, 1992). The requirement for IGF-1 can also be demonstrated by the use of IGF-1 peptide analogues, that inhibit the activation of the IGF-1 receptor and, simultaneously, cellular proliferation of both normal and cancer cells (Pietrzkowski et al., 1992c and 1993). The requirement for a functional IGF-1 receptor can

Current Directions in Insulin-Like Growth Factor Research,
Edited by D. LeRoith and M.K. Raizada, Plenum Press, New York, 1994

105

be instead demonstrated by using antisense oligodeoxynucleotides to the IGF-1 receptor RNA: the antisense strategy was successful in inhibiting cellular proliferation in fibroblasts (Pietrzkowski et al., 1992a and b, Porcu et al., 1992), in interleukin-2 stimulated T lymphocytes and in HL-60 cells (Reiss et al., 1992), in prostatic cancer cells (Pietrzkowski et al., 1993) and in ovarian carcinoma cells (Raphael Rubin, personal communication). Finally, kinetic experiments indicate that the cell cycle clock starts with the addition of IGF-1 (Cristofalo et al., 1989, Yoshinouchi and Baserga, 1993).

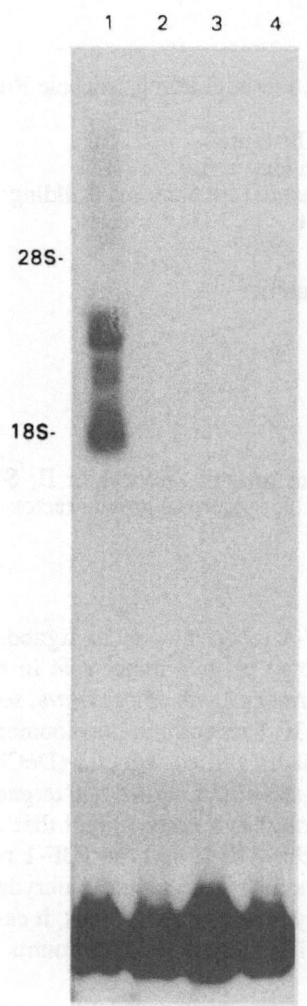

Figure 1. Northern Blot of IGF-1 Expression in SV40-transformed 3T3 Cells.

RNA was extracted from two cell lines: a) Balb58 cells, which are Balb/c 3T3 cells stably transfected with a temperature-sensitive mutant of the SV40 T antigen, tsA58 (Porcu et al., 1992); and b) the parental cell line, Balb/c 3T3 cells. The blot was hybridized with a full length mouse IGF-1 cDNA (Bell et al., 1986). Lanes: 1 and 2, Balb58 cells, at 34° and 39.6°, respectively; 3 and 4: 3T3 cells, at 34° and 39.6°, respectively.

Thus, the evidence is strong, and grows steadily stronger, that the activation of the IGF-1 receptor by its ligands is crucially important in regulating the extent of cellular proliferation in many types of cells. A recent review by Macaulay (1992) gives a list of human cancers in which the expression of the IGF-1 receptor or of the IGF's is altered.

INDUCTION OF IGF-1 EXPRESSION BY CELLULAR AND VIRAL ONCOGENES

In the past two years, our laboratory has been instrumental in defining a new mechanism for cellular transformation. We have found that both the proto-oncogene c-myb (Travali et al., 1991) and the viral DNA oncogene, the SV40 large T antigen (Porcu et al., 1992) markedly increase the expression of IGF-1 and IGF-1 mRNA. A striking example is shown in Fig. 1, which is a Northern blot of RNA's from Balb58 cells. These cells (described in detail by Porcu et al., 1992) are 3T3 derivatives stably transfected with the tsA58 T antigen, a thermo-sensitive (ts) T antigen. At 34°C, these cells are transformed, but at the restrictive temperature of 39.6°C, the T antigen is not functional, and the cells revert to an untransformed phenotype (Radna et al., 1989). The Northern blot was hybridized to an IGF-1 probe: at 34°C, there is a large amount of IGF-1 mRNA, while at 39.6°C, under these conditions, it is not even detectable. The other two lanes show RNA from 3T3 cells, in which IGF-1 mRNA (again under these conditions) is undetectable at either temperature. This dramatic increase in the expression of IGF-1 mRNA occurs largely at the transcriptional level (Porcu et al., 1992).

Although it has been known for some time that certain proto-oncogenes are actually growth factors or growth factors receptors (see for instance, the review by Hunter, 1991), our findings with c-myb and SV40 T antigen define a new mechanism for oncogenesis, i.e. the lowering of growth factor requirements by the induction of specific growth factors (and also of their receptors). In agreement with these reports is the observation that overexpression of the IGF-1 receptor results in increased transformability of NIH 3T3 cells (Kaleko et al., 1990). If oncogenes activate the transcription of the IGF's genes, one could predict that tumor suppressor genes may have the opposite effect. Indeed, this is the case: the Wilms' tumor suppressor gene WT1 represses the transcription of the IGF-II gene (Drummond, et al., 1992), and preliminary results from Jennifer Swantek in our laboratory indicate that the Retinoblastoma gene product inhibits the expression of the IGF-1 gene.

INDUCTION OF IGF-1 EXPRESSION BY OTHER GROWTH FACTORS

It has been known for many years that growth factors like PDGF and EGF increase the secretion of IGF-1 and IGF-1 like substances into the medium by cells in culture (Clemmons and Shaw, 1983, Clemmons, 1984). We have confirmed and extended these findings and shown that exposure of 3T3 cells to EGF markedly increases the expression of IGF-1 and IGF-1 mRNA (Pietrzkowski et al., 1992b). We have also found that some cancer cell lines in culture secrete remarkably large amounts of IGF-1, for instance prostatic cancer cells (Pietrzkowski et al., 1993) and glioblastoma cells (unpublished data). Indeed, we suspect that many cell lines that can grow in serum free medium, do so through an autocrine mechanism based on the activation of the IGF-1 receptor by its ligands (especially IGF-1 and IGF-II).

REGULATION OF THE EXPRESSION OF THE IGF-1 RECEPTOR.

Under physiological conditions, the levels of IGF-1 receptor in a given type of cell vary very little. For instance, certain growth factors, like EGF and PDGF, do increase the number of IGF-1 binding sites, but by not more than 2-3 fold (Van Wyk et al., 1981, Clemmons and Van Wyk, 1981, Clemmons et al., 1986). The levels of IGF-1 receptor RNA are also increased in Interleukin-2-stimulated T lymphocytes (Reiss et al., 1992). High levels of IGF-1 receptor (and IGF-1 receptor RNA) can be observed, though, in cells stably transfected with an expression plasmid in which the human IGF-1 receptor cDNA is under the control of a viral promoter (Kaleko et al., 1990, McCubrey et al., 1991, Pietrzkowski et al., 1992a, and b). This suggests that the levels of IGF-1 receptor and IGF-1 receptor RNA may be autoregulated.

Preliminary results in our laboratory, in collaboration with the laboratory of Derek LeRoith have shown that PDGF can induce transcription from the IGF-1 receptor promoter. In these experiments, we used a 2.6 kb rat promoter driving the luciferase reporter gene. PDGF increases the expression of luciferase mRNA to a maximum of 4-fold over unstimulated cells. The expression of the endogenous mouse IGF-1 receptor RNA is also increased after stimulation with PDGF. Indeed, even a short promoter (about 100 bp upstream of the transcription initiation site) is responsive to the stimulation by PDGF (Rubini et al., submitted). The ability of PDGF or serum to increase transcription from the IGF-1 receptor promoter is illustrated in Fig. 2, where we used a short promoter, the SmaI promoter, comprising about 350bp upstream of the transcription initiation site. Clearly, both PDGF and 10% serum increase transcription; by densitometry, PDGF produced a 4.8-fold increase, and serum a 9.1-fold increase.

Interestingly, the expression of the IGF-1 receptor is not affected by the SV40 T antigen (Ferber et al., submitted).

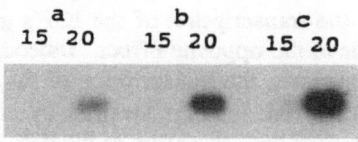

Figure 2. Expression of Luciferase RNA in SmaI/Luc cells after Stimulation with PDGF or serum. SmaI/Luc cells are 3T3 cells stably transfected with a construct in which the luciferase reporter gene is under the control of a rat IGF-1 receptor promoter (Werner et al., 1990). The SmaI restriction site is located about 350bp upstream of the transcription initiation site. RNA was extracted and the luciferase RNA levels determined by RT-PCR, with the methodology, amplimers and probe described by Porcu et al., (1992). Lanes: a) quiescent cells; b) 4 hrs. after PDGF-BB (5 ng/ml); c) 4 hrs. after 10% serum.

THE DOUBLE LIFE OF THE IGF-1 RECEPTOR

One of the most intriguing features of the IGF-1 receptor is the ambiguity of its response when activated by its ligands. As an illustration of this phenomenon we shall take 3T3 cells, which require both PDGF and IGF-1 for optimal growth (Scher et al., 1979). Addition of IGF-1 only to quiescent 3T3 cells fails to elicit a mitogenic response. But if the cells are previously or simultaneously incubated with PDGF (which, by itself, is also non mitogenic), then the addition of IGF-1 results in maximal stimulation of cellular proliferation.

The problem is that quiescent cells have IGF-1 receptors (Clemmons et al., 1986, Pietrzkowski et al., 1992a), that these receptors are autophosphorylated by the appropriate ligands (Pietrzkowski et al., 1992b) and do transmit a signal to the nucleus, since specific genes are activated (Baserga and Rubin, 1993). Thus, in quiescent 3T3 cells (and in other cell types), the IGF-1 receptor can be activated by its ligands, but is not mitogenic. It becomes mitogenic (after IGF-1) when overexpressed (McCubrey et al., 1991, Pietrzkowski et al. 1992 a and b), or when the cells are primed by another growth factor. Since the possible explanations and supporting evidence for each of them have been discussed at length in a recent review by Baserga and Rubin (1993), we will limit ourselves to summarize that review as follows: there are 3 large categories of alternative explanations for the shift of the IGF-1 receptor from a non-mitogenic to a mitogenic mode. 1) IGF-1 and the other growth factor (PDGF-EGF, etc.) have independent pathways both of which have to be completed for the cells to enter S phase and divide. 2) the first growth factor induces an increase in the number of IGF-1 receptors, i.e. a quantitative change that makes the activated IGF-1 receptor mitogenic. 3) the other growth factors induce a qualitative change in the IGF-1 receptor, or its substrates or its signal transduction pathway. We believe that the identification of the mechanism (whatever the mechanism is) that changes the IGF-1 receptor from a non-mitogenic to a mitogenic mode means to identify the most important single step that controls the proliferation of animal cells.

IS THERE LIFE AFTER THE IGF-1 RECEPTOR?

The answer is yes   at least to the extent of 30%. We have mentioned above the experiments of Efstratiadis and co-workers on the targeted disruption of both the IGF-1 receptor and the IGF-II genes in knockout mice. We have grown the cells from these embryos in our laboratory, and the results are intriguing in many respects. We shall mention here only two findings with these cells that are of general interest and are summarized in Table 1. The first point is that the mouse embryo cells lacking the IGF-1 receptor genes do grow in 10% serum, but at a rate that is roughly 35% the rate of wild type mouse embryo cells. In mice, the embryos with targeted disruption of the IGF-1 receptor genes, reach a weight of about 30% that of wild type embryos (see above), and it is comforting that the results in vivo and in vitro are very similar. The second point is that these mouse embryo cells were stably transfected with the SV40 T antigen. Wild type cells became transformed and grew vigorously; the cells from the homozygous mutant embryos grew a little better than the parental cells, but still less than the untransformed wild type cells. What Table 1 does not show is that the SV40-transfected mutant cells are contact-inhibited and, in fact, they cannot grow in soft agar, whereas the SV40-transformed wild type cells produced more than 300 colonies. It seems therefore that in mouse embryo cells with a targeted disruption of the IGF-1 receptor genes, the SV40 T antigen cannot induce transformation. In this respect, our several studies with antisense oligodeoxynucleotides to the IGF-1 receptor have consistently demonstrated that cancer cells are more susceptible to an antisense than normal cells (Pietrzkowski et al., 1992c, 1993, Porcu et al., 1992). In general, an antisense oligodeoxynucleotide to the IGF-1 receptor RNA causes a 60% inhibition of growth (in agreement with the growth rate of mutant cells lacking IGF-1 receptors): cancer cells are more strongly inhibited and, even more important, become incapable to grow in soft agar (Sell et al., in preparation).

Table 1. Growth of Mouse Embryo Cells with a Targeted Disruption of the IGF-1 Receptor Genes

Cell line	number of cells/cm^2/x10^3
C2(wild type embryo cells)	55
K10 (mutant embryo cells)	25
C2tA58 (transformed with Tag)	150
K10tA58(T antigen carrying)	35

The number of cells was determined after 5 days of growth in 10% serum. The cells transfected with the SV40 T antigen were grown at 34°C, because we used a ts mutant of the T antigen (Radna et al., 1989, Porcu et al., 1982). The k10 cells derive from embryos homozygous for a targeted disruption of the IGF-1 receptor genes. In all cases, the cells had reached confluence, but in C2tA58 cells, there were foci and piling-up.

CONCLUSIONS

The IGF-1 receptor and the IGF-1/IGF-1 receptor autocrine loop are emerging as playing a pivotal role in the control of cellular proliferation in animal cells. The findings that the proto-oncogene c-myb and the viral oncogene SV40 T antigen can alter the growth factor requirements of cells by inducing the expression of IGF-1 and IGF-1 mRNA define a novel mechanism by which oncogenes can transform cells through the production of growth factors (and possibly their receptors). On the other hand, the inability of the SV40 T antigen to transform cells that are null for the IGF-1 receptor, points out that the expression of this receptor, at least in certain types of cells, is obligatory for transformation. It will be interesting to see whether other oncogenes need the receptor for transformation, and also whether they can alter growth by modifying the IGF-1/IGF-1 receptor autocrine loop. It also seems quite likely that in the near future both the IGF-1 and its receptor will become targets for the inhibition of abnormal growth.

ACKNOWLEDGMENTS

This work was supported by grants CA 56309 and 53484 from the National Institutes of Health and CB 48 from the American Cancer Society.

REFERENCES

Baserga, R. and Rubin, R., 1993, Cell cycle and growth control. *Crit. Rev. Eukar. Gene Expr.* 3: 47-61.
Bell, G.I., Stempien, M.M., Fong, N.M., and Rall, L.B, 1986, Sequences of liver cDNAs encoding two different mouse insulin-like growth factor I precursors. *Nucl. Acids Res.* 14:7873-7882.
Clemmons, D.R., 1984, Multiple hormones stimulate the production of somatomedin by cultured human fibroblasts. *J. Clin. Endocrin. & Metab.* 58:850-856.
Clemmons, D.R. and Shaw, D.S., 1983, Variables controlling somatomedin production by cultured human fibroblasts. *J. Cell. Physiol.* 115:137-142.

Clemmons, D.R. and Van Wyk, J.J., 1981, Somatomedin: physiological control and effects on cell proliferation, *in*: "Tissue Growth Factors," R. Baserga, ed., Springer-Verlag KG, Berlin.

Clemmons, D.R., Elgin, R.G., and James, P.E., 1986, Somatomedin-C binding to cultured human fibroblasts is dependent on donor age and culture density. *J. Clin. Endocrinol. Metab.* 63:996-1001.

Cristofalo, V.J., Phillips, P.D., Sorger, T. and Gerhard, G., 1989, Alterations in the responsiveness of senescent cells to growth factors. *J. Gerontol.* 44:55-62.

DeChiara, T.M., Efstradiatis, A. and Robertson, E.J., 1990, A growth-deficiency phenotype in heterozygous mice carrying an insulin-like growth factor II gene disrupted by targeting. *Nature* 345:78-80.

Drummond, I.A., Madden, S.L., Rohwer-Nutter, P., Bell, G.I., Sukhatme, V.P. and Rauscher, F.J., III, 1992, Repression of the insulin-like growth factor II gene by the Wilms tumor suppressor WT1. *Science* 257:674-678.

Goldring M.B., and Goldring, S.R., 1991, Cytokines and cell growth control. *Eukar Gene Express.* 1:301-326.

Huang, S. and Terstappen, L.W.M.M., 1992, Formation of haematopoietic microenvironment and haematopoietic stem cells from single human bone marrow stem cells. *Nature* 360: 745-749.

Hunter, T., 1991, Cooperation between oncogenes. *Cell* 64:249-270.

Kaleko, M., Rutter, W.G. and Miller, A.D., 1990, Overexpression of the human insulin-like factor I receptor promotes ligand dependent neoplastic transformation. *Mol. Cell. Biol.* 10:464-473.

Lowe, W.L., Jr., 1991, Biological actions of the insulin-like growth factors. *in*: "Insulin-like Growth Factors: Molecular and Cellular Aspects," D. LeRoith, ed., CRC Press.

Macaulay, V.M., 1992, Insulin-like growth factors and cancer. *Brit.J. Cancer* 65:311-320.

McCubrey, J.A., Stillman, L.S., Mayhew, M.W., Algate, P.A., Dellow, R.A. and Kaleko, M., 1991, Growth promoting effects of insulin-like growth factor I (IGF-1) on hematopoietic cells. Overexpression of introduced IGF-1 receptor abrogates interleukin-3 dependency of murine factor dependent cells by ligand dependent mechanism. *Blood* 78:921-929.

Pietrzkowski, Z., Lammers, R., Carpenter, G., Soderquist, A.M., Limardo, M., Phillips, P.D., Ullrich, A. and Baserga, R., 1992, Constitutive expression of insulin-like growth factor 1 and insulin-like growth 1 receptor abrogates all requirements for exogenous growth factors. *Cell Growth and Diff.* 3:199-205.

Pietrzkowski, Z., Sell, C., Lammers, R., Ullrich, A. and Baserga, R., 1992, Roles of insulin-like growth factor 1 (IGF-1) and the IGF-1 receptor in epidermal growth factor-stimulated growth of 3T3 cells. *Mol. Cell. Biol.* 12:3883-3889.

Pietrzkowski, Z., Wernicke, D., Porcu, P., Jameson, B.A. and Baserga, R., 1992, Inhibition of cell proliferation by peptide analogs of IGF-1. *Cancer Res.* 52:6447-6451.

Pietrzkowski, Z., Mulholland, G., Gomella, L., Jameson, B.A., Wernicke, D. and Baserga, R., 1993, Inhibition of growth of prostatic cancer cell lines by peptide analogs of IGF-1. *Cancer Res.* 53:1102-1106

Porcu, P., Ferber, A., Pietrzkowski, Z., Roberts, C.T., Adamo, M., LeRoith, D. and Baserga, R., 1992, Roles of Insulin-like growth factor 1(IGF-1) and the IGF-1 receptor in epidermal growth factor-stimulated growth of 3T3 cells. *Mol. Cell. Biol.* 12:3883-3889.

Radna, R.L., Caton, Y., Jha, K.K., Kaplan, P., Li, G., Traganos, F. and Ozer, H.L., 1989, Growth of immortal simian virus 40 tsA transformed human fibroblasts is temperature dependent. *Molec. Cell. Biol.* 9:3093-3096.

Reiss, K., Porcu, P., Sell, C., Pietrzkowski, Z. and Baserga, R., 1992, The insulin-like growth factor 1 receptor is required for the proliferation of hemopoietic cells. *Oncogene* 7:2243-2248.

Scher, C.D., Shephard, R.C., Antoniades, H.N., and Stiles, C.D., 1979, Platelet derived growth factor and the regulation of the mammalian fibroblasts cell cycles. *Biochim. Biophys, Acta* 560:217-241.

Travali, S., Reiss, K., Ferber, A., Petralia, S., Mercer, W.E., Calabretta, B. and Baserga, R., 1991, Constitutively expressed c-myb abrogates the requirement for insulin-like growth factor 1 in 3T3 fibroblasts. *Mol. Cell. Biol.* 11:731-736.

Van Wyk, J.J., Underwood, L.E., D'Ercole, A.J., Clemmons, D.R., Pledger, W.J., Wharton, W.R. and Leof, E.B., 1981, Role of somatomedin in cellular proliferation, *in*: "The Biology of Normal Human Growth," M. Ritzen et al., eds., Raven Press, N.Y.

Werner, H., Stannard, B., Bach, M.A., LeRoith, D. and Roberts, C.T., Jr., 1990, Cloning and characterization of the proximal promoter region of the rat insulin-like growth factor 1 (IGF-1) receptor gene. *Biochem. Biophys. Res. Comm.* 169:1021-1027.

Werner, H., Woloschak, M., Stannard, B., Shen-Orr, Z., Roberts, C.T. Jr. and LeRoith, D., 1991, The insulin-like growth factor receptor: molecular biology, heterogeneity and regulation. *in*: "Insulin-like Growth Factors: Molecular and Cellular Aspects," D. LeRoith, ed., CRC Press, Boca Raton.

Yoshinouchi, M. and Baserga, R. The role of the IGF-1 receptor in the stimulation of cells by short pulses of growth factors. *Cell Proliferation*, in press.

THE INSULIN RECEPTOR FAMILY

Karen A. Seta, Kristina S. Kovacina, and Richard A. Roth

Department of Pharmacology
Stanford University School of Medicine
Stanford, CA 94305

INTRODUCTION

The insulin receptor family in mammals includes the receptors for insulin, insulin-like growth factor I (IGF-I), and the insulin receptor-related receptor (IRR), a receptor whose sequence is homologous to the sequences of the other two receptors but whose ligand is unknown (Fig. 1) (1). Another potential receptor in this family is the receptor for relaxin, a hormone whose structure is related to that of insulin (2). In addition, other members of this receptor family could exist for IGF-I and the highly related IGF-II. Although the cDNA and gene for one receptor which binds IGF-I with high affinity has been isolated (3), several pieces of data suggest that there may be other related IGF receptors. For example, although the receptor expressed from this cDNA had high affinity for IGF-I and IGF-II but almost 1000-fold weaker affinity for insulin (4), several reports in the literature have indicated that there are IGF receptors with a much higher affinity for insulin or a much weaker affinity for IGF-II (5, 6). Some of these studies may have been affected by the presence of hybrid insulin-IGF-I receptors (7). In addition, expression of the apparently same cDNA for the IGF-I receptor in another cell type led to the formation of a receptor which had high affinity for IGF-I and weak affinity for IGF-II when binding studies were performed on whole cells but which bound IGF-I and II almost equally when binding studies were performed in cell lysates (8). In addition, several

Current Directions in Insulin-Like Growth Factor Research,
Edited by D. LeRoith and M.K. Raizada, Plenum Press, New York, 1994

113

monoclonal antibodies to the IGF-I receptor were found to stimulate an increase in the affinity of the IGF-I receptor for insulin almost to the level of the insulin receptor (9). These results suggest that the same IGF receptor can bind the IGFs and insulin with different relative affinities depending on the environment of the receptor in the cell membrane. Thus, it is not clear at the present time whether another IGF receptor exists or whether these different binding data can be explained by the present IGF-I receptor interacting with other molecules in different cell backgrounds.

In addition to the mammalian receptors, related receptors exist in lower organisms. A putative insulin receptor in Drosophila has been described biochemically (10). In addition, a partial cDNA clone for the Drosophila receptor has been reported (11). Insulin related molecules also exist in the silkworm and mollusc (12, 13). It will be of interest to see whether the receptors for these molecules are related to the mammalian insulin receptor family.

The sequences for the three known members of the mammalian insulin receptor family all predicted that they are synthesized as a single polypeptide that is cleaved to yield the mature α and β subunits. For the insulin and IGF-I receptors, this has been confirmed by biochemical studies. However, a native insulin receptor-related receptor has yet to be described. The extracellular α subunit of all three receptors contains a cysteine-rich region. Since the sequence of this region diverges to a greater extent than surrounding residues (Fig. 1), it was predicted that this region might confer the distinct ligand specificities to the different receptors. A chimeric receptor containing the cysteine-rich region of the IGF-I receptor in the backbone of the insulin receptor was in fact found to bind both IGF-I and II with affinities comparable to that of the native IGF-I receptor (14). However, the cysteine-rich region of the insulin receptor was not found to confer high affinity insulin binding on the IGF-I receptor (15, 16). Instead, residues both on the amino and carboxy-side of the cysteine-rich region were found to contribute to the high affinity insulin binding (15-17). These results suggest that multiple regions of the receptors are required for high affinity ligand binding and that the higher divergence in sequence in the cysteine-rich region may indicate a lower need to conserve these particular residues. In support of this hypothesis is the finding that the cysteine-rich region of the rat insulin receptor also diverges from that of the human insulin receptor to a greater degree than the surrounding regions do (Fig. 1).

The most highly conserved region of the three receptors is the portion of the β subunit which encodes for tyrosine kinase activity (Fig. 1). In contrast, the carboxy-tail of the β subunits of the three receptors exhibits the lowest degree of sequence identity (Fig. 1), with this region of IRR being only 19% identical to the corresponding sequence of the insulin receptor. Some studies have suggested that

114

this carboxy-tail region of the insulin receptor is important in signaling (18). The high divergence in sequence of this region of the receptors could therefore explain the different biological functions mediated by insulin and IGF-I, with the insulin receptor mediating metabolic responses and the IGF-I receptor mediating more proliferative responses. However, other studies have shown that the three receptors are quite similar in their abilities to stimulate different responses when expressed in the same cell type (4, 19). In addition, the substrates and signaling systems elicited by the three receptors appear quite similar (4, 19). Thus, the high divergence in the sequences of the carboxy-tails of the three receptors (like that of the cysteine-rich regions) could arise from a lack of importance in conserving the particular amino acids in this region. In support of this hypothesis is the finding that truncated receptors lacking much of the carboxy-tail function normally (20) and that this region shows greater divergence than other regions between the human and rat insulin receptors (Fig. 1).

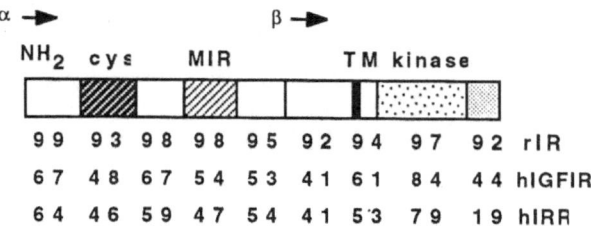

Figure 1. Schematic of the insulin receptor family. Shown are the % sequence identities with the human insulin receptor for different regions of the rat insulin receptor (rIR), human insulin-like growth factor I receptor (hIGFIR), and human insulin receptor-related receptor (hIRR). cys: cysteine-rich region, MIR: major immunogenic region, TM: transmembrane region, and kinase: kinase region.

In the present manuscript, we will present some recent data we have generated on the insulin receptor-related receptor. This receptor was first identified by Shier and Watt from a screen of a genomic library (1). Identification of cells and tissues which express this receptor will help in determining the function of this receptor. Studies on the signal transducing abilities of this receptor may help in determining the role of different residues of these receptors in eliciting a particular response or in phosphorylating a particular substrate.

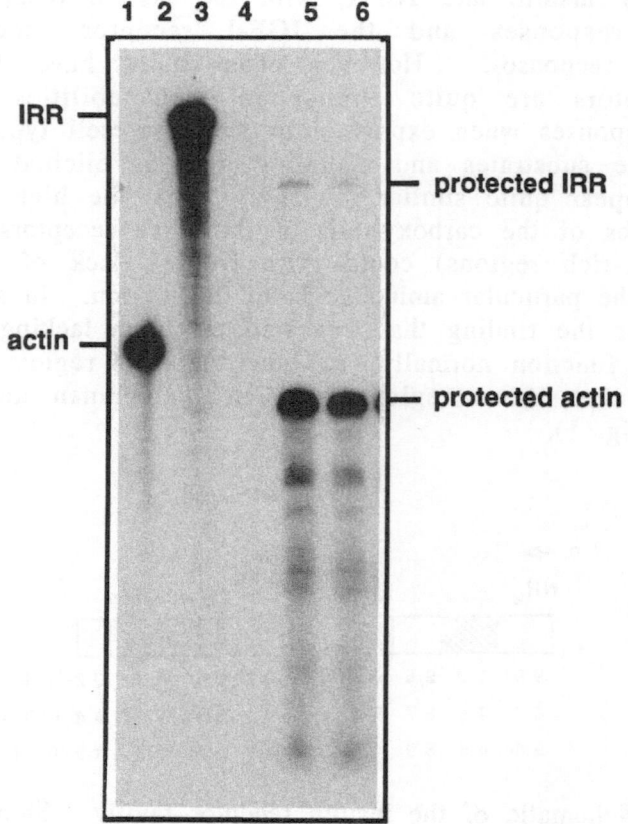

Figure 2: RNase assay for IRR mRNA. Rat IRR exon 3 (E3) was subcloned into Bluescript^TM KS and linearized with HindIII. The control template for mouse pT3-β-actin (MAXIscript^TM kit, Ambion Inc., Austin, TX) was digested with DdeI to make it distinguishable from the IRRE3 probe. Antisense RNA probes labeled with $[\alpha\text{-}^{32}P]$ UTP were constructed according to directions in the MAXIscript^TM kit. Probes were gel purified and annealed to 20 µg rat kidney total RNA. The RNase protection assay was performed with the RPA II^TM kit (Ambion) according to kit directions. Lanes 1 and 3 contain undigested actin and IRRE3 probes, respectively. Lanes 2 and 4 contain the same probes and yeast total RNA and show that the digestion was complete. Lanes 5 and 6 contain rat total RNA and both probes and was digested with RNase. The IRRE3 probe gives a protected fragment of 300 bp and the actin probe gives a protected fragment of 140 bp.

MEASUREMENT OF THE INSULIN RECEPTOR-RELATED RECEPTOR mRNA BY RNase PROTECTION

To better quantitate the mRNA for the insulin receptor-related receptor in different tissues under various physiological conditions, a RNase protection assay was developed. An antisense probe to the cysteine-rich region of the rat receptor (exon 3) was labeled and gel purified. When total RNA from cells and tissues expressing the IRR mRNA were incubated with this probe, a fragment of the expected size (300 bp) was protected from RNase digestion (Fig. 2). The amount of this protected fragment was proportional to the amount of input RNA. To quantitate the amount of total RNA present in the different samples, a control probe for actin mRNA was also included which was designed to give a band of a different size (140 bp) (Fig. 2). When this assay was utilized to measure the amount of IRR mRNA in different rat tissues, kidney gave the greatest signal of the tissues tested (with weaker signals observed with thymus and stomach). The amount of IRR mRNA in kidney in weanling and adult rat kidney was found to be approximately the same. In streptozotocin-induced diabetic rats , the amount of IRR mRNA was found to be within 2-fold that of non-diabetic rats. The amount of IRR mRNA was also found not to greatly change in the remnant kidney of a unilateral nephrectomized rat. These results differ from those previously reported for the insulin and IGF-I receptors and emphasize the different roles for these receptors (21, 22).

PRODUCTION AND CHARACTERIZATION OF AN ANTISERUM TO THE INSULIN RECEPTOR-RELATED RECEPTOR

To better study the insulin receptor-related receptor protein, mice were injected with CHO cell lines overexpressing a chimeric receptor containing the complete extracellular domain of the human insulin receptor-related receptor and the cytoplasmic domain of the insulin receptor (IRR/IRK) (19). The antisera from these mice were found to precipitate the chimeric receptor from both CHO cells overexpressing this chimeric receptor (Fig. 3) as well as from NIH 3T3 cells overexpressing the same chimera (data not shown). Moreover, these antisera were capable of activating the intrinsic tyrosine kinase activity of this receptor (Fig. 4). These results indicate that these antisera contain antibodies directed against an epitope in the extracellular domain of IRR. These antisera and monoclonal antibodies derived from the mice producing them should be useful reagents for characterizing the IRR protein found in different tissues. In addition, if monoclonal antibodies which act as agonists can be isolated, it should be possible to study the function of the native insulin receptor related receptor without the identification of a ligand for this receptor.

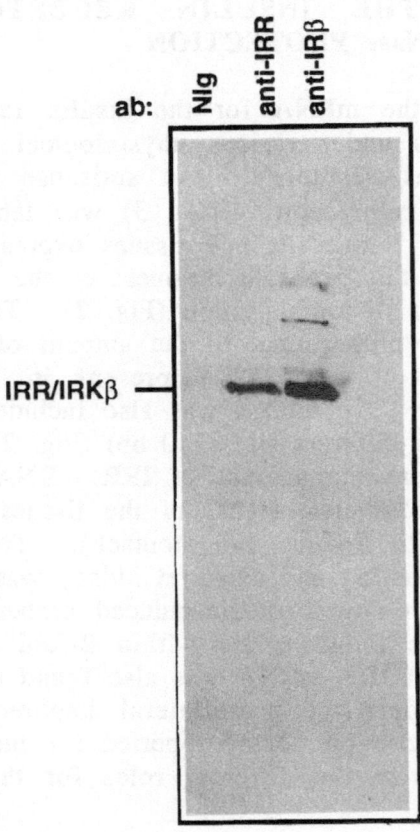

Figure 3. Immunoprecipitation of IRR/IRK by an antiserum to IRR. CHO cells expressing IRR/IRK were lysed and the lysates were precipitated with either control mouse sera (NIg), sera from a mouse which was injected with cells overexpressing IRR/IRK (anti-IRR) or a monoclonal antibody to the intracellular domain of the β subunit of the insulin receptor (anti-IRβ). The precipitates were then analyzed by SDS gel electrophoresis and Western blotting with an antibody to the β subunit of the IR.

IDENTIFICATION OF SHC AS A SUBSTRATE FOR THE INSULIN RECEPTOR-RELATED RECEPTOR

A chimeric receptor with the extracellular domain of the insulin receptor and the kinase domain of IRR (IR/IRRK) has allowed studies of the signal transducing abilities of the IRR kinase in the absence of an activating ligand for IRR (19). In prior studies, this kinase, like those of the insulin and IGF-I receptors, was found to phosphorylate IRS-1 and the GAP-associated p62 protein in response to insulin (19).

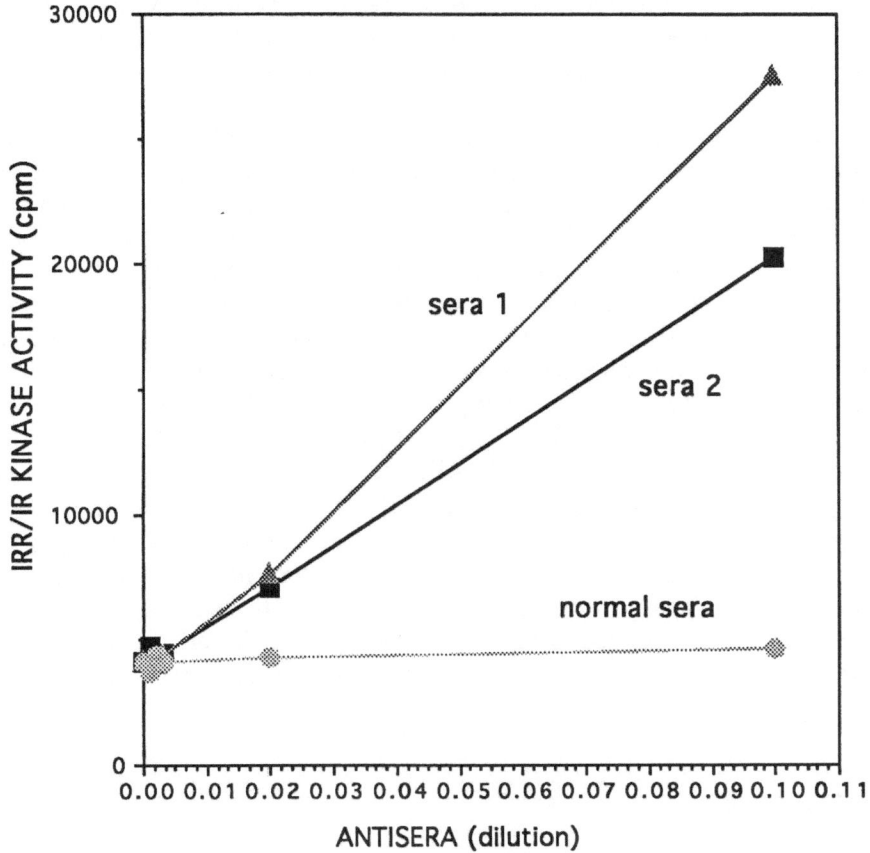

Figure 4. Activation of the IRR/IR kinase activity by antisera to IRR. Antisera from two mice injected with CHO.IRR/IRK cells, and normal mouse serum were tested for their ability to stimulate the IR kinase activity of IRR/IRK. CHO.IRR/IRK cells were plated in 24-well tissue culture plates and allowed to reach confluence. After treatment for 1 hour with serum-free medium, cells were stimulated with various concentrations of sera for 10 min at 37°C. The cells were then washed and lysed. IRR/IRK was captured on microtiter wells that had been coated with 29B4, a monoclonal antibody to the IR kinase domain that does not interfere with kinase activity. A kinase reaction mixture containing [γ-^{32}P] ATP and poly (Glu:Tyr) 4:1 as a substrate was added to the wells and allowed to incubate for 1 hr at room temperature. The reaction mixture was spotted onto strips of Whatman paper and TCA precipitated. Strips were air dried and counted by the Cerenkov method.

Recently, the SH-2 containing protein called SHC has been found to be phosphorylated in response to the activation of the insulin receptor kinase (23). To see if this protein was also a substrate of the IRR kinase, cells overexpressing the above described chimeric receptor were treated with insulin in the presence or absence of the phosphatase inhibitor, vanadate. The cell lysates were immunoprecipitated with either antibodies to SHC or control immunoglobulin. The precipitates were electrophoresed on an SDS gel, transferred to nitrocellulose, and reacted with an antibody to phosphotyrosine. Both the 52 and 66 kDa SHC bands were observed to be tyrosine phosphorylated by the chimeric receptor in response to insulin (Fig. 5). This phosphorylation was potentiated in the presence of the tyrosine phosphatase inhibitor, vanadate, as was previously observed with the insulin receptor (23). The only other band observed in the anti-SHC precipitates which became tyrosine phosphorylated in response to insulin is a band of 90 kDa, a band which presumably represents the β subunit of the chimeric receptor. This band was observed in the control immunoglobulin precipitations at about the same levels (Fig. 5), indicating that the IRR kinase is not specifically associating with SHC. These results are similar to prior studies of the insulin receptor, which was also found to mediate the tyrosine phosphorylation of SHC but not to specifically associate with it.

CONCLUSIONS

Recent studies have identified low levels of the IRR mRNA in various tissues including kidney, stomach, thymus and muscle by PCR and Northern analyses (19,24,25). The present studies confirm the expression of the IRR mRNA in rat kidney by the RNase protection assay. This assay should allow quantitation of this mRNA under different physiological conditions. However, preliminary studies have not found any large changes in the IRR mRNA in rat kidney during development, in diabetes, or after unilateral nephrectomy. More recent in situ studies have found that the highest levels of IRR mRNA are in the sensory neurons of the trigeminal and dorsal root ganglia of the rat (26). This finding suggests that the ligand for IRR may be more like the insulin related peptides in the silkworm and mollusc since these peptides act as neuropeptides (12,13). It should be possible to measure the levels of the IRR mRNA in various ganglia by the RNase protection assay under different physiological conditions.

We also show in the present studies that mice injected with cells overexpressing a chimeric receptor with the extracellular domain of the insulin receptor-related receptor readily produce antisera which precipitate this receptor and activate its intrinsic kinase activity. It should therefore be possible to obtain hybridomas producing monoclonal antibodies to this receptor as has previously been done for the insulin and IGF-I receptors (9). These antibodies will be valuable reagents for studying the native IRR protein as well as potentially being used to study the physiological function of this receptor in the absence of its ligand.

Finally, we show that the newly described substrate of the insulin receptor, SHC, is also tyrosine phosphorylated in response to activation of the IRR kinase. Thus, all three identified substrates of the insulin receptor kinase (IRS-1, the GAP p62 associated protein and SHC) appear to also be substrates of IRR even though the carboxy-tails of these two receptors only exhibit 19% sequence identity. These results are consistent with the finding that these two receptors stimulate similar biological responses (19). These results would suggest that the different roles played by different members of this family of receptors may be more determined by the levels of expression of the different receptors in various tissues and the levels and availability of their ligands than by differences in their intrinsic signaling capabilities.

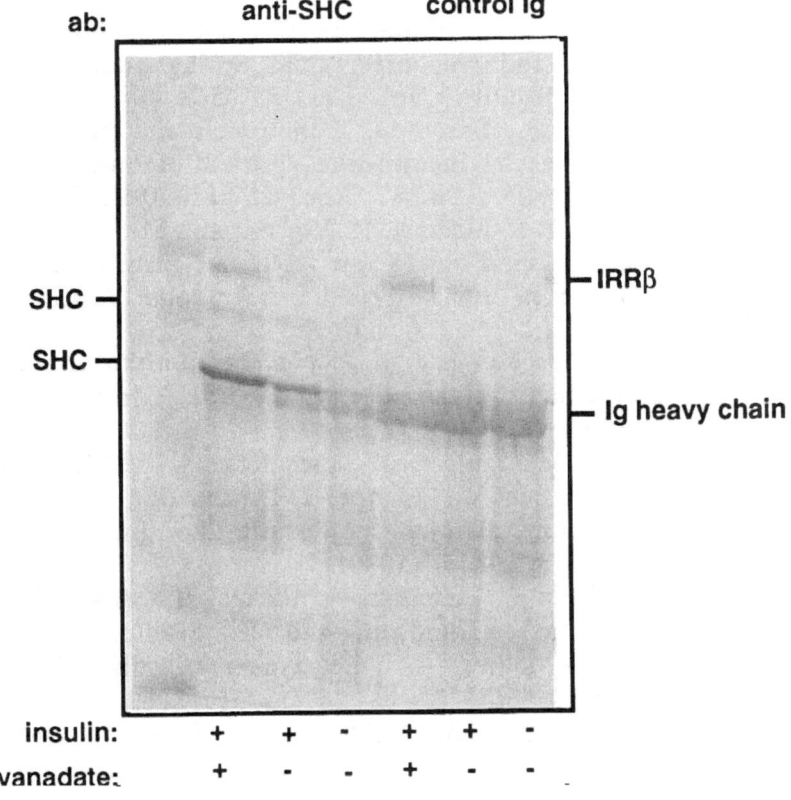

Figure 5. Tyrosine phosphorylation of SHC after activation of the kinase domain of IRR. CHO cells overexpressing IR/IRRK were treated as indicated with either buffer, insulin or vanadate plus insulin. The cells were then lysed and the lysates precipitated with either control immunoglobulin or an antibody to SHC. The precipitates were analyzed by SDS gel electrophoresis and Western blotting with an anti-phosphotyrosine antibody.

ACKNOWLEDGMENTS

The authors thank Drs. Ralph Rabkin, Derek LeRoith, Susan Mulroney, and Fredric Kraemer for RNA samples. This work was supported by National Institutes of Health grant DK 45652.

REFERENCES

1. P. Shier and V.M. Watt, Primary structure of a putative receptor for a ligand of the insulin family, *J. Biol. Chem.* 264:14605-14608 (1989).
2. C. Schwabe and E.E. Bullesbach, Relaxin, *Comp. Biochem. Physiol.* 96B:15-21 (1990).
3. A. Ullrich, A. Gray, A.W. Tam, T. Yang-Feng, M. Tsubokawa, C. Collins, W. Henzel, T. Le Bon, S. Kathuria, E. Chen, S. Jacobs, U. Francke, J. Ramachandran, and Y. Fujita-Yamaguchi, Insulin-like growth factor I receptor primary structure: comparison with insulin receptor suggests structural determinants that define functional specificity, *EMBO J.* 5:2503-2512 (1986).
4. R.A. Roth, G. Steele-Perkins, J. Hari, C. Stover, S. Pierce, J. Turner, J.C. Edman, and W.J. Rutter, Insulin and insulin-like growth factor receptors and responses, Cold Spring Harbor Symposia on Quantitative Biology, Vol. LIII:537-543 (1988).
5. R.S. Garofalo and B. Barenton, Functional and immunological distinction between insulin-like growth factor I receptor subtypes in KB cells, *J. Biol. Chem.* 267:11470-11475 (1992).
6. G. Milazzo, C.C.Yip, B.A. Maddux, R. Vigneri, and I.D. Goldfine, High-affinity insulin binding to an atypical insulin-like growth factor-I receptor in human breast cancer cells, *J. Clin. Invest.* 89:899-908 (1992).
7. C.P. Moxham, V. Duronio, and S. Jacobs, Insulin-like growth factor I receptor β-subunit heterogeneity, *J. Biol. Chem.* 264:13238-13244 (1989).
8. E.L. Germain-Lee, M. Janicot, R. Lammers, A. Ullrich, and S.J. Casella, Expression of a type I insulin-like growth factor receptor with low affinity for insulin-like growth factor II, *Biochem. J.* 281:413-417 (1992).
9. M.A. Soos, C.E. Field, R. Lammers, A. Ullrich, B. Zhang, R.A. Roth, A.S. Andersen, T. Kjeldsen, and K. Siddle, A panel of monoclonal antibodies for the type I insulin-like growth factor receptor, *J. Biol. Chem.* 267:12955-12963 (1992).
10. R. Fernandez-Almonacid and O.M. Rosen, Structure and ligand specificity of the *Drosophila melanogaster* insulin receptor, *Molec. Cell. Biol.* 7:2718-2727 (1987).
11. Y. Nishida, M. Hata, Y. Nishizuka, W.J. Rutter, and Y. Ebina, Cloning of a drosophila cDNA encoding a polypeptide similar to the human insulin receptor precursor, *Biochem. Biophys. Res. Comm.* 141:474-481 (1986).

12. K. Maruyama, H. Hietter, H. Nagasawa, A. Isogai, S. Tamura, A. Suzuki, and H. Ishizaki, Isolation and primary structure of bombyxin-IV, a novel molecular species of bombyxin from the silkworm, *Bombyx mori*, *Agric. Biol. Chem.* 52:3035-3041 (1988).

13. W.P.M. Geraerts, A.B. Smit, K.W. Li, E. Vreugdenhil, and H. van Heerikhuizen, Neuropeptide Gene Families that Control Reproductive Behaviour and Growth in Molluscs, *in*: Current Aspects of the Neurosciences, N.N. Osborne, ed., The Macmillan Press Ltd, New York (1991).

14. B. Zhang, and R.A. Roth, Binding properties of chimeric insulin receptors containing the cysteine-rich domain of either the insulin-like growth factor-I receptor or the insulin receptor-related receptor, *Biochemistry* 30:5113-5117 (1991).

15. A.S. Andersen, T. Kjeldsen, F.C. Wiberg, H. Vissing, L. Schaffer, J.S. Rasmussen, P. De Meyts, and N.P.H. Moller, Identification of determinants that confer ligand specificity on the insulin receptor, *J. Biol. Chem.* 267:13681-13686 (1992).

16. R. Schumacher, M.A. Soos, J. Schlessinger, D. Brandenburg, K. Siddle, and A. Ullrich, Signaling-competent receptor chimeras allow mapping of major insulin receptor binding domain determinants, *J. Biol. Chem.* 268:1087-1094 (1993).

17. B. Zhang and R.A. Roth, A region of the insulin receptor important for ligand binding (residues 450-601) is recognized by patients' autoimmune antibodies and inhibitory monoclonal antibodies, *Proc. Natl. Acad. Sci. USA* 88:9858-9862 (1991).

18. Y. Takata, N.J.G. Webster, and J.M. Olefsky, Mutation of the two carboxyl-terminal tyrosines results in an insulin receptor with normal metabolic signaling but enhanced mitogenic signaling properties, *J. Biol. Chem.* 266:9135-9139 (1991).

19. B. Zhang and R.A. Roth, The insulin receptor-related receptor, *J. Biol. Chem.* 267:18320-18328 (1992).

20. K. Yonezawa, S. Pierce, C. Stover, M. Aggerbeck, W.J. Rutter, and R.A. Roth, Endogenous substrates of the insulin receptor: studies with cells expressing wild-type and mutant receptors, *in*: Molecular Biology and Physiology of Insulin and Insulin-like Growth Factors, M.K. Raizada and D. LeRoith, eds., Plenum Press, New York (1991).

21. S.E. Mulroney, A. Haramati, H. Werner, C. Bondy, C.T. Roberts, Jr., and D. LeRoith, Altered expression of insulin-like growth factor-I (IGF-I) and IGF receptor genes after unilateral nephrectomy in immature rats, *Endocrinology* 130:249-256 (1992).

22. H. Werner, Z. Shen-Orr, B. Stannard, B. Burguera, C.T. Roberts, Jr., and D. LeRoith, Experimental diabetes increases insulin-like growth factor I and II receptor concentration and gene expression in kidney, *Diabetes* 39:1490-1497 (1990).

23. K.S. Kovacina and R.A. Roth, Identification of SHC as a substrate of the insulin receptor kinase distinct from the GAP-associated 62 kDa tyrosine phosphoprotein, *Biochem. Biophys. Res. Comm.*, in press.

24. H. Kurachi, K. Jobo, M. Ohta, T. Kawasaki, and N. Itoh, A new member of the insulin receptor family, insulin receptor-related receptor, is expressed preferentially in the kidney, *Biochem. Biophys. Res. Comm.* 187:934-939 (1992).

25. P. Shier and V.M. Watt, Tissue-specific expression of the rat insulin receptor-related receptor gene, *Molec. Endocrinology* 6:723-729.

26. R.R. Reinhardt, E. Chin, B. Zhang, R.A. Roth, and C.A. Bondy, Insulin receptor-related receptor mRNA is focally expressed in sensory neurons and renal distal tubule, *Endocrinology*, in press.

IRR: A NOVEL MEMBER OF THE INSULIN RECEPTOR FAMILY

Valerie M. Watt, Peter Shier, Joanne Chan,
Bradley A. Petrisor, and Swarna K. Mathi

Department of Physiology
University of Toronto
Toronto, Ontario, Canada M5S 1A8

INTRODUCTION

Insulin's effects, including its pivotal regulation of blood glucose levels, are mediated by a cell-surface receptor (reviewed in Olefsky, 1990; Ullrich and Schlessinger, 1990). The structure of the insulin receptor, defined using recombinant DNA techniques, exhibits a high degree of overall similarity with the receptor for the structurally related insulin-like growth factor (IGF), IGF-I (for review, see Czech, 1989). These heterotetrameric glycoproteins consists of extracellular α-subunits containing the insulin-binding region disulfide-bonded to β-subunits which span the membrane and contain a cytoplasmic tyrosine kinase activated by insulin binding (Ebina et al., 1985; Ullrich et al., 1985; Ullrich et al., 1986). The α- and β-subunits are derived by proteolytic cleavage of the proreceptor.

In my laboratory we have identified a third, novel member of the insulin receptor family, the insulin receptor-related receptor, IRR (Shier and Watt, 1989). The IRR gene, in contrast to the insulin and IGF-I receptor genes, exhibits a limited tissue-specific expression and a single transcriptional start site. Chimeric receptors between IRR and the insulin receptor, although demonstrating functional conservation of the Cys rich region, have not yet identified the IRR ligand.

IDENTIFICATION AND CHARACTERIZATION OF THE IRR GENE

We used insulin receptor DNA to isolate homologous DNA encoding this novel receptor from human and guinea pig genomic libraries (Fig.1; Shier and Watt, 1989). Human IRR genomic DNA was then used to map the IRR gene to human chromosome 1 (Shier et al., 1990) distinct from those of the insulin and IGF-I receptor genes which are located on human chromosomes 19 and 15 (Yang-Feng et al., 1985; Ullrich et al., 1986). The IRR gene probably evolved from an ancestor common to the insulin and IGF-I receptor genes since the overall organization of all three genes is strikingly similar (IRR, Fig.1 and Shier and Watt, 1989; insulin receptor, Seino et al., 1989; IGF-I receptor, Abbott et al., 1992). Indeed, of the 42 identifiable intron/exon boundaries in the IRR gene (Table 1) and insulin receptor gene, 36 are at analogous positions and the remaining 6 accommodate deletions or insertions of no more than 5 amino acids. Thus, the IRR gene does not exhibit the common pseudogene structure arising from reverse transcription of a processed RNA (Vanin, 1985). The absence of in-frame stop codons in any of the 22 human and guinea pig IRR exons even though these

Current Directions in Insulin-Like Growth Factor Research,
Edited by D. LeRoith and M.K. Raizada, Plenum Press, New York, 1994

125

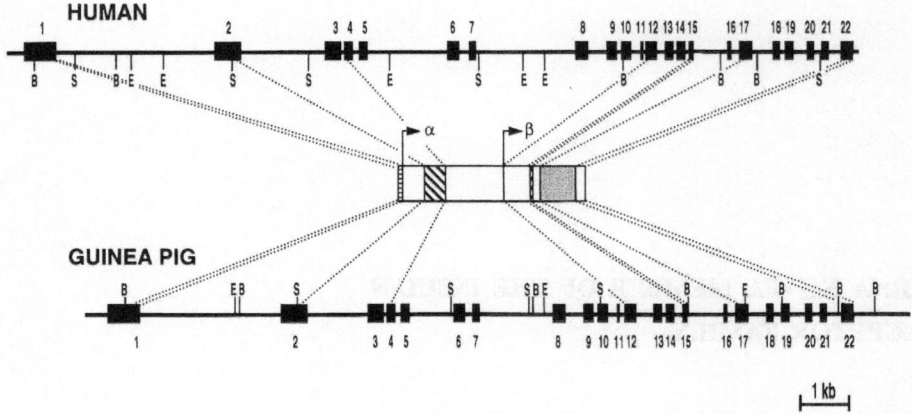

Figure 1. Partial restriction maps of genomic DNA encoding the human and guinea pig IRR. Exons are indicated by *black boxes*. Exon boundaries were determined by the presence of consensus splice donor and acceptor sites at identical positions in both species and comparison of the predicted primary translation product of the IRR genes with the human insulin receptor and IGF-I receptor amino acid sequences. The sequence of the human IRR signal peptide (MAVPSLWPWGACLPVIFLSLGF-GLDT) has been determined both from cDNA and genomic clones. *Arrows* in the predicted prereceptor indicate the putative signal peptide and precursor processing sites; ▤, the signal peptide; ▨ the Cys-rich domain; ▧, the transmembrane domain; and ▨, the tyrosine kinase domain. B, *Bam*HI; E, *Eco*RI; S, *Sst*I.

two IRR amino acid sequences have diverged significantly from each other and from those of the insulin and IGF-I receptors also suggested that the IRR gene is not a pseudogene.

The predicted IRR protein is strikingly similar to the insulin and IGF-I receptors throughout their entire length, with highest identity in the tyrosine kinase domain (79-84%; Fig. 2; Shier and Watt, 1989). Conservation of specific extracellular features, including the α/β-cleavage site, Cys residues, and N-linked glycosylation sites suggested that post-translational processing of all three receptors is similar. Based on its high degree of sequence similarity to the insulin and IGF-I receptors, we concluded that the IRR gene probably encodes a receptor protein-tyrosine kinase that exhibits functional characteristics similar to those of the insulin and IGF-I receptors (Shier and Watt, 1989).

TISSUE-SPECIFIC EXPRESSION OF THE IRR GENE

Since the determination of the tissue-specific expression of the IRR gene was crucial to our understanding of the function of IRR, we initially used the guinea pig and human

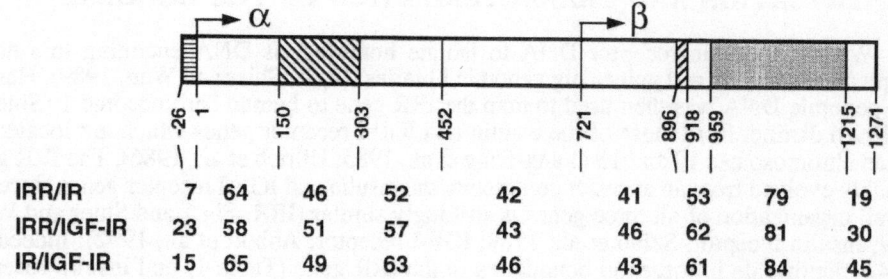

	−26 1	150	303	452	721	896 918	959	1215 1271
IRR/IR	7	64	46	52	42	41	53	79	19
IRR/IGF-IR	23	58	51	57	43	46	62	81	30
IR/IGF-IR	15	65	49	63	46	43	61	84	45

Figure 2. Comparison of the amino acid sequence of human IRR with those of closely related members of the insulin receptor family. The scheme of the predicted prereceptor is described in Fig.1. *Vertical numbers* indicate the positions of the initial amino acids in each region or the terminal residue of the IRR precursor. *Horizontal numbers* indicate percent identity between corresponding domains.

Table I. Exon/Intron Organization of Human and Guinea Pig IRR Genes.

Exon no.	Exon size (bp)	Sequence at exon/intron junction[a] 5'Splice donor.....3'Splice Acceptor			Amino acid interrupted
1	667	GAG G	gtgagtcccc.....atgcccacag	TG TGC	Val-3
	625	A.T .	......t.t......g.a.......	.. ...	
2	552	AGA G	gtgggcactg.....ctccccacag	TG TGC	Val-187
		C.. .	......g.........g.g.......	.. ..T	
3	301	AGC AG	gtgagtgtag.....ctcctgctag	C ATA	Ser-288
		... ..	..a..ca........gc.....c..	. ...	
4	146	GGC T	gtcagtacct.....ccatccctag	AC AAC	Tyr-336
		... .	........a.c....t.....tc..	.. ...	
5	145	GAT GG	gtaagggtta.....cctcccccag	G AAC	Gly-384
		... ..	.....t.a.t.............t..	. ...	
6	215	GCC T	gtgagacccc.....ccccaccag	GC CAG	Cys-456
		..T .	......c.t.t......g..cg....	.. ...	
7	127	GAG TC	gtgagtgccg.....gggtgttcag	C CCA	Ser-498
		... ..	......agg.................	. ...	
8	239	GCA G	gtaggcattt.....acactcctag	CT CCC	Ala-578
		... .	......g........tg.....	.G ..T	
9	168	CGC G	gtgcgcaggg.....gcggcgccag	GC TTG	Gly-634
		... .	........t......tgctg.....	.. ...	
10	196	CCC AT	gtgcgaagag.....gcctccccag	A TCC	Ile-699
		... .A	.....gccg................	. CCC	Lys-699
11	42	CAA AG	gtgagcagga.....ctctcccag	G GAC	Arg-713
		..G ..	......c..g......gc....t...	. ...	
12	221	CAC A	gtaggtgatc.....cttcccgcag	GA GAG	Arg-787
		.G. .	......gc.........t......	.. ...	
13	137	GGA GAG	gtaggtgccc.....ttactctcag	GAG GCC	Glu-832
		... ...	...a.ca..a......c....ct....	... ...	
14	163	CCA G	gtatacacag.....tgactgccag	AG GAG	Glu-887
		... .	...cg..g.t......a..cat...	.A ..A	
15	106	AAG AG	gtgatgataa.....cccgttgcag	A AAC	Arg-922
		... ..	......c.........t.......	. ..T	
16	53	GAT A	gtaggtctgg.....ctccttgcag	TG TAT	Met-940
		..C .	................t.cc...	C	
17	230	CAT GTG	gtaagggaga.....tctgtcccag	GTG CGT	Val-1016
		... ...	......a..t......ca......	... ...	
18	111	GCA GAG	gtaggacca.....ttggctctag	AAC AAC	Glu-1053
		... ...	..g....t................	..T ...	
19	160	GGG G	gtacagaggg.....cactttccag	AC TTC	Asp-1107
		... .	.................c..t...	.. ..T	
20	130	GTC TG	gtggggccga.....atgcccccag	G TCC	Trp-1150
		... ..	..a.....tg.....tgt..t....	. ...	
21	135	CAG CT	gtgagtcacc.....cacccaccag	G CAG	Leu-1195
		... ..	........tt........tt.....	. ...	
22	>337				
	>238				

a Human nucleotide sequence at exon/intron junctions is given in full; guinea pig differences only are indicated below. Exon sequences are in *uppercase*, intron sequences are in *lowercase*. Exon sizes are identical between species with the exception of exons 1 and 22.

genomic DNA to probe Northerns of RNA from various tissues and species. However, we were unable to detect the presence of IRR transcripts (Shier and Watt, 1989). Subsequently, we decided to optimize detection of IRR transcripts using rat IRR DNA to probe rat mRNA. Same-species DNA probes enabled use of more stringent hybridization conditions and thus reduced non-specific hybridization. We chose the rat as the experimental model since rat tissue is more available than human tissue; and unlike the guinea pig (King and Kahn, 1981), binding by insulin, IGF-I, and IGF-II is typical of other mammals, and the gestation period is short.

For these studies (Shier and Watt, 1992), we isolated rat genomic DNA which Southern blot and sequence analyses revealed contained exons 1 to 6 of the rat IRR gene. The rat IRR DNA probe encoding part of exon 2 detected hybridizing transcripts only in rat stomach and kidney and much more faintly in RNA from thymus (Fig.3; Shier and Watt, 1992). The two mRNAs of ~2 and ~6kb detected in RNA from stomach and kidney both probably encode IRR, since both transcripts were also detected by Northern analysis, using as probe rat DNA encoding a nonoverlapping fragment of exons 2 and 3. The ~6kb transcript, but not the ~2kb transcript, is large enough to encode the entire predicted IRR protein. Under these conditions, no hybridizing transcripts were visible in many other tissues including skeletal muscle, brain, intestine, and uterus (Fig.3; Shier and Watt, 1992). This very limited tissue distribution for IRR mRNA is in distinct contrast to the widespread distribution of the larger insulin and IGF-I receptor mRNAs.

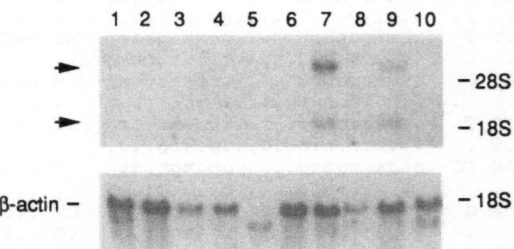

Figure 3. Northern blot analysis of the tissue distribution of IRR transcripts. Poly(A) RNA (2 µg) extracted from rat fetal head (~day 21; lane 1), neonatal brain (lane 2), adult brainstem (lane 3), brain (lane 4), skeletal muscle (lane 5), placenta (lane 6), stomach (lane 7), heart (lane 8), kidney (lane 9), and testis (lane 10) was hybridized with rat IRR exon 2 DNA (rIRRx2p) and rehybridized with rat β-actin DNA. *Arrows* indicate IRR-specific transcripts.

To confirm the identity of the transcripts which hybridized to the IRR probe, we used polymerase chain reaction (PCR) amplification and found that the relative amount of IRR DNA amplified from several tissues corresponded to the relative amounts of IRR mRNA detected using Northern blot analyses (Shier and Watt, 1992). Sequence analysis of the amplified DNA confirmed that we had amplified IRR cDNA and that we had correctly predicted the intron/exon junctions between exons 1 and 6. Confirmation of all the intron/exon boundaries has been obtained by construction of a full-length human IRR cDNA isolated by cDNA library screening and PCR amplification from human kidney RNA (Chan and Watt, unpublished).

Using a Northern blot of poly(A) RNA (Clontech) as well as quantitative PCR analyses of total RNAs from human tissues, we have demonstrated that the tissue distribution of IRR transcripts in the human is limited and similar to that in the rat (Chan and Watt, unpublished). Recently, our results were confirmed using non-quantitative PCR amplification and a variety of rat tissues (Kurachi et al., 1992); IRR transcripts were present at much higher levels in kidney than in all other tissues including heart, spleen, and liver. Much less was detected in thymus, and stomach RNA was not used (Kurachi et al., 1992). In contrast, using a Northern blot of RNA from human tissues, Zhang and Roth (1992) detected IRR transcripts in many tissues and with a different distribution (heart > liver, kidney). Our results indicate that this widespread distribution of human IRR likely arises from riboprobe hybridization to the similarly sized 28S ribosomal RNA (Chan and Watt, unpublished).

In studies focusing on IRR's role in the kidney, we have used Northern blot, in situ hybridization, and PCR analyses to localize IRR mRNA to specific kidney cortical cells (Mathi and Watt, unpublished).

IDENTIFICATION OF THE IRR GENE TRANSCRIPTION START SITE

Nucleotide sequence analysis of ~800bp 5' to the putative initiator methionine of the rat IRR gene identified consensus DNA promoter sequences involved in the control of gene transcription. These included a TATA-box, SP1 binding sites, and a CCAAT box (Shier and

Watt, 1992). Primer extension analysis using RNA from rat stomach revealed a single transcriptional start site 29bp down-stream from the putative TATA box and 544bp up-stream of the initiator methionine. The presence of these potential promoter elements in the IRR DNA 5' to the initiator methionine codon as well as the presence of a single transcription start site contrasts with constitutively expressed genes such as the insulin receptor gene, which lack both TATA and CCAAT regulatory sites and have multiple start sites (Araki et al., 1987; Mamula et al., 1988). Analysis of the regions 5' to the initiator methionine of the human and guinea pig IRR genes have also identified a putative TATA-box in an analogous position (Petrisor, Shier, and Watt, unpublished observations).

LIGAND BINDING TO IRR AND INSULIN RECEPTOR/IRR CHIMERAS

The comparable similarities between the extracellular, putative ligand-binding regions of the IRR, insulin, and IGF-I receptors suggests that the ligands which bind these receptors are also similar. Related ligands bind members of the same receptor protein-tyrosine kinase families, including the families of the insulin receptor, EGF receptor, FGF receptor, PDGF receptor, and trk (NGF receptor). Assuming that the tissue distribution of the IRR protein parallels that of IRR transcripts, it is unlikely that IRR is the sole physiological receptor for any known member of the insulin family. The distribution of IRR mRNA (kidney>>brain, heart, and uterus) contrasts markedly with that for IGF-II mRNA (brain>heart>uterus> kidney; Murphy et al., 1987), which suggest that IRR is unlikely to be a receptor that mediates the paracrine actions of IGF-II. Similarly, IRR is probably not the relaxin receptor, since we were unable to detect IRR transcripts in the uterus, a tissue where relaxin is known to bind (Osheroff et al., 1990). IRR may, however, be an additional receptor for a known member of the insulin family or, alternatively, a receptor for a related but currently unknown ligand. Insulin family members also include altered forms of known members: different forms of the IGFs exhibit altered biological activity (Carlsson-Skwirut et al., 1989) and proinsulin (Rubenstein et al., 1972) as well as the insulin C-peptide (Zierath et al., 1991) have been suggested to exhibit biological activity. The divergent insulins of hystricomorph rodents such as guinea pig (Chan et al., 1984; Watt, 1985; Graur et al., 1991), may also represent additional types of insulin present in other mammals (Rosenzweig et al., 1983) that have arisen after duplication and divergence of a primitive mammalian gene (Watt 1985). In addition, vertebrate homologues of the invertebrate insulin-like molecules, such as molluscan insulin-related peptide (MIP, Smit et al., 1988) and bombyxins (4K-prothoracicotropic hormone; Adachi et al., 1989), may also exist.

Initial experiments to determine which ligand bound this putative receptor have made use of chimeric receptors between IRR and the insulin receptor. We substituted part of the ligand binding region of the insulin receptor with the analogous region of IRR. The chimeric receptor (IRRx3/IR) contained residues Ser^{210} to Met^{294} encoded by exon 3 of the human insulin receptor replaced with analogous residues of IRR. This region of both the insulin and IGF-I receptors was suggested to be a major ligand binding determinant since a switch in ligand binding specificity was exhibited by chimeras IGFIR225IR and IGFIR286IR (Gustafson and Rutter, 1990). [^{125}I]Insulin binding to the IRRx3/IR chimeric receptor did not differ from the wild type insulin receptor in terms of its affinity for insulin or its inability to be competed by IGF-I, IGF-II, or relaxin (Fig.4A and B). Similar binding results have been obtained using another exon 3 IRR/insulin receptor chimera (Zhang and Roth, 1991). That chimeric receptors containing IRR exons 1, 2, and 3 (Shier and Watt, unpublished observations) or exons 2 and 3 (Zhang and Roth, 1991) replacing the analogous region of the insulin receptor do not bind insulin, IGF-I, IGF-II, or relaxin, suggested that IRR is not a receptor for one of these ligands. This is further supported by lack of binding of insulin, IGF-I, or IGF-II to the full-length IRR (Chan and Watt, unpublished observations) or to a chimeric IRR/IR receptor containing the extracellular domain of the IRR (Zhang and Roth, 1992). Neither these nor some other known ligands of the insulin family including relaxin, bombyxins II and IV, and molluscan insulin-like peptide stimulated the tyrosine kinase activity of this IRR/IR chimera (Zhang and Roth, 1992). IR/IRR chimeras with the extra-cellular and transmembrane domains of IR and the intracellular domain of IRR did exhibit insulin-stimulated tyrosine kinase activity as well as elevated thymidine incorporation and 2-deoxyglucose uptake (Zhang and Roth, 1992). Thus the IRR kinase domain is functional and can modulate cellular growth and metabolism.

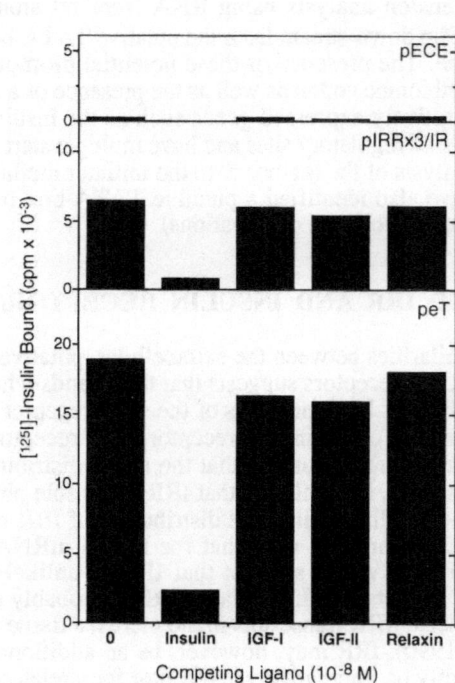

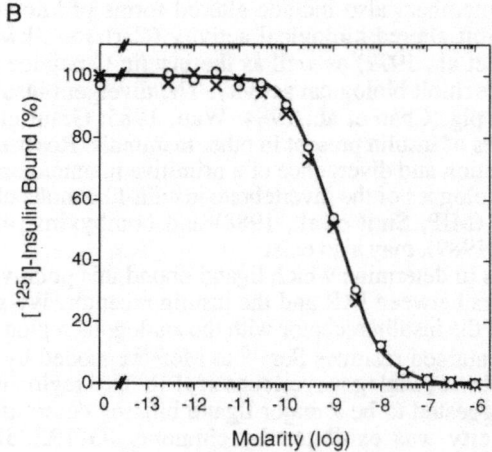

Figure 4. A. Insulin binding to the exon 3 IRR/insulin receptor chimera and to the insulin receptor. COS-1 cells were transfected with the vector (pECE), the exon 3 IRR/insulin receptor chimera (pIRRx3/IR), or the wild type human insulin receptor (peT). [^{125}I]-insulin (100 pM) binding to monolayer cultures was displaced with 10^{-8} M insulin, IGF-I, IGF-II, or relaxin. Values are averages of triplicate determinations that differ by less than 12 %.
B. Binding competition of [^{125}I]insulin (100 pM) by unlabeled insulin. COS-1 cells were transfected with the exon 3 IRR/insulin receptor chimera (pIRRx3/IR,o) or the wild type human insulin receptor (peT, x). Cells transfected with the vector (pECE) exhibited <3 % of the binding of cells transfected with pIRRx3/IR. Non-specific binding was <1 % of total binding. Values are averages of duplicate or triplicate determinations from three independent experiments.

CONCLUSIONS

The surprising identification of the gene encoding IRR, a novel member of the insulin receptor family, was the first evidence that the insulin receptor family extended beyond the well-known insulin and IGF-I receptors. Elucidation of the function of IRR should further our understanding of the role of both the families of insulin receptors and their ligands, as well as determine if an altered IRR could contribute to a pathophysiological state. IRR may play a role in diabetes mellitus, not only because IRR is similar to the insulin receptor, but also because of its similarity with the IGF-I receptor; IGF-I has been implicated as a mitogen in diabetic nephropathy (Hostetter, 1991), the single most important cause of renal failure in adults in the Western world (Mauer et al., 1989).

ACKNOWLEDGEMENTS

We thank S. B. Runciman for technical assistance. This work was funded by the Medical Research Council of Canada and the Canadian Diabetes Association. Student scholarships from the Canadian Diabetes Association were awarded to P.S. and J.C.; a student scholarship from the Quebec Diabetes Association was awarded to P.S.; summer studentships from the Juvenile Diabetes Foundation Canada were awarded to J.C. and B.A.P.; and the Banting and Best Diabetes Centre, University of Toronto awarded a studentship to J.C. and a Hugh Sellers post-doctoral fellowship to S.K.M.

REFERENCES

Abbot, A.M., Bueno, R., Pedrini, M.T., Murray, J.M., and Smith R.J., 1992, Insulin-like growth factor I receptor gene structure, *J. Biol. Chem.* 267:10759.

Adachi, T., Takiya, S., Suzuki, Y., Iwami, M., Kawakami, A., Takahashi, S.Y., Ishizaki, H., Nagasawa, H., and Suzuki, A., 1989, cDNA structure and expression of bombyxin, an insulin-like brain secretory peptide of the silkmoth *bombyx mori*, *J. Biol. Chem.* 264:7681.

Araki, E., Shimada, F., Uzawa, H., Mori, M., and Ebina, Y., 1987, Characterization of the promoter region of the human insulin receptor gene, *J. Biol. Chem.* 262:16186.

Carlsson-Skwirut, C., Lake, M., Hartmanis, M., Hall, K., and Sara, V.R., 1989, A comparison of the biological activity of the recombinant intact and truncated insulin-like growth factor 1 (IGF-1), *Biochim. Biophys. Acta.* 1011:192.

Chan, S.J., Episkopou, V., Zeitlin, S., Karathanasis, S.K., MacKrell, A., Steiner, D.F., and Efstratiadis, A., 1984, Guinea pig preproinsulin gene: an evolutionary compromise?, *Proc. Natl. Acad. Sci. USA.* 81: 5046.

Czech, M.P., 1989, Signal transmission by the insulin-like growth factors, *Cell.* 59:235.

Ebina, Y., Ellis, L., Jarnagin, K., Edery, M., Graf, L., Clauser, E., Ou, J., Masiarz, F., Kan, Y.W., Goldfine, I.D., Roth, R.A., and Rutter, W.J., 1985, The human insulin receptor cDNA: the structural basis for hormone-activated transmembrane signalling, *Cell.* 40:747.

Graur, D., Hide, W.A., and Li, W.-H., 1991, Is the guinea-pig a rodent?, *Nature.* 351:649.

Gustafson, T.A., and Rutter, W.J., 1990, The cysteine-rich domains of the insulin and insulin-like growth factor I receptors are primary determinants of hormone binding specificity, *J. Biol. Chem.* 265:18663.

Hostetter, T.H., 1991, Diabetic Nephropathy, in: "The Kidney Vol. 2," B.M. Brenner and F.C Rector, eds., W.B. Saunders Company, Philadelphia.

King, G.L., and Kahn, C.R., 1981, Non-parallel evolution of metabolic and growth-promoting functions of insulin, *Nature.* 292:644.

Kurachi, H., Jobo, K., Ohta, M., Kawasaki, T., and Itoh, N., 1992, A new member of the insulin receptor family, insulin receptor-related receptor, is expressed perferentially in the kidney, *Biochem. Biophys. Res. Commun.* 187:934.

Mamula, P.W., Wong, K.-Y., Maddux, B.A., McDonald, A.R., and Goldfine, I.D., 1988, Sequence and analysis of promoter region of human insulin-receptor gene, *Diabetes.* 37:1241.

Mauer, S.M., Ellis, E.J., Bilous, R.W., and Steffes, M.W., 1989, The pathology of diabetic nephropathy, *in*: "Complications of Diabetes Mellitus," B. Draznin, S. Melmed, and D. LeRoith, eds., Alan R. Liss Inc., New York.

Murphy, L.J., Bell, G.I., and Friesen, H.G., 1987, Tissue distribution of insulin-like growth factor I and II messenger ribonucleic acid in the adult rat, *Endocrinology.* 120:1279.

Olefsky, J.M., 1990, The Insulin receptor. A multifunctional protein, *Diabetes.* 39:1009.

Osheroff, P.L., Ling, V.T., Vandlen, R.L., Cronin, M.J., and Lofgren, J.A., 1990, Preparation of biologically active ^{32}P-labeled human relaxin; displaceable binding to rat uterus, cervix, and brain, *J. Biol. Chem.* 265:9396.

Rosenzweig, J.L., LeRoith, D., Lesniak, M.A., MacIntyre, I., Sawyer, W.H., and Roth, J., 1983, Two distinct insulins in the guinea pig: the broad relevance of these findings to evolution of peptide hormones, *Fed. Proc.* 42:2608.

Rubenstein, A.H., Melani, F., and Steiner, D.F., 1972, Circulating proinsulin: immunology, measurement, and biological activity. *Handbook of Physiology, Section 7 Endocrinology.* 1:515.

Seino, S., Seino, M., Nishi, S., and Bell, G.I., 1989, Structure of the human insulin receptor gene and characterization of its promoter, *Proc. Natl. Acad. Sci. USA.* 86:114.

Shier, P., and Watt, V.M., 1989, Primary structure of a putative receptor for a ligand of the insulin family, *J. Biol. Chem.* 264:14605.

Shier, P. and Watt, V.M., 1992, Tissue-specific expression of the rat insulin receptor-related receptor gene, Mol. Endocrinol. 6:723.

Shier, P., Willard, H.F. and Watt, V.M., 1990, Localization of the insulin receptor-related receptor gene to human chromosome 1, *Cytogenet. Cell Genet.* 54:80.

Smit, A.B., Vreugdenhil, E., Ebberink, R.H.M., Geraerts, W.P.M., Klootwijk, J., and Joosse, J., 1988, Growth-controlling molluscan neurons produce the precursor of an insulin-related peptide, *Nature.* 331:535.

Ullrich, A., Bell, J.R., Chen, E.Y., Herrera, R., Petruzzelli, L.M., Dull, T.J., Gray, A., Coussens, L., Liao, Y.-C., Tsubokawa, M., Mason, A., Seeburg, P.H., Grunfeld, C., Rosen, O.M., and Ramachandran, J., 1985, Human insulin receptor and its relationship to the tyrosine kinase family of oncogenes, *Nature.* 313:756.

Ullrich, A., Gray, A., Tam, A.W., Yang-Feng, T., Tsubokawa, M., Collins, C., Henzel, W., Le Bon, T., Kathuria, S., Chen, E., Jacobs, S., Francke, U., Ramachandran, J., and Fujita-Yamaguchi, Y., 1986, Insulin-like growth factor I receptor primary structure: comparison with insulin receptor suggests structural determinants that define functional specificity, *EMBO J.* 5:2503.

Ullrich, A., and Schlessinger, J., 1990, Signal transduction by receptors with tyrosine kinase activity, *Cell.* 61:203.

Vanin E.F., 1985, Processed pseudogenes: characteristics and evolution, *Ann. Rev. Genet.* 19:253.

Watt, V.M., 1985, Sequence and evolution of guinea pig preproinsulin DNA, *J. Biol. Chem.* 260:10926.

Yang-Feng, T.L., Francke, U., and Ullrich, A., 1985, Gene for human insulin receptor: localization to site on chromosome 19 involved in pre-B-cell leukemia, *Science.* 228:728.

Zhang, B., and Roth, R.A., 1991, Binding properties of chimeric insulin receptors containing the cysteine-rich domain of either the insulin-like growth factor I receptor or the insulin receptor related receptor, *Biochemistry.* 30:5113.

Zhang, B., and Roth, R.A., 1992, The insulin receptor-related receptor: tissue expression, ligand binding specificity, and signaling capabilities, *J. Biol. Chem.* 267:18320.

Zierath, J.R., Galuska, D., Johansson, B.-L., and Wallberg-Henriksson, H., 1991, Effect of human C-peptide on glucose transport in in vitro incubated human skeletal muscle, *Diabetologia.* 34:899.

MOLECULAR PROPERTIES OF INSULIN/IGF-1 HYBRID RECEPTORS
4th INTERNATIONAL SYMPOSIUM ON INSULIN, IGFs AND THEIR RECEPTORS

Jeffrey E. Pessin

Department of Physiology & Biophysics
The University of Iowa
Iowa City, Iowa, USA

INTRODUCTION

Insulin and IGF-1 are two highly related growth factors which mediate pleiotropic cellular responses[1-3]. These hormones regulate their specific actions via binding to specific high affinity receptors on the plasma membrane of target cells. Similar to these growth factors, the receptors for insulin and IGF-1 share a high degree of structural and functional properties[4-7]. Both these receptors are classified as Type I receptors being composed of two α subunits and two β subunits which are disulfide-linked into an $\alpha_2\beta_2$ heterotetrameric complex (Figure 1). The receptor subunits are initially synthesized as $\alpha\beta$ polypeptide fusion precursor proteins which undergoes extensive co- and post-translational modifications[8-12]. The receptor Mr 155,000 $\alpha\beta$ precursors undergoe co-translational acylation and Asn-linked glycosylation. As the precursor is processed from the endoplasmic reticulum through the trans-Golgi network, intramolecular disulfide bonds are formed linking the α subunit to the β subunit. During this time frame, two $\alpha\beta$ fusion precursors non-covalently dimerize followed by proteolytic cleavage at a furin-like tetrabasic sequence which separates the α subunit from the β subunit. Finally, the dimerized $\alpha\beta$ half-receptors become disulfide-linked through the α subunits coincident with the addition of terminal sialic acid residues and exposure of the native $\alpha_2\beta_2$ heterotetrameric receptor on the cell surface membrane.

Recent studies have begun to dissect the structural features of these receptors responsible for their physiochemical properties. The insulin and IGF-1 receptor α subunits contain the high affinity binding sites for insulin and IGF-1, respectively[13-24]. Biochemical and molecular biological approaches have identified various α subunit domains responsible for high affinity ligand binding specificity. The primary determinant of IGF-1 binding to the IGF-1 receptor appears to specifically reside within the cysteine-rich domain[13-17]. In contrast, the insulin binding domain appears to be somewhat more complex having ligand

Current Directions in Insulin-Like Growth Factor Research,
Edited by D. LeRoith and M.K. Raizada, Plenum Press, New York, 1994

133

binding determinants located over several discrete sub domains. These include the amino-terminus, a portion of the cysteine-rich domain as well as sequences carboxy terminal to the cysteine-rich domain[9-24]. The cysteine residue responsible for disulfide-linkage of the α subunit to the β subunit has been identified as amino acid 674, however, the corresponding β subunit cysteine nor the cysteine residues responsible for disulfide-linkage of the α subunits have not yet been determined[25].

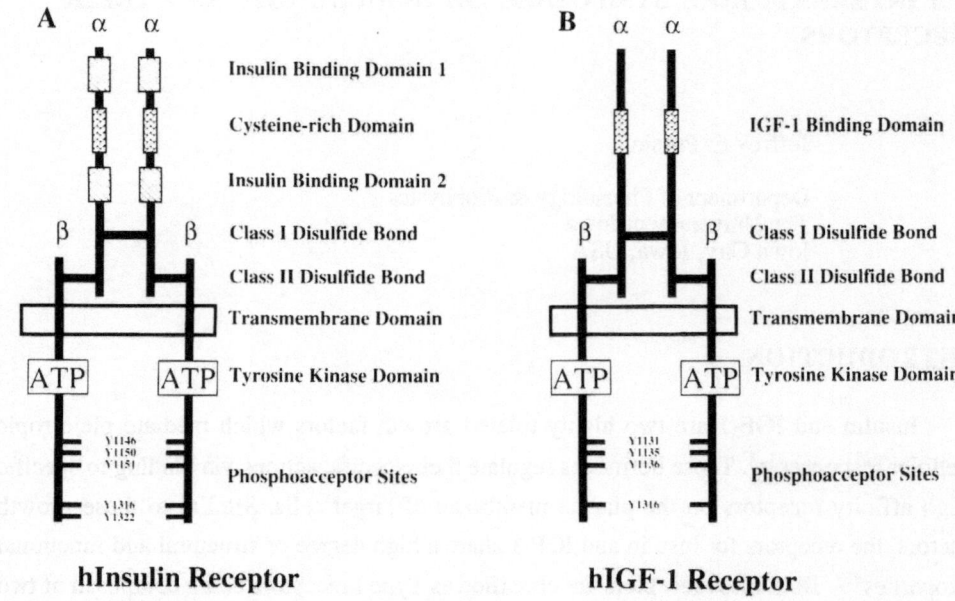

Figure 1. Schematic model of the human insulin and IGF-1 receptors. Various functional domains of the insulin (A) and IGF-1 (B) receptors are indicated.

The receptor α subunits are spatially localized to the extracellular face of the plasma membrane where exposure to circulating insulin or IGF-1 occurs. This extracellular binding event is then transmitted as an intracellular signal by activation of the transmembrane β subunit. The intracellular region of each b subunit contains an intracellular tyrosine-specific protein kinase domain, ATP binding sites, several phosphotyrosine as well as phosphoserine/threonine acceptor sites[26,27]. Upon binding ligand, the receptor rapidly undergoes autophosphorylation at several tyrosine residues on the β subunit. Phosphorylation of the 1146-1151 region in the insulin receptor activates the receptor kinase toward intermediary cellular substrates[28-32]. Although the specific tyrosine autophosphorylation sites of the IGF-1 receptor have not been experimentally determined, based upon the high degree of sequences identity it is likely that the corresponding conserved tyrosine residues function in a similar manner. Several studies have also demonstrated that the carboxy terminal tyrosine autophosphorylation sites (1316 and 1322) may also play an important role in insulin signaling. Rat 1 cells expressing a carboxyl terminal deletion displays enhanced mitogenic properties with impaired metabolic responses[33-35]. Similarly, mutations in the carboxy terminal tyrosine residues of the insulin receptor β subunit were found to enhance mitogenic signaling in CHO cells[36]. However, this observation is not universally accepted as another study in CHO cells did not indicate any specific role for the insulin receptor β subunit carboxy terminus[37]. Nevertheless, it is interesting to note that the IGF-1 receptor β subunits contain only the equivalent of tyrosine 1316 of the insulin receptor and, in general, function as a more potent effector of mitogenesis.

To date, there remains a distinct controversy with regard to the role of receptor kinase activity in mediating biological responses. The majority of studies are consistent with an obligatory role of receptor tyrosine kinase activity for the regulation of both acute metabolic and chronic mitogenic actions of insulin[2,38-40]. However, several studies have suggested that the insulin receptor can function in the absence of tyrosine kinase activation and/or tyrosine phosphorylation of down stream effector molecules[41-44]. Whether or not the insulin and IGF-1 receptors require tyrosine autophosphorylation to mediate all their diverse biological actions, it is clear that tyrosine kinase activity is intrinsic to the receptor β subunits and that down stream tyrosine phosphorylation is required for a variety of post-receptor signaling events.

INSULIN/IGF-1 HYBRID RECEPTORS

During our studies on insulin and IGF-1 receptor mediated transmembrane signaling, previous studies have suggested the presence of insulin and IGF-1 receptors that were immunological and functionally different from classical insulin and IGF-1 receptors[45-50]. This receptor diversity is greater than could now be accounted for by expression of the known receptor genes and potential splice variants. Initially the existence of a receptor species composed of an insulin αβ half-receptor disulfide-linked with an IGF-1 αβ half-receptor (insulin/IGF-1 hybrid receptor) was suggested by Siddle and colleagues[51,52]. This was based upon the observation that highly specific anti-insulin receptor monoclonal antibodies would precipitated IGF-1 binding from solubilized extracts of human placenta and cultured cell membranes. This cross-immunoreactivity strongly suggested the presence of heterotypic receptors having properties of both insulin and IGF-1 receptors. In this regard, these receptors may account for the previously identified atypical insulin receptors found in human placenta membranes[46]. More recently, immunoaffinity purification has been successfully used to isolated insulin/IGF-1 hybrid receptors from human placenta membranes[53].

Several other studies have documented the existence of insulin/IGF-1 hybrid receptors. Jacobs and colleagues demonstrated the presence of hybrid receptors in NIH3T3 cells and HepG2 cells both by immunological criteria as well as by phosphopeptide mapping[54].In these studies, the presence of insulin/IGF-1 hybrid receptors was readily detected owing to the differences in apparent molecular of the insulin and IGF-1 receptor β subunits. Other studies have demonstrated that human and rodent insulin receptors can also readily assembly into a hybrid receptor complex and that insulin/IGF-1 hybrid receptors are also present in neuronal cell lines[55,56].

Taken together, these data strongly support the existence of insulin/IGF-1 hybrid receptors in both human placenta and cultured cell lines. Although the presence and functional properties of insulin/IGF-1 hybrid receptors in typical insulin target tissues such as adipose and skeletal muscle has not yet been demonstrated, it is likely that these additional receptor complexes will contribute to the diversity of cellular signaling by these ligands. Thus, a detailed description of hybrid receptor signaling is not only an important issue for our understanding of normal receptor function but also to evaluate the linkage between insulin resistance and the presence of wild type/mutant hybrid insulin receptors in target tissues.

FUNCTIONAL PROPERTIES OF HYBRID RECEPTORS

Detailed analysis of the functional properties of insulin/IGF-1 hybrid receptors has been difficult due to the simultaneous presence of both classical insulin and IGF-1 receptors. Nevertheless, several studies have indicated that the insulin/IGF-1 hybrid receptors have a greater degree of specificity for IGF-1 than insulin and that activation of β subunit autophosphorylation is more sensitive to IGF-1 than insulin[54,56]. Recently, Siddle and colleagues have successfully purified insulin/IGF-1 hybrid receptors away from the endogenous classical homotypic receptors[53]. In addition, we have used the vaccinia virus transient transfection system, coupled with selective immunoprecipitation to assess the binding characteristics of purified insulin/IGF-1 hybrid receptors[57]. Consistent with the early studies, ligand binding analysis demonstrated a significantly greater binding affinity for IGF-1 than insulin in the insulin/IGF-1 hybrid receptors. Together, these data suggest that cells expressing insulin/IGF-1 hybrid receptors will display kinase activation properties more consistent with classical IGF-1 receptors than insulin receptors.

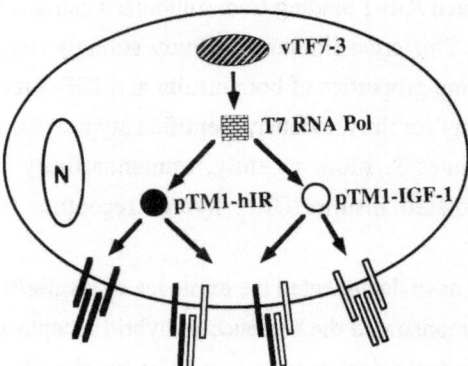

Figure 2. Schematic model of insulin/IGF-1 hybrid receptor assembly using the vaccinia virus/T7 bacteriophage expression system. The cells are initially infected with T7 bacteriophage expressing recombinant virus (vTF7-3), followed by transfection with a plasmid (pTM1) containing either the insulin or IGF-1 receptor cDNA. The resulting receptor precursors can then form various combinations of homotypic and heterotypic hybrid receptors depending upon the ratio of plasmids used in the transfection.

During the course of these studies, we recognized that specific ligand binding and immunological properties of insulin/IGF-1 hybrid receptors could be used to dissect the mechanism of ligand-dependent transmembrane signaling. To generate define populations of homotypic and heterotypic insulin/IGF-1 receptors, we took advantage of the vaccinia virus/bacteriophage T7 transient expression system[58]. Previous studies have documented the reproducible and highly efficient nature of this system to produce fully mature and functional receptors[59]. The assembly of heterotypic insulin/IGF-1 hybrid receptors by this method is schematically illustrated in Figure 2.

To assess the molecular events in ligand stimulated transmembrane signaling, we initially confirmed that these insulin/IGF-1 hybrid receptors display a greater apparent binding affinity for IGF-1 than insulin[57]. Although these receptors will preferentially bind IGF-1 over insulin, at low hormone concentrations each ligand is specific for its cognizant α subunit. Using the vaccinia virus/T7 expression system, we prepared heterotypic insulin/IGF-1 hybrid receptor composed of a truncated, kinase-active insulin half-receptor and a wild type IGF-1 half-receptor ($\alpha\beta_{IR.\Delta43}$-$\alpha\beta_{IGFR.WT}$). In these experiments the $\Delta43$ insulin half-receptor was used rather than the wild type insulin receptor to accentuate the molecular weight difference between the β subunits detected by SDS-polyacrylamide gels (Fig. 3). In the absence of ligands, the $\alpha\beta_{IR.\Delta43}$-$\alpha\beta_{IGFR.WT}$ insulin/IGF-1 hybrid receptor displayed specific autophosphorylation of both the 90 kDa and 102 kDa β subunits (Fig. 3, lane 1). In the presence of 10 nM insulin (Fig. 3, lane 2) or 10 nM IGF-1 (Fig. 3, lane 3) autophosphorylation of both β subunits occurred to a similar extent. As expected ligand specificity was also observed for β subunit autophosphorylation of both the homotypic insulin and IGF-1 receptors (Fig. 3, lanes 4-9).

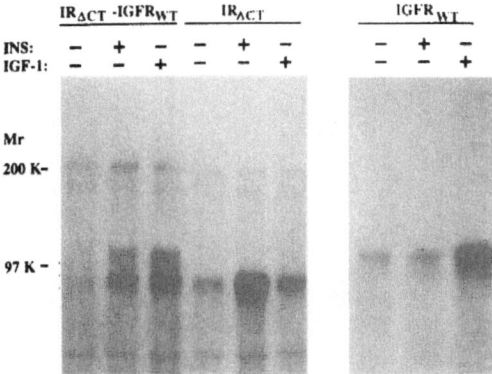

Figure 3. Ligand stimulation of intramolecular β subunit autophosphorylation within kinase functional insulin/IGF-1 hybrid receptors. 3T3442A fibroblasts were infected with vTF3-7 and transfected with 5 µg/150 mm dish of either pTM1-IR.$\Delta43$, pTM1-IGF.WT or both pTM1-IR.$\Delta43$ plus pTM1-IGF.WT. Homotypic and hybrid receptors were partially purified as previously described (57). The insulin/IGF-1 hybrid receptor $\alpha\beta_{IR.\Delta43}$-$\alpha\beta_{IGFR.WT}$ (lanes 1-3), homotypic insulin receptor $\alpha\beta_{IR.\Delta43}$-$\alpha\beta_{IR.\Delta43}$ (lanes 4-6), and homotypic IGF-1 receptor $\alpha\beta_{IGFR.WT}$-$\alpha\beta_{IGFR.WT}$ were then subjected to autophosphorylation in the absence of ligand (lanes 1, 4 and 7), in the presence of 10 nM insulin (lanes 2, 5 and 8), or in the presence of 10 nM IGF-1 (lanes 3, 6 and 9). The samples were immunoprecipitated with the insulin receptor monoclonal antibody, 83-7 (lanes 1-6) or with the IGF-1 receptor monoclonal antibody, αIR-3 (lanes 7-9) and subjected to reducing SDS-polyacrylamide gel electrophoresis and autoradiography. Reprinted with permission from Frattali & Pessin (1993) J. Biol. Chem. 268: 7393-7400 (57).

Since the homotypic insulin and IGF-1 receptors as well as the insulin/IGF-1 hybrid receptors maintained ligand binding specificity, we next utilized these receptors to address the spatial relationship between α subunit ligand occupancy and β subunit autophosphorylation. This was initially examined by determining the β subunit autophosphorylation pattern between two different mutant insulin/IGF-1 hybrid receptors each composed of one active and one kinase-defective αβ half-receptor (Fig. 4). In these experiments, we initially utilized a hybrid receptor consisting of a truncated kinase-inactive insulin αβ half-receptor ($\alpha\beta_{IR.A/K.\Delta43}$) assembled with a wild type IGF-1 αβ half-receptor (Fig. 4, lanes 1-3). In this hybrid receptor, labeling of the 90 kDa subunit was indicative of an intramolecular trans-autophosphorylation reaction pathway whereas labeling of the 102 kDa β subunit represents intramolecular *cis*-autophosphorylation. Under basal conditions both the 90 kDa and 102 kDa β subunits were found to be autophoshorylated (Fig. 4, lane 1). In contrast, insulin primarily stimulated the labeling of the 90 kDa kinase-defective insulin receptor β subunit with little increase in labeling of the 102 kDa kinase-active IGF-1 receptor β subunit (Fig. 4, lane 2). Surprisingly, IGF-1 treatment also resulted in the predominant labeling of the 90 kDa kinase-defective insulin receptor β subunit (Fig. 4, lane 3).

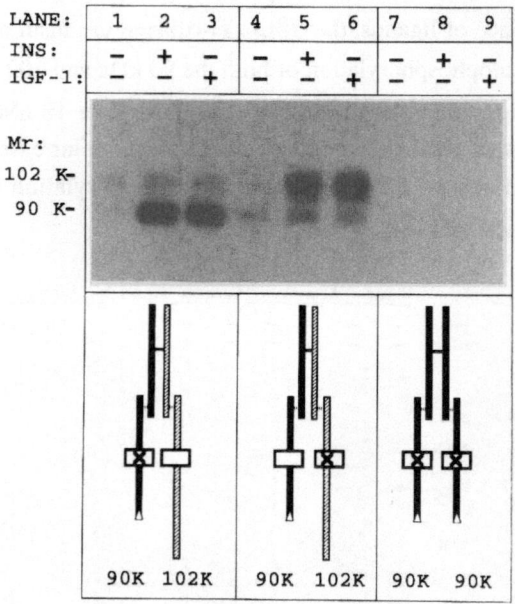

Figure 4. Ligand stimulation of intramolecular *trans*-autophosphorylation in mutant/wild type insulin/IGF-1 hybrid receptors. 3T3442A fibroblasts were infected with vTF3-7 and transfected with 9 μg/150 mm dish pTM1-IR.A/K.Δ43 plus 3 μg/150 mm dish pTM1-IGFR.WT (lanes 1-3), 6 mg/150 mm dish of both pTM1-IR.Δ43 plus pTM1-IGFR.A/K (lanes 4-6), and 9 μg/150 mm dish pTM1-IR.A/K.Δ43 (lanes 7-9). The homotypic and insulin/IGF-1 hybrid receptors were partially purified as previously described (57). Insulin/IGF-1 hybrid receptor, $\alpha\beta_{IR.A/K.\Delta43}$-$\alpha\beta_{IGFR.WT}$, (lanes 1-3), insulin/IGF-1 hybrid receptor, $\alpha\beta_{IR.\Delta43}$-$\alpha\beta_{IGFR.A/K}$, (lanes 4-6), and homotypic insulin receptor, $\alpha\beta_{IR.A/K.\Delta43}$-$\alpha\beta_{IR.A/K.\Delta43}$, were immunoprecipitated with the insulin receptor monoclonal antibody, 83-7, and subjected to autophosphorylation in the absence of ligand (lanes 1, 4 and 7), in the presence of 80 nM insulin (lanes 2, 5 and 8), or in the presence of 80 nM IGF-1 (lanes 3, 6 and 9). The receptor subunits were resolved by reducing SDS-polyacrylamide gel electrophoresis and autoradiography. Reprinted with permission from Frattali & Pessin (1993) J. Biol. Chem. 268: 7393-7400 (57).

In a complementary approach we also examined the relationship between α subunit ligand binding with β subunit autophosphorylation in a hybrid receptor composed of a truncated kinase-active insulin αβ half-receptor ($\alpha\beta_{IR.\Delta43}$) assembled with a kinase-inactive IGF-1 αβ half-receptor ($\alpha\beta_{IGFR.A/K}$). In this situation, labeling of the full length 102 kDA β subunit represents intramolecular trans-autophosphorylation; whereas, labeling of the 90 kDa kinase-active insulin receptor β subunit represents intramolecular cis-autophoshorylation. In this hybrid receptor, both insulin (Fig. 4, lane 5) and IGF-1 (Fig. 4, lane 6) increased autophosphorylation of the kinase-inactive 102 kDa β subunit with little effect on the kinase-active 90 kDa β subunit. As controls, the homotypic, kinase-defective insulin receptor demonstrated no significant β subunit autophosphorylation in either the absence or presence of insulin or IGF-1 (Fig. 4, lanes 7-9). Similarly, no significant β subunit autophosphorylation was observed within homotypic kinase-defective IGF-1 receptors in either the basal state or in response to either ligand (data not shown). Taken together, these data strongly support a model in which insulin and IGF-1 activate β subunit autophosphorylation by an intramolecular trans-autophosphorylation mechanism independent of which a subunit was ligand occupied.

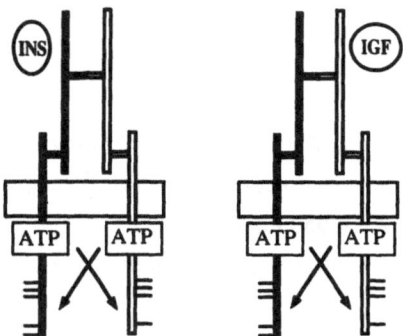

Figure 5. Schematic representation of ligand-stimulated transmembrane autophosphorylation of insulin/IGF-1 hybrid receptors. In this model, ligand occupancy of either α subunit results in the activation of both β subunits such that a series of intramolecular autophosphorylation events occur. Under these conditions, both β subunits will become autophosphorylated followed by activation of receptor substrate kinase activity.

Thus our studies to date, strongly support a mechanism by which ligand occupancy of either α subunit results in the propagation of an intramolecular signal that can activate both β subunits to function as tyrosine-specific protein kinases. These activated β subunits initial function to trans-phosphorylate each other resulting in the tyrosine phosphorylation of both β subunits. This model for intramolecular trans-autophosphorylation is schematically illustrated in Figure 5. In addition, recent studies by Shoelson and Pilch utilizing a photoreactive insulin derivative have also demonstrated that ligand occupancy of one insulin receptor α subunit activates the autophosphorylation of both β subunits roughly in a 60% to 40% ratio[60]. These data are consistent with our findings in the insulin/IGF-1 hybrid receptors and strongly support a molecular mechanisms requiring two intramolecular trans-autophosphorylation events in the native wild type holoreceptor complex.

CONCLUSIONS

During the past several years, the molecular characterization of insulin and IGF-1 receptor structure/function has demonstrated a surprising complexity. Although we and others are currently investigating several properties of insulin/IGF-1 hybrid receptors substantially more studies will be required to develop a precise molecular understanding of receptor transmembrane signaling. For example, previous studies have demonstrated that hybrid holoreceptors composed of a kinase-active $\alpha\beta$ half-receptor assembled with a kinase-defective $\alpha\beta$ half-receptor are substrate kinase inactive[59,61]. Thus, a pressing questions is whether two functional ATP catalytic sites are required for substrate kinase activity or is one ATP catalytic site sufficient if both β subunit are appropriately phosphorylated? In addition, the molecular basis for the surprising ligand binding specificity of the insulin/IGF-1 hybrid receptors should provide a valuable approach to dissect the precise determinants involved in ligand binding specificity and the non-linear binding properties of the homotypic insulin receptors. Future studies directed at developing a detailed understanding of these hybrid receptors will undoubtedly provide important new insights into the molecular basis of insulin and IGF-1 receptor transmembrane signaling.

REFERENCES

1. M.M. Rechler and S.P. Nissley, The nature and regulation of the receptors for insulin-like growth factors, *Annu. Rev. Physiol.* 47:425-442 (1985).

2. I.D. Goldfine, The insulin receptor; Molecular biology and transmembrane signaling, *Endocrinol. Rev.* 8: 235-255 (1987).

3. S. Jacobs and P. Cuatrecasas, Insulin Receptors, *Annu. Rev. Pharmacol. Toxicol.* 23: 461-479 (1983).

4. Y. Ebina, L. Ellis, K. Jarnagin, M. Edery, L. Graf, E. Clauser, J.-H. Ou, R. Masiarz, Y.W. Kan, I.D. Goldfine, R.A. Roth, and W.J. Rutter, The insulin receptor; Molecular biology and transmembrane signaling, *Cell* 40: 747-758 (1985).

5. A. Ulrich, J.R. Bell, E.Y. Chen, R. Herrera, L.M. Petruzzelli, T.J. Dull, A. Gray, L. Coussens, Y.-C. Liao, M. Tsubokawa, A. Mason, P.H. Seeburg, C.L. Grunfeld, O.M. Rosen, and J. Ramachandran, Human insulin receptor and its relationship to the tyrosine kinase family of oncogenes, *Nature* . 313: 756-761 (1985).

6. G. Steele-Perkins, J. Turner, J.C. Edman, J. Hari, S.B. Pierce, C. Stover, W.J. Rutter, and R.A. Roth, Expression and characterization of a functional human insulin-like growth factor I receptor, *J. Biol. Chem.* 263: 11486-11492 (1988).

7. A. Ullrich, A. Gray, A.W. Tam, T. Yang-Feng, M. Tsubokawa, C. Collins, W. Henzel, T. LeBon, S. Kathuria, E. Chen, S. Jacobs, U. Francke, J. Ramachandran, and Y. Fujita-Yamaguchi, Insulin-like growth factor I receptor primary structure: Comparison with insulin receptor suggests structural determinants that define functional specificity, *EMBO J.* 5: 2503-2512 (1986).

8. E. Van Oberghen, M. Kasuga, M. LeCam, J.A. Hedo, A. Itin, and L.C. Harrison, Biosynthetic labeling of insulin receptor: Studies of subunits in cultured human IM-9 lymphocytes, *Proc. Natl. Acad. Sci. USA* . 78: 1052-1056 (1981).

9. J.B. Forsayeth, B. Maddux,and I.D. Goldfine, Biosynthesis and processing of the human insulin receptor, *Diabetes* . 35: 837-846 (1986).

10. J.A. Hedo, E. Collier, and A. Watkinson, Myristyl and palmityl acylation of the insulin receptor, *J. Biol. Chem.* 262: 954-957 (1987).

11. P.J. Deutsch, C.F. Wan, O.M. Rosen, and C.S.Rubin, Latent insulin receptors and possible receptor precursors in 3T3-L1 adipocytes, *Proc. Natl. Acad. Sci. USA* . 80: 133-136 (1983).

12. G.V. Ronnett, P. Knutson, R.A. Kohanski, T.L. Simpson, and M.D. Lane, Role of glycosylation in the processing of newly translated insulin proreceptor in 3T3-L1 adipocytes, *J. Biol. Chem.* 259: 4566-4575 (1984).

13. A.S. Andersen, T. Kjeldsen, F.C. Wiberg, P.M. Christensen, J.S. Rasmussen, K. Norris, K.B. Moller, and N.P.H. Moller, Changing the insulin receptor to possess insulin-like growth factor I ligand specificity, *Biochemistry*. 29: 7363-7366 (1990).

14. T.A. Gustafson, and W.J. Rutter, The cysteine-rich domains of the insulin and insulin-like growth factor I receptors are primary determinants of hormone binding specificity, *J. Biol. Chem.* 265: 18663-18667 (1990).

15. T. Kjeldsen, A.S. Andersen, F.C Wiberg, J.S. Rasmussen, L. Schaffer, P. Balschmidt, D.B. Moller, and N.P.H. Moller, The ligand specificities of the insulin receptor and the insulin-like growth factor I receptor reside in different regions of a common binding site, *Proc. Natl. Acad. Sci. USA*. 88: 4404-4408 (1991).

16. R. Schumacher, L. Mosthaf, J. Schlessinger, D. Brandenburg, and A. Ullrich, Insulin and IGF1 binding specificity is determined by distinct regions of their cognate receptors, *J . Biol. Chem.* 266: 19288-19295 (1991).

17. B. Zhang, and R.A. Roth, A region of the insulin receptor important for ligand binding, *Proc. Natl. Acad. Sci. USA* . 88: 9858-9862 (1991).

18. C.C. Yip, C. Grunfeld, and I.D. Goldfine, Identification and characterization of the l igand-binding domain of insulin receptor by use of an anti-peptide antiserum against amino acid sequence 241-251 of the alpha subunit, *Biochemistry*. 30: 695-701 (1991).

19. C.C. Yip, H. Hsu, R.G. Patel, H.M. Hawley, B.A. Maddux, and I.D. Goldfine, Localization of the insulin-binding site to the cysteine-rich region of the insulin receptor alpha-subunit, *Biochem. Biophys. Res. Commun.* 157: 321-329 (1988).

20. R. Rafaeloff, R. Patel, C.C. Yip, I.D. Goldfine, and D.M. Hawley, Mutation of the high cysteine region of the human insulin receptor α-subunit increases insulin receptor binding affinity and transmembrane signalling, *J. Biol. Chem.* 264: 15900-15904 (1989).

21. F. Wedekind, K. Baer-Pontzen, S. Bala-Mohan, D. Choli, H. Zahn, and D. Brandenburg, Hormone binding site of the insulin receptor: Analysis using photoaffinity-mediated avidin complexing, *Biol. Chem. Hoppe-Seyler.* 370: 251-258 (1989).

22. S.M. Waugh, E.E. DiBella, and P.F. Pilch, Isolation of a proteolytically derived domain of the insulin receptor containing the major site of cross-linking/binding, *Biochemistry.* 28: 3448-3455 (1989).

23. M. Fabry, E. Schaefer, L. Ellis, E. Kojro, Fahrenholz and D. Brandenburg, Detection of a new hormone contact site within the insulin receptor ectodomain by the use of a novel photoreactive insulin, *J. Biol. Chem.* 267: 8950-8956 (1992).

24. P. DeMeyts, J.-L. Gu, R.M. Shymko, B.E. Kaplan, G.I. Bell, and J.Whittaker, Identification of a ligand-binding region of the human insulin receptor encoded by the second exon of the gene, *Mol. Endocrinol.* 4: 409-416 (1990).

25. B. Cheathman, and C.R. Kahn, Cyseine 647 in the insulin receptor is required for normal covalent interaction between α and β subunits in signal transduction, *J. Biol. Chem.* 267: 7108-7115 (1992).

26. H.E. Tornqvist, J.R. Gunsalus, R.A. Nemenoff, A.R. Frackelton, M.W. Pierce, and J. Avruch, Identification of the insulin receptor tyrosine residues undergoing insulin-stimulated phosphorylation in intact rat hepatoma cells, *J. Biol. Chem.* 263: 350-359 (1988).

27. M.F. White, S.E. Shoelson, H. Keutmann, and C.R.Kahn, A cascade of tyrosine autophosphorylation in the β-subunit activates the phosphotransferase of the insulin receptor, *J. Biol. Chem.* 263: 2969-2980 (1988).

28. H.Maegawa, J.M. Olefsky, S. Thies, D. Boyd, A. Ullrich, and D.A. McClain, Insulin receptors with defective tyrosine kinase inhibit normal receptor function at the level of substrate phosphorylation, *J. Biol. Chem.* 263: 12629-12637 (1988).

29. K.-T. Yu, and M.P. Czech, Tyrosine phosphorylation of insulin receptor β subunit activates the receptor tyrosine kinase in intact H-35 hepatoma cells, *J. Biol. Chem.* 261: 4715-4722 (1986).

30. O.M. Rosen, R. Herrera, Y. Olowe, L.M. Petruzzelli, and M.H. Cobb, Phosphorylation activates the insulin receptor tyrosine protein kinase, *Proc. Natl. Acad. Sci. USA.* 80: 3237-3240 (1983).

31. R.A. Kohanski, and M.D. Lane, Kinetic evidence for activating and non-activating components of autophosphorylation of the insulin receptor protein kinase, *Biochem. Biophys. Res. Commun.* 134: 1312-1318 (1986).

32. L.J.Sweet, B.D. Morrison, P.A. Wilden, and J.E. Pessin, Insulin dependent intermolecular subunit communication, *J. Biol. Chem.* 262: 16730-16738 (1987).

33. D.A. McClain, H. Maegawa, J. Levy, T. Huecksteadt, T.J. Dull, A. Ullrich, and J.M. Olefsky, Properties of a human insulin receptor with a COOH-terminal truncation, *J. Biol. Chem.* 263: 8904-8911 (1988).

34. H. Maegawa, D.A. McCalin, G. Freidenberg, J.M. Olefsky, M. Napier, T. Lapari, T.J. Dull, J. Lee, and A. Ullrich, Properties of a human insulin receptor with a COOH-terminal truncation, *J. Biol. Chem.* 263: 8912-8917 (1988).

35. R.S. Thies, A. Ullrich, and D.A. McClain, Augmented mitogenesis and impaired metabolic signaling mediated by a truncated insulin receptor, *J. Biol. Chem.* 264: 12820-12825 (1989).

36. A. Ando, K. Momomura, K. Tobe, R. Yamamoto-Honda, H. Sakura, Y. Tamori, Y. Kaburagi, O. Koshio, Y. Akanuma, Y. Yazaki, M. Kasuga, and T. Kadowaki, Enhanced insulin-induced mitogenesis and mitogen-activated protein kinase activities in mutant insulin receptors with substitution of two COOH-terminal tyrosine autophosphorylation sites by phenylalanine, *J. Biol. Chem.* 267: 12788-12796 (1992).

37. M.G. Myers, Jr., J.M. Backer, K. Siddle, and M.F. White, The insulin receptor functions normally in Chinese hamster ovary cells after truncation of the C terminus, *J. Biol. Chem.* 266: 10616-10623 (1991).

38. O.M. Rosen, After Insulin Binds, *Science* 237: 1452-1458 (1987).

39. C.R. Kahn, The molecular mechanism of insulin action, *Ann, Rev. Med.* 36: 429-451 (1985).

40. M.F. White, and C.R. Kahn, The Enzymes, Vol. XVII, pp. 456-502, P.D. Boyer and E.G. Krebs, Eds., Academic Press, Orlando, FL (1986).

41. C.K. Sung, B.A. Maddux, Hawley, D.M. and I.D. Goldfine, Monoclonal antibodies mimic insulin activation of ribosomal protein S6 kinase without activation of insulin receptor tyrosine kinase, *J. Biol. Chem.* 264: 18951-18959 (1989).

42. A. Debant, G. Ponzio, E. Clausen, and B. Rossi, Receptor cross-linking restores an insulin metabolic effect altered by mutation on tyrosine 1162 and tyrosine 1163, *Biochemistry.* 28: 14-17 (1989).

43. D.E. Moller, H. Benecke, and J.S. Flier, Biologic activities of naturally occurring human insulin receptor mutations. Evidence that metabolic effects of insulin can be mediated by a kinase-deficient insulin receptor mutant, *J. Biol. Chem.* 266:10995-11001 (1991).

44. W. Gottschalk, The pathway mediating insulin's effects on pyruvate dehydrogenase bypasses the insulin receptor tyrosine kinase, *J. Biol. Chem.* 266: 8814-8819 (1991).

45. R.L.Hintz, A.V. Thorsson, G. Enberg, and K, Hall, IGFII binding on human lymphoid lines: Demonstration of a common high affinity receptor for insulin like peptides, *Biochem. Biophys. Res. Commun.* 127: 929-936 (1984).

46. H.A. Jonas, J.D. Newman, and L.C. Harrison, An atypical insulin receptor with high affinity for insulin-like growth factors copurified with placental insulin receptors, *Proc. Natl. Acad. Sci. USA.* \83: 4124-4128 (1986).

47. P.Misra, R.L. Hintz, and R. Rosenfeld, Structure and immunological characterization of insulin like growth factor II binding to IM-9 cells., *J. Clin. Endocrinol. Metab.* 63: 1400-1405 (1986).

48. S.E.Tollefson, K. Thompson, and D.J. Petersen, Separation of the high affinity IGF I receptor from low affinity binding sites by affinity chromatography, *J. Biol. Chem.* 262: 16461-16469 (1987).

49. T.K. Alexandrides, and R.J. Smith, A novel fetal insulin-like growth factor (IGF) I receptor, *J. Biol. Chem.* 264: 12922-12930 (1989).

50. S.J.Casella, V.K. Han, A.J. D'Ercole, M.E. Svoboda, and J.J. Van Wyk, Insulin-like growth factor II binding to the type I somatomedin receptor: Evidence for two high affinity binding sites, *J. Biol. Chem.* 261: 9268-9273 (1986).

51. M.A. Soos, and K. Siddle, Immunological relationships between receptors for insulin and insulin-like growth factor I, *Biochem. J.* 263: 553-563 (1989).

52. M.A. Soos, J. Whittaker, R. Lammers, A. Ullrich, and K. Siddle, Receptors for insulin and insulin-like growth factor-I can form hybrid dimers, *Biochem. J.* 270: 383-390 (1990).

53. M.A. Soos, C.E. Field, and K. Siddle, Purified hybrid insulin/insulin-like growth factor-I receptors bind insulin-like growth factor-I, but not insulin, with high affinity, *Biochem. J.* 290: 419-426 (1991).

54. C.P. Moxham, V. Duronio, and S. Jacobs, Insulin-like growth factor I receptor β-subunit heterogeneity, *J. Biol. Chem.* 264: 13238-13244 (1989).

55. J.E. Chin, J.M. Tavare, L. Ellis, and R.A. Roth, Evidence for hybrid rodent and human insulin receptors in transfected cells, *J. Biol. Chem.* 266: 15587-15590 (1991).

56. R.S. Garofalo, and B. Barenton, Functional and immunological distinction between insulin-like growth factor I receptor subtypes in KB cells, *J. Biol. Chem.* 267: 11470-11475 (1992).

57. A.L. Frattali and J.E. Pessin, Relationship between α subunit ligand occupancy and β subunit autophosphorylation in insulin/insulin-like growth factor-1 hybrid receptors, *J. Biol. Chem.* 268: 7393-7400 (1993).

58. B. Moss, O. Elroy-Stein, T. Mizukami, W.A. Alexander, and T.R. Fuerst, New mammalian expression vectors, *Nature.* 348: 91-92 (1990).

59. A.L. Frattali, J.L. Treadway, and J.E. Pessin, Transmembrane signaling by the human insulin receptor kinase, *J. Biol. Chem.* 267: 19521-19528 (1992).

60. J. Lee, T. O'Hare, P.F. Pilch, and S.E. Shoelson, Insulin receptor autophosphorylation occurs asymmetrically, *J. Biol. Chem.* 268: 4092-4098 (1993).

61. J.L. Treadway, B.D. Morrison, M.A. Soos, K. Siddle, J. Olefsky, A. Ullrich, D.A. McClain, and J.E. Pessin, Transdominant inhibition of tyrosine kinase activity in mutant insulin/inulin-like growth factor I hybrid receptors, *Proc. Natl. Acad. Sci., USA.* 88: 214-218 (1991).

IMMUNOLOGICAL STUDIES OF TYPE I IGF RECEPTORS AND INSULIN RECEPTORS: CHARACTERISATION OF HYBRID AND ATYPICAL RECEPTOR SUBTYPES

Maria A. Soos, Barbara.T. Navé, and Kenneth Siddle

Department of Clinical Biochemistry
University of Cambridge
Addenbrooke's Hospital
Cambridge CB2 2QR, U.K.

INTRODUCTION

The insulin receptor (IR) and type I IGF receptor (IGFR) are widely distributed in mammalian tissues, though varying in concentration between different cell types. The receptors show considerable similarity in primary sequence within a common disulphide-linked $(\alpha\beta)_2$ subunit structure. The insulin receptor-related receptor (IRR), a third member of this receptor family for which a ligand has yet to be identified, apparently has a much more restricted tissue distribution (Shier and Watt, 1992). The IR and IGFR mediate overlapping biological responses for which the intrinsic tyrosine-specific kinase activity of the receptors appears to be essential. These tyrosine kinases are highly homologous and differences in signalling capacities of the IR and IGFR in a given cell background have not been clearly defined.

Although each is the product of a single gene, both the IR and IGFR exhibit structural and functional heterogeneity. This diversity can arise in several ways. We (Soos and Siddle, 1989; Soos et al., 1990) and others (Moxham et al., 1989; Treadway et al., 1989) have demonstrated the existence of hybrid receptors, containing disulphide-linked $\alpha\beta$ halves of both the IR and IGFR in an asymmetric $(\alpha\beta)(\alpha^*\beta^*)$ structure. Hybrids were identified in human placenta and a variety of cell lines by using IR-specific antibodies. Such antibodies react with a significant fraction of ^{125}I-IGF-1 binding sites (Soos and Siddle, 1989; Soos et al., 1990) and precipitate autophosphorylated receptor from IGF-1-stimulated cells (Moxham et al., 1989). Hybrid receptors have also been generated *in vitro* from isolated half-receptors (Treadway et al., 1989). Analogous hybrids have been demonstrated for other families of receptors with tyrosine kinase activity, including the α and β isoforms of the PDGF receptor (Heidaran et al., 1991), and the EGF receptor and neu proto-oncogene (Qian et al., 1992), although in these cases hybrid formation is ligand-dependent and non-covalent.

Current Directions in Insulin-Like Growth Factor Research,
Edited by D. LeRoith and M.K. Raizada, Plenum Press, New York, 1994

145

Heterogeneity of both the IR and IGFR also results from alternative splicing of mRNA. Two isoforms of the IR have been described, differing in the presence or absence of a 12-amino acid sequence close to the carboxyl-terminus of the α-subunit which is encoded by a discrete exon. The isoforms show a small difference in affinity for insulin (Mosthaf et al., 1990; Yamaguchi et al., 1991), and a larger difference in extent of cross-reaction with IGF-1 (Yamaguchi et al., 1993). Two isoforms of the IGFR have also been reported (Yee et al., 1989), in this case arising from differential use of adjacent splice acceptor sites, and resulting in sequence variation in the extracellular portion of the β-subunit close to the transmembrane domain. The extent to which the assembly and properties of hybrids may depend on IR or IGFR isoforms is not known, although both isoforms of the IR form hybrids in transfected fibroblasts (Yamaguchi et al., 1993).

Tissue-specific patterns of glycosylation may also influence receptor size (Heidenreich et al., 1983; McElduff et al., 1985; Ota et al., 1988). It is not known whether there are significant functional differences as a consequence of such glycosylation differences, although mutational analysis suggests that gross alteration of IR β-subunit glycosylation does impair function (Leconte et al., 1992).

Other types of heterogeneity have been described for which the molecular basis is less clear. Several groups have reported that the IGFR β-subunit in various cell types appears as a doublet following stimulation of autophosphorylation by IGF-1 (Alexandrides and Smith, 1989; Garofalo and Rosen, 1989; Hainaut et al., 1991; Garofalo and Barenton, 1992). Together with immunoreactivity data, these observations have been taken to indicate the existence of two distinct IGFRs. The structural relationship between the two forms is unclear, although some of the observations are reminiscent of those which were interpreted as evidence for IR/IGFR hybrid receptors (Moxham et al., 1989).

Functional heterogeneity of receptors, in terms of ligand binding specificity, has been reported for both the IR and IGFR. In various cell types, subfractions of IR with relatively high affinity for IGFs (Jonas et al., 1989; Jonas and Cox, 1990), or of IGFR with relatively high affinity for insulin (Burant et al., 1987; Waldbillig and Chader, 1988; Milazzo et al., 1992) have been detected. The structural basis for this heterogeneity is not known, but it is unlikely to reflect known splice variants or differences in primary sequence. In fact, atypical IR have been described in cells transfected with cloned IR cDNA (Jonas et al., 1990).

The physiological significance of receptor subtypes remains unclear, although hybrid and atypical receptors are of potential importance for the sensitivity and specificity of cellular responses to insulin and IGF-1. As yet, little is known concerning the tissue distribution and functional properties of these receptor subtypes, and the aim of the present work was to provide such information.

PRODUCTION AND PROPERTIES OF IGFR MONOCLONAL ANTIBODIES

A panel of IR-specific monoclonal antibodies (Soos et al., 1986) was originally used to identify hybrid receptors in human tissues and transfected cells (Soos and Siddle, 1989; Soos et al., 1990). To complement the IR antibodies as probes for the further study of classical, hybrid and atypical receptors, we produced a similar panel of IGFR-specific monoclonal antibodies (Soos et al., 1992, 1993). These antibodies were obtained using as immunogen either transfected mouse fibroblasts expressing high levels of human IGFR (IGF-1R/3T3 cells) or conjugates of a peptide corresponding to the carboxyl-terminal 15 amino acids of the human IGFR β-subunit. Antibodies recognising at least 8 distinct epitopes were obtained, with various effects on receptor function (Table 1). The epitopes were localised within different receptor domains by testing reactivity of antibodies with chimeric constructs, in which portions of IR sequence were replaced with corresponding IGFR sequence or *vice versa* (Soos et al., 1993; Schumacher et al., 1993).

Table 1. Properties of the principal IGFR monoclonal antibodies

Antibody	Epitope	Binding of ^{125}I-IGF-1	Species cross-reaction		
			Rat	Rabbit	Pig
4-52	62-184	↑↑	-	Nd	Nd
16-13	62-184	↑↑	-	Nd	Nd
24-60	184-283	↓↓	-	±	±
24-31	283-240	0	-	±	Nd
26-3	283-440	↑↑	-	-	Nd
24-55	440-514	↓	-	-	Nd
24-57	440-514	↓↓	-	+	±
17-69	514-586	↓	-	+	+
1-2	1323-1337	0	+	Nd	Nd

Species and receptor specificity of antibodies

The IGFR antibodies, like the IR antibodies, exhibit considerable species specificity, and none of the antibodies recognising extracellular epitopes cross-reacts with rat receptors. This specificity presumably arises because extracellular structures common to human and rat/mouse receptors are not readily immunogenic in mice. Sequence differences between human and mouse receptors therefore play a large part in determining the epitopes for which monoclonal antibodies may be obtained. Indeed, the precise antibody binding sites must include residues which differ between human and rodent receptors within the broad regions defined by studies with chimeric receptors. The full sequence of the rat IGFR has not been published, so that detailed comparison with the human IGFR is not possible. However, the number of differences between human and rodent IR sequences is small, there being 95% sequence identity in the extracellular portion (Flores-Riveros et al., 1989; Goldstein and Dudley, 1990), and it would be anticipated that human and rodent IGFRs share a similarly high level of sequence identity.

Notwithstanding the lack of reaction with rat receptors, some IGFR antibodies do react with receptors from other species. Thus antibody 17-69 cross-reacts well with both rabbit and pig receptors, while 24-57 reacts well with rabbit but only poorly with pig receptors (Table 1). Similarly, we previously identified IR antibodies such as 83-7 which react with rabbit and pig receptors (Soos et al., 1986). Furthermore, the anti-peptide antibody 1-2 which recognises the carboxyl-terminal sequences of the IGFR β-subunit does cross-react fully with rat receptors (Soos et al., 1993). An analogous anti-peptide antibody CT-1, directed against the carboxyl-terminus of the human IR β-subunit, likewise cross-reacts with rat IR (Ganderton et al., 1993).

The IGFR antibodies showed no reaction with human IR expressed in NIH3T3 HIR3.5 transfected cells (Soos et al., 1992). Similarly, IR antibodies were unreactive with human IGFR in IGF-1R/3T3 transfected cells (Soos et al., 1990). Although both panels of antibodies are receptor-specific, the epitopes for most antibodies are accessible in IR/IGFR hybrids (Soos et al., 1990, 1993). However, the IGFR antibodies 24-55 and 25-57 react only poorly with hybrids compared to classical type I IGFR, a property which has proved useful in the isolation of hybrid receptors (see below). The epitopes for these antibodies (within the region 440-514) may be close to the interface between adjacent α-subunits which is distinct in hybrid receptors. Mutational analysis of the IR has also implicated the region around Lys 460 in interactions between α-subunits (Kadowaki et al., 1990), and Cys 524 has been proposed as a site of inter-subunit disulphide cross-linking (Schaffer and Ljungqvist, 1992).

Effects of antibodies on ligand binding

Anti-receptor antibodies have a variety of effects on ligand binding, depending on individual epitopes (Table 1). Antibodies interacting with several distinct regions of the IGFR inhibit IGF-1 binding (Soos et al., 1992; Schumacher et al., 1993). The epitope for the strongly inhibitory antibody 24-60 is located in the cysteine-rich domain between residues 184-283. This antibody is very similar in properties and epitope to the previously described monoclonal antibody αIR-3 (Kull et al., 1983; Gustafson and Rutter, 1990). A number of studies have shown that the cysteine-rich domain (exon 3) of the IGFR is necessary for high affinity IGF-1 binding, indicating that residues within this domain make an important contribution to the IGF-1 binding site (Gustafson and Rutter, 1990; Kjeldsen et al., 1991; Zhang and Roth, 1991a; Schumacher et al., 1991). It would be consistent with such studies if antibodies such as 24-60 and αIR-3 inhibited IGF-1 binding by direct competition for a common binding site. However, the inhibitory effect of antibody 24-60 on IGF-1 binding is appreciably less for solubilised receptors than for intact cells (Soos et al., 1992). Moreover, αIR-3 does not inhibit binding of IGF-2 to the IGFR (Steele-Perkins and Roth, 1990), nor does it inhibit IGF-1 binding to a chimeric construct consisting of the IGFR exon 3 in place of the corresponding IR sequence (Zhang and Roth, 1991a). Such data suggest that these antibodies act indirectly and selectively to decrease IGF-1 binding affinity, perhaps by changing the conformation of the receptor, rather than by direct steric blockade.

Interaction of antibodies with the carboxyl-terminal half of the IGFR α-subunit can also inhibit IGF-1 binding (Table 1). For instance, the epitope for the stongly inhibitory antibody 24-57 lies within the region 440-514, while weakly inhibitory antibodies such as 24-55 and 17-69 have epitopes in the same or adjacent regions. A similar region of the IR, between residues 450-601, contains the epitopes for all the IR antibodies which inhibit insulin binding (Prigent et al., 1990; Gustafson and Rutter, 1990; Zhang and Roth, 1991b). This region may therefore contribute to high affinity ligand binding in both receptors, as suggested also by studies with IR/IGFR chimeras (Schumacher et al., 1993). However an IR construct in which residues 486-569 have been deleted binds insulin with relatively high affinity (Kadowaki et al., 1990), indicating that the primary determinants of ligand binding lie elsewhere.

Several IGFR antibodies stimulate IGF-1 binding, especially to intact cells (Table 1). Two such antibodies, 4-52 and 16-13, recognise epitopes close to the amino-terminus of the α-subunit (residues 62-184), while another, 26-3, reacts in the region 283-440 (Soos et al., 1992; Schumacher et al., 1993). Although these regions are not contiguous in the primary sequence they may be close to each other in the native molecule, since they largely correspond to the L1 and L2 domains which in the IR have been suggested to associate as a pseudosymmetrical dimer (Bajaj et al., 1987; Schaefer et al., 1992). Antibodies which interact at corresponding regions of the IR have not been identified. However, in the case of the IR, antibodies 83-7 and 18-146 which stimulate insulin binding (Soos et al., 1986) react with the cysteine-rich region of the receptor (Schaefer et al., 1990; Zhang and Roth, 1991b). The effect of such antibodies contrasts markedly with the inhibition of ligand binding produced by antibodies recognizing the cysteine-rich region of the IGFR. Thus the IR and IGFR show several differences with respect to the relationship between antibody epitopes and effects on ligand binding. These observations lend support to the notion that different aspects of the ligand binding sites of the IR and IGFR are responsible for high affinity binding of insulin and IGF-1 respectively (Kjeldsen et al., 1991; Schumacher et al., 1991). However it has to be said that the number of antibodies available for study is limited, and as yet no epitopes have been mapped in sufficient detail to be certain that the IR and IGFR respond differently to binding of antibodies at precisely equivalent sites.

INSULIN/IGF RECEPTOR HYBRIDS

Although hybrid receptors have been detected in cell lines derived from various tissues, there is as yet no information regarding their presence in normal tissues apart from placenta. Moreover, it has not so far been possible to study the properties of hybrid receptors in the absence of contaminating IR and IGFR. The availability of large panels of IR and IGFR antibodies has allowed us to select antibodies suitable for use in studying the distribution of hybrids in animal tissues, and to purify hybrids from human sources.

Tissue Distribution of Hybrid Receptors

To circumvent difficulties in obtaining fresh human tissues, we have used IR- and IGFR-specific antibodies which cross-react with rabbit receptors to study the occurrence of hybrids in normal rabbit tissues (Fig. 1). In all tissues studied, IR antibodies precipitated an appreciable fraction of the total receptor-bound IGF-1, ranging from 40% in liver and spleen to 90% in heart and skeletal muscle. Since there is little difference in the binding affinity of hybrids and type I IGF receptors for IGF-1 (see below), these fractions probably approximate to the proportion of total IGFR halves incorporated into hybrids. Thus we conclude that in many tissues the concentration of hybrids may actually be greater than that of classical type I receptors. The fraction of total IR halves which are incorporated into hybrids has yet to be established, as hybrids bind insulin with a lower affinity than normal insulin receptors (see below).

The proportion of IGF-1 binding attributable to hybrid receptors did not correlate well with the relative amounts of ^{125}I-insulin and ^{125}I-IGF-1 binding in different tissues. The highest ratio of ^{125}I-insulin to ^{125}I-IGF-1 binding was in liver, which had the lowest proportion of hybrids. It might have been expected that an excess of IR would favour

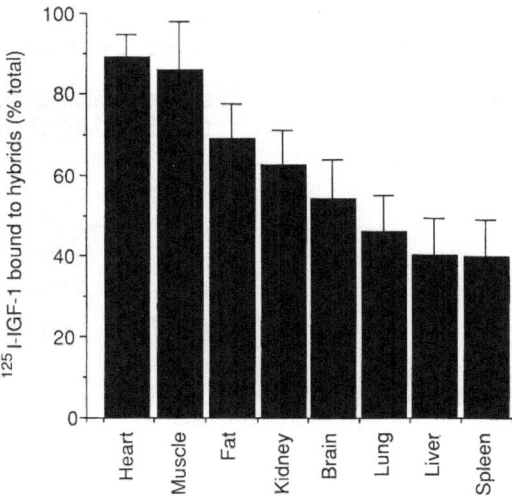

Figure 1.   Fresh rabbit tissues were homogenised in 50 mM Hepes/150 mM NaCl buffer pH 7.6. Microsomal membrane fractions were prepared and solubilised in 1% Triton X-100. Solubilised membranes were incubated with ^{125}I-IGF-1 before addition of monoclonal antibody and precipitation with sheep anti-(mouse IgG) immunoadsorbent. ^{125}I-IGF-1 bound to hybrids (determined using antibody IR 83-7) is expressed as a percentage of the total ^{125}I-IGF-1 binding to hybrids and type I receptors (determined using antibody IGFR 17-69).

incorporation of IGFR into hybrids, so at first sight this observation implies that the formation of hybrids is not a simple function of the relative concentrations of IR and IGFR. However many of the tissues examined contain multiple cell types, and the distribution of receptors amongst these is unlikely to be uniform. More sophisticated studies on pure cell populations will be necessary to determine whether hybrid formation depends only on the concentration of individual receptors, and to estimate the relative efficiency of formation of hybrid compared to classical receptors. Studies of ligand-induced hybrid formation *in vitro* suggest that IR and IGFR halves combine near randomly, with no marked preference for homologous rather than heterologous assembly (Treadway et al., 1989). The simplest model for IR/IGFR hybrid formation *in vivo* is that assembly occurs during postranslational modification and maturation of the individual receptors. It is believed that dimerisation and disulphide formation between IR halves occurs in the endoplasmic reticulum, perhaps facilitated by protein disulphide isomerase (Olson et al., 1988). Alternatively, the possibility cannot be ruled out that disulphide and subunit interchange is also possible between mature receptors in the plasma membrane, where ligand binding and other regulatory factors might influence hybrid formation. It has been reported that free half-receptors are present in plasma membranes (Koch et al., 1986; Fujita-Yamaguchi and Harmon, 1988; Tollefsen and Thompson, 1988), although these have not been apparent in other studies (Boni-Schnetzler et al., 1986; Morrison et al., 1988). Further work is required to determine the site of hybrid assembly *in vivo*, and to investigate the possible regulation of this process.

Purification and Binding Properties of Hybrid Receptors

To study the properties of hybrids without interference from other receptor types, we have purified hybrids from detergent solubilised placental membranes using a two-step immuno-affinity method (Soos et al., 1993). Classical type I IGF receptors were first removed using the IGFR antibody 24-55, which reacts poorly with hybrids, leaving insulin receptors and the bulk of hybrids in the supernatant. Hybrid receptors were then further purified by binding to the IGFR-specific anti-peptide antibody 1-2, followed by elution with free peptide (Fig 2). For comparison, classical type I IGF receptors and insulin receptors

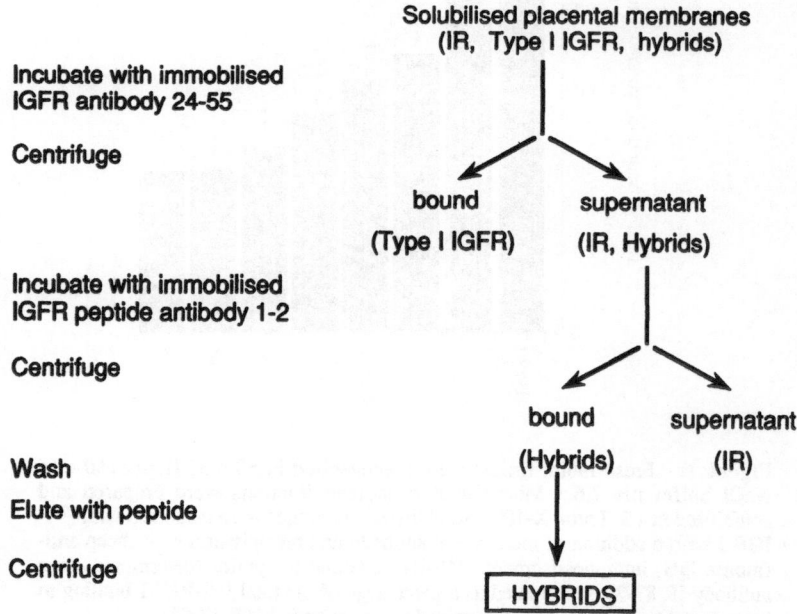

Figure 2. Hybrid receptors were purified from solubilised placental microsomal membranes by sequential use of two monoclonal antibodies.

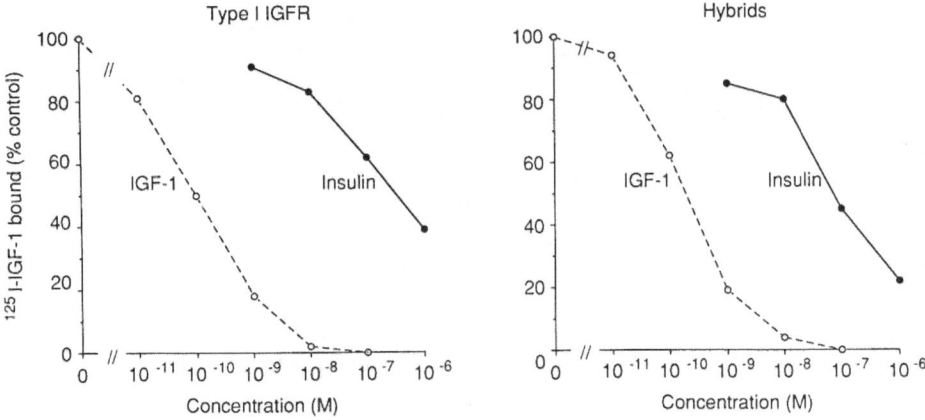

Figure 3. Binding of ^{125}I-IGF-1 to purified type I IGF receptors and hybrids was measured in the presence of the indicated concentrations of unlabelled IGF-1 or insulin. Specific receptor-bound ^{125}I-IGF-1 is expressed as a percentage of that in the absence of unlabelled ligand.

were isolated by similar procedures using appropriate antibodies. The purified hybrid receptor preparation was essentially free of classical receptors as judged by the ability of both IR-specific and IGFR-specific antibodies to deplete binding of both ^{125}I-IGF-1 and ^{125}I-insulin.

We found that hybrids purified from human placenta bind IGF-1 but not insulin with high affinity (Soos at al., 1993). Hybrids bind IGF-1 with comparable affinity to classical type I IGF receptors, as judged by the IC$_{50}$ for displacement of ^{125}I-IGF-1 binding (Fig. 3). However, the affinity of hybrids for insulin is about 10-fold lower than that of insulin receptors for insulin (Fig. 4). As a result, hybrids bind approx. 20 times as much IGF-1 as insulin at tracer concentrations. Since the concentration of insulin necessary to inhibit insulin binding to hybrids (Fig. 4) is much lower than that required to inhibit IGF-1 binding (Fig. 3), it is likely that insulin is binding to the IR half of hybrids, albeit with low affinity.

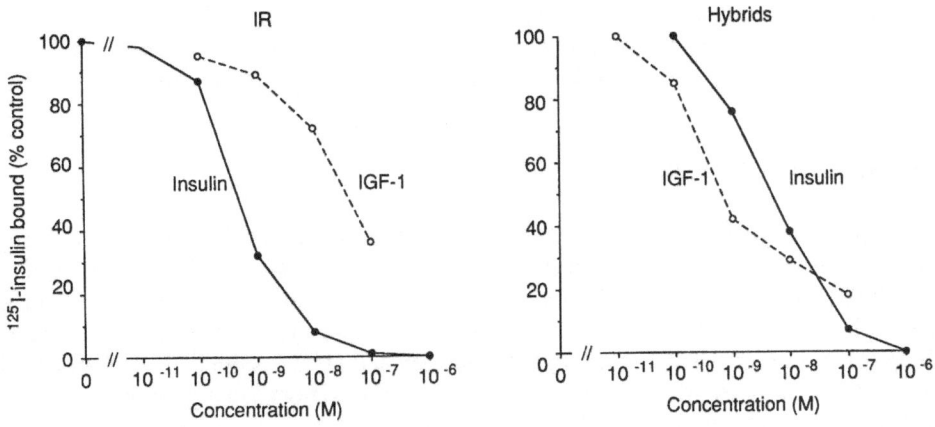

Figure 4. Binding of ^{125}I-insulin to purified insulin receptors and hybrids was measured in the presence of the indicated concentrations of unlabelled IGF-1 or insulin. Specific receptor-bound ^{125}I-insulin is expressed as a percentage of that in the absence of unlabelled ligand.

A further indication of the low affinity of insulin binding to hybrids is provided by the observation that ^{125}I-insulin binding increased 1.8-fold when hybrids were subjected to mild reducing agents to liberate individual half receptors (Soos et al., 1993). By contrast, binding to purified IR decreased by 75% after disulphide reduction. Thus, the affinity of hybrids for insulin is even less than that of isolated IR halves. These observations provide further evidence that the presence of two IR α-subunits is necessary for high affinity insulin binding. It remains to be determined whether the association of α-subunits provides an optimal conformation of binding sites wholly contained within each subunit, or whether high affinity binding requires simultaneous contact of insulin with both subunits. In either case, it is clear that an IGFR α-subunit cannot substitute for an IR α-subunit in providing the necessary interactions. On the other hand, binding of IGF-1 to the IGFR shows very little dependence on α-subunit associations, the affinity being similar for type I receptors, hybrids and half-receptors.

Although in our hands purified hybrids clearly bind insulin with lower affinity than IGF-1, it has been suggested that hybrids reconstituted *in vitro* from isolated receptor halves bind comparable amounts of IGF-1 and insulin (Treadway et al., 1989). It is difficult to reconcile these disparate findings at present although they may, at least in part, reflect the use of different ligand concentrations, since the use of higher concentrations would diminish the effects of any affinity difference. It is also possible that hybrids assembled *in vitro* differ from purified hybrids in, for example, the formation of disulphide bonds.

Heterologous binding competition studies reveal further interesting properties of hybrids. Binding of ^{125}I-insulin is displaced by low concentrations of IGF-1, appropriate for interaction with the IGFR half receptor (Fig. 4). This observation implies that the affinity of the IR half receptor within a hybrid is modulated by occupancy of the IGFR half, analogous to negative cooperativity within the IR itself. By contrast, very high concentrations of insulin are required to displace binding of ^{125}I-IGF-1 to hybrids and to type I receptors (Fig. 3), suggesting the IGFR half receptor is not susceptible to trans-modulation by the IR half. This is consistent with the lack of reports of negative cooperativity of IGF-1 binding to the type I receptor.

The autophosphorylation of purified hybrids in response to IGF-1 and insulin remains to be examined. In intact cells, it appears that low concentrations of IGF-1 induce autophosphorylation of both IR and IGFR β-subunits within hybrids (Moxham et al., 1989). This is consistent with the mechanism of autophosphorylation within the IR itself, which involves predominantly an intramolecular "trans" reaction between adjacent β-subunits (Frattali et al., 1992). However, in the IR this reaction appears to be asymmetric, phosphorylation occurring preferentially on the unoccupied αβ half of a receptor with a single molecule of bound ligand (Lee et al., 1993). It is possible therefore that, although hybrid receptors preferentially bind IGF-1 rather than insulin, this binding will result in preferential autophosphorylation and activation of the insulin half-receptor rather than the IGF half receptor. The availability of purified hybrids will allow us to test this possibility.

ATYPICAL RECEPTORS

Atypical receptors which bind both IGF-1 and insulin with relatively high affinity have been described in several cell types (Misra et al., 1986; Jonas et al., 1986; Burant et al., 1987; Waldbillig and Chader, 1988; Jonas et al., 1989; Jonas and Cox, 1990; Milazzo et al., 1992). The detection of atypical receptors frequently depends on differences in apparent affinity for insulin or IGF-1 according to the radioligand (^{125}I-insulin or ^{125}I-IGF) used in competition binding assays. In such instances, atypical receptors can represent only a small subfraction of total receptors. Although the binding properties of atypical receptors in some respects resemble those of IR/IGFR hybrids, atypical and hybrid receptors are clearly distinct by immunological criteria. While hybrids react with both IR-specific and IGFR-specific antibodies, atypical receptors should show specificity in their reaction with receptor antibodies. In fact, atypical binding specificity has been reported for subfractions of both the IR and IGFR in various cell types.

Atypical receptors in placenta and MCF-7 cells

We have confirmed the presence of atypical insulin receptors in purified insulin receptor preparations (Soos et al 1993), as previously reported by others (Jonas et al., 1986). Insulin receptors purified from placenta bind more ^{125}I-IGF-1 than expected from the cross-reactivity of insulin receptors with IGF-1. Moreover binding of ^{125}I-IGF-1 is displaced by low concentrations of insulin (IC_{50} 0.2 nM) as well as IGF-1 (IC_{50} 3 nM) and is depleted by IR-specific but not by IGFR-specific antibodies. These receptor preparations therefore show the characteristic properties of atypical insulin receptors (Jonas et al., 1989; Jonas and Cox, 1990; Jonas et al., 1990).

In a human breast cancer cell line MCF-7 the majority of insulin binding has been reported to be to atypical IGF receptors (Milazzo et al., 1992). We confirmed that ^{125}I-insulin binding to solubilised MCF-7 cells is effectively displaced by both IGF-1 (IC_{50} 0.1 nM) and insulin (IC_{50} 2 nM), and that insulin binding activity in these cells is removed by IGFR-specific but not by IR-specific antibodies. These observations indicate that most of the insulin binding in these cells is to atypical IGF receptors and not to hybrid or insulin receptors. However, in our hands the level of insulin binding to intact MCF-7 cells was very low, and the atypical IGFR binding specificity was only apparent after solubilisation (M.A. Soos and K. Siddle, unpublished).

We conclude therefore that subfractions of both the IR and IGFR can exhibit atypical binding specificity, with anomalously high affinity for heterologous ligand. These atypical receptors are distinct from hybrids in their recognition only by IR-specific or IGFR-specific antibodies but not both.

Atypical IGF receptors IGF-1R/3T3 cells

The immunological data imply that atypical receptors are structurally very closely related to the corresponding classical receptors. Furthermore, atypical insulin receptors have been detected in transfected fibroblasts as a product of cloned IR cDNA (Jonas et al., 1990). We have similarly detected atypical behaviour of type I IGF receptors solubilised from IGF-1R/3T3 cells (mouse fibroblasts transfected with human IGFR cDNA). We reported that binding of ^{125}I-insulin to receptors solubilised from IGF-1R/3T3 cells was displaced by low concentrations of IGF-1 (IC_{50} 0.1 nM) as well as insulin (IC_{50} 1 nM) and that the binding activity was removed by two IGFR antibodies (Soos et al., 1990). At that time we interpreted these data as evidence for hybrids between human IGFR and endogenous mouse IR, which bind both insulin and IGF-1 with high affinity. Unfortunately, we were unable to use our IR antibodies to distinguish between hybrids and atypical IGFR in this case, as these antibodies do not react with mouse insulin receptors. As with MCF-7 cells, a significant level of ^{125}I-insulin binding was detected only for solubilised receptors and not intact cells. The similarities in behaviour of receptors from IGF-1R/3T3 and MCF-7 cells lead us now to believe that most of the insulin binding detected with solubilised IGF-1R/3T3 cells is to atypical IGF receptors rather than to hybrid receptors.

This view is reinforced by effects of IGFR antibodies on insulin binding to both cell types. Insulin binding to intact IGF-1R/3T3 cells is dramatically increased by some IGFR antibodies such as 16-13 and 26-3, so that the binding affinity is then comparable to that of *bona fide* insulin receptors (Soos et al., 1992). Moreover in the presence of the antibodies, insulin binding is displaced by low concentrations of both IGF-1 (IC_{50} 0.2 nM) and insulin

(IC$_{50}$ 2 nM) and the binding properties thus closely resemble those of atypical IGF receptors. The antibodies have a much smaller effect on insulin binding to solubilised IGF-1R/3T3 receptors, suggesting that the effect on intact cells is mimicking that of solubilisation. MCF-7 cells behave in all respects in a qualitatively similar way to IGF-1R/3T3 cells in their responses to binding-stimulatory IGFR antibodies (M.A. Soos and K. Siddle, unpublished). However, as far as we have determined, none of our IR antibodies affect IGF-1 binding to the IR in a comparable fashion.

The structural basis for these phenomena is completely obscure, and it cannot yet be assumed that the same molecular mechanisms are responsible for atypical binding properties of insulin receptors and IGF receptors. Possibly a subfraction of receptors is modified by limited proteolysis, differential glycosylation or disulphide reduction in a way which directly alters binding properties or renders them susceptible to modulation by antibody or solubilisation. It is interesting that certain IR/IGFR chimeras are also capable of binding both insulin and IGF-1 with high affinity, suggesting that distinct residues are involved in interactions with the two ligands (Kjeldsen et al., 1991; Schumacher et al., 1991). It may be that receptor specificity is dependent not only on the presence of essential positive interactions but also on negative constraints which impede the binding of cross-reacting ligand. The changes in binding specificity seen in atypical receptors could result from the loss of such inhibitory constraints as much as the provision of additional positive interactions.

CONCLUSIONS

Hybrid and atypical receptors are of undoubted interest because of the insights they provide into the structure and function of normal receptors. An equally important issue, however, is whether such receptor subtypes are physiologically important. We have shown that IR/IGFR hybrids are widely distributed in mammalian tissues, and that in some tissues they may be more abundant than classical type I IGF receptors. It will be a more difficult task to document the occurrence of atypical receptors, which may be more abundant than previously suspected but masked by classical receptors. Thus for example, insulin binding to atypical IGF receptors will only be detected in cells with few classical insulin receptors.

The binding properties of insulin receptors and IGF receptors as classically described are such that each would be expected to show near absolute specificity for its own ligand at physiological concentrations of insulin and IGF-1. It is tempting to speculate that hybrid and atypical receptors provide mechanisms to facilitate "specificity spillover", permitting insulin to activate IGFR signalling pathways or IGF-1 to exert insulin-like effects. For example, such "specificity spillover" has been postulated to underly paradoxical hyper-responsiveness of certain tissues to insulin in the face of generally prevailing insulin resistance (Geffner and Golde, 1988; Poretsky, 1991). In this context, an important unanswered question is whether there are in fact significant differences in signalling capacity of the IR and IGFR intracellular domains. Some difference in relative potency for mediating growth-promoting effects has been reported (Lammers et al., 1989), but otherwise the two receptors seem to mediate remarkably similar effects in a given cell background (Shimizu et al., 1986; Weiland et al., 1991). Further studies in this area are clearly warranted.

The situation is further complicated by the observation that the properties of hybrid receptors are not simply the sum of their component halves. As purified from placenta, hybrids have a relatively low affinity for insulin, suggesting that they would be poorly responsive to insulin at physiological concentrations. Rather than providing a mechanism for insulin to activate IGF-like responses, it may be therefore that formation of hybrids serves to sequester insulin receptors in a relatively unresponsive form.

For atypical receptors also, the situation is not straightforward. At least for IGF receptors, atypical binding properties were only apparent in the present study after receptor solubilisation or addition of anti-receptor antibodies. The effect of several different monoclonal antibodies to dramatically stimulate insulin binding to IGFR in intact cells invites the speculation that there may be other serum or tissue proteins capable of inducing such modulation *in vivo*. Certainly there remain many intriguing aspects of the structure, function and physiological role of both hybrid and atypical receptors which require further research.

Acknowledgments

We are grateful to the Wellcome Trust and the British Diabetic Association for financial support, and to Dr. J. Whittaker and Dr. A. Ullrich for the generous gift of transfected cells.

REFERENCES

Alexandrides, T.K., and Smith, R.J., 1989, A novel fetal insulin-like growth factor (IGF) I receptor: mechanism for increased IGF I- and insulin-stimulated tyrosine kinase activity in fetal muscle, *J. Biol. Chem.*, 264:12922-12930.

Bajaj, M., Waterfield, M.D., Schlessinger, J., Taylor, W.R., and Blundell, T., 1987, On the tertiary structure of the extracellular domains of the epidermal growth factor and insulin receptors, *Biochim Biophys Acta* 916:220-226.

Boni-Schnetzler, M., Scott, W., Waugh, S.E., DiBella, E., and Pilch, P.F., 1987, The insulin receptor: structural basis for high affinity ligand binding, *J. Biol. Chem.* 262:8395-8401.

Burant, C.F., Treutelaar, M.K., Allen, K.D., Sens, D.A., and Buse, M.G., 1987, Comparison of insulin and insulin-like growth factor I receptors from rat skeletal muscle and L-6 myocytes, *Biochem. Biophys. Res. Comm.* 147:100-107.

Flores-Riveros, J.R., Sibley, E., Kastelic, T., and Lane, M.D., 1989, Substrate phosphorylation catalyzed by the insulin receptor tyrosine kinase; kinetic correlation to autophosphorylation of specific sites in the β subunit, *J. Biol. Chem.*, 264:21557-21572.

Frattali, A.L., Treadway, J.L., and Pessin, J.E., 1992, Transmembrane signaling by the human insulin receptor kinase: relationship between intramolecular β-subunit *trans*- and *cis*-autophosphorylation and substrate kinase activation, *J. Biol. Chem.* 267:19521-19528.

Fujita-Yamaguchi, Y., and Harmon, J.T., 1988, A monomer-dimer model explains the results of radiation activation: binding characteristics of insulin receptor purified from human placenta, *Biochemistry* 27:3252-3260.

Ganderton, R.H., Stanley, K.K., Field, C.E., Coghlan, M.P., Soos, M.A., and Siddle, K., 1992, A monoclonal anti-peptide antibody reacting with the insulin receptor β-subunit: characterization of the antibody and its epitope and use in immunoaffinity purification of intact receptors, *Biochem. J.* 288:195-205.

Garofalo, R.S., and Rosen, O.M., 1989, Insulin and insulinlike growth factor 1 (IGF-1) receptors during central nervous system development: expression of two immunologically distinct IGF-1 receptor β subunits, *Mol. Cell. Biol.*, 9:2806-2817.

Garofalo, R.S., and Barenton, B., 1992, Functional and immunological distinction between insulin-like growth factor I receptor subtypes in KB cells, *J. Biol. Chem.*, 267:11470-11475.

Geffner, M.E., and Golde, D.W., 1988, Selective insulin action on skin, ovary and heart in insulin-resistant states, *Diabetes Care* 11:500-505.

Goldstein, B.J., and Dudley, A.L., 1990, The rat insulin receptor: primary structure and conservation of tissue-specific alternative messenger RNA splicing, *Molec. Endocrinol.*, 4:235-244.

Gustafson, T.A., and Rutter, W.J., 1990, The cysteine-rich domains of the insulin and insulin-like growth factor I receptors are primary determinants of hormone binding specificity: evidence from receptor chimeras, *J. Biol. Chem.* 265:18663-18667.

Hainaut, P., Kowalski, A., Giorgetti, S., Baron, V., and Van Obberghen, E., 1991, Insulin and insulin-like-growth-factor-I (IGF-I) receptors in *Xenopus laevis* oocytes: comparison with insulin receptors from liver and muscle, *Bichem. J.* 273:673-678.

Heidaran, M.A., Pierce, J.H., Yu, J.C., Lombardi, D., Artrip, J.E., Fleming, T.P.,Thomason, A., and Aaronson, S.A., 1991, Role of αβ receptor heterodimer formation of β platelet-derived growth factor receptor activation by PDGF-AB, *J. Biol .Chem.* 266:20232-20237.

Heidenreich, K.A., Zahiser, N.R., Berhanu, P., Brandenburg, D., and Olefsky, J.M., 1983, Structural differences between insulin receptors in the brain and peripheral target tissues, *J. Biol. Chem.* 258:8527-8530.

Jonas, H.A., and Cox, A.J., 1990, Insulin-like growth factor binding to the atypical insulin receptors of a human lymphoid-derived cell line (IM-9), *Biochem. J.* 266:737-742.

Jonas, H.A., Newman, J.D., and Harrison, L.C., 1986, An atypical insulin receptor with high affinity for insulin-like growth factors copurified with placental insulin receptors, *Proc. Natl. Acad. Sci. U.S.A.* 83:4124-4128.

Jonas, H.A., Cox, A.J., and Harrison, L.C., 1989, Delineation of atypical insulin receptors from classical insulin and Type I insulin-like growth factor receptors in human placenta, *Biochem. J.* 257:101-107.

Jonas, H.A., Eckardt, G.S., and Clark, S. 1990, Expression of atypical and classical insulin receptors in Chinese hamster ovary cells transfected with cloned cDNA for the human insulin receptor, *Endocrinology* 127:1301-1309.

Kadowaki, H., Kadowaki, T., Camra, A., Marcus-Samuels, B., Rovira, A., Bevins, C.L., and Taylor, S., 1990, Mutagenesis of lysine 460 in the human insulin receptor: effects upon receptor recycling and cooperative interactions among binding sites, *J. Biol. Chem.* 265:21285-21296.

Kjeldsen, T., Andersen, A.S., Wiberg, F., Rasmussen, J.S., Schafer, L., Balschmidt, P., Moller, K.B., and Moller, N.P., 1991, The ligand specificities of the insulin receptor and the insulin-like growth factor-1 receptor reside in different regions of a common binding site, *Proc. Natl. Acad. Sci. U.S.A.* 88:4404-4408.

Koch, R., Deger, A., Jack, J.M., Klotz, K.N., Schenzle, D., Kramer, H., Kelm, S., Muller, G., Rapp, R., and Weber, U., 1986, Characterization of solubilized insulin receptors from rat liver microsomes: existence of two receptor species with different binding properties, *E. J. Biochem.* 154:281-287.

Kull, F.C., Jacobs, S., Su, Y.-F., Svoboda, M.E., van Wyk, J.J., and Cuatrecasas, P., 1983, Monoclonal antibodies to receptors for insulin and somatomedin-C, *J. Biol. Chem.* 258:6561-6566.

Leconte, I., Auzan, C., Debant, A., Rossi, B., and Clauser, E., 1992, N-linked oligosaccharide chains of the insulin receptor β subunit are essential for transmembrane signalling, *J. Biol. Chem.*, 267:17415-17423.

Lee, J., O'Hare, T., Pilch, P.F., and Shoelson, S.E.,1993, Insulin receptor autophosphorylation occurs asymmtrically, *J. Biol. Chem.* 268:4092-4098.

McElduff, A., Grunberger, G., and Gorden, P., 1985, An alteration in apparent molecular weight of the insulin receptor from the human monocyte cell line U-937, *Diabetes* 34:686-690.

Milazzo, G., Yip, C.C., Maddux, B., Vigneri, R., and Goldfine, I.D., 1992, High-affinity insulin binding to an atypical insulin-like growth factor-I receptor in human breast cancer cells, *J. Clin. Invest.* 89:899-908.

Misra, P., Hintz, R.L., and Rosenfeld, R., 1986, Structure and immunological characterization of insulin-like growth factor II binding to IM-9 cells, *J. Clin. Endocrinol. Metab.* 63:1400-1405.

Mosthaf, L., Grako, K., Dull, T.J., Coussens, L., Ullrich, A., and McClain D.A., 1990, Functionally distinct insulin receptors generated by tissue-specific alternative splicing, *EMBO J.*, 9:2409-2413.

Morrison, B.D., Swanson, M.L., Sweet, L.J., and Pessin, J., 1988, Insulin-dependent covalent reassociation of isolated $\alpha\beta$ heterodimeric insulin receptors into an $\alpha_2\beta_2$ heterotetrameric disulfide-linked complex, *J. Biol. Chem.* 263:7806-7813.

Moxham, C.P., Duronio, V., and Jacobs, S., 1989, Insulin-like growth factor I receptor β-subunit heterogeneity: evidence for hybrid tetramers composed of insulin-like growth factor I and insulin receptor heterodimers, *J. Biol. Chem.* 264:13238-13244.

Olson, T.S., Bamberger, M.J., and Lane, M.D., 1988, Post-translational changes in tertiary and quaternary structure of the insulin pro-receptor: correlation with acquisition of function, *J. Biol. Chem.* 263:7342-7352.

Ota, A., Wilson, G.L., and LeRoith, D., 1988, Insulin-like growth factor I receptors on mouse neuroblastoma cells: two β subunits are derived from differences in glycosylation, *Eur. J. Biochem.* 174:521-530.

Poretsky, L., 1991, On the paradox of insulin-induced hyperandrogenism in insulin-resistant states, *Endocr. Rev.* 12:3-13.

Prigent, S.A., Stanley, K.K., and Siddle, K., 1990, Identification of epitopes on the human insulin receptor reacting with rabbit polyclonal antisera and mouse monoclonal antibodies., *J. Biol. Chem.* 265:9970-9977.

Qian, X., Decker, S.J., and Greene, M.I., 1992, p185[c-neu]and epidermal growth factor receptor associate into a structure composed of activated kinases, *Proc. Natl. Acad. Sci. USA.* 89:1330-1334.

Schaffer, L., and Ljungqvist, L., 1992, Identification of a disulfide bridge connecting the α-subunits of the extracellular domain of the insulin receptor, *Biophys. Res. Comm.* 189:650-653.

Schaefer, E.M., Siddle, K., and Ellis, L., 1990, Deletion analysis of the human insulin receptor ectodomain reveals independently folded soluble subdomains and insulin binding by a monomeric α-subunit, *J. Biol. Chem.* 265:13248-13253.

Schaefer, E.M., Ericjon, H.P., Federwisch, M., Wollmer, A., and Ellis, E., 1992, Structural organisation of the human insulin receptor ectodomain, *J. Biol. Chem.* 267:23393-23402.

Schumacher, R., Mosthaf, L., Schlessinger, J., Brandenburg, D., and Ullrich, A., 1991, Insulin and insulin-like growth factor-1 binding specificity is determined by distinct regions of their cognate receptors, *J. Biol. Chem.* 266:19288-19295.

Schumacher, R., Soos, M.A., Schlessinger, J., Brandenburg, D., Siddle, K., and Ullrich, A., 1993, Signaling-competent receptor chimeras allow mapping of major insulin receptor binding domain determinants, *J. Biol. Chem.* 268:1087-1094.

Shier, P., and Watt, V.M., 1992, Tissue-specific expression of the rat insulin receptor-related receptor gene, *Molec. Endocrinol.* 6:723-729.

Shimizu, M., Webster, C., Morgan, D.O., Blau, H., and Roth, R.A., 1986, Insulin and insulin-like growth factor receptors and responses in cultured human muscle cells, *Am. J. Physiol.* 251:E611-E615.

Soos M.A., and Siddle K., 1989 Immunological relationships between receptors for insulin and insulin-like growth factor-I: evidence for structural heterogeneity of insulin-like growth factor-I receptors involving hybrids with insulin receptors, *Biochem.J* 263:553-563.

Soos, M.A., Siddle, K., Baron, M.D., Heward, J.M., Luzio, J.P., Bellatin, J., and Lennox, E.S., 1986, Monoclonal antibodies reacting with multiple epitopes on the human insulin receptor, *Biochem. J.* 235:199-208.

Soos M.A., Whittaker J., Lammers R., Ullrich A., and Siddle K., 1990, Receptors for insulin and insulin-like growth factor-I can form hybrid dimers: characterization of hybrid receptors in transfected cells, *Biochem. J.* 270:383-390.

Soos, M.A., Field, C.E., Lammers, R., Ullrich, A., Zhang, B., Roth, R.A., Andersen, A., Kjeldsen, T., and Siddle, K., 1992, A panel of monoclonal antibodies for the type I insulin-like growth factor receptor: epitope mapping, effects on ligand binding and biological activity, *J. Biol. Chem.* 267:12955-12963.

Soos, M.A., Field, C.E., and Siddle, K., 1993, Purified hybrid insulin/insulin-like growth factor-I receptors bind insulin-like growth factor-I, but not insulin, with high affinity, *Biochem. J.* 290:419-426.

Steele-Perkins, G., and Roth, R.A., 1990, Monoclonal antibody αIR-3 inhibits the ability of insulin-like growth factor II to stimulate a signal from the type I receptor without inhibiting its binding, *Biochem. Biophys. Res. Comm.* 171:1244-1251.

Tollefsen, S.E., and Thompson, K., 1988, The structural basis for insulin-like growth factor I receptor high affinity binding, *J Biol Chem* 263:16267-16273.

Treadway J.L., Morrison B.D., Goldfine I.D., and Pessin J.E., 1989, Assembly of insulin/insulin-like growth factor-1 hybrid receptors in vitro, *J. Biol. Chem.* 264:21450-21453.

Waldbillig, R.J., and Chader, G.J., 1988. Anomalous insulin-binding activity in the bovine neural retina: a possible mechanism for the regulation of receptor binding specificity, *Biochem. Biophys. Res. Comm.* 151:1105-1112.

Weiland, M., Bahr, F., Hohne, M., Schurmann, A., Ziehm, D., and Joost, H.G., 1991, the signalling potential of the receptors for insulin and insulin-like growth factors 1 in 3T3-L1 adipocytes, *J. Cell. Physiol.* 149:428-435.

Yamaguchi, Y., Flier, J.S., Yokota, A., Benecke, H., Backer, J.M., and Moller, D.E., 1991, Functional properties of two naturally occurring isoforms of the human insulin receptor in Chinese hamster ovary cells, *Endocrinology* 129:2058-2066.

Yamaguchi, Y., Flier, J.S., Benecke, H., Ransil, B.J., and Moller, D.E., 1993, Ligand binding properties of the two isoforms of the human insulin receptor, *Endocrinology* 132:1132-1138.

Yee, D., Lebovic, G.S., Marcus, R.R., and Rosen, N., 1989, Identification of an alternate type I insulin-like growth factor receptor β subunit mRNA transcript, *J. Biol. Chem.* 264:21439-21441.

Zhang, B., and Roth, R. A., 1991a, Binding properties of chimeric insulin receptors containing the cysteine-rich domain of either the insulin-like growth factor receptor or the insulin receptor related receptor, *Biochemistry* 30:5113-5117.

Zhang, B.,and Roth, R. A., 1991b, A region of the insulin receptor important for ligand binding (residues 450-601) is recognized by patient's autoimmune antibodies and inhibitory monoclonal antibodies, *Proc. Natl. Acad. Sci. USA.* 88:9858-9862.

INSULIN LIKE GROWTH FACTOR 1 RECEPTOR SIGNAL TRANSDUCTION TO THE NUCLEUS

Steven A. Rosenzweig,[1,4] Barry S. Oemar,[2] Norman M. Law,[3]
Uma T. Shankavaram,[1] and Bradley S. Miller[1]

[1]Department of Cell and Molecular Pharmacology and
Experimental Therapeutics
Medical University of South Carolina
Charleston, SC 29425
[2]Department of Research
Kantonsspital Basel, University of Basel
CH-4031 Basel, Switzerland
[3]Oncology and Virology Research Department
Pharmaceuticals Division
CIBA-GEIGY, Ltd.
CH-4002 Basel, Switzerland
[4]Corresponding Author

INTRODUCTION

The IGF-1 receptor (IGF-1R) is a member of the tyrosine kinase class of cell surface receptors which become autophosphorylated on tyrosyl residues upon ligand binding (Czech, 1989). It has striking homology to the insulin receptor; however, each receptor maintains a unique specificity for its own ligand (Schumacher et al., 1991). The mechanism by which the binding of IGF-1 to its receptor elicits a cellular effect has been the subject of considerable research with a singular cohesive model having yet to be defined. The autophosphorylation of the IGF-1R via subunit transphosphorylation is clearly a necessary requisite for transmission of an intracellular signal, as found for the insulin receptor (Sweet et al., 1987; Ullrich and Schlessinger, 1990). In the case of other growth factor receptors with tyrosine kinase domains such as the EGF and PDGF receptors, receptor activation results in phospholipase C activation leading to 1,2-diacylglycerol and inositol 1,4,5-trisphosphate production and a corresponding increase in protein kinase C (pkC) and calcium mobilization, respectively (Berridge, 1993). Recently, evidence for a direct association between the EGF and PDGF receptors and a number of

Current Directions in Insulin-Like Growth Factor Research,
Edited by D. LeRoith and M.K. Raizada, Plenum Press, New York, 1994

159

key substrates such as phospholipase C-γ, PI-3 kinase and GAP (p21^{ras} GTPase activating protein, Cantley et al., 1991) has been demonstrated. This direct link has also been established for the IGF-1R and PI-3 kinase interactions (Cantley et al., 1991; Yamamoto et al., 1992; Lavan et al., 1992). It was recently reported that in cells stimulated with insulin in the presence of the protein tyrosine phosphatase inhibitor, phenylarsine oxide, 5-10% of the cellular GAP associates with the insulin receptor (Pronk et al., 1992). This implies that p21^{ras} may be a downstream effector of this receptor. Whether this is also true for the IGF-1R is not known. The best characterized phosphotyrosyl substrate of the insulin/IGF-1 receptors is a cytoplasmic protein of ~165-195 kDa collectively referred to as pp185 (White et al., 1985). Pp185 has recently been purified (Rothenberg et al., 1991), cloned and sequenced (Sun et al., 1991). Of interest is the observation that this protein, termed IRS-1 (insulin receptor substrate-1), contains several potential binding sites for SH2- (*src* homology domain-2) containing proteins, but no SH2 or SH3 domains itself (SH3 domains may be potential sites of cytoskeleton (actin) binding or protein:protein interactions; Koch et al., 1991). Suggesting that it may serve as a "docking" protein for other cellular effectors (Sun et al., 1991).

Thus, it would appear that IGF-1 may stimulate nuclear events by initiating a protein effector network, or scaffold at the plasma membrane, which in turn, activates downstream effectors. This signaling mechanism is an area of active investigation and should provide important new insights into the mechanisms whereby the IGF-1R can stimulate cell proliferation. What is still lacking is an understanding of how these upstream effectors regulate nuclear events leading to cell proliferation. To this end, work in our laboratory has demonstrated that the IGF-1R can activate nuclear protein phosphorylation events with one of the targets of this activation being the proto-oncogene protein, c-Jun (Oemar et al., 1991b). This has provided us with a system with which we can work our way upstream from the nucleus through the signal transduction cascade in order to define the effectors involved in IGF-1 signal transduction to the nucleus.

IGF-1R REGULATION AND FUNCTION IN MESANGIAL CELLS

An unexpected finding stemming from our studies on IGF-1R expression in mesangial cells isolated from diabetic mice (*db/db*) was the observation of a discordant effect of glucose on IGF-1R mRNA and protein levels (Oemar et al., 1991a). These analyses revealed that incubation of the *db/db* cells in high glucose (28 mM) containing medium resulted in an increase in the levels of IGF-1R and insulin receptor mRNA (Fig. 1). Whereas insulin receptor protein levels coordinately increased under these conditions, IGF-1R protein levels were found to decrease (Fig. 2). This was not due to selective protein loss as mesangial cells isolated from non-diabetic littermates exhibited lower levels of IGF-1Rs than the diabetic cells (Oemar et al., 1991a). The reason for this dual regulation of IGF-1R mRNA and protein levels by glucose is still not understood. One explanation for this discrepancy is that the *db/db* cells are producing high levels of IGF-1 itself, which, in turn, is stimulating a rapid turnover or down-regulation of IGF-1R protein. This may be even more pronounced in diabetes, as IGF-1 levels are increased (Werner et al., 1990). Consistent with this idea is the increased proliferative response of mesangial cells in type 1 diabetes.

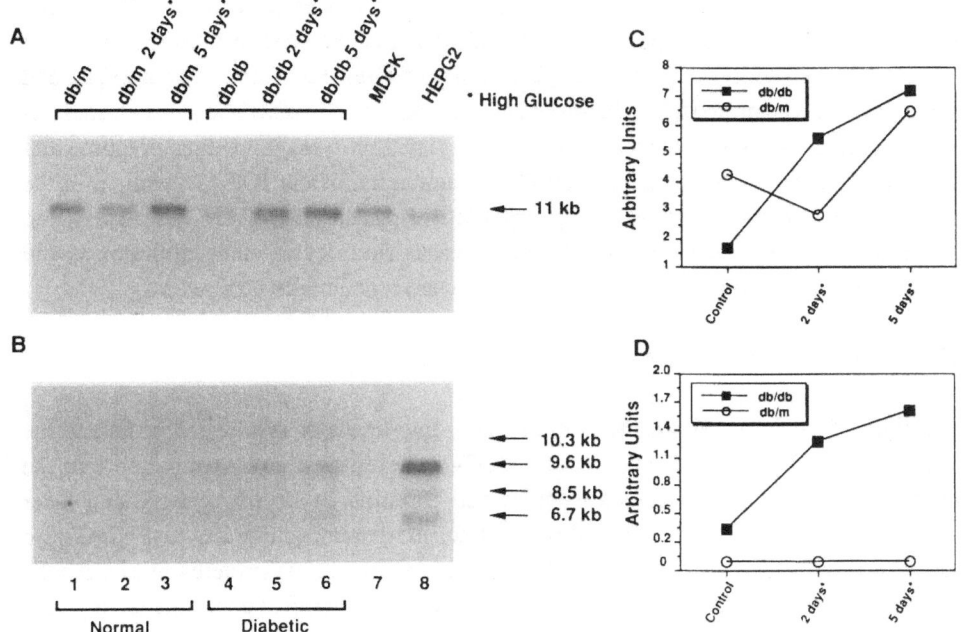

Figure 1. Northern blot analysis of IGF-1R (A) and insulin receptor (B) mRNAs. 20 μg of total cellular RNA was resolved on a 1% formaldehyde-agarose gel and transferred to Zetabind filters. The filters were probed with the appropriate receptor cDNA probe. Quantification of the autoradiographs obtained for each receptor are provided in C and D. Madin-Darby canine kidney (MDCK) cells; *db/db*, diabetic mesangial cells; *db/m*, control mesangial cells. [From Oemar et al., *J. Biol. Chem.* **266**, 2369-2373, (1991), after Figure 3, with permission of the American Society for Biochemistry and Molecular Biology, Inc., Bethesda].

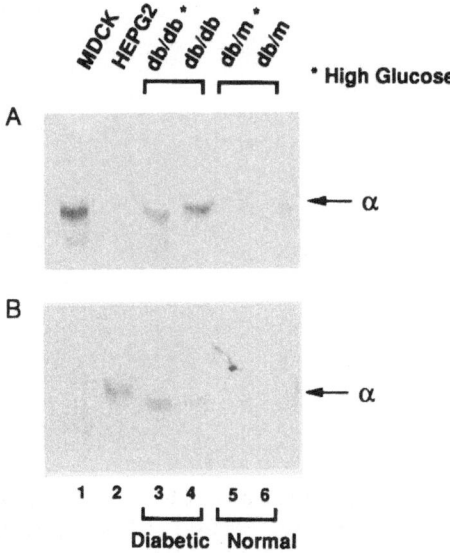

Figure 2. Immunoblot analysis of IGF-1R (A) and insulin receptor (B) proteins. 50 μg of wheat germ agglutinin-agarose purified membrane proteins were resolved on 7.5% SDS gels, transferred to nitrocellulose and probed with site-specific anti-peptide IgG (Rosenzweig et al., 1990). [From Oemar et al., *J. Biol. Chem.* **266**, 2369-2373, (1991), after Figure 2, with permission of the American Society for Biochemistry and Molecular Biology, Inc., Bethesda]. α, migration position of receptor α subunits.

An important question raised by these observations relates to whether IGF-1Rs are functionally compromised or perhaps on the other end of the spectrum, overactivated in the *db/db* cells. To address this question, the ability of IGF-1Rs to autophosphorylate and mediate the subsequent phosphorylation of cellular substrates in *db/db* cells was tested. As shown in Figure 3A, IGF-1Rs were autophosphorylated on tyrosyl residues in a time and dose-dependent (not shown) manner. Similarly, activation of the IGF-1R resulted in the time and dose-dependent tyrosyl phosphorylation of pp185 (IRS-1) in these cells (Fig. 3B). These results were consistent with the previous findings of other laboratories and thus indicated that the IGF-1Rs in the *db/db* cells were functionally competent.

IGF-1 Stimulated Nuclear Protein Tyrosyl Phosphorylation

In light of the large complement of IGF-1Rs, low levels of insulin receptors and the well characterized proliferative response to IGF-1, we initiated an investigation into the immunolocalization of tyrosyl phosphorylated proteins in *db/db* cells. Using laser scanning confocal microscopy employing anti-phosphotyrosine antibodies (α–Py IgG), we observed a rapid and transient tyrosyl phosphorylation of nuclear proteins in *db/db* cells (Oemar et al., 1991b). These data were corroborated in biochemical experiments in which nuclear fractions were prepared from ^{32}P-labeled *db/db* cells stimulated with IGF-1 and subjected to immunoprecipitation with anti-phosphotyrosine antibodies (Fig. 4). The results obtained indicated that a subset of proteins in the nucleus becomes tyrosyl phosphorylated in response to IGF-1 stimulation.

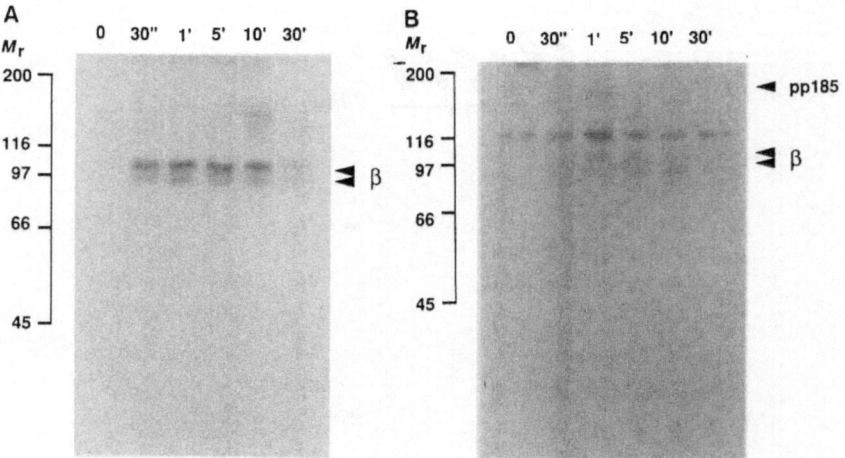

Figure 3. IGF-1-induced tyrosyl phosphorylation of IGF-1R β subunits and cytoplasmic proteins. *db/db* cells were serum-starved (18 h), labeled with ^{32}Pi for 2 h followed by stimulation with IGF-1 (10^{-7}M) for the various times indicated. Cells were lysed and the detergent extracts were sequentially immunoprecipitated with α–Py IgG followed by α–IGF-1R IgG (A). For the second round of immunoprecipitations, antigens were eluted with 10 mM O-phospho-L-tyrosine prior to addition of α–IGF-1R IgG. Analysis of cytoplasmic proteins was carried out by immunoprecipitating lysates with α–Py IgG alone (B). These data demonstrate the time-dependency of IGF-1 on the tyrosyl phosphorylation of receptor and pp185 (i.e., IRS-1). β, migration position of IGF-1R β subunit.

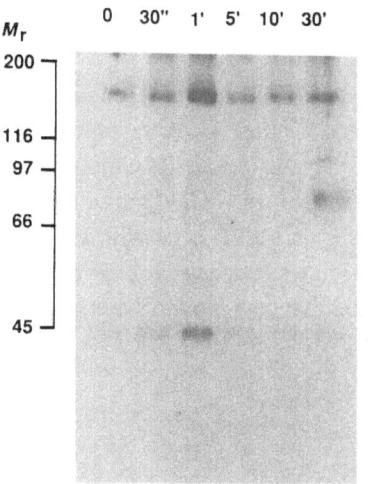

M_r 0   30"   1'   5'   10'   30'

200 —

116 —
97 —

66 —

45 —

Figure 4. IGF-1-induced tyrosyl phosphorylation of nuclear proteins. *db/db* cells prelabeled with $^{32}P_i$ were stimulated for the indicated times with 10^{-7} M IGF-1. Nuclear fractions were isolated and immunoprecipitated with α–Py IgG and analyzed by SDS-PAGE and autoradiography of the dried gel. [From Oemar et al., *J. Biol. Chem.* **266**, 24241-24244, (1991), Figure 2, with permission of the American Society for Biochemistry and Molecular Biology, Inc., Bethesda].

Physiological Significance of Nuclear Protein Tyrosyl Phosphorylation: Regulation of Activator Protein-1 Complexes

The most prominent nuclear protein to become tyrosine phosphorylated in response to IGF-1 action had an apparent mass of ~43 kDa. (Fig. 4). A candidate protein having this mass is the product of the *c-jun* proto-oncogene. Using a combined sequential immunoprecipitation protocol, we established that c-Jun is tyrosine phosphorylated within 1 min of IGF-1 treatment in *db/db* cells, as well as in the human epidermoid carcinoma cell line, CaSki (Oemar et al., 1991b). c-Jun is a member of the immediate early gene family so designated because their transcription is activated within minutes of growth factor addition (Sheng and Greenberg, 1990). This increase is rapid, transient and independent of new protein synthesis. mRNA for these components typically are present at low to undetectable levels in quiescent cells. In the nucleus, c-Jun, via leucine zipper interaction (Bos et al., 1988) either forms a homodimer with a second molecule of c-Jun or forms a heterodimer with c-Fos. The heterodimeric complex is more stable and as a result more transcriptionally active (Nakabeppu et al., 1988). This functional dimer is referred to as the AP-1 (activator protein 1) complex (Curran and Franza, 1988). Activated AP-1 binds to a specific DNA sequence, the AP-1 binding motif, also called the TRE sequence for TPA Response Element. This domain derives its name from the fact that this consensus sequence was initially defined as a phorbol ester inducible promoter (Angel et al., 1987).

To examine the physiological role of tyrosyl phosphorylation in regulating c-Jun/AP-1 function we have begun measuring changes in AP-1 DNA binding activity in IGF-1 treated cells using mobility shift assays (Singh et al., 1986). For these studies, we use an oligonucleotide containing the AP-1 binding site/TRE sequence (Nakabeppu et al., 1988):

```
5'-AGCTTGGTGACTCATCCG-3'
3'-TCGAACCACTGAGTAGGC-5'
```

(with the consensus TRE sequence in boldface). Cells are treated with human recombinant IGF-1 for different times and nuclear extracts prepared according to the method of Andrews and Faller, 1991. The typical results of a mobility shift assay are shown in Fig. 5. These results clearly demonstrate that IGF-1 stimulates the rapid activation of AP-1 complex DNA binding activity in CaSki cells with kinetics that are consistent with the tyrosyl phosphorylation of c-Jun we previously observed. Cycloheximide treatment had no effect on this profile indicating that new protein synthesis was not required.

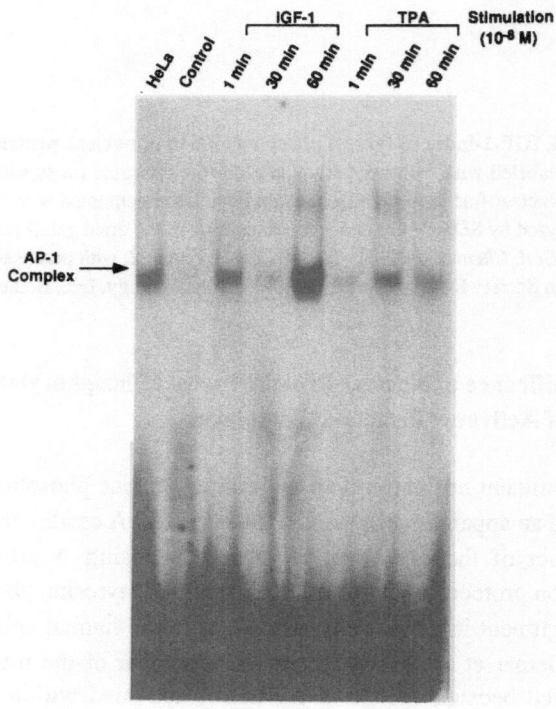

Figure 5. IGF-1-stimulated increase in AP-1 complex DNA binding activity.
Nuclear extracts from serum starved CaSki cells treated with 10^{-8} M IGF-1 or TPA for the indicated times were incubated with the 5'-end labeled TRE sequence containing oligonucleotide probe. Protein-DNA complexes were resolved by non-denaturing gel electrophoresis in Tris-borate-EDTA running buffer and visualized by autoradiography of the dried gel. HeLa, nuclear extract from serum stimulated HeLa cells was run as a positive control; Control, unstimulated CaSki cells.

Identification of IGF-1 Regulated Nuclear Protein Kinases

Foremost in establishing that IGF-1 regulates nuclear events by changes in phosphorylation of c-Jun, is the need to demonstrate the presence of nuclear kinases responsive to IGF-1 activation. The identification and isolation of nuclear kinases will provide us with an important tool with which to work in reverse order and trace the signal transduction pathway to the plasma membrane. To this end, we have examined nuclear extracts from cells treated with IGF-1 for 1 min for tyrosine and ser/thr kinase activities. Our previous observations indicated that only part of the anti-phosphotyrosine immunoprecipitable 43 kDa material (Fig. 4) was capable of being further immunoprecipitated with anti-c-Jun antibodies (Oemar et al., 1991b). Since MAP kinase is tyrosyl phosphorylated (Ray and Sturgill, 1988), 41-44 kDa in mass, is known to be activated by IGF-1, and may translocate to the nucleus (Chen et al., 1992) its candidacy as a c-Jun kinase was implicit. We have explored this possibility in *db/db* cells, CaSki cells and in NIH 3T3 cells overexpressing the hIGF-1R (c4 cells; Kaleko et al., 1990). To date we have obtained data indicating that c4 cell nuclei express both tyrosine and MBP (myelin basic protein) kinase activities which are increased by IGF-1 stimulation. It is the identification of these nuclear kinase activities that forms the basis for our goal of isolating IGF-1 regulatable nuclear protein kinases.

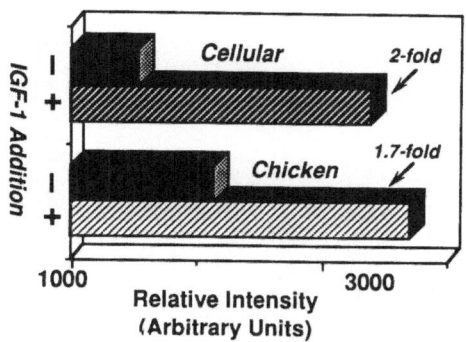

Figure 6. *In vitro* **nuclear c-Jun kinase assay.** Nuclear extracts were prepared from untreated cells or cells stimulated with IGF-1 (10^{-8} M) for 1 min according to the method of Andrews and Faller, 1991. Five µg of recombinant chicken c-Jun was added to each tube and the assay initiated by addition of [γ-^{32}P]ATP. The incubations were heated to stop the reaction and inactivate the kinase(s). The samples were immunoprecipitated with α-c-Jun IgG, resolved by SDS-PAGE, transferred to PVDF paper and the paper exposed to x-ray film. Shown is the quantification of the autoradiograph indicating a 1.7- and 2-fold increase in chicken and cellular c-Jun phosphorylation, respectively, in extracts from IGF-1 treated C4 cells. Phosphoamino acid analyses of the *in vitro* phosphorylated c-Jun are in progress.

One way we have begun to examine these kinase activities is by determining whether nuclear extracts from IGF-1 stimulated cells will phosphorylate exogenously added recombinant c-Jun. In preliminary studies, we have assayed c4 cell nuclear extracts for kinases capable of phosphorylating recombinant c-Jun. We are now characterizing a kinase activity that is increased in nuclear extracts isolated from IGF-1 stimulated cells. In this regard, Adler et al., 1992, isolated a 67 kDa Jun kinase by affinity chromatography of cell extracts on a column containing an N-terminal fragment of c-Jun. It is conceivable that this kinase may also remain in close contact with c-Jun during immunoprecipitation. This possibility is presently under investigation. As shown in Fig. 6, CaSki cell nuclear extracts possess a c-Jun kinase whose activity is increased ~2-fold *vs.* either endogenous (cellular) or exogenously added recombinant c-Jun (chicken).

Although the literature is replete with data suggesting a role for pp44/42^{erk} (i.e., MAP kinase) activation in the phosphorylation-activation of c-Jun/AP-1, we have been unable to detect significant quantities of p44/42^{erk} in nuclear extracts from quiescent c4 or CaSki cells by immunoblot analysis. Similarly, nuclear pp44/42^{erk} activity was not detectable following stimulation of cells with 10^{-7}M IGF-1 for 1, 15, or 60 min. by either immunoblot or by immunocytochemical analyses (Fig. 7, Miller and Rosenzweig, manuscript in preparation). Given the overwhelming evidence from *in vitro* analyses, that indicate pp44/42^{erk} phosphorylates c-Jun on ser 243 resulting in inhibition of DNA binding activity (Trejo et al., 1992, Smeal et al., 1992) in conjunction with our own findings, it is evident that c-Jun kinases other than pp44/42^{erk} likely exist in the nucleus.

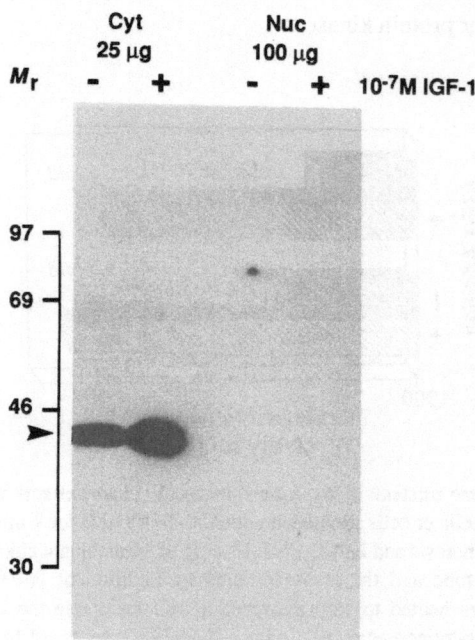

Figure 7. Anti-ERK immunoblot. Cytoplasmic (Cyt) and nuclear (Nuc) extracts from untreated cells or cells stimulated for 1 min with IGF-1 (10^{-7} M) were prepared according to the method of Bell et al., 1987. Proteins were resolved by SDS-PAGE, transferred to nitrocellulose and immunoblotted with monoclonal anti-ERK IgG (Zymed Laboratories). Immunoreactive pp42/44^{erk} was not detectable in nuclear extracts from quiescent or stimulated CaSki cells. Arrowhead indicates migration position of pp42^{erk}. Note: a four-fold excess of nuclear over cytoplasmic protein was used in this blot.

Summary

The mechanism by which IGF-1Rs regulate the growth and maintenance of cells in normal and disease states provides an important setting for studies addressing signal transduction events at the nuclear level. With the identification of c-Jun/AP-1 as a nuclear target of IGF-1 action we are provided with a model system for pursuing the molecular mechanisms triggered by IGF-1 action.

Acknowledgements

We wish to thank Dr. Nick Tonks, Cold Spring Harbor Laboratory, for recombinant PTPase, Dr. Peter Vogt, University of Southern California, for the recombinant c-Jun plasmid, and Pauline Brown for excellent technical assistance. This work was supported by NIH grants DK-34386 and EY-06581 and a research grant from the Juvenile Diabetes Foundation International (SAR).

REFERENCES

Adler, V., Polotskaya, A., Wagner, F. and Kraft, A.S. (1992). Affinity-purified c-Jun amino-terminal protein kinase requires serine/threonine phosphorylation for activity. *J. Biol. Chem.* **267**, 17001-17005.

Andrews, N.C. and Faller, D.V. (1991). A rapid micropreparation technique for extraction of DNA-binding proteins from limiting numbers of mammalian cells. *Nuc. Acids Res.* **19**, 2499.

Angel, P., Imagawa, M., Chiu, R., Stein, B., Imbra, R.J., Rahmsdorf, H.J., Jonat, C., Herrlich, P. and Karin, M. (1987). Phorbol ester-inducible genes contain a common *cis* element recognized by a TPA-modulated *trans*-activating factor. *Cell* **49**, 729-739.

Bell, J.C., Mahadevan, L.C., Colledge, W.H., Frackleton Jr., A.R., Sargent, M.G. and Foulkes, J.G. (1987). Abelson-transformed fibroblasts contain nuclear phosphotyrosyl proteins which preferentially bind to murine DNA. *Nature* **325**, 552-554.

Berridge, M.J. (1993). Inositol trisphosphate and calcium signalling. *Nature* **361**, 315-325.

Bos, T.J., Bohmann, D., Tsuchle, H., Tjian, R. and Vogt, P.K. (1988). v-jun encodes a nuclear protein with enhancer binding properties of AP-1. *Cell* **52**, 705-712.

Cantley, L.C., Auger, K.R., Carpenter, C., Duckworth, B., Graziani, A., Kapeller, R. and Soltoff, S. (1991). Oncogenes and signal transduction. *Cell* **64**, 281-302.

Chen, R.H., Sarnecki, C. and Blenis, J. (1992). Nuclear localization and regulation of erk-encoded and rsk-encoded protein kinases. *Mol. Cell. Biol.* **12**, 915-927.

Czech, M.P. (1989). Signal transmission by the insulin-like growth factors. *Cell* **59**, 235-238.

Kaleko, M., Rutter, W.J. and Miller, A.D. (1990). Overexpression of the human insulinlike growth factor I receptor promotes ligand-dependent neoplastic transformation. *Mol. Cell. Biol.* **10**, 464-73.

Koch, C.A., Anderson, D., Moran, M.F., Ellis, C. and Pawson, T. (1991). SH2 and SH3 domains - Elements that control interactions of cytoplasmic signaling proteins. *Science* **252**, 668-674.

Lavan, B.E., Kuhne, M.R., Garner, C.W., Anderson, D., Reedijk, M., Pawson, T. and Lienhard, G.E. (1992). The association of insulin-elicited phosphotyrosine proteins with src homology 2 domains. *J. Biol. Chem.* **267**, 11631-6.

Nakabeppu, Y., Ryder, K. and Nathans, D. (1988). DNA binding activities of three murine jun proteins: Stimulation by fos. *Cell* **55**, 907-915.

Oemar, B.S., Foellmer, H.G., Hodgdon-Anandan, L. and Rosenzweig, S.A. (1991a). Regulation of Insulin-Like Growth Factor-I Receptors in diabetic mesangial cells. *J. Biol. Chem.* **266**, 2369-2373.

Oemar, B.S., Law, N.M. and Rosenzweig, S.A. (1991b). Insulin-Like Growth Factor-I induces tyrosyl phosphorylation of nuclear proteins. *J. Biol. Chem.* **266**, 24241-24244.

Pronk, G.J., Polakis, P., Wong, G., Devriessmits, A.M.M., Bos, J.L. and Mccormick, F. (1992). Association of a tyrosine kinase activity with gap complexes in v-src transformed fibroblasts. *Oncogene* **7**, 389-394.

Ray, L.B. and Sturgill, T.W. (1988). Insulin-stimulated microtubule-associated protein kinase is phosphorylated on tyrosine and threonine in vivo. *Proc. Natl. Acad. Sci. USA* **85**, 3753-3757.

Rosenzweig, S.A., Zetterström, C. and Benjamin, A. Identification of retinal insulin receptors using site-specific antibodies to a carboxyl-terminal peptide of the human insulin receptor α-subunit: Up-regulation of neuronal insulin receptors in diabetes. J. Biol. Chem. 265:18030-18034, 1990.

Rothenberg, P.L., Lane, W.S., Karasik, A., Backer, J., White, M. and Kahn, C.R. (1991). Purification and partial sequence analysis of pp185, the major cellular substrate of the insulin receptor tyrosine kinase. *J. Biol. Chem.* **266**, 8302-8311.

Schumacher, R., Mosthaf, L., Schlessinger, J., Brandenburg, D. and Ullrich, A. (1991). Insulin and Insulin-Like Growth Factor-1 binding specificity is determined by distinct regions of their cognate receptors. *J. Biol. Chem.* **266**, 19288-19295.

Sheng, M. and Greenberg, M.E. (1990). The regulation and function of c-fos and other immediate early genes in the nervous system. *Neuron* **4**, 477-485.

Singh, H., Sen, R., Baltimore, D. and Sharp, P.A. (1986). A nuclear factor that binds to a conserved sequence motif in transcriptional control elements of immunoglobulin genes. *Nature* **319**, 154-158.

Smeal, T., Binetruy, B., Mercola, D., Groverbardwick, A., Heidecker, G., Rapp, U.R. and Karin, M. (1992). Oncoprotein-mediated signalling cascade stimulates c-Jun activity by phosphorylation of serine-63 and serine-73. *Mol. Cell. Biol.* **12**, 3507-3513.

Sun, X.J., Rothenberg, P., Kahn, C.R., Backer, J.M., Araki, E., Wilden, P.A., Cahill, D.A., Goldstein, B.J. and White, M.F. (1991). Structure of the insulin receptor substrate IRS-1 defines a unique signal transduction protein. *Nature* **352**, 73-77.

Sweet, L.J., Morrison, B.D., Wilden, P.A. and Pessin, J.E. (1987). Insulin-dependent intermolecular subunit communication between isolated $\alpha\beta$ heterodimeric insulin receptor complexes. *J. Biol. Chem.* **262**, 16730-16738.

Trejo, J., Chambard, J-C., Karin, M. and Brown, J.H. (1992). Biphasic increase in *c-jun* mRNA is required for induction of AP-1-mediated gene transcription: Differential effects of muscarinic and thrombin receptor activation. *Mol. Cell Biol.* **12**, 4742-4750.

Ullrich, A. and Schlessinger, J. (1990). Signal transduction by receptors with tyrosine kinase activity. *Cell* **61**, 203-212.

Werner, H., Shenorr, Z., Stannard, B., Burguera, B., Roberts, C.T. and Leroith, D. (1990). Experimental diabetes increases Insulinlike Growth Factor-I and Factor-II receptor concentration and gene expression in kidney. *Diabetes* **39**, 1490-1497.

White, M.F., Maron, R. and Kahn, C.R. (1985). Insulin rapidly stimulates tyrosine phosphorylation of a Mr 185,000 protein in intact cells. *Nature* **318**, 183-186.

Yamamoto, K., Altschuler, D., Wood, E., Horlick, K., Jacobs, S. and Lapetina, E.G. (1992). Association of phosphorylated Insulin-like Growth Factor-I receptor with the SH2 domains of phosphatidylinositol 3-kinase p85. *J. Biol. Chem.* **267**, 11337-43.

MOLECULAR CLONING OF pp120/ ECTO-ATPase, AN ENDOGENOUS SUBSTRATE OF THE INSULIN RECEPTOR KINASE

Sonia M. Najjar, Neubert Philippe,
Simeon I. Taylor, and Domenico Accili

The Diabetes Branch
National Institute of Diabetes and Digestive and Kidney Diseases
National Institutes of Health, Bethesda, MD 20892

INTRODUCTION

In recent years, a great deal of attention has been devoted to dissecting intracellular pathways of insulin action following activation of the insulin receptor kinase (1). Several laboratories have reported the identification of protein substrates of the insulin receptor tyrosine kinase (2). Here, we review our studies of pp120/ecto-ATPase, a liver-specific glycoprotein of M_r-120,000 which was originally identified by us as a substrate of the insulin receptor tyrosine kinase in cell-free phosphorylation assays (3). Independently, other laboratories have identified the same protein as a hepatic enzyme capable of hydrolyzing extracellular ATP and GTP. Recently, cDNA and genomic cloning studies have contributed to elucidate the structure of this protein and to formulate hypotheses on its function. This chapter will focus on the potential role of pp120/ ecto-ATPase in mediating actions of insulin in the liver.

Historical background

In studies of the effects of dexamethasone on the kinase activity of insulin receptors isolated from rat liver, we detected a glycoprotein of apparent molecular mass 120,000, the phosphorylation of which was stimulated by insulin in a cell-free system (3). Based on immunological criteria, as well as a survey of the tissue distribution and subcellular fractionation of this protein, it was concluded that the M_r-120,000 phosphoprotein (henceforth referred to as pp120) was distinct from the insulin receptor and acted as a substrate for phosphorylation by the insulin receptor kinase (3, 4). Indeed, phosphorylation of pp120 *in vitro* occurred on tyrosine residues and fulfilled the requirements of an enzymatic reaction mediated by the activated, autophosphorylated insulin receptor kinase. Furthermore, it was possible to show that pp120 was a substrate for phosporylation induced by other receptor-type tyrosine kinases including the IGF-1 receptor (5) and the EGF receptor (6). Finally, studies of of intact rat hepatoma cells indicated that insulin induced phosphorylation of a similar (probably identical) protein. As was observed previously with cell-free extracts, the M_r-120,000 glycoprotein in intact cells was phosphorylated on tyrosine residues, and was immunoprecipitated by a polyclonal antiserum raised against pp120 (7).

Current Directions in Insulin-Like Growth Factor Research,
Edited by D. LeRoith and M.K. Raizada, Plenum Press, New York, 1994

169

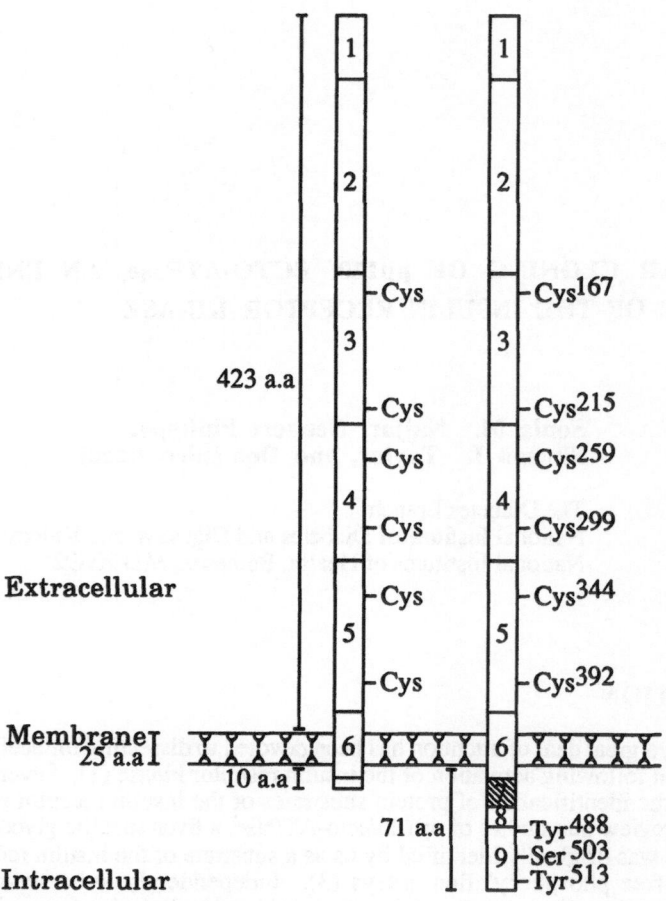

Fig.1 Primary structure of pp120/ecto-ATPase. The primary sequence of pp120/ecto-ATPase was deduced from cDNA clones isolated from a rat liver library. The cDNAs are predicted to encode a polypeptide of M_r 47,000. The extracellular domain consists of 423 amino acids and contains 15 potential sites for N-linked glycosylation. The intracellular domain contains potential sites for tyrosine as well as serine phosphorylation. To the left, a diagram of the form encoded by the short cDNA isoform is shown. As a result of alternative splicing, 60 of the 71 amino acids of the cytoplasmic domain are truncated. The shorter isoform is therefore predicted to lack all potential intracellular sites of phosphorylation. To the right, the full length pp120/ecto-ATPase is shown

Strikingly, expression of pp120 is restricted to liver, and the protein product localizes to the plasma membane fraction of rat liver extracts. Because of these features, we attempted to identify pp120 by immunodetection using a panel of antibodies raised against rat liver proteins. Interestingly, a monoclonal antibody raised against HA4, a liver glycoprotein associated with the bile canalicular side of the hepatocyte plasma membrane, immunoprecipitated the same glycoprotein detected by the polyclonal antiserum raised against pp120 (8). In fact, HA4 and pp120 share the same molecular mass, tissue distribution and subcellular localization. The physiological function of HA4 in vivo is uncertain; however, it has been thought to participate in bile acid transport across the hepatic plasma membrane (9).

Based on these findings, a purification scheme was devised in order to isolate pp120. Rat liver microsomes were solubilized and subjected to a first enrichment step by lectin chromatography. Glycoproteins eluted from the lectin column were then affinity purified on an anti-HA4 antibody column. A pure preparation of pp120 was obtained, and subjected to Edman degradation in order to determine the amino acid sequence. Partial sequence data revealed that pp120/ HA4 was the same protein as ecto-ATPase (10), a liver-specific glycoprotein, the cDNA sequence of which had just been reported (11, 12).

Ecto-ATPase had been independently identified as a Ca/Mg-dependent ATPase (11, 12). It is therefore intriguing that three groups, working independently on three seemingly unrelated projects, converged in the identification of the same protein.

Cloning of the gene encoding pp120/ecto-ATPase: identification of variably spliced isoforms.

Cloning of cDNA encoding ecto-ATPase (12) revealed an open reading frame of 1560 nucleotides, predicted to encode a peptide of M_r-47,000. The difference between the predicted molecular mass of the peptide and the observed mobility on denaturing SDS-polyacrylamide gels is readily explained by some of our studies. Indeed, we had shown that pp120 is exquisitely sensitive to endoglycosidase digestion, and that the apparent

Cloning of the pp120/Ecto-ATPase Gene

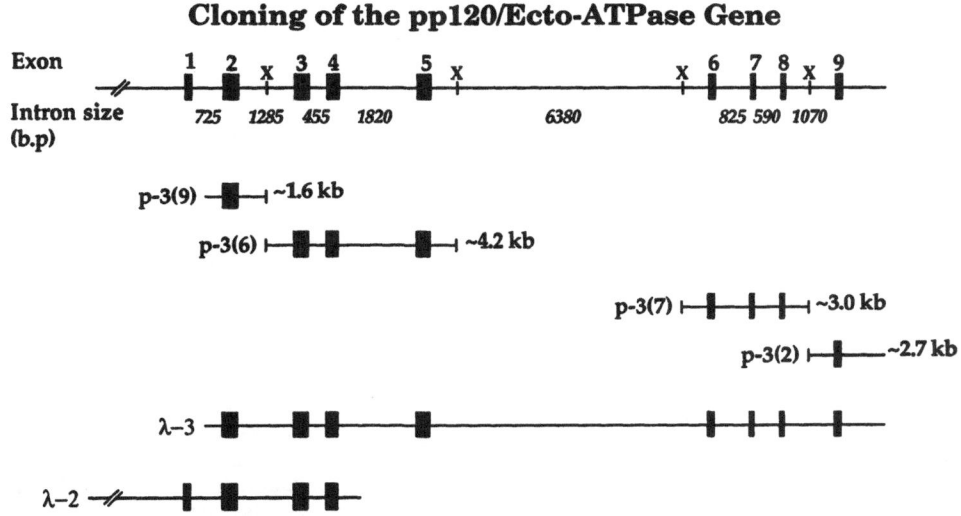

Fig.2 Genomic organization of the pp120/ecto-ATPase gene. A rat genomic library constructed in a lambda replacement vector of the EMBL family was screened using a rat complementary DNA clone. Individual recombinant phages were plaque-purified and analyzed by a combination of restriction mapping, PCR amplification, and sequence analysis. A PCR-based strategy was employed in order to determine the size of the introns. Intron-exon boundries were then determined by sequence analysis according to the AG/TG rule.

molecular mass of 120,000 is due to substantial N-linked glycosylation (8). Upon inspection of the predicted primary sequence, 15 potential sites of asparagine-linked glycosylation are present in the extracellular domain. Other structural features are also noteworthy. The hydropathy plot of the peptide predicts a single hydrophobic domain with features consistent with those of a membrane spanning domain. As shown in Fig.1, pp 120 /ecto-ATPase would then consist of a 423-amino acid extracellular domain, a single transmembrane domain, and a 71 amino acid cytoplasmic tail. The extracellular domain has enzymatic activity as phosphohydrolase for which ATP and GTP can serve as substrates (12). In the intracellular domain, a potential site for phosphorylation by cyclic AMP-dependent protein kinase (KRPTS, amino acids 499-503) can be identified (12), as well as two tyrosines in the

171

context of negatively charged residues (13). They appear to be candidate sites for tyrosine phosphorylation induced by receptor-type kinases (Y^{488} and Y^{513}).

We used the polymerase chain reaction to clone the cDNA encoding rat liver pp120/ecto-ATPase. We detected two populations of cDNA clones encoding pp120/ecto-ATPase in rat liver (Figs. 1 and 3) (13).These two forms differ by the absence or presence of a 53 bp fragment between bp 1362 and 1414. The explanation of this deletion became apparent upon cloning of the gene encoding pp120 (Fig. 2). The gene consists of nine exons spread over 15 kbp of sequence. The 53 bp fragment that is absent from some of the cDNA clones is encoded by exon 7. This exon is alternatively spliced in mRNA molecules, giving rise to the two isoforms of pp120/ecto-ATPase cDNA. Splicing of the 53 nucleotides corresponding to exon 7 generates a different open reading frame and introduces a chain termination codon at nucleotide 1374. A truncated protein product, lacking 61 of the 71 amino acids of the cytoplasmic domain, is thus generated (Fig. 1). Interestingly, the shorter isoform lacks all cytoplasmic phosphorylation sites. Expression studies are currently underway in our laboratory to address the question as to the role of the two isoforms.

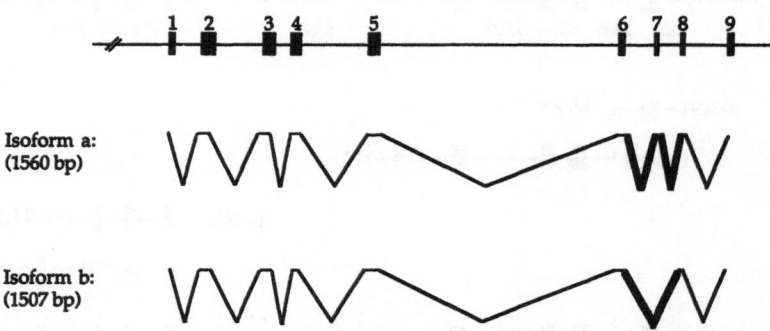

Fig. 3 Alternative splicing of pp120/ecto-ATPase. This diagram depicts the splicing patterns of two different mRNA isoforms derived from trnascription of the pp120/ ecto-ATPase gene. The absence of 53 nucleotides encoded by exon 7 generates a protein product of 458 amino acid. This short isoform lacks 61 of 71 amino acids of the cytoplasmic domain of pp120/ ecto-ATPase. All potential intracellular phosphorylation sites are thus absent from the shorter isoform.

Interestingly, pp120 had occasionally been visualized after metabolic labeling as a doublet of 120/110 kDa on SDS/ployacrylamide gels. However, the shorter isoform was not detected in our phosphorylation assays, suggesting that it does not undergo phosphorylation. These findings are consistent with the notion that both alternatively spliced isoforms are translated into protein products; however, only full-length isoforms act as substrates for growth factor receptor tyrosine kinases. Alternative explanations, however, have not been formally ruled out (e.g., the possibility that the two isoforms are produced by proteolytic digestion during the purification procedure, or that they represent products of different glycosylation patterns).

pp120/ecto-ATPase in the mechanism of action of insulin.

The primary structure of pp120/ ecto-ATPase shares considerable homology with other cell surface glycoproteins of the immunoglobulin superfamily, such as carcinoembryonic antigen, pregnancy-specific β-glycoprotein, and biliary glycoprotein, all of which are known to exist in multliple isoforms derived from alternative splicing of mRNA molecules (12). Interestingly, exon 7 of the pp120/ ecto-ATPase gene is identical to the corresponding exon of CEA (12), which is also alternatively spliced (Fig. 4).

The physiological function of pp120/ ecto-ATPase is unclear. The long isoform is a substrate for tyrosine phosphorylation by at least three receptor tyrosine kinases, while the short isoform appears not to be. Furthermore, the extracellular domain of the protein has been shown to contain enzymatic activity as an ATP/GTP hydrolase. Whether this activity represents the major physiological role of pp120 is also to be resolved. Hydrolysis of extracellualr ATP has been postulated to play a role in chemotaxis and T lymphocyte-mediated cytotoxicity (14). Cytotoxic T-lymphocytes may utilize ecto-ATPases in order to protect themselves from the toxic actions of ATP (14). In this respect, it has been proposed that ecto-ATPases and ecto-kinases may act so as to modulate the functions of extracellular ATP. P-glycoprotein, the protein product of the multidrug resistance gene, has also been shown to possess phosphohydrolase activity. Other authors have suggested that pp120 may act as a bile acid transporter, based on its subcellular and tissue distribution.

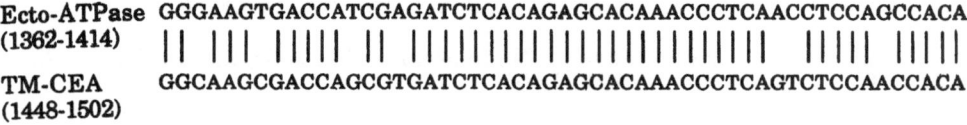

Ecto-ATPase (1362-1414)

TM-CEA (1448-1502)

Fig. 4. Exon 7 of the pp120/ecto-ATPase gene is identical to the homologous exon of CEA gene. The primary sequence of exon 7 of pp120/ecto-ATPase gene and the homologous gene of the carcinoembryonic antigen are reported. CEA gene undergoes alternative splicing of this exon, and is thought to play a role in mechanisms of cell adesion. Its unregulated expression has been implicated in the onset of the metastatic phase of cancer. pp120 shares homoogy with other members of the family of cell adhesion molecules, like NCAM (11).

Among many intracellular substrates of the insulin receptor kinase identified to date, pp120/ ecto-ATPase has unique features. In fact, unlike other proteins that are phosphorylated by or bind to the intracellular domain of the insulin receptor, pp120/ecto-ATPase appears to exert its enzymatic activity by way of an extracellular catalytic domain. It is our goal to establish whether phosphorylation by the insulin receptor kinase regulates the enzymatic activity of pp120. Further studies will address the role of this protein in the mechanism of action of insulin in the liver.

REFERENCES

1. Taylor, S. I., Najjar, S., Cama, A., and Accili, D. (1991) *Insulin- Like Growth Factors: Molecular and Cellular Aspects* (LeRoith, D., ed) pp. 221-244, CRC press, Inc., Boca Raton.

2. Kasuga, M., Izumi, T., Tobe, K., Shiba,T., Momomura, K., Tashiro-Hashimoto, Y., Kadowaki, T.(1990) *Diabetes Care* **13**, 317-26.

3.   Rees-Jones, R. W., and Taylor, S. I. (1985) *J. Biol. Chem.* **260**, 4461-7.

4.   Accili, D., Perrotti, N., Rees-Jones, R., and Taylor, S. I. (1986) *Endocrinology* **119**, 1274-80.

5.   Fanciulli, M., Paggi, M.G., Mancini, A., Del Carlo, C., Floridi, A., Taylor, S.I., and Perrotti, N. (1989) *Biochem. Biophys. Res. Commun.*, **160**: 168-73.

6.   Phillips, S. A., Perrotti, N., and Taylor, S., I. (1987) *FEBS Lett.* **212**, 141-4.

7.   Perrotti, N., Accili, D., Marcus-Samuels, B., Rees-Jones, R.W., and Taylor, S. I. `(1987) *Proc. Natl. Acad. Sci. U. S. A.* **84**, 3137-40.

8.   Margolis, R.N., Taylor, S. I., Seminara, D., and Hubbard, A. L. (1988) *Proc. Natl. Acad. Sci. U. S. A.* **85**, 7256-9.

9.   Sippel CJ ; Suchy FJ ; Ananthanarayanan M ; Perlmutter DH (1993) *J. Biol Chem*, **268**:2083-91.

10.   Margolis, R. N., Schell, M. J., Taylor, S. I., and Hubbard, A. L. (1990) *Biochem. Biophys. Res. Commun.* **166**, 562-6·

11.   Lin, S.H. (1989) *J. Biol. Chem.* **264**, 14403-407·

12.   Lin, S. H., and Guidotti, G. (1989) *J. Biol. Chem.* **264**, 14408-14·

13.   Najjar, S.M., Accili, D. Philippe, N., Jernberg, J., Margolis, R., and Taylor, S.I. (1992) *J. Biol. Chem.* **268**, 1201-06·

14.   Filippini, A., Taffs, R.E., Agui, T., Sitkovsky, M.V. (1990) *J Biol Chem* **265**, 334-40.

THE INSULIN-LIKE GROWTH FACTOR-II/MANNOSE-6-PHOSPHATE RECEPTOR: STRUCTURE, FUNCTION AND DIFFERENTIAL EXPRESSION.

Wieland Kiess[1], Andreas Hoeflich[1], Yi Yang[2], Ulrike Kessler[1], Allan Flyvbjerg[3], and Bruno Barenton[4]

[1]Cell Biology Laboratory, Dept. Pediatric Endocrinology
Children´s Hospital, University of Munich
Lindwurmstr. 4, D-8000 Munich 2
Germany

[2]Children´s Hospital
Medical University
Feng Li Rd., Shanghai
People´s Republic of China

[3]Institute of Experimental Clinical Research
Kommunehospitalet
University of Aarhus, DK-8000 Aarhus
Denmark

[4]INRA, Laboratoire de Croissance et Differenciation Cellulaire
Place Pierre Viala, F-Montpellier Cedex
France

INTRODUCTION

The insulin like growth factor-II/mannose-6-phosphate (IGF-II/M6P) receptor is a bifunc-tional binding protein that binds lysosomal enzymes bearing the M6P recognition marker and IGF-II at distinct binding sites (45, 52). In addition, transforming growth factor (TGF) beta precursor, thyroglobulin and proliferin, a protein which is expressed in rapidly proliferating cells are also recognized by this receptor (Table 1). In avian and amphibian cells the receptor lacks the binding site for IGF-II but serves as a binding protein for M6P-bearing ligands (7,9,76). Almost all mammalian cells described until today express IGF-II/M6P receptors that bind both classes of ligands, namely M6P-containing glycoproteins and IGF-II (55-59).

The genes that code for the IGF-II/M6P receptor map to the centromeric third of chromo-some 17 in the mouse, and on the long arm of human chromosome 6, region 6q25-

Current Directions in Insulin-Like Growth Factor Research,
Edited by D. LeRoith and M.K. Raizada, Plenum Press, New York, 1994

175

Table 1. Ligands of the IGF-II/M6P receptor
(for review see 22, 40, 55, 56).

1. IGF-II binding site:

_____

insulin-like growth factor-II

(insulin-like growth factor-I)

2. M6P-binding sites:

_____

proliferin
transforming growth factor-beta1 precursor
thyroglobulin
unteroferrin
prorenin
acid hydrolases (approx. 50 lysosomal
enzymes bearing the M6P recognition marker)

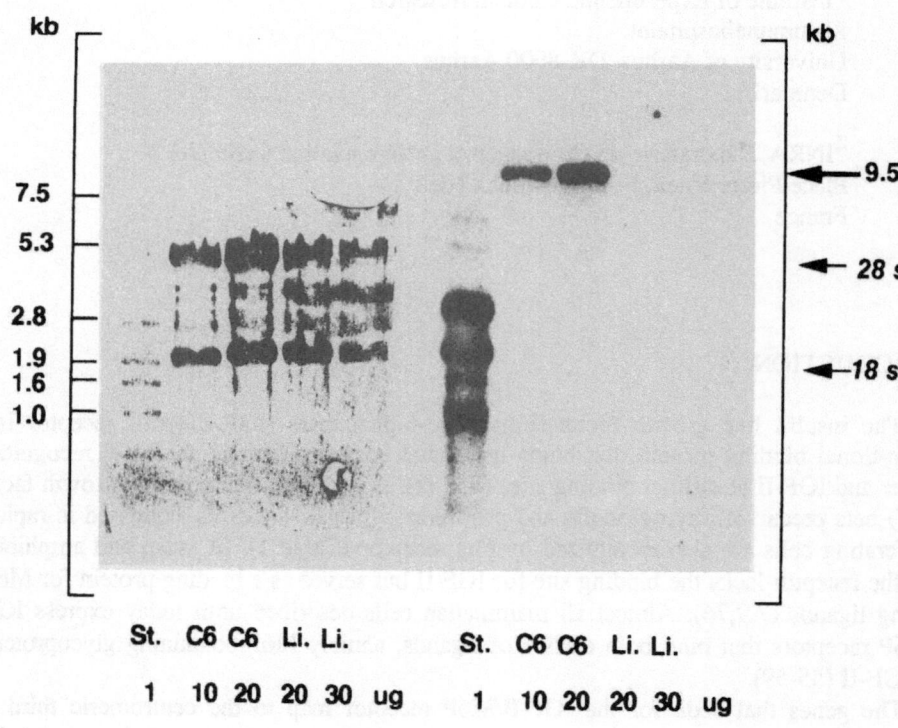

Figure 1

q27 (41). The aligned sequences of the full-length human and bovine receptors that are derived from the cDNA sequences are 80% identical. Homology between the rat and the bovine amino acid sequences also infers to approximately 80% (45,52,69). A 9 kb RNA transcript of the IGF-II/M6P receptor is routinely detected in Northern blotting experiments using RNA from cells or tissues that express this receptor (Figure 1). Interestingly, the gene for the IGF-II/ M6P receptor is maternally imprinted, whereas the IGF-II gene is paternally imprinted (1,20, 75). It has been suggested that the differential imprinting of the IGF-II and the IGF-II/M6P receptor genes has important functional consequence for fetal growth and development (30).

The mammalian IGF-II/M6P receptor encomprises a large extracellular domain which consists of 15 repeats with an average size of approximately 147 amino acids. These repeats share between 16 and 38% homology and are also homologous to the extracellular domain of the so-called cation-dependent mannose-6-phosphate receptor. The cation-dependent mannose-6-phosphate receptor is a 40 kDa glycoprotein which also binds M6P-containing ligands but does not recognize IGF-II. 16 potential N-linked glycosylation sites have been identified upon examining the cDNA sequence of the IGF-II/M6P receptor. Differential glycosylation of the receptor in different tissues has been described. The function of the carbohydrate chains of the receptor molecule remains unclear. However, it has become evident that glycosylation of the receptor is not necessary for binding of IGF-II or M6P-containing ligands.

A truncated form of the receptor lacking the intracellular domain is present in the serum of the rat, sheep, monkey and human (36,46). In the sheep the serum form of the IGF-II/M6P receptor seems to be an important carrier of IGF-II during fetal life (27). The function of the truncated IGF-II/M6P receptor form has not been elucidated However, it has been suggested that release of the extracellular part of the receptor represents the major degradative pathway of the IGF-II/M6P receptor (11).

INSULIN-LIKE GROWTH FACTOR-II BINDING SITE

Equilibrium binding experiments have revealed that the IGF-II/M6P receptor binds 1 mol of IGF-II. IGF-II binds to the receptor with high affinity (Kd 2×10^{-10} M). The receptor seems to be rather specific for IGF-II. Isulin, IGF-I and IGF-II variants do not recognize the receptor at all or with much lower affinity (2,5,40,55-59). It has been suggested that IGF-II binds to a sequence which localizes close to repeat number 13 which is close to the transmembrane spanning region of the receptor and which shares some homology with fibronectin. The exact IGF-II binding site of the IGF-II/M6P receptor has yet to be defined.

It is a matter of debate whether or not the IGF-II/M6P receptor serves a signaling role for the IGFs. A number of different laboratories have described biologic effects of IGF-II that are being mediated through the IGF-II/M6P receptor (Table 2). Evidence for a signaling role of the receptor has been brought forward in these studies by the fact that very low concentrations of IGF-II were required to elicit a biologic response in a certain cell type. Alternatively, blocking or stimulating antibodies to the receptor were used to support the notion that the IGF-II/M6P receptor indeed mediates some biologic effects of IGF-II. Other workers using blocking anti IGF-II/M6P receptor antibodies could not identify any IGF-II/M6P receptor mediated biologic responses of IGF-II (for review see 40, 55-59, 68-69). Interestingly, the cell types and tissues which were used to show specific IGF-II/M6P receptor mediated effects of IGF-II were all of embryonal origin or tumor derived (Table 2). This finding could supplement data which suggest a role of IGF-II for fetal growth and development. Indeed, targeted disruption of the IGF-II gene leads to a growth-deficiency phenotype in mice heterozygous for the gene deletion (19). However, no data are available

Table 2. IGF-II/M6P-receptor mediated biologic effects (for review see 40, 50, 55-59, 68)

cell type/tissue	biologic effect	dose effect	effect of antibody
HepG2 hepatoma cells	glycogen synthesis	II > I	yes
fetal chondrocytes	clonal growth	II > I	no
K 562 erythro-leukemia cells	clonal growth	II > I	no
3T3 fibro-blasts	Ca^{2+} flux DNA synthesis	II > I	yes
renal tubuli	Na$^+$/H$^+$, IP3, DAG	II > I	no
MCF7 mammary-carcinoma cells	proliferation	II > I	yes
adrenal cells	DNA synthesis	II > I	no
embryonal limb buds	DNA synthesis glucose transport glycogen synthesis	II > I	no
neuronal cells	DNA synthesis	II > I, M6P	yes
cardiomyo-cytes	PI turnover	M6P	no
pheochromo-zytoma cells	DNA synthesis	II > I	no
3T3 cells	G$_{ialpha}$ coupling	II > I	yes
HepG2 hepatoma cells	insulin-inhibitable protein degradation	II > I	yes
rhabdomyo-sarcoma cells	cell motility	II > I	yes
skeletal myoblasts	differentiation	II > I	no

which would suggest that the IGF-II effect during fetal growth was mediated via the IGF-II/M6P receptor.

SIGNALING AND SIGNAL TRANSDUCTION MECHANISMS

In a series of publications Nishimoto and coworkers presented evidence that the IGF-II/M6P receptor was coupled to a calcium gating system and to a GTP-binding protein, Gialpha (39). Induction of calcium fluxes by IGF-II in 3T3 fibroblasts was mediated via activation of Gialpha.and lead to increased thymidine incorporation into DNA in such cells. A simple structure of the IGF-II/M6P receptor encodes the G protein-activating function of the IGF-II/M6P receptor (62-63). Interestingly, distinctive regulation of the interaction of the receptor with Gialpha by M6P and IGF-II has been shown (53-54). It is still controversial whether or not the linkage between G-proteins and the IGF-II/M6P receptor plays a role in signaling IGF-II growth promoting effects in cell types other than 3T3 cells. Another rather attractive hypothesis which could explain the linkage between the IGF-II/M6P receptor and G-proteins might be that small molecular weight G-proteins are involved in transport throught the Golgi stack (49). Linkage of the IGF-II/M6P receptor with G-proteins might then play an important role in targeting the receptor and its ligands throught the Golgi apparatus towards prelysosomal compartments in the cell.

A possible linkage of the IGF-II/M6P receptor with the phosphoinositol (PI) pathway of second messengers has also been suggested. In canine kidney proximal tubules picomolar and nanomolar concentrations of IGF-II and M6P enhanced the production of inositol trisphosphate and diacylglycerol. In contrast, in primary cultures of rat cardiac myocytes IGF-I and M6P but not IGF-II induced inositol 1,4,5, trisphosphate formation. The funtional consequences of the interrelationship between the receptor and the PI breakdown are not known (29, for review see 56).

MANNOSE-6-PHOSPHATE BINDING SITES

The mannose-6-phosphate binding sites of the IGF-II/M6P receptor localize to repeats 1-3 and 7-11 of the extracytoplasmic region (74). Equilibrium dialysis experiments revealed that the receptor binds 2 mol of M6P or 1 mol of beta-galactosidase or equivalent lysosomal enzymes via their M6P-residues (22,23,40). The process whereupon M6P residues are attached to lysosomal enzymes in the Golgi involves two enzymes, N-acetylglucosamine-1-phosphotransferase and alpha-N-acetylglucosaminyl-phosphodiesterase (23,40,64). Only processed lysosomal enzymes and ligands bearing the M6P recognition marker bind to the M6P binding sites of the receptor and are targeted to lysosomes.

The receptor functions to target lysosomal enzymes bearing the M6P recognition marker to lysosomes by binding M6P-bearing ligands in the Golgi network and delivering them to a prelysosomal compartment with acidic pH. In the trans Golgi network cytoplasmic domains of the receptor molecules are being recognized by the Golgi-specific adaptor complex HA-2 whereupon specific kinases associate with the receptor (23,40). Phosphorylation of the receptor leads to translocation to different cellular compartments. The receptor exits from the trans Golgi network via clathrin-coated pits. The degree of phosphorylation of receptor molecules pretermines their localization within the cell (13,23,40). In the acidic prelysosomal compartment the ligands dissociate from the receptor and are transported to the lysosome via capillary movement (15,40).

The majority of cellular IGF-II/M6P receptors are localized intracellularly whereas only about 10% of the total receptor pool is cell-surface expressed. Cell-surface receptors

Table 3. Interaction of IGF-II and lysosomal enzymes at the IGF-II/M6P receptor.

1. effect of IGF-II on lysosomal enzyme binding/uptake

acid hydrolase	cell type	reference
beta-galactosidase	C6 glial, BRL 3A2 liver cells purified receptors	(37)
beta-glucuronidase	L cells	(61)
arylsulphatase	hepatocytes	(31)
beta-hexosaminidase B	C6 glial cells	(35)
cathepsin D	MCF7 mammary carcinoma	(47)

2. effect of lysosomal enzymes on IGF-II binding

acid hydrolase	cell type	reference
beta-galactosidase	purified receptors	(38)
beta-glucuronidase	L cells	(61)
arylsulphatase	hepatocytes	(31)
cathepsin D	MCF7 mammary carcinoma	(47)

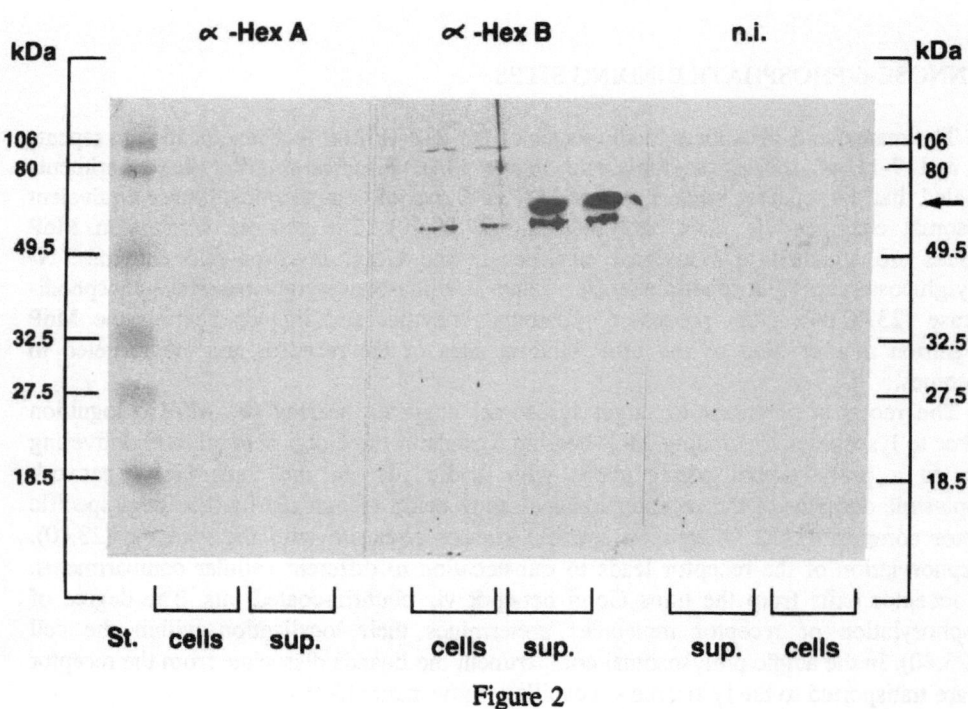

Figure 2

180

serve to internalize extracellular lysosomal enzymes. It is important to note that in many cell types a distinctive pattern of subcellular distribution is found: in gastrointestinal cells IGF-II/M6P receptors are predominantly found close to the apical membranes. In proximal tubule cells of rat kidney apical localization of the IGF-II/M6P receptor has also been described (14). There is constant recycling of receptors from the cell surface to intracellular compartments (15,40,42).

MODULATION OF RECEPTOR RECYCLING AND CELLULAR TRANSLOCATION

A number of chemical compounds, enzymes and hormones have been found to modulate recycling and routing of the IGF-II/M6P receptors (Table 4). Some of these compounds such as Brefeldin A, phorbol esters and polymyxin B have been useful tools for the elucidation of transport pathways of the lysosomal system. Others such as the IGFs, insulin and EGF might well play a physiologic role in modulating receptor localization and receptor function in vivo. It is known that insulin stimulation of fat cells induces an increased surface expression of IGF-II/M6P receptors. This increase of IGF-II/M6P receptor surface expression coincides with increased glucose transporter translocation to the cell surface which occurs after insulin stimulation (10,12,43-44). Brefeldin A treatment of human skin fibroblasts also induces a redistribution of the receptor to the cell surface and increased both binding and internalization of M6P-containing ligands (16,18). A number of kinases and phosphatases are being recognized to participate in the molecular mechanisms which lead to IGF-II/M6P receptor translocation and redistribution (4,17).

INTERACTION OF LIGANDS

IGF-II has been shown to inhibit the cellular uptake of a number of lysosomal enzymes (Table 3). In one series of experiments secretions from Tay Sachs disease fibroblasts were biosynthetically labeled and partially purified. These secretions contain high amounts of beta-hexosaminidase B (Figure 2). When such secretions were used to study the cellular uptake of naturally occuring M6P-containing ligands via the IGF-II/M6P receptor, IGF-II prevented the uptake of the radiolabeled secretions partially (35). The uptake of artificial M6P-containing substrates is less affected by IGF-II (3,40,55-59). The most likely explanation for the IGF-II induced inhibition of cellular uptake of lysosomal enzymes is a direct effect of IGF-II on binding of the other class of ligands to the receptor: IGF-II has indeed been shown to prevent the binding of beta-galactosidase to purified IGF-II/M6P receptors (37). Since IGF-II and M6P-containing ligands bind to different binding sites on the receptor, sterical hindrance or conformational changes of the receptor molecule after binding of IGF-II represent the most probable causes of the IGF-II effect on lysosomal enzyme binding. It is known that low pH and a number of reagents can cause rapid conformational changes of the IGF-II/M6P receptor molecule (57-58).

Several lysosomal enzymes inhibit the binding of IGF-II to purified IGF-II/M6P receptors or interfere with the cellular uptake of the growth factor (Table 3). Again, sterical hindrance or conformational changes of the receptor upon binding of one class of ligands seem to be the most probable mechanisms which cause the reciprocal inhibition of binding of the two classes of ligands that bind to the IGF-II/M6P receptor (38). It is clear that IGF-II and M6P-bearing ligands reciprocally interfere with the extracellular pathway of ligand internalization of the IGF-II/M6P receptor.

The question whether or not the intracellular sorting of newly synthesized lysosomal enzymes containing M6P is affected by the presence of IGF-II in intracellular compartments is under intense investigation. In a first report, Braulke et al. reported that overexpression of IGF-II in NIH 3T3 cells did not affect the synthesis and the sorting of

Table 4. Modulators of IGF-II/M6P receptor recycling/routing.

agents:	reference
brefeldin A	(16, 18)
phorbol esters	(17, 33)
IGFs	(17)
EGF	(4)
insulin	(10, 12, 44)
polymyxin B	(12)
phosphatases	(4)
kinases	(13, 23)
M6P	(4, 67)
G-proteins (?)	(12)

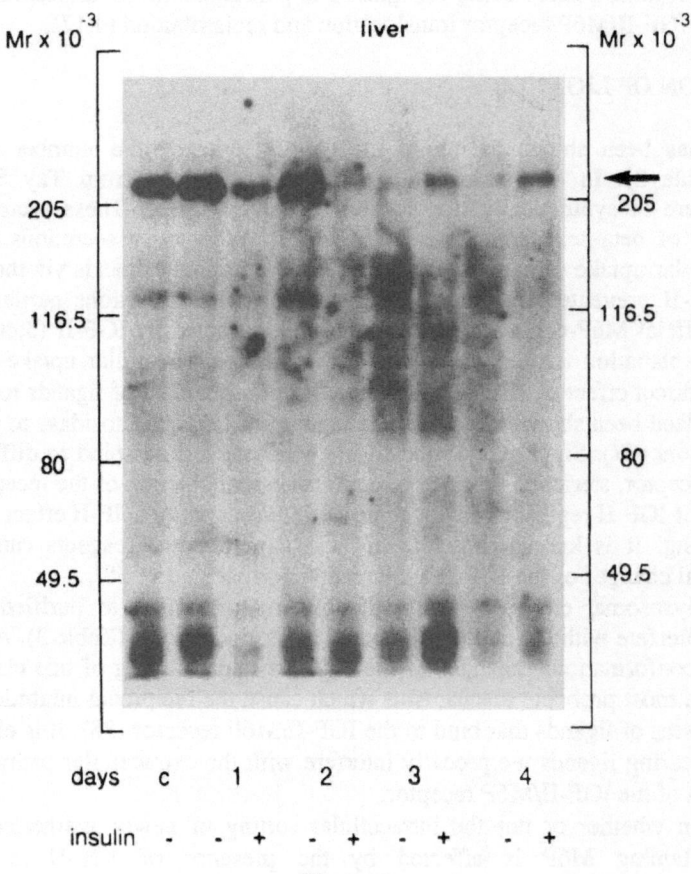

Figure 3

M6P-containing proteins. In the same model situation, overexpression of IGF-II did not affect the binding and uptake of arylsulfatase A, an enzyme bearing the M6P recognition marker (3). Several other groups are currently investigating whether or not the expression of IGF-II plays a role in modulating the sorting of lysosomal enzymes from the Golgi to the lysosome.

RECEPTOR EXPRESSION

The expression of the IGF-II/M6P receptor varies considerably (Table 5). Variable levels of the IGF-II/M6P receptor are measured in different tissues: in the rat and human, heart,kidney and thymus are among the organs which contain high levels of the receptor. Brain expresses only small amounts of IGF-II/M6P receptor (26, 60, 70). There is a strong developmental pattern of receptor expression in the rat and sheep, with high concentrations of IGF-II/M6P receptor found during fetal life. After birth, receptor concentrations in all tissues decline dramatically (60). In the human, the developmental regulation of receptor levels is less evident (26). It is unclear whether the high levels of receptor protein and mRNA expression during fetal life reflects the involvement of this receptor in growth processes. Alternatively, the receptor could function in tissue remodeling or play a role in tissue differentiation. For example, it has been reported that both IGF-II and IGF-II/M6P receptor levels increase during skeletal muscle differentiation (24,72). Compensatory growth of hepatic, thyroid or renal tissue is accompanied by increased levels of receptors in such tissues (6,65,66). In addition, a number of pathologic states such as the onset of diabetes (25,73) or fetal starvation (27) seem to modulate IGF-II/M6P receptor expression. In addition, a number of hormones are thought to be capable of regulating IGF-II/M6P receptor levels at least in some tissues (10,12,43,48). The physiologic role of such differential expression of the receptor has not been elucidated.

ALTERNATIVE RECEPTORS

It is clear that the IGF-II/M6P receptor binds IGF-II with high affinity. Recently, the existence of additional high affinity receptors for IGF-II have been proposed. Such receptors are thought to be heterotetrameric receptors sharing high homology with the insulin and the IGF-I receptor but no structural similarity with the IGF-II/M6P receptor (8,21,28,51,71). These alpa2beta2 IGF receptors could represent insulin/IGF-I receptor hybrids with high affinity for IGF-II. Alternatively, such receptors could be insulin or IGF-I receptor subtypes having acquired exceptional binding characteristics involving IGF-II (34,51,59). The purification and molecular characterization of such putative IGF-II selective receptors has been difficult mainly because of the very low abundance of atypical IGF-II binding that is found in most tissues and cells.

CONCLUSIONS

The IGF-II/M6P receptor is a bifunctional binding protein which binds IGF-II and M6P containing ligands at distinct binding sites. While its role in targeting lysosomal enzymes to lysosomes has been elucidated, the physiologic role of IGF-II binding to this receptor remains a mystery. The variable expression during development and in different cells and tissues and the delicate modulation of receptor expression by many factors that are linked to growth and development make it most likely that the receptor serves a key role in tissue growth and development. Differential localization within a cell might not only be important for lysosomal enzyme sorting but also for any putative role of IGF-II exerting a signaling function via the recptor.

Table 5. Modulation of IGF-II/M6P receptor expression

regulatory factor	reference
in vivo:	
pre/postnatal development	(26, 32, 60, 70))
muscle differentiation	(24, 72)
diabetes onset	(25, 73)
liver regeneration	(6)
hyperplasia of the thyroid	(66)
fetal starvation	(27)
glucose infusion	(27)
compensatory growth of the kidney	(65)
in vitro:	
insulin	(10, 12, 43)
GH	(43)
estradiol	(48)

ACKNOWLEDGEMENT

This work was supported by Deutsche Forschungsgemeinschaft, DFG Ki 365/1.1-3, Bonn, Germany, PROCOPE, Paris-Bonn, and the Danish Diabetes Assoziation.

REFERENCES

1 Barlow DP, Stoeger R, Herrmann BG, Saito K, Schweifer N. The mouse IGF-2 receptor is imprinted and closely linked to the Tme locus. Nature 349:84-87 (1991)

2 Beukers MW, Oh Y, Zhang H, Ling N, Rosenfeld RG. (Leu 27)-IGF-II is highly selective for the type II IGF receptor in binding, crosslinking and thymidine incorporation experiments. Endocrinology 128:1201-1203 (1991)

3 Braulke T., Bresciani R., Buergisser D.M., K. von Figura. IGF-II overexpression does not affect sorting of lysosomal enzymes in NIH 3T3 cells. Biochem Biophys Res Commun 179: 108-115 (1991)

4 Braulke T, Mieskes G. Role of protein phosphatases in IGF-II-stimulated M6P/IGF-II receptor redistribution. J Biol Chem. 267:17347-17353 (1992)

5 B_rgisser DM, Roth BV, Giger R. Mutants of human IGF-II with altered affinities for the type I and type II IGF receptor. J Biol Chem. 266:1029-1033 (1991)

6 Burguera B, Werner H, Sklar M, Shen-Orr Z, Stannard B, Roberts CT, Nissley SP, Vore SJ, Caro JF, LeRoith D. Liver regeneration is associated with increased expression of the IGF-II/M6P receptor. Molec. Endocrinol 4:1539-1545 (1990)

7 Canfield WM, Kornfeld S. The chicken liver cation-independent M6P receptor lacks the high affinity binding site for IGF-II. J Biol Chem. 264: 7100-7103 (1989)

8 Cassella SJ, Han VK, D'Ercole AJ, Svoboda ME, Van Wyk JJ. IGF-II binding to the type I somatomedin receptor. J. Biol. Chem. 261:9268-9273 (1986)

9 Clairmont KB, Czech MP. Chicken and Xenopus M6P-receptors fail to bind IGF-II. J Biol Chem. 264: 16390-16392 (1989)

10 Clairmont KB, Czech MP. Insulin injection increases the levels of serum receptors for transferrin and IGF-II/M6P in intact rats. Endocrinology 127:1568-1573 (1990)

11 Clairmont KB, Czech MP. Extracellular release as the major degradative pathway of the IGF-II/M6P receptor. J Biol Chem. 266:12131-12134 (1991)

12 Cormont M, Gremeaux T, Tanti JF, Van Obberghen E, Le Marchand-Brustel Y. Polymyxin B inhibits insulin-induced glucose transporter and IGF-II receptor translocation in isolated adipocytes. Eur. J. Biochem. 207:185-193 (1992)

13 Corvera S, Folander K, Clairmont KB, Czech MP. A highly phosphorylated subpopulation of IGF-II/M6P receptors is concentrated in a clathrin-enriched plasma membrane fraction. Proc. Natl. Acad. Sci. USA 85:7567 (1988)

14 Cui S, Flyvbjerg A, Nielsen S, Kiess W, Christensen EI. Distribution of IGF-II/M6P receptors in rat kidney: evidence for apical localization in proximal tubule cells. Kidney Intern. 43:796-807 (1993)

15 Dahms N.M., P. Lobel, S. Kornfeld. Mannose-6-phosphate receptors and lysosomal enzyme targeting. J Biol Chem 264:12115-12118 (1989)

16 Damke H., Klumpermann J., von Figura K., Braulke T. Effects of Brefeldin A on the endocytic route. Redistribution of M6P/IGF-II receptors to the cell surface. J Biol Chem. 266:24829-24833 (1991)

17 Damke H, von Figura K, Braulke T. Simultaneous redistribution of M6P and transferrin receptors by IGFs and phorbol ester. Biochem J. 281:225-229 (1992)

18 Damke H, Klumperman J, von Figura K, Braulke T. Brefeldin A affects the cellular distribution of endocytic receptors differentially. Biochem Biophys. Res Commun. 185:719-727 (1992)

19 DeChiara T.M., A. Efstradiadis , E.J. Robertson . A growth-deficiency phenotype in heterozygous mice carrying an IGF-II gene disrupted by targeting. Nature 345: 78-80 (1990)

20 DeChiara TM, Robertson EJ, Efstradiadis A. Parental imprinting of the mouse IGF-II gene. Cell 64:849-859 (1991)

21 Domeyne A, Pinset C, Montarras D, Garandel V, Rosenfeld RG, Barenton B. Preferential binding of IGF-II to a putative alpha2beta2 IGF-II receptor type in C2 myoblasts. Eur J Biochem. 208:273-279 (1992)

22 Figura von K., A. Hasilik. Lysosomal enzymes and their receptors.Ann. Rev. Biochem. 55:167-193 (1986)

23 Figura von K. Molecular recognition and targeting of lysosomal proteins. Curr Opinion in Biol 3:642-646 (1991)

24 Florini J, Magri KA, Ewton ZA, James PL, Grindstaff K, Rotwein PS. Spontaneous differentiation of skeletal myoblasts is dependent upon autocrine secretion of IGF-II. J. Biol. Chem. 266:15917-15923 (1991)

25 Flyvbjerg A, Kessler U, Funk B, Dorka B, Orskov H, Kiess W. Transient increase of IGF-II/M6P receptor protein content in diabetic rat kidney and liver tissue. Diabetologia (submitted) (1993)

26 Funk B, Kessler U, Eisenmenger W, Hansmann A, Kolb H, Kiess W. Expression of the human IGF-II/M6P receptor during fetal life and early infancy. J. Clin Endocrinol. Metab. 75:424-431 (1992)

27 Gallaher BW, Oliver MH, Eichorn K, Kessler U, Kiess W, Harding JE, Gluckman PD, Breier BH. The circulating IGF-II/M6P receptor and IGFBPs in fetal sheep plasma are regulated by glucose and insulin. Endocrinology (submitted) (1993)

28 Garafalo RS, Barenton B. Functional and immunological distinction between IGF-I receptor subtypes in KB cells. J Biol Chem 275: 7735-7738 (1992)

29 Guse A, Kiess W, Funk B, Kessler U, Berg I, Gercken G. Identification and charcterization of IGF receptors on adult rat cardiac myocytes: linkage to inositol 1,4,5 trisphosphate formation. Endocrinology 130:145-151 (1992)

30 Haig D, Graham C. Genomic imprinting and the strange case of the IGF-II receptor. Cell 64:1045-1046 (1991)

31 Hartmann H, Meyer-Alber A, Braulke T. Metabolic actions of IGF-II in cultured adult rat hepatocytes are not mediated through the IGF-II receptor. Diabetologia 35:216-223 (1992)

32 Harvey MB, Kaye PL. IGF-II receptors are first expressed at the 2-cell stage of mouse development. Development 111:1057-1060 (1991)

33 Hu K-Q, Backer JM, Sahagian G, Feener EP, King GL. Modulation of the IGF-II/M6P receptor in microvascular endothelial cells by phorbol ester via protein kinase C. J Biol Chem 265:13864-13870 (1990)

34 Jonas HA, Cox AJ. IGF binding to the atypical insulin receptors of a human lymphoid-derived cell line (IM-9). Biochem. J. 266:737-742 (1990)

35 Kessler U, Aumeier S, Funk B, Kiess W. Biosynthetic labeling of beta-hexosaminidase B: inhibition of the cellular uptake of ^{3}H hexosaminidase B by IGF-II in rat C6 glial cells. Mol Cell. Endocrinol 90:147-153 (1992)

36 Kiess W, Greenstein LA, White RM, Lee L, Rechler MM, Nissley SP. Type II IGF receptor is present in rat serum. Proc Natl Acad Sci USA 84:7720-7724 (1987)

37 Kiess W, Thomas CL, Greenstein L, Lee L, Sklar MM, Rechler MM, Sahagian GG, SP Nissley SP. IGF-II inhibits both the cellular uptake of beta-galactosidase and the binding of beta-galactosidase to purified IGF-II/M6P receptor. J.Biol.Chem. 264:4710-4714 (1989)

38 Kiess W, Thomas CL, Sklar MM, Nissley SP. Beta-galactosidase decreases the binding affinity of the IGF-II/M6P receptor for IGF-II. Eur J. Biochem 190:71-77 (1990)

39 Kojima I, Nishimoto I., Iiri T, Ogata E, Rosenfeld RG. Evidence that the type II IGF receptor is coupled to calcium gating system. Biochem. Biophys. Res. Commun. 154:9 (1988)

40 Kornfeld S. Structure and function of the M6P/IGF-II receptors. Ann.Rev.Biochem. 61:307-330 (1992)

41 Laureys G, Barton DE, Ullrich A, Francke U. Chromosomal mapping of the gene for the type II IGF receptor/ cation-independent M6P receptor in man and mouse.Genomics 3:224-229 (1988)

42 Lobel P, Fujimoto K, Ye RD, Griffiths G, Kornfeld S. Mutations in the cytoplasmic domain of the 215 kDa M6P receptor differentially alter lysosomal enzyme sorting and endocytosis.Cell 57:787 (1989)

43 Lonroth P, Assmundsson K, Eden S, Enberg G, Gause I, Hall K, Smith U. Regulation of IGF-II receptors by GH and insulin in rat adipocytes. Proc. Natl. Scad. Sci. USA 84:3619 (1987)

44 Lonroth P, Appell KC, Wesslau C, Cushman SW, Simpson IA, Smith U. Insulin-induced subcellular redistribution of IGF-II receptors in the rat adipose cell. Counterregulatory effects of isoproterenol, adenosine, and cAMP analogues. J Biol Chem. 263:15386 (1988)

45 MacDonald R., Pfeffer ,S.R., Coussens, L., Tepper, M.A., Brocklebank,C.M., Mole, J.E., Anderson, J.K., Chen, E., Czech, M.P., Ullrich , A. A single receptor binds both IGF-II and mannose-6-phosphate. Science 239: 1134-1136 (1988)

46 MacDonald RG, Trepper MA; Clairmont KB, Perregaux SB, Czech MP. Serum form of the rat IGF-II/M6P receptor is truncated in the carboxyl-terminal domain. J Biol Chem. 264:3256-3261 (1989)

47 Mathieu M, Rochefort H, Barenton B, Prebois C, Vignon F. Interactions of cathepsin D and IGF-II on the IGF-II/M6P receptor in human breast cancer cells and possible consequences on mitogenic activity of IGF-II. Mol.Endocrinol. 4:1327-1335 (1990)

48 Mathieu M. Vignon F, Capony F, Rochefort H. Estradiol down-regulates the M6P/IGF-II receptor gene and induces cathepsin-D in breast cancer cells: a receptor saturation mechanism to increase the secretion of lysosomal proenzymes. Mol. Endocrinology 4: 1327-1335 (1991)

49 Melancon P, Glick BS, Malhotra V, Weidman PJ, Serafini T, Gleason ML, Orci L, Rothman JE. Involvement of GTP-binding `G´proteins in transport through the Golgi stack. Cell 51:1053-1062 (1987)

50 Minniti CP, Kohn EC, Grubb JH, Sly WS, Y. Oh, HL M_ller, RG Rosenfeld, LJ Helman. The IGF-II/M6P receptor mediates IGF-II induced motility in human rhabdomyosarcoma cells. J.Biol.Chem. 267:9000-9004 (1992)

51 Misra P, Hintz R, Rosenfeld RG. Structural and immunological characterization of IGF-II binding to IM-9 cells. J Clin Endocrinol Metab. 63:1400 (1986)

52 Morgan, D.O., J. C. Edman, D.N. Standring, V.A. Fried, M.C. Smith, R.A. Roth, W.J. Rutter. Insulin-like growth factor-II receptor as a multifunctional binding protein. Nature 329:301-307 (1987)

53 Murayama Y, Okamoto T, Ogata E, Asano T, Iiri T, Katada T, Ui M, Grubb JH, Sly WS, Nishimoto I.Distinctive regulation of the functional linkage between the human cation-independent mannose-6-phosphate receptor and GTP-binding proteins by IGF-II and M6P. J Biol Chem 265:17456-17462 (1990)

54 Nishimoto I, Ogata E, Okamoto T. Guanine nucleotide-binding protein interacting but unstimulating sequence located in IGF-II receptor. J Biol Chem 266:12747-12751 (1991)

55 Nissley SP, Kiess W, Sklar MM. The IGF-II/M6P receptor. In: LeRoith D, Raiszada M, (Eds.), Molecular and cellular biology of IGFs and their receptors. Plenum Press, New York, pp 359-368.(1990)

56 Nissley SP, Kiess W, Sklar MM. The IGF-II/mannose-6-phosphate receptor. in: IGFs: molecular and cellular aspects. Ed. D.LeRoith . CRC Press, Boca Raton, pp 111-150 (1991)

57 Nissley SP, Kiess W. Reciprocal modulation of binding of lysosomal enzymes and IGF-II to the M6P/IGF-II receptor. In: Molecular biology of insulin, IGFs and their receptors. Eds. D.LeRoith, MK Raizada, Plenum Press, New York, (1991)

58 Nissley SP, Kiess W. Binding of IGF-II and lysosomal enzymes to the IGF-II/M6P receptor.In: Modern concepts of IGFs. Ed. M. Spencer, pp 419-430, Elsevier Publisher, Amsterdam (1991)

59 Nissley SP, Lopeszinsky W. IGF receptors. Growth factors 5:29-43 (1991)

60 Nissley P, Kiess W, Sklar MM The developmental expression of the IGF-II/M6P receptor. Molecular Reproduction & Developm (in press) (1993)

61 Nolan CM , JW Kyle , H Watanabe, WS Sly. Binding of IGF-II by human cation-independent mannose-6-phosphate receptor/IGF-II receptor expressed in receptor-deficient mouse L cells. Cell Regulation 1:197-213 (1990)

62 Okamoto T , Katada T , Murayama Y , Ui M , Ogata E , Nishimoto I   A simple structure encodes G protein-activating function of the IGF-II/mannose-6-phosphate receptor. Cell 62:709-717 (1990)

63 Okamoto T, Ohkuni Y, Ogata E, Nishimoto I. Distinct mode of G protein activation due to single residue substitution of active IGF-II receptor peptide Arg^{2410}-Lys^{2423}: evidence for stimulation acceptor region other than C-terminus of G_{ialpha}. Biochem Biophys Res Commun. 179:10-16 (1991)

64 Pfeffer SR. Mannose-6-phosphate receptors and their role in targeting proteins to lysosomes. J Membr Biol 103:7 (1988)

65 Polychronakos C, Guyda HJ, Posner BI. Increase in the type 2 IGF receptors in the rat kidney during compensatory growth. Biochem Biophys Res. Commun. 132: 418-423 (1985)

66 Polychronakos C, Guyda HJ, Patel B, Posner BI.Increase in the number of type II IGF receptors during propylthiouracil-induced hyperplasia in the rat thyroid. Endocrinology 119: 1204 (1986)

67 Polychronakos C, Guyda HJ, Janthly U, Posner BI. Effects of M6P on receptor-mediated endocytosis of IGF-II. Endocrinology 127:1861-1866 (1990)

68 Rechler MM , Nissley SP .Insulin-like growth factors. In: Sporn MB, Roberts AB,eds. Peptide growth factors and their receptors I. Handbook of Pharmacology, Heidelberg, New York, Springer Publisher; 263-367 (1990)

69 Roth RA. Structure of the receptor for IGF-II: the puzzle amplified. Science 239:1269-1271 (1988)

70 Senior PV, Byrne S, Brammar WJ, Beck F. Expression of the IGF-II/M6P receptor mRNA and protein in the developing rat. Development 109:67-73 (1990)

71 Shier P, Watt VM. Primary structure of a putative receptor for a ligand of the insulin family. J Biol Chem. 264:14605-14608 (1989)

72 Tollefsen SE, Sadow JL, Rotwein P. Coordinate expression of IGF-II and its receptor during muscle differentiation. Proc. Natl Acad Sci USA 86:1543-1547 (1989)

73 Werner H, Shen-Orr Z, Stannard B, Burguera B, Roberts CT, LeRoith D. Experimental diabetes increases IGF-I and IGF-II receptor concentration and gene expression in kidney. Diabetes 39:1490-1497 (1990)

74 Westlund B, Dahms NM, Kornfeld S. The bovine M6P/IGF-II receptor: localization of M6P binding sites to domains 1-3 and 7-11 of the extracytoplasmic region. J. Biol. Chem. 266:23233-23239 (1991)

75 Willison K. Opposite imprinting of the mouse IGF-II and IGF-II receptor genes. Trends in Genetics 7:107-108 (1991)

76 Yang YWH, Robbins AR, Nissley SP, Rechler MM. The chick embryo fibroblast cation-independent M6P receptor is functional and immunologically related to the mammalian IGF-II/M6P receptor but does not bind IGF-II. Endocrinology 126:1177-1189 (1991)

PARENTAL IMPRINTING OF THE GENES FOR IGF-II AND ITS RECEPTOR

Constantin Polychronakos

Associate Professor, Department of Pediatrics,
Division of Endocrinology, McGill University
Molecular Endocrinology Laboratory
Montreal Children's Hospital Research Institute, H3H 1P3

INTRODUCTION

One of the basic tenets of Mendelian genetics is that autosomal traits are equivalently transmitted from each of the two parents. In recent years the molecular corollary of this has also been demonstrated for the majority of the genes studied: Both copies of each autosomal gene are equivalently expressed at the mRNA level, regardless of the parent from which each was derived.

The phenomenon of parental genomic imprinting is an exception to this rule. It is now clear that a small number of genes preferentially produce mRNA transcripts from the gene copy derived from the parent of a specific sex. Four murine genes with this property have so far been identified, two exclusively transcribed from the maternal copy, and two from the paternal (Table 1). The significance of the phenomenon for the insulin-like growth factor (IGF) field lies in the fact that two of the four genes are those coding for IGF-II and the IGF-II/mannose 6-phosphate receptor (IGF-II/M6P-R).

THE PHENOMENON OF PARENTAL IMPRINTING

Genomic imprinting was first directly demonstrated with pronuclear transplantation experiments involving mouse zygote constructs carrying pronuclei that were both derived from oocytes (gynogenotes) or sperm (androgenotes). Neither zygote resulted in viable mice, and this was shown to be due to nuclear rather than cytoplasmic factors[1-3], indicating that genome copies derived from both parents are necessary for normal development. More interestingly, the two lethal phenotypes are drastically different[1]. The embryo proper is relatively spared in gynogenotes but the placenta and extraembryonic membranes are atrophic, while androgenotes have relatively well developed placenta but very hypoplastic embryo proper. Complete hydatiform moles, abnormal products of conception carrying two copies of the paternal genome are the human equivalent of androgenotes and display similar features[4].

Current Directions in Insulin-Like Growth Factor Research,
Edited by D. LeRoith and M.K. Raizada, Plenum Press, New York, 1994

189

Table 1. Mammalian genes known to be imprinted.

Gene	Expressed copy	Function	Species (chromosome)
IGF2	paternal	Fetal growth factor	mouse (distal 7) human (11p15.5)
IGF2R	maternal	Sorting of lysosomal enzymes. (?) IGF-II degradation.	mouse (17)
H19	maternal	Unknown. Functions as RNA	mouse (distal 7) human (11p15.5)
SNRPN	paternal	RNA splicing in brain and heart	mouse (central 7)

A finer characterization of the non-equivalence of the two parental sets of genes was possible using mice carrying balanced chromosomal translocations. When two mice carrying the same balanced translocation are mated, a certain fraction of their progeny will lose a chromosomal fragment, but will be rescued by an excess copy of the identical fragment from the other parent. Such animals have a normal gene complement but carry a chromosomal segment both copies of which are derived from the same parent (paternal or maternal disomy). They are usually viable but often have gross abnormalities that differ according to parental origin of the segment involved[5,6,7]. Sometimes the phenotypes of maternal vs. paternal uniparental disomy have reciprocal features[7], attributable to absence vs. double dose of a gene. Thus, certain areas of the mouse genome have been characterized as carrying imprinted genetic material, while others do not, at least at the gross phenotypic level[6].

Thus, although the basic information contained in the DNA sequence of the genome is identical in the oocyte or the sperm, the two copies derived from each clearly behave differently. Passage through the male or female germline must, then, leave an "imprint" on the basis of which somatic cells can recognize gene copies as maternal or paternal even after repeated cell divisions and DNA replications.

These observations provide the basis for a mechanistic explanation of several apparently non-Mendelian patterns of inheritance in human disease phenotypes: The severity and/or age of onset of several dominant diseases, including neurofibromatosis[8] spinocerebellar ataxia[9], Huntington's disease[10], and myotonic dystrophy[11], is substantially affected by the sex of the transmitting parent. The Prader-Willi and Angelman syndromes provide an even more striking and direct example of imprinting in human disease: These two distinct dysmorphic phenotypes, both involving central nervous system dysfunction, can be caused by the same chromosomal deletion at 15q11-13, the former appearing when the deletion is on the paternal chromosome, and the latter when on the maternal one[12]. Complementation with identical genetic material from the other parent has no effect, as cases of each syndrome have been identified with 15q11-13 uniparental disomy from the reciprocal parental sex but no net deletion[13,14]. Another imprinted human phenotype is the Beckwith-Wiedemann syndrome, which involves parental-origin specific abnormalities of the tip of the short arm of chromosome 11, the exact locus where *IGF2* maps. Because of the mounting evidence that *IGF2* is the gene involved in this disease, it will be discussed in greater detail below.

A final line of evidence implicating imprinting in human disease is the non-random loss of chromosomal material of a specific (usually maternal) origin in malignant tumors. This is discussed in more detail in reference to the relation between *IGF2* imprinting and Wilms' tumor.

IMPRINTING OF THE IGF II GENE

Imprinting of Igf2, the mouse IGF-II gene, at the phenotypic and molecular level was first demonstrated by DeChiara et al., using a transgenic mouse model carrying an IGF-II gene disrupted by homologous recombination. As expected, homozygous animals failed to express IGF-II and were small at birth with normal post-natal growth, thus confirming the long-held view that IGF-II is a fetal growth factor[16]. Unexpectedly, however, an identical phenotype was seen in heterozygous mice inheriting the disrupted gene from their father[17]. These animals failed to produce any IGF-II mRNA. Mice inheriting the same disrupted gene from their mother were phenotypically normal and expressed IGF-II at levels indistinguishable from controls.

The same paper demonstrated that the phenomenon was repeated over several successive generations, showing that the imprint was completely reversible: maternally inherited *Igf2* copies that were inactive in male animals were fully expressed in their offspring.

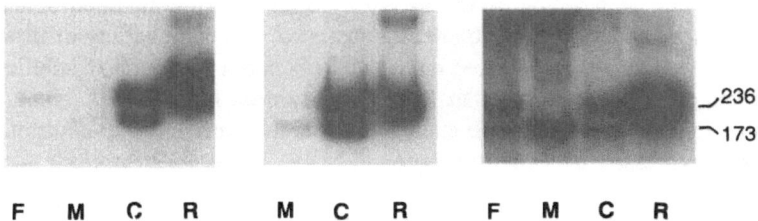

Figure 1. Monoallelic expression of *IGF2* in human placenta. An exon 9 fragment, containing a polymorphic *Apa*I site was PCR-amplified from the genomic DNA of each parent (F or M) or the child (C), as well as from reverse-transcribed placental RNA (R). After digestion with *Apa*I, two alleles can be seen in the child's genomic DNA, but only one is expressed. The expressed allele can, in each case, be identified as paternal as the mother is homozygous for the other allele.

Interestingly, IGF-II expression in the central nervous system, the only tissue where significant amounts of IGF-II are expressed post-natally, appeared to be exempt from imprinting, as it was observed at a normal level in mice whose only functional *Igf2* copy was maternal[17]. This clearly indicates that the phenotypic expression of gene imprinting is subject to tissue-specificity and, possibly, developmental regulation.

We have recently demonstrated that *IGF2*, the human homologue gene, is imprinted in an identical fashion, using a different approach[18]: In order to distinguish between transcripts derived from each copy of the gene, we used a transcribed *Apa*I RFLP in the 3' untranslated region (3' UTR) of the gene. We examined RNA from kidney and placenta of fetuses heterozygous for the polymorphism by reverse-transcription-PCR. Digestion of the PCR products with *Apa*I allowed us to distinguish between transcripts derived from each gene copy. In all eight fetuses thus studied, RNA transcripts from only one of the two genomic copies could be detected. The phenomenon was not specific for a particular allele, as each of the two alleles was found to be exclusively transcribed in different individuals. In all three cases where parental DNA was informative, the transcribed allele was shown to be paternal, as the mother was a homozygote for the non-transcribed allele (Figure 1). This observation was simultaneously and independently confirmed by three other laboratories[19-21].

IMPRINTING OF THE MANNOSE 6-PHOSPHATE/IGF-II RECEPTOR GENE

IGF-II binds to two membrane receptors. The type I receptor recognizes both IGFs with high affinity and is believed to be the transducer of their mitogenic actions via a tyrosine phosphorylation pathway[22]. In addition, IGF-II also binds to the mannose 6-phosphate- IGF-II receptor (M6P/IGF-II-R), a transmembrane protein with no apparent role in transducing IGF-II mitogenic signals, and whose only demonstrated biologic action is the targeting of lysosomal enzymes to the lysosomes, through its separate binding site for M6P. [23]. It appears, indeed, likely, that the main purpose of the IGF-II binding site on the M6P/IGF-II-R is lysosomal targeting of IGF-II for degradation. In the rodent fetus, this receptor is expressed at levels that are more than twenty times higher than post-natally[24].

A considerable amount of interest was, therefore, generated when Barlow et al.[25] showed that the murine gene for this receptor (*Igf2r*) is also imprinted, following an independent line of investigation based on the study of *Tme*, a locus on mouse chromosome 17. Deletion of *Tme* is lethal at 15 days of gestation in *mus musculus*, but only when the deletion is on the maternally derived chromosome. Animals carrying the deletion on the paternal chromosome are phenotypically normal. Four genes, including *Igf2r*, are known to map at *Tme* and mRNA from all but *Igf2r* can be detected in 15-day embryos carrying the maternal deletion. *Igf2r* mRNA, on the other hand, normally abundantly expressed in a wide variety of tissues at this gestational age, is totally absent in those embryos. This is not the result of allelic variation, as embryos identical by descent for that locus fully express *Igf2r* mRNA when paternally derived[25]. Thus, *Igf2r* is imprinted with exclusive maternal expression. In addition, although other maternally expressed genes may be present in the locus, *Igf2r* appears necessary for fetal survival.

It is not known whether the human *IGF2R* gene is also imprinted. To address this question, we used an approach similar to that described above for *IGF2*. Two transcribed CA repeat polymorphisms, both in the 3'UTR, one published[26] and one recently identified in our laboratory, were used. In RNA form term placenta of heterozygous newborns, carefully dissected to eliminate any contamination with decidua or maternal blood, both alleles were detected by RT-PCR. It appears, therefore, that the human *IGF2R* is not imprinted, at least in term placenta. Such species differences may be important in the elucidation of the molecular mechanisms and the biologic purpose of imprinting.

THE BIOLOGIC PURPOSE OF IMPRINTING

It is difficult to imagine what evolutionary advantage could be gained by expressing only the paternal or only the maternal copy of a gene. The occurrence of this phenomenon in genes coding for a functional pair of proteins, in copies derived from opposite parental sexes, provides the first opportunity to attempt insight into this question. As mentioned above, if it is accepted that the function of M6P/IGF-II-R in relation to IGF-II is to target it to the lysosomes for degradation, then these two proteins would serve opposing purposes. This has prompted Haig and Graham[27] to put forward the following conjecture: Mutations that enhance the expression or activity of fetal growth factors result in larger newborns, better fit for survival. However, where food availability is limiting, excessive transfer of nutrients to the fetus diminishes the mother's fitness, compromising her ability to survive child-bearing and to reproduce again, and compromising the viability of littermates carrying different alleles from the same or a different father. Therefore, enhanced allelic forms of a fetal growth factor might best serve their own propagation by remaining silent when transmitted from the mother, and being fully expressed when transmitted from the father. By the reciprocal reasoning, genes acquiring properties that tip the balance of nutrients in the mother's favor would promote their own survival by not being expressed when paternally transmitted. The extraordinarily high levels of expression of both IGF-II and M6P/IGF-II-R in the rodent fetus[24,28,29] are certainly consistent with such spiraling competitive evolutionary drive.

The situation in the human is different. IGF-II is expressed at high levels in the fetus[30,31], but measurement of M6P/IGF-II-R levels by quantitative immunoblotting in many tissues from stillborns and post-natal autopsy material failed to show in the human anything approaching the dramatically increased fetal expression seen in the rat[32]. We have corroborated this by measuring the soluble form of the receptor in amniotic fluid and serum from living, healthy fetuses and newborns[33]. Thus IGF-II seems to be regulated similarly in rodents and humans, in terms of high-level fetal expression and repression of the maternal gene copy. Unlike the mouse, however, the human fetus does not appear to either express M6P/IGF-II-R at levels higher than post-natally or repress the paternal copy. In Homo Sapiens, a monogamous species with single-fetus gestations, the evolutionary pressure driving the events postulated by Haig and Graham would be less intense. As a result, high fetal expression of *IGF2R* and the need to imprint its expression may have been lost or never developed.

Haig's conjecture makes the testable prediction that imprinted genes discovered in the future will code for proteins whose function, in a situation of competition for nutrients, favors the fetus if paternally expressed, and the mother if maternally expressed. The two additional imprinted genes found since the conjecture was made may be examined in this light.

1. *H19*, codes for a polyadenylated cytoplasmic RNA which does not appear to be translated into protein, and whose function is obscure. Its short physical distance from *IGF2* (<200 kb in the human, <90 kb in the mouse) and its reciprocal imprinting[34-36] has led to the speculation that its transcription, in and by itself, suppresses *IGF2* expression in cis[37]. If this is its main function, then its exclusive maternal expression would be consistent with the conjecture.

2. *SNRPN* (small nuclear ribonucleoprotein N), expressed only form the paternal copy, codes for a small peptide found in a ribonucleoprotein complex in particles involved in the splicing of RNA in brain and heart[38]. Its exclusive paternal expression in the mouse[39], as well as its status as the only gene mapping to the critical Prader-Willi region[40] makes it a highly likely candidate for this disease. It is not obvious how such a gene product could in any way fit Haig's conjecture, suggesting that the role of imprinting involves functions other than transplacental nutrient partition.

Whatever the biologic purpose of imprinting, it must involve functions related to parent-offspring interactions, as it is difficult to imagine any other context in which functional distinction of parental origin would have the biologic effects required to drive evolutionary emergence of the phenomenon. Such interactions need not be intrauterine, but may be related to lactation or genetic determinants of social interchanges important for gene survival, such as food consumption or mating. In this respect, it is interesting that many of the features of Prader-Willi syndrome, quite likely due to total failure to express *SNRPN*, suggest disorders of such functions: Severe neonatal hypotonia, seriously interfering with spontaneous lactation, followed by uncontrollable hyperphagia later in life, and hypothalamic hypogonadism.

IMPRINTING AND DNA METHYLATION

The chemical nature of the imprinting modification has not been firmly established, but the transcriptional behavior of imprinted genes allows inferences about it: a) the modification must be reproduced through many rounds of DNA replication, allowing the transcription machinery of somatic cells to recognize the imprinted gene copy many cell divisions after fertilization. b) it must be reversible on passage through the germline. This practically rules out that the modification is a structural mutation of the DNA sequence. Such non-sequence change of DNA has been termed an epigenetic modification. c) finally, the imparting of the new imprint, as well as the erasure of the old one, must be happening in germline cells, as it is only there that the chromosomes are segregated by the sex of the transmitting parent.

Cytosine methylation is the only covalent DNA modification that satisfies these criteria, and recent evidence supports the widely held expectation that it is the basis of gene imprinting. Direct evidence, discussed below, has firmly established that imprinted genes are differentially methylated according to parental origin. The remaining question is whether methylation is a cause or a consequence of imprinting.

Cytosine methylation at position 5 is a very common DNA modification. In mammals it occurs almost exclusively in a 5'-CG-3' dinucleotide, often referred to as a CpG. It is estimated that as much as 80-90% of the CGs in the mammalian genome are thus methylated[41]. Methylation is found most consistently and abundantly on DNA sequences that are not actively transcribed, such as outside genes or in genes subject to long-term inactivation as a result of tissue specificity, developmental regulation of their expression, or random X chromosome inactivation in females[41]. Such methylation can be the initiating event in gene inactivation or a consequence of it. There is good evidence, reviewed by Hergerzberg[41], that either can be the case the mammalian genome.

Practically all methylation in the mouse is performed by a single methyltransferase[42]. The specificity of methylation of particular genes must therefore be determined by molecules that interact with both DNA and the enzyme. These may be proteins that either bind to DNA in a dynamic way, such as transcription factors and repressors, or are more stably associated with it, such as histones and other proteins determining chromatin structure[43-45] (reviewed by Hergersberg[41], and Szyf[46]).

Specific proteins whose DNA binding properties are affected by methylation have been identified. Methylation usually inhibits the binding of such proteins, including tanscription factors[47,48], while the ubiquitous Sp1 site, which contains a CG dinucleotide is unaffected[49]. Proteins that bind methylated DNA preferentially have also been described[50], and they appear to inhibit transcription[51].

The imprinting modification may be on the active gene copy (positive imprint) or on the inactive one (negative imprint). Because of the association of imprinting, methylation, and gene repression, it was widely assumed that parental imprints are negative. Recent evidence, reviewed below, seriously challenges this concept.

Transgene methylation

Some information about the molecular mechanisms involved in imprinting became available by studying the phenomenon of methylation imprinting in certain transgenes. A substantial percentage of transgenes used for overexpression of specific coding sequences and randomly integrated into the genome, can be shown to be methylated only when transmitted by the mother[52-55]. Although this methylation is not always reversible when the gene passes again through the male germline, and affects expression of the gene only occasionally[53,54], the observation helped establish a firm connection between methylation and imprinting.

THE MOLECULAR BASIS OF *Igf2* AND *Igf2r* IMPRINTING

The frequency of CG dinucleotides in the genome is much lower than expected by a random arrangement of bases. An exception are CpG islands, sequences found around the transcription initiation sites of many genes, which contain many CG dinucleotides, and which are kept actively demethylated[56]. As CG island methylation is found in the inactivated X

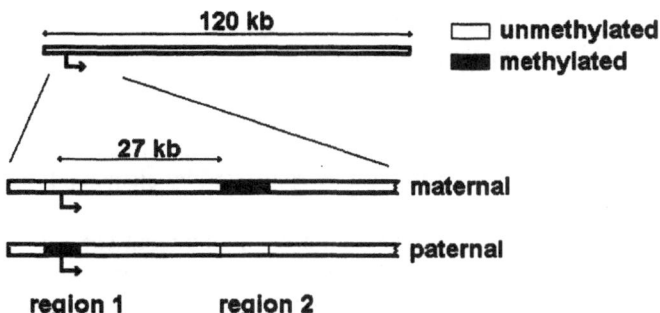

120 kb

☐ unmethylated
■ methylated

27 kb

maternal

paternal

region 1   region 2

Figure 2. Schematic representation of *Igf2r* methylation, as demonstrated by Stöger et al. Methylation of region 1, encompassing the transcription initiation site (arrow), is present in the inactive paternal copy, and appears to be secondary to gene inactivity. The primary imprint is in region 2, contained in an intron, 27 kb downstream, and it is a positive imprint, as it is found only on the active maternal copy.

chromosome[57], it is one candidate mechanism for chromosome-specific gene inactivation, such as seen in imprinting. In this respect it is interesting that both *Igf2* and *Igf2r* have CG islands around their promoters[58,59]. These islands were recently examined in the mouse genes for evidence of parental-origin specific methylation, with the use of methylation-sensitive endonucleases. A typical example of such approach is the use of *Hpa*II and *Msp*I. The two enzymes both cut at CCGG, a recognition site frequently found in CpG islands. As *Hpa*II but not *Msp*I is inhibited by methylation, Southern blots following digestion of genomic DNA with each enzyme will show fragments of different length if one or more sites covered by the probe are methylated.

Sasaki et al.[60] used this approach to compare *Igf2* methylation in normal mice to that in animals with maternal uniparental disomy of distal chromosome 7, the locus to which *Igf2* maps. As one might expect, these mice express IGF-II at very low levels and are born smaller than their normal littermates[61]. The two major promoters from which transcription is known to be initiated in the fetal mouse were studied, to test the hypothesis that hypermethylation of the maternal copy underlies its repression. Contrary to this expectation, neither was found to be methylated in either normal or maternally isodisomic animals. Methylation was, however, found that was different in isodisomic mice compared to controls, several kilobases 5' to the promoters. The significance of this finding remains to be established, especially considering that the studies were done on a disrupted chromosome. It does demonstrate, however, that promoter methylation is not necessary for imprinting, and that distal regulatory elements may be involved instead.

These conclusions are supported by the findings of Stöger et al., who studied methylation of *Igf2*r in mouse embryos heterozygous for a chromosomal deletion involving this gene[59]. This approach allowed isolated study of only one gene copy, identifiable as paternal or maternal, in each animal. The study found methylation in two regions of the 120 kb spanned by the primary transcript sequence (Figure 2). Region 1, surrounding the transcription initiation sites and containing a 1 kb CpG island, was methylated only in the repressed paternal copy. Region 2, found in an intron approximately 27 kb 3' to the transcription initiation site and containing another CpG island, was methylated only in the active, maternal copy.

Promoter methylation in the inactive copy, however, did not turn out to be the primary event: Study of germ cells and early embryos demonstrated that region 1 methylation was not present in sperm or early gestation embryos, and gradually appeared long after the failure to express the gene was fully manifest. The imprint appeared, instead, to be a positive one, located in region 2. This region was methylated in oocytes and early embryos, and the pattern did not change through gestation, to post-natal life. This time course suggests that methylation of the paternal *Igf2r* promoter is a consequence rather than the cause of the gene Tcopy repression. Methylation as a primary event was only found in region 2, 27 kb downstream. It is known that transcription can be controlled by elements lying that far from its initiation site[62].

These observations suggest that the activation of the maternal copy by methylation is, indeed, the primary imprint. If so, then methylation has a positive effect on transcription. This could be mediated either by inhibiting DNA interaction with a repressor, or by enhancing the binding of a trans activator.

More work is obviously needed to completely define these molecular interactions. Defining parental imprinting at the molecular level has just begun, and future findings may have relevance to our understanding of gene regulation that is not be confined to the few genes that are imprinted. Studies of the genes coding for IGF-II and M6P/IGF-II-R are likely to continue to play a central role in the field.

IGF2 IMPRINTING IN HUMAN DISEASE

Our knowledge of the effects of gene imprinting on human disorders has, so far, relied on the study of disease phenotypes (Reviewed by Hall[63] and Reik[64]). The availability of molecular information on imprinted genes of known function now allows a more powerful examination of these effects.

What can be predicted about the relevance of *IGF2* and *IGF2R* imprinting in human disease? Imprinting effectively diminishes to half the amount of gene product that would otherwise be expressed. A malfunction of the mechanisms involved could result in a double dose of the gene in the case of failure to silence one gene copy, referred to as "relaxation of imprinting". On the other hand, failure to erase a negative imprint or to impose a positive one would result in the absence of gene expression. In the case of *IGF2*, recent reports provide the first direct evidence that that the first mechanism may be involved in two human diseases, Beckwith- Wiedemann syndrome and Wilms' tumor. A double dose of IGF-II seems to play an role in the pathogenesis of both.

Beckwith-Wiedemann Syndrome

Beckwith-Wiedemann syndrome (BWS) is characterized by several features in the fetus and newborn, the most striking of which can be explained by a double dose of a fetal growth factor: Fetal macrosomia, excessive size of the placenta, muscle mass, tongue and kidney, umbilical hernia probably as a result of large viscera, and pancreatic hyperplasia, both exocrine and endocrine, resulting in transient neonatal hypoglycemia[65,66]. A considerable proportion of patients with BWS develop embryonic tumors in early life, of which Wilms' tumor is the most common[67]. Although most cases are sporadic, familial occurrence is known to exist and, in many pedigrees, the disease behaves as a dominant trait transmitted exclusively by the mother, although individuals of either sex can be affected[68].

By linkage analysis, familial BWS maps to 11p15, near the tip of the short arm of chromosome 11 where *IGF2* also maps[68]. The majority of BWS patients, however, are sporadic. When the 11p15 locus of some of these patients was analyzed with the use of polymorphic markers, it was found that a considerable proportion of them (3/8) had paternal uniparental disomy of 11p15. Because of *IGF2* imprinting, paternal uniparental disomy would result in a double dose of IGF-II. However, the chromosomal abnormalities resulting in excess *IGF2* expression may also cause dysfunction of other, nearby, genes which are really responsible for the disease, and for which *IGF2* is only a marker. The expression of such gene(s) may be affected by the same factor(s) that determine *IGF2* imprinting. It is currently not known whether imprinted chromosomal domains exist that encompass more than one gene. It is certainly not the case with three genes closely linked to *Igf2r*[25]. The close physical distance of *IGF2* and *H19*[37], two genes that are imprinted albeit in opposite parental sexes, suggests that such a domain can extend over at least two genes, even if it may affect the expression of each in opposite directions. BWS, therefore, may be due to excess IGF-II coupled with failure to express *H19* plus, conceivably but unlikely, dysfunction of other genes whose expression is tightly coupled to the molecular mechanisms controlling *IGF2* imprinting.

Additional insight into BWS may be gained from a mouse model, developed using mice with uniparental disomy of distal 7, a locus syntenic to 11p15[69]. Maternally isodisomic mice fail to express *IGF2* and their birth size is smaller than normal[5]. Mice chimeric for paternal uniparental disomy, which would be expected to have increased expression of the peptide, are larger than normal and display phenotypic features analogous to BWS[69]. Again, it is not clear whether *IGF2* is necessarily the only gene involved, and further study of this model will complement exploration of molecular pathology in BWS patients.

The Molecular Pathology of Wilms' Tumor

A double dose of IGF-II appears also to be involved in the causation of Wilms' tumor, a kidney malignancy appearing in the first five years of life. This tumor has long been known to overexpress IGF-II[70,71]. Mutations at WT1, a putative tumor suppressor gene at 11p13, are found in some Wilms' tumors[72, 73]. WT1 codes for a zinc-finger protein that binds to the EGR1 consensus response element sequence[74]. Interestingly, the promoter region from which *IGF2* is expressed in human fetal kidney has three EGR1 response elements, and *IGF2* is the only gene on whose expression WT1 has been shown to have dramatic suppressive effects in co-transfection experiments[75]. *IGF2* overexpression may be causally involved in Wilms' tumorigenesis or it may be simply a marker of WT1 loss of function. Recent evidence supports a causative role: *IGF2* overexpression, generated by another two, completely independent, molecular events has also been found with high frequency in Wilms' tumors.

More than half of all Wilms' tumors cannot be explained by WT1 mutations[81]. About half of these have loss of heterozygosity at another locus which, almost certainly by coincidence, is also on the short arm of chromosome 11 and appears to independently cause the same tumor phenotype[76-81].

Losses of heterozygosity (LOH) in tumors have been very useful in identifying genes involved in tumorigenesis. The clonal genotype of many tumors often involves gross deletions at specific loci, believed to inactivate tumor suppressor genes[82]. Both copies of a tumor suppressor gene must be inactivated for tumor to occur. When the remaining copy of such a gene has been inactivated by a point mutation, the phenomenon is detected as a loss of heterozygosity (LOH) for the locus, by finding only one allele in a tumor from a heterozygous patient.

In the case of Wilms' tumor, the overlapping regions of the various 11p15 LOHs observed in different tumors all involve a small critical area at 11p15.5, which includes the *IGF2-H19* complex. These LOHs have two unusual features. First, the allele retained is almost always paternal[83]. Second, by DNA dosage studies, a considerable proportion of these LOHs are due to paternal uniparental disomy rather than simple deletion of the maternal copy[76-81].

It is difficult to find a precedent where double dose of a normal growth factor could cause a malignant tumor. Also, the presence of 11p15.5 LOH not associated with paternal duplication suggests the involvement, in these deletions, of a tumor suppressor gene. Indeed, evidence for such a tumor suppressor gene has been found, using complementation with subchromosomal transferable fragment (STF) technology[84,85]. A fragment very close to, but distinct from, the *IGF2-H19* locus on 11p15.5 abolished the malignant phenotype in a cell line derived from rhabdomyosarcoma, a tumor that shows the same parental-origin selective LOH at 11p15.5.

If the tumors are primarily caused by loss of this gene, then *IGF2* may drive the selection of the paternal allele by playing a permissive role. In other words, *IGF2* expression is necessary for the development of the tumor. If it is the paternal copy of the putative tumor suppressor that is lost through a deletion gross enough to be detected as LOH, *IGF2* is very likely to be also deleted, because of the physical proximity, and the tumor will not occur. The problem with this model is that it does not explain the increased expression of *IGF2*, a feature almost invariably seen in Wilms' tumors. That increased expression of *IGF2* is important in Wilms' tumorigenesis is supported by two recent papers, showing that up to two thirds of those tumors that retain both parental alleles show relaxation of imprinting and express both gene copies[19,20], while normal kidney expresses only the paternal one[18-20].

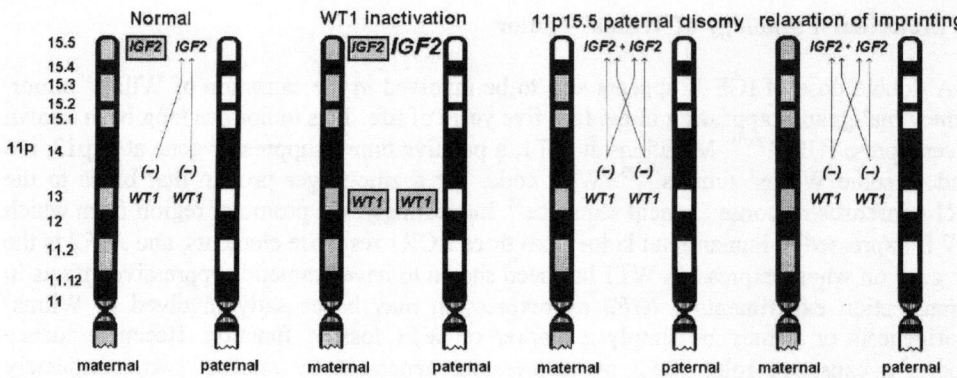

Figure 3 three distinct mechanisms by which *IGF2* overexpression has been shown to arise in Wilms' tumors. Shaded boxes represent inactivated genes.

Increased *IGF2* expression in Wilms' tumors may be: (a) a causative event in the development of the tumor (b) a marker for other causative molecular events or (c) a consequence of the tumor. The studies reviewed above show that increased *IGF2* expression can result from three totally independent and distinct molecular mechanisms: WT1 inactivation, paternal uniparental disomy, and relaxation of imprinting (Figure 3). It would be an unlikely coincidence if all three of these mechanisms happened to result from the tumor phenotype or were, independently of each other, markers of something else causing Wilms' tumor.

To sum up, it is obvious that more than one molecular lesions are involved in Wilms' tumorigenesis (the era of the single-gene cancer probably began and ended with retinoblastoma). The available evidence at this time supports a permissive (necessary but not sufficient) role for *IGF2* overexpression, as subtle as two-fold. Further work in the area, actively pursued in many laboratories, will allow a more precise understanding of the role of *IGF2* in Wilms' tumor and, more generally, of gene imprinting in carcinogenesis.

REFERENCES

1. M.A. Surani, S.C. Barton, and M.L. Norris, Development of reconstituted mouse eggs suggests imprinting of the genome during gametogenesis. *Nature* 308:548 (1984).
2. J. McGrath and D. Solter, Completion of mouse embryogenesis requires both the maternal and paternal genomes. *Cell* 37:179 (1984).
3. J.R. Mann and R.H. Lovell-Badge, Inviability of parthenogenones is determined by pronuclei, not egg cytoplasm. *Nature* 310:66 (1984).
4. K.D. Bagshawe and S.D. Lawler, Unmasking moles. *Br J Obstet Gynaecol* 89:255 (1982).
5. A.G. Searle and C.V. Beechey, Genome imprinting phenomena on mouse chromosome 7. *Genet* 56:237 (1991).
6. A.G. Searle, J. Peters, M.F. Lyon, J.G. Hall, E.P. Evans, J.H. Edwards, and V.H. Buckle, Chromosome maps of man and mouse. *IV Ann. Hum Genet* 53:89 (1989).
7. B.M. Cattanach and M. Kirk, Differential activity of maternally and paternally derived chromosome regions in mice. *Nature* 315:496 (1985).
8. M. Miller and J.G. Hall, Possible maternal effect on severity of neurofibromatosis. *Lancet* 2:1071 (1978).
9. A.E. Harding, Genetic aspects of autosomal dominant late onset cerebellar ataxia. *J Med* 18:436 (1981).
10. R.M. Ridley, C.D. Frith, L.A. Farrer, and P.M. Conneally, Patterns of inheritance of the symptoms of Huntington's disease suggestive of an effect of genomic imprinting. *J Med Genet* 28:22 (1991).
11. M.C. Koch, T. Grimm, H.G. Harley, and P.S. Harper, Genetic risks for children of women with myotonic dystrophy. *Am J Hum Genet* 48:1084 (1991).
12. I. Kennerknecht, A genetic model for the Prader-Willi syndrome and its implication for Angelman syndrome. *Hum Genet* 90:91 (1992).
13. R.D. Nicholls, J.H.M. Knoll, M.G. Butler, S. Karam, and M. Lalande, Genetic imprinting suggested by maternal heterodisomy in non-deletion Prader-Willi syndrome. *Nature* 342:281 (1989).
14. S. Malcolm, J. Clayton-Smoth, M. Nochols, S. Robb, T. Webb, J.A.L. Armour A.J. Jeffreys, et al., Uniparental paternal disomy in Angelman's sydrome. *Lancet* 337:694 (1991).
15. W.P. Robinson, A. Bottani, Y.G. Xie, J. Balakrishman, F. Binkert, M. Machler, A. Prader, and A. Schinzel, Molecular, cytogenetic, and clinical investigations of Prader-Willi syndrome patients. *Am J Hum Genet* 49:1219 (1991).
16. T.M. DeChiara, A. Efstratiadis, and E.J. Robertson, A growth-deficiency phenotype in heterozygous mice carrying an insulin-like growth factor II gene disrupted by targeting. *Nature* 345:78 (1990).
17. T.M. Dechiara, E.J. Robertson, and A. Efstratiadis, Parental imprinting of the mouse insulin-like growth factor II gene. *Cell* 64:849 (1991).
18. N. Giannoukakis, C. Deal, C.G. Goodyer, J. Paquette, and C. Polychronakos, Parental genomic imprinting of the human *IGF2* gene. *Nature Genetics* 4:98 (1993).
19. S. Rainier, L.A. Johnson, C.J. Dobry, A.J. Ping, P.E. Grundy, and A.P. Feinberg, Relaxation of imprinted genes in human cancer. *Nature* 362:747 (1993).
20. O. Ogawa, M.R. Eccles, J. Szeto, L.A. McNoe, K. Yun, M.A. Maw, P.J. Smith, and A.E. Reeve, Relaxation of insulin-like growth factor II gene imprinting implicated in Wilms' tumour. *Nature* 362:749-751 (1993).

21. R. Ohlsson, A. Nyström, S. Pfeifer-Ohlsson, V. Töhönen, F. Hedborg, P. Schofield, Flam, F., and T.J. Edström, *IGF2* is parentally imprinted during human embryogenesis and in the Becwith-Wiedemann syndrome. *Nature Genetics* 4:94 (1993).

22. P. Nissley and W. Lopaczynski, Insulin-like growth factor receptors. *Growth Factors* 5:29-43 (1991).

23. C. Polychronakos, The mannose 6-phosphate/IGF-II receptor. in: Molecular and Cellular Biology of the IGFs M. Raizada and D. LeRoith, eds, Plenum Press, New York, (1989).

24. M.M. Sklar, W. Kiess, C.L. Thomas, and S.P. Nissley, Developmental expression of the tissue insulin-like growth factor II/mannose 6-phosphate receptor in the rat. Measurement by quantitative immunoblotting. *J Biol Chem* 264:16733 (1989).

25. D.P. Barlow, R. Stöger, B.G. Herrmann, K. Saito, and N. Schweifer, The mouse insulin-like growth factor type-2 receptor is imprinted and closely linked to the Tme locus. *Nature* 349:84 (1991).

26. J. Goto, D.A. Figlewicz, C. Marineau, N. Khodr, and G.A. Rouleau, Dinucleotide repeat polymorphism at the GF2R locus. *Nucleic Acids Res* 20:923 (1992).

27. D. Haig and C. Graham, Genomic imprinting and the strange case of the insulin- like growth factor II receptor. *Cell* 64:1045 (1991).

28. S.O. Adams, S.P. Nissley, S. Handwerger, and M. Rechler, Developmental patterns of of IGF-I and II synthesis and regulation in rat fibroblasts *Nature* 302:150-153 (1993).

29. P.K. Lund, B.M. Moats-Staats, M.A. Hynes, J.G. Simmons, M. Jansen, A.J. D'Ercole, and J.J. Van Wyk, Somatomedin C/IGF-I and IGF-II mRNAs in rat fetal and adult tissues. *J Biol Chem* 261:14539 (1986).

30. V.K.M. Han, A.J. D'Ercole, P.K. Lund, Cellular localization of somatomedin (IGF) mRNA in the human fetus. *Science* 236:193 (1987).

31. A. Gray, A.W. Tam, T.J. Dull, J. Hayflick, J. Pintar, W.K. Cavenee, A. Koufos, A. Ullrich, Tissue-specific and developmentally regulated transcription of IGF-II. *DNA* 6:283 (1987).

32. B. Funk, U. Kessler, W. Eisenmenger, A. Hansmann, H.J. Kolb, and W. Kiess, Expression of the M6P/IGF-II receptor in multiple human tissues during fetal life and early infancy. *J Clin Endocrinol Metab* 75:431 (1992).

33. Y. Xu and C. Polychronakos, A soluble form of the M6P/IGF-II receptor in human amniotic fluid. 74th Annual Meeting, Endocrine Society, San Antonio. Abstract #1443, (1992).

34. M.S. Bartolomei, S. Zemel, and S.M. Tilghman, Parental imprinting of the mouse *H19* gene. *Nature* 351:153 (1991).

35. Y. Zhang and B. ycko, Monoallelic expression of the human *H19* gene. *Nature Genetics* 1:40 (1992).

36. J. Rachmilewitz, R. Goshen H. Ariel, T. Schneider, N. de Groot, and A. Hochberg, Parental imprinting of the human *H19* gene. *FEBS* 309:25 (1992).

37. S. Zemel, M.S. Bartolomei, and S.M. Tilghman, Physical linkage of two mammalian imprinted genes, *H19* and *igf2*. *Nature Genetics* 2:61-65 (1992).

38. D. Fischer, D. Weisenberger, and U. Scheer, Assigning functions to nucleolar structures. *Chromosoma* 101:133 (1991).

39. S.E. Leef, C.I. Brannan, M.I. Reed, T. Ozcelik, U. Francke, N.G. Copeland, and N.A. Jenkins, Maternal imprinting of the mouse Snrpn gene and conserved linkage homology with the human Prader-Willi syndrome region. *Nature Genetics* 2:259 (1992).

40. T. Ozcelik, S. Leff, W. Robinson, T. Donlon, M. Lalande, E. Sanjines, A. Schinzel, and U. Francke, Small nuclear ribonucleoprotein polypeptide N (SNRPN), an expressed gene in the Prader-Willi syndrome critical region. *Nature Genetics* 2:265 (1992).

41. M. Hergersberg, Biological aspects of cytosine methylation in eukaryotic cells. *Experientia* 47:1171 (1991).

42. E. Li, T.H. Bestor, and R. Jaenisch, Targeted mutation of the DNA methyltransferase gene results in embryonic lethality. *Cell* 69:9151 (1992).

43. I. Keshet, J. Lieman-Hurwitz, and H. Cedar, DNA methylation affects the formation of active chromatin. *Cell* 4:535 (1986).

44. D.J. Ball, D.S. Gross, and W.T. Garrard, 5-methylcytosine is localized in nucleosomes that contain histone H1. *Proc Natl Acad Sci USA* 80:5490 (1983).

45. J. Tazi and A. Bird, Alternative chromatin structure at CpG islands. *Cell* 60:909 (1990).

46. M. Szyf, DNA methylation patterns: an additional level of information. *Biochem Cell* 69:764 (1991).

47. R. Hermann and W. Doerfler, Interference with protein binding at AP2 sites by sequence-specific methylation in the late E2A promoter of adenovirus type 2 DNA. *FEBS Lett* 29:238 (1991).

48. M. Comb and H.M. Goodman, CpG mehtylation inhibits proenkephalin gene expression and binding of the transcription factor AP-2. *Nucleic Acids Res* 18:3975 (1990).

49. M.A. Harrington, P.A. Jones, M. Imagawa, and M. Karin, Cytosine methylation does not affect binding of transcription factor Sp1. *Proc Natl Acad Sci USA* 85:2066 (1988).

50. J.D. Lewis, R.R. Meehan, W.J. Henzel, I. Maurer-Fogy, P. Jeppesen, F. Klein, and A. Bird, Purification, sequence, and cellular localization of a novel chromosomal protein that binds to methylated DNA. *Cell* 69:905 (1992).

51. J. Boyes and A. Bird, DNA methylation inhibits transcription indirectly via a methyl-CpG binding protein. *Cell* 64:1123 (1991).

52. C. Sapienza, T.H. Tran, J. Paquette, R. McGowan, and A. Peterson, Degree of methylation of transgenes is dependent on gamete of origin. *Nature* 328:251 (1987).

53. J.L. Swain, T.A. Stewart, and P. Leder, Parental legacy determines methylation and expression of an autosomal transgene: a molecular mechanism for parental imprinting. *Cell* 50:719 (1987).

54. C. Pourcel, Maternal inhibition of hepatitis B surface antigen gene expression in transgenic mice correlates with de novo methylation. *Nature* 329:454 (1987).

55. W. Reik, A. Collick, M.L. Norris. S.C. Barton, and M.A.H. Surani, Genomic imprinting determines methylation of parental alleles in trnasgenic mice. *Nature* 328:248 (1987).

56. A.B. Kolsto, G. Kollias, V. Giguere, K.I. Isobe, H. Prydz, and F. Grosveld, The maintenance of methylation-free islands in transgenic mice. *Nucleic Acids Res* 14:9667 (1987).

57. S.G. Grant and V.M. Chapman, Mechanism of X-chromosome regulation. *A Rev Genet* 22:199 (1988).

58. P. Rotwein and L.J. Hall, Evolution of insulin-like growth factor II: Characterization of the mouse IGF-II gene and identification of two pseudo-exons. *DNA Cell Biol* 9:725 (1990).

59. R. Stöger, P. Kubicka, C.G. Liu, T. Kafri, A. Razin, H. Cedar, and D.P. Barlow, Maternal-specific methylation of the imprinted mouse *Igf2r* locus identifies the expressed locus as carrying the imprinting signal. *Cell* 73:61 (1993).

60. H. Sasaki, P.A. Jones, J.R. Chaillet, A.C. Ferguson-Smith, S.C. Barton, W. Reik, and M.A. Surani, Parental imprinting: potentially active chromatin of the repressed maternal allele of the mouse insulin-like growth factor II (*Igf2*) gene. *Genes & Development* 6:1843 (1992).

61. A.G. Searle and C.V. Beechey, Genome imprinting phenomena on mouse chromosome 7. *Genet Res* 56:237 (1990).

62. O. Hanscombe, D. Whyatt, P. Fraser, N. Yannoutsos, D. Greaves, N. Dillon, and F. Grosvelt, Importance of globin gene order for correct gene expression. *Genes Dev.* 5:1387 (1991).

63. J.G. Hall, Genomic imprinting: review and relevance to human diseases. *American J of Human Genetics* 46:857-873 (1990).

64. W. Reik, Genomic imprinting and genetic disorders in man. *Trends Genet* 5:331 (1989).

65. H.R. Wiedemann, Complexe malformatif familial avec hernie ombilicale et macroglossie - un "syndrome nouveau". *J Genet Hum* 13:223 (1964).

66. J.B. Beckwith, Macroglossia, omphalocele, adrenal cytomegaly, gigantism, and hyperplastic visceromegaly. *Birth Defects* 5:188 (1969).

67. C.L. Clericuzio, Clinical phenotypes and Wilms tumor. *Med Pediatr Oncol* 21:182 (1993).

68. A. Koufos, P. Grundy, K. Morgan, K. et al., Familial Wiedemann-Beckwith syndrome and a second Wilms' tumor locus both map to 11p15.5. *Am J Hum Genet* 44:711 (1989).

69. A.C. Ferguson-Smith, B.M. Cattanach, S.C. Barton, C.V. Beechey, and M.A. Surani, Embryologic and molecular investigations of parental imprinting on mouse chromosome 7. *Nature* 351:667 (1991).

70. A.E. Reeve, M.R. Eccles, R.J. Wilkins, G.I. Bell, and L.J. Millow, Expression of insulin-like growth factor-II transcripts in Wilms tumour. *Nature* 317:258 (1985).

71. J. Scott, J. Cowell, M.E. Robertson, L.M. Priestley, R. Wadey, B. Hopkins, J. Pritchard, G.I. Bell, L.B. Rall, C.F. Graham, et al., Insulin-like growth factor II gene expression in Wilms' tumor and embryonic tissues. *Nature* 317 (1985).

72. K.M. Call, T. Glaser, C.L. Ito, et al., Isolation and characterization of a zinc finger polypeptide gene at the human chromosome 11 Wilms' tumor locus. *Cell* 60:509 (1990).

73. M. Gessler, A. Poustka, W. Cavanee, et al., Homozygous deletions in Wilms' tumour of a zinc-finger gene identified by chromosome jumping. *Nature* 343:774 (1990).

74. F.J. Rauscher III, J.F. Morris, O.E. Tournay, et al., Binding of the Wilms' tumor locus zinc finger protein to the EGR-1 consensus sequence. *Science* 250:1259 (1991).

75. I.A. Drummond, S.L. Madden, P. Rohwer-Nutter, G.I. Bell V.P. Sukhatme, F.J. Rauscher III, Repression of the IGF-II gene by the Wilms tumor suppressor WT1 *Science* 257:674 (1992).

76. M. Mannens, R.M. Slater, C. Heyting, J. Bliek, J. de Kraker, N. Coad, P. de Pagter-Holthuizen, and P.L. Pearson, Molecular nature of genetic changes resulting in loss of heterozygosity of chromosome 11 in Wilms' tumours. *Human Genetics* 81:41 (1989).

77. S.H. Orkin, D.S. Goldman, and S.E. Sallan, Development of homozygosity for chromosome 11p markers in Wilms' tumour. *Nature* 309:172 (1984).

78. A.E. Reeve, P.J. Housiaux, R.J.M. Gardner, W.E. Chewings, R.M. Grindley, and L.J. Millow, Loss of a Harvey ras allele in sporadic Wilms' tumour. *Nature* 309:174 (1984).

79. E.R. Fearon, B. Vogelstein, and P. Feinberg, Somatic deletion and duplication of genes on chromosome 11 in Wilms' tumours. *Nature* 309:176 (1984).

80. A.M. Raizis, D.M. Becroft, R.L. Shaw, A.E. Reeve, A mitotic recombination in Wilms tumor occurs between the parathyroid hormone locus and 11p13. *Human Genetics* 70:344 (1985).
81. R.M. Slater and M. Mannens, Cytogenetics and molecular genetics of Wilms' tumor of childhood. *Cancer Genet Cytogenet* 61:111 (1992).
82. B. Ponder, Gene losses in human tumours. *Nature* 335:400 (1988).
83. W.T. Schroeder, L.Y. Chao, D.D. Dao, L.C. Strong, S. Pathak, V. Riccardi, V.H. Lewis, and G.F. Saunders, Nonrandom loss of maternal chromosome 11 alleles in Wilms tumors. *Am J Hum Genet* 40:413 (1987).
84. S.F. Dowdy, C.L. Fasching, D. Araujo, et al., Suppression of tumorigenicity in Wilms tumor by the p15.5-p14 region of chromosome 11. *Science* 254:293 (1991).
85. M. Koi, L.A. Johnson, L.M. Kalikin, P.F.R. Little, Y. Nakamura, and A.P. Feinberg, Tumor cell growth arrest caused by subchromosomal transferable DNA fragments from chromosome 11. *Science* 260:361 (1993).

MULTIHORMONAL REGULATION OF IGFBP-1 PROMOTER ACTIVITY

David R. Powell, Phillip D. K. Lee and Adisak Suwanichkul

Department of Pediatrics
Baylor College of Medicine
Houston, Texas 77030

INTRODUCTION

Insulin-like growth factor binding protein-1 (IGFBP-1) is a 25 kiloDalton protein which can compete with IGF receptors for binding of IGF-I and IGF-II peptides. Such high affinity binding allows IGFBP-1 to influence IGF action; IGFBP-1 can inhibit or potentiate the effects of IGF peptides depending on experimental conditions and on post-translational modifications of this binding protein (1). IGFBP-1 is expressed in a tissue-specific manner, with significant expression essentially limited to liver and uterus in most individuals (2,3).

IGFBP-1 is readily detected in human serum and is most likely synthesized by hepatocytes (1). Although the role of circulating IGFBP-1 is not clear, it is clear that serum IGFBP-1 levels are highly regulated and may fluctuate by more than 15-fold in a matter of hours. Such fluctuations appear to depend on the nutritional status of the individual; levels rise with fasting and fall rapidly after a meal (1,4,5). The activity of gluconeogenic enzymes such as phosphoenolpyruvate carboxykinase (PEPCK) is regulated in a similar manner, suggesting that serum IGFBP-1 may play a role in glucose counterregulation. Hypothetically, IGFBP-1 levels increase during fasting in order to bind IGF peptides, thereby blocking the insulin-like effects of these growth factors during the period of substrate deficiency (1,6). Some support for this hypothesis is provided by a recent study showing that acute infusion of IGFBP-1 into rats results in a small but significant increase in serum glucose levels (6).

In the case of the cytosolic form of hepatic PEPCK, activity is stimulated by cAMP (which serves as second messenger for glucagon) and glucocorticoids, and is inhibited in a dominant fashion by insulin (7). Glucocorticoids, cAMP and insulin, alone and in combination, regulate PEPCK expression primarily at the level of transcription, and the effect of these regulators is confered to the PEPCK gene within the span of 460 basepairs (bps) just 5' to the mRNA cap site (8-10). Considering the regulatory pattern of serum IGFBP-1 levels, numerous studies have been performed to determine the role of glucocorticoids, cAMP and insulin in hepatic IGFBP-1 expression (1). *In vivo* studies find that glucocorticoids stimulate (11) while insulin inhibits IGFBP-1 expression; the glucocorticoid effect is mediated at the level of mRNA abundance (12), while the insulin effect is mediated at the level

Current Directions in Insulin-Like Growth Factor Research,
Edited by D. LeRoith and M.K. Raizada, Plenum Press, New York, 1994

205

of transcription (13). *In vitro* studies confirm these observations, show that cAMP stimulates IGFBP-1 expression, and suggest that all three regulate IGFBP-1 expression at the level of transcription (14-19). In addition, the inhibitory effect of insulin on IGFBP-1 expression may be dominant over the stimulatory effects of cAMP and glucocorticoids (1,16-18).

Recent studies have begun to characterize the *cis* elements and *trans*-acting factors responsible for multihormonal regulation of PEPCK transcription (9,10,20,21). This manuscript will discuss similar investigations of the multihormonal regulation of IGFBP-1 promoter activity, and will compare the regulation of the two promoters by glucocorticoids, cAMP and insulin.

MATERIALS AND METHODS

Plasmid constructs

The construction of plasmids p357CAT and p103CAT, which contain 357 and 103 bp of human IGFBP-1 (hIGFBP-1) promoter sequence, respectively, 5' to the chloramphenicol acetyltransferase(CAT) reporter gene, has been described previously (22).

Cell culture and DNA transfection

HEP G2 human hepatoma cells were maintained, plated and transfected as described previously (22,23). Each plate was transfected with 10 μg of CAT plasmid; 1 μg of pRSVL plasmid, which contains the RSV LTR upstream to the luciferase reporter gene, was cotransfected to control for transfection efficiency (19,23). Cells were then incubated with or without experimental additives for 18 hours. Cellular protein was collected from each plate and assayed for CAT and luciferase activity as described previously (22-24).

DNaseI Protection Assay

The DNaseI protection assay was described previously, as was the HEP G2 nuclear extract (19,22). Recombinant cAMP-responsive element binding protein (CREB) was kindly provided by Dr. William Roesler (University of Saskatchewan, Saskatchewan, Canada) (21).

Alignment of IGFBP-1 Promoter Sequences

Human, rat and mouse IGFBP-1 promoter sequences were aligned with the help of the Molecular Biology Computer Resource Center at Baylor College of Medicine. This analysis used the GAP program of the Sequence Analysis Software Package provided by the Genetics Computer Group (GCG), Inc. The GAP program employs the algorithm of Needleman and Wunsch (25) to maximize the number of matches and minimize the number of gaps. Human and rat IGFBP-1 promoter sequences were from references 22 and 26, respectively. The mouse IGFBP-1 promoter sequence, currently unpublished, was from the Genbank™/EMBL Data Bank under accession number X67493.

RESULTS AND DISCUSSION

Utility of hIGFBP-1 Promoter Studies in HEP G2 Cells

HEP G2 cells express many proteins characteristically expressed by hepatocytes *in vivo*, and have proven useful in identifying *cis* elements and *trans*-acting factors responsible for producing the hepatic phenotype. Initial transient transfection studies in HEP G2 cells found that a 1205 bp fragment of the hIGFBP-1 promoter directed expression of the reporter gene CAT when inserted in the sense, but not antisense, orientation (22). Further transfection of CAT constructs containing progressive deletions of the IGFBP-1 promoter found that basal promoter activity was directed by a *cis* element located from -81 to -53 bp of the IGFBP-1 promoter (Figure 1) (22). This sequence contains the consensus motif found in other *cis* elements which bind hepatic nuclear factor 1 (HNF1) (27). The following studies indicate that HNF1 directs basal expression of IGFBP-1 in HEP G2 cells: i) mutations which decrease binding of purified HNF1 binding domain (HNF1bd) decrease basal IGFBP-1 promoter activity, and mutations which increase HNF1 binding increase basal IGFBP-1 promoter activity (22); ii) HNF1bd produces the exact footprint of the -81 to -53 bp *cis* element as does HEP G2 nuclear extract (22); and iii) transfection of an HNF1 expression vector into HeLa cells activates the native IGFBP-1 promoter construct but not the construct containing the mutated *cis* element which binds HNF1bd poorly (28). The importance of the HNF1 *cis* element in IGFBP-1 expression is also suggested by the fact that this element is highly conserved among the human, rat and mouse IGFBP-1 promoters (Figure 1).

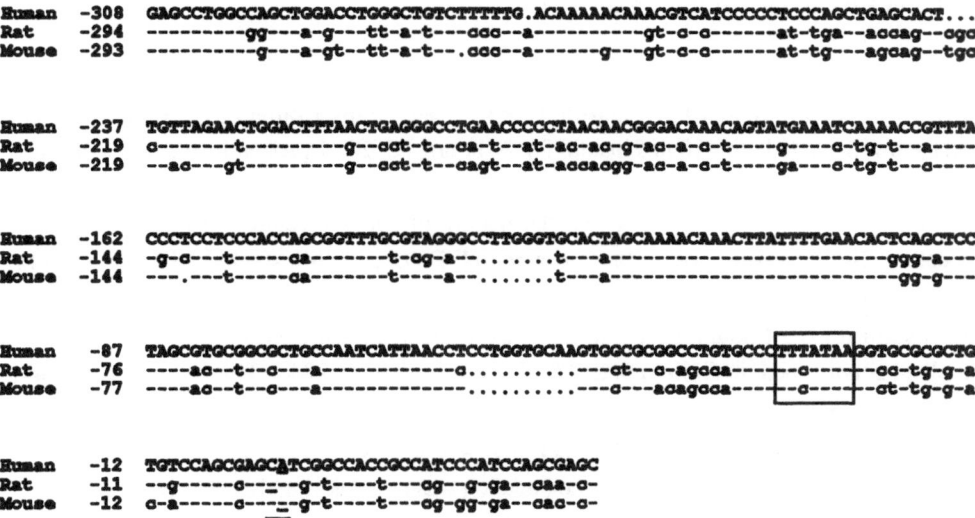

Figure 1. Comparison of IGFBP-1 gene promoters. Sequences of human (22), rat (26) and mouse (GenBank/ EMBL accession #X67493) IGFBP-1 promoter sequences were aligned by the method of Needleman and Wunsch (25). The entire human sequence is presented as upper case letters; conserved rat and mouse bases are presented as dashes, and nonconserved bases presented as lower case letters. Numbers refer to the distance 5' to the mRNA capsite, which is underlined for each promoter. The TATA element of each sequence is boxed.

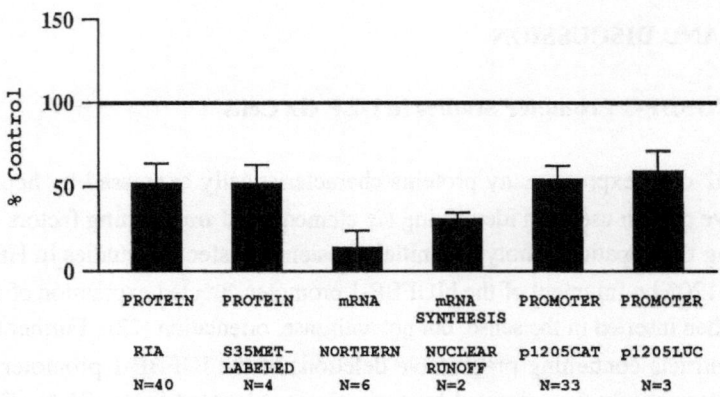

Figure 2. Effect of insulin on IGFBP-1 expression in HEP G2 cells. Data were compiled on insulin inhibition of a number of measures of IGFBP-1 expression (16,34,35). The effects of insulin are presented as % control value, with control value emphasized as the black line at 100%. Insulin effects represent the mean ± standard deviation of N experiments.

HNF1 is a homeodomain protein expressed in a tissue-specific manner, with highest levels found in epithelial cells of liver and kidney (27,29,30). In conjunction with other *trans*-acting factors, HNF1 appears to play an important role in directing the transcription of albumin, α1-antitrypsin and other proteins expressed primarily in liver but also in kidney (29,31). The fact that HNF1 and IGFBP-1 are expressed in roughly the same tissue-restricted pattern *in vivo* suggests that HNF1 is indeed important in directing IGFBP-1 expression, and suggests that HEP G2 cells may be a useful model for identifying other factors which participate in the regulation of hepatic IGFBP-1 expression.

Effect of Insulin on IGFBP-1 Expression

A number of studies suggest that insulin is the major regulator of serum IGFBP-1 levels (1,5,32), probably by inhibiting hepatic IGFBP-1 gene transcription (13). When HEP G2 cells were incubated with 10 nM insulin for 24 hours, IGFBP-1 protein levels in conditioned medium fell to 47% of control values; this inhibition was blocked by anti-insulin receptor antibody MC-51 (33), indicating that the insulin effect is mediated through the insulin receptor. Similar results were obtained when HEP G2 cells transfected with the 1205 bp IGFBP-1 promoter/CAT reporter gene construct p1205CAT were incubated with 10 or 100 nM insulin for 6-24 hours; in a total of 40 experiments, insulin lowered IGFBP-1 protein levels in conditioned medium to 51±13% of control values (16,34). This effect was due to a fall in the IGFBP-1 protein synthesis rate, since HEP G2 cells incorporated 49% less [^{35}S] methionine into IGFBP-1 during 4-6 hour incubations with 100 nM insulin (Figure 2), and since no effect of insulin on IGFBP-1 degradation was noted during this time interval (16).

The effect of insulin resulted from a decrease in the rate of IGFBP-1 mRNA synthesis. IGFBP-1 mRNA levels fell to 45 ± 11% of control values after a 2 hour incubation, and to 14±10% of control values after 4-14 hour incubations with 100 nM insulin. In the presence of 100 nM insulin, IGFBP-1 mRNA synthesis rates also fell by nuclear runoff assay, measuring 31% of control values after a 2 hour incubation (16) (Figure 2). Similar effects of insulin on IGFBP-1 protein and mRNA levels, and on IGFBP-1 transcription rate, have been noted by other investigators using cultured hepatocytes or fetal hepatic tissue (1,3,14,15,17).

After transient transfection into HEP G2 cells, constructs containing the first 1205 bp of the IGFBP-1 promoter showed decreased activity when exposed to 10 or 100 nM insulin for 14-24 hours. In 33 experiments using p1205CAT, insulin lowered promoter activity to 54±8% of control values; similar data resulted from 3 experiments with p1205LUC, an IGFBP-1 promoter construct with luciferase replacing CAT as reporter gene (figure 2) (16,35).

Early studies tested the ability of insulin to inhibit activity of CAT constructs containing the first 529 bp (p529CAT) and 103 bp (p103CAT) of the IGFBP-1 promoter. Insulin inhibited activity of p529CAT to 52% of control, but actually stimulated activity of p103CAT, suggesting that the insulin-responsive element (IRE) was located between -529 and -103 bp of the IGFBP-1 promoter (16). Recent studies used progressive deletion, internal deletion and site-directed mutations of the IGFBP-1 promoter to demonstrate that the IRE is in fact located between -120 and -96 bp of the IGFBP-1 promoter (35). This element, which is located in the most highly conserved region of the IGFBP-1 promoter (Figure 1), contains a palindromic sequence; each half of the palindrome bears some similarity to IREs functionally mapped in the PEPCK and α-amylase promoters (10,36). The IGFBP-1 IRE can confer insulin responsiveness to the normally insulin-nonresponsive thymidine kinase promoter, and by gel shift assay this IRE can also specifically bind at least one factor present in HEP G2 nuclear extract (35); studies are currently underway to identify and characterize the proteins which confer the insulin effect to the IGFBP-1 promoter. The above work strongly suggests that the effect of insulin on hepatic IGFBP-1 expression *in vivo* is confered at the level of transcription by proteins which bind to this -120 to -96 bp element.

Effect of cAMP on IGFBP-1 Expression

The probable stimulation of serum IGFBP-1 levels by glucagon (37) suggests a probable role for the glucagon second messenger cAMP in stimulating hepatic IGFBP-1 expression. Indeed, early studies in hepatocytes and fetal hepatic tissues showed that either glucagon, cAMP or theophylline can stimulate a modest accumulation of IGFBP-1 protein and mRNA, and that the inhibitory effect of insulin is dominant to the stimulatory effect of cAMP (14,16,17). Studies in HEP G2 cells found that 5 mM dibutyrylcAMP (cAMP) could stimulate a ~1.5- to 2-fold increase in IGFBP-1 protein levels and p1205CAT promoter activity, and a ~6-fold increase in IGFBP-1 mRNA levels; further studies found that the combination of 2 mM theophylline and 0.5 mM cAMP also increased IGFBP-1 protein and mRNA levels, but in addition stimulated activity of p1205CAT ~5-fold (19). Consistent with past studies, insulin inhibited the effect of theophylline and cAMP on IGFBP-1 protein and mRNA accumulation, but in addition insulin inhibited the increase in IGFBP-1 promoter activity, suggesting that the effect of insulin is dominant and that the insulin and cAMP pathways converge in the proximal IGFBP-1 promoter to regulate IGFBP-1 expression (16). The ~5-fold increase in p1205CAT activity with the combination of 2 mM theophylline and 0.5 mM cAMP was used to localize *cis* elements confering the cAMP effect to the IGFBP-1 promoter. HEP G2 cells transfected with a series of IGFBP-1 promoter deletion constructs

Figure 3. Binding of CREB and HEP G2 nuclear extract to native or mutated forms of the IGFBP-1 promoter CRE. The -357 to -148 bp fragment of the native IGFBP-1 promoter containing the native CRE, and the same fragment containing the mutant (TAGCA) CRE, were independently labeled on the coding strand. They were then incubated with DNaseI either without nuclear factors (0), with recombinant CREB, or with HEP G2 nuclear extract (N.E.). HEP G2 nuclear extract footprints protected regions 2 through 5 (P2-P5); their location and extent are shown on the right, with numbers representing the number of bp 5' to the mRNA cap site. P2 represents the CRE. Sequencing ladders (T+C;T) were derived from the labeled fragment as described (22).

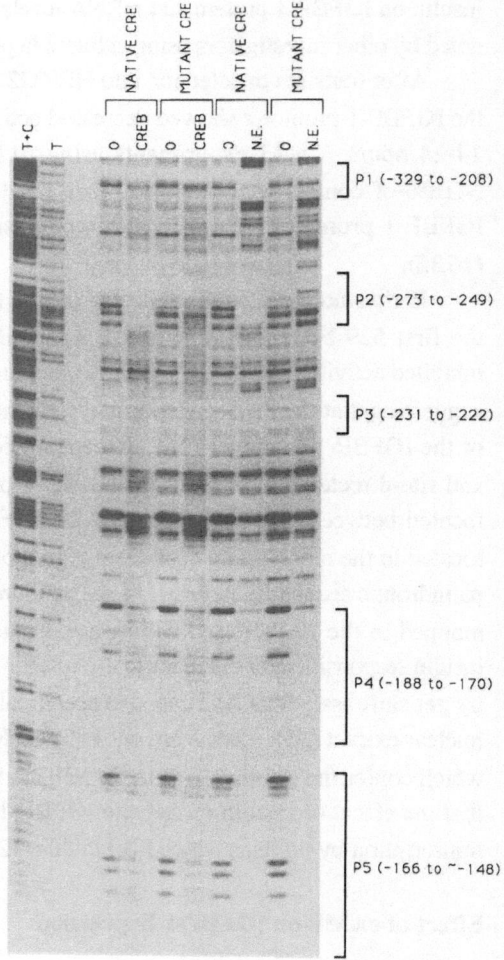

were incubated with the theophylline/ cAMP combination. This analysis found that the majority of the ~5-fold increase was confered between -269 and -207 bp of the IGFBP-1 promoter. Within this region, at -263 to -259 bp, is a CGTCA motif which confers the effects of cAMP to a number of other gene promoters (19,38). When the CGTCA motif of p1205CAT was mutated to TAGCA, the theophylline/cAMP combination stimulated activity of the mutant construct to only 49% of the stimulation obtained with the native construct (19). Additional studies showed that the catalytic subunit of cAMP-dependent protein kinase A, which mediates most intracellular effects of cAMP, increased activity of native p1205CAT by ~290% but could only increase activity of the construct containing the TAGCA mutant by ~90% (19).

The CGTCA motif of many cAMP responsive elements (CREs) can bind a number of proteins, the most extensively studied of which is CRE binding protein or CREB (38,39). CREB apparently constitutively binds to these CREs, and the cAMP effect is not confered

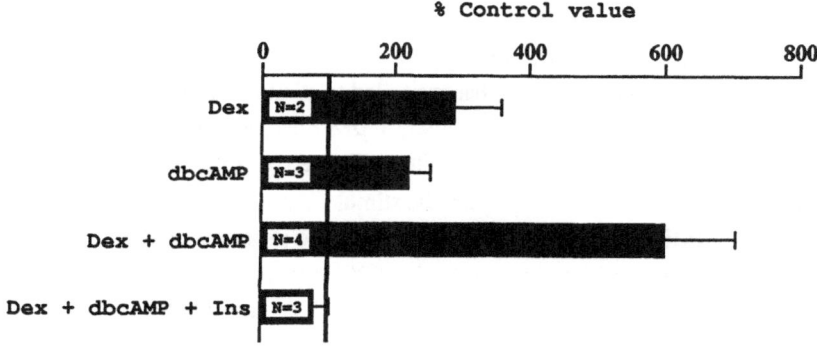

A   Promoter Activity: p357CAT

% Control value

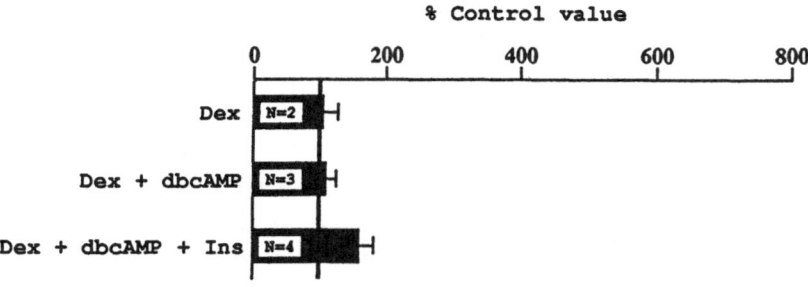

B   Promoter Activity: p103CAT

% Control value

Figure 4. Effect of dexamethasone, dbcAMP and insulin on activity of IGFBP-1 promoter constructs p357CAT and p103CAT. HEP G2 cells transiently transfected with IGFBP-1 promoter constructs were incubated with dexamethasone (1 μM), dbcAMP (0.5 mM) and insulin (100 nM) for 18 hours and then assayed for CAT activity. Data are presented as % control (no additives) value and represent the mean ± standard deviation of N experiments.

until cAMP phosphorylates CREB, thereby altering this protein to a form which enhances transcription (38,39). The ability of recombinant CREB and HEP G2 nuclear extract to bind to native (CGTCA) and mutated (TAGCA) CREs of the IGFBP-1 promoter are presented in Figure 3. Clearly, both CREB and the nuclear extract produce an identical footprint over the CGTCA motif (P2 region) on the IGFBP-1 promoter coding strand, whereas neither can footprint the same region of the IGFBP-1 promoter containing the TAGCA mutant, indicating that the TAGCA mutation is associated with loss of function and with loss of ability to bind a *trans*-acting factor, CREB, known to confer the cAMP effect to other cAMP-responsive gene promoters.

Of interest, the CGTCA motif of the human promoter is not conserved in the rat and mouse promoters (Figure 1), although cAMP does stimulate IGFBP-1 expression in a rat hepatocyte cell line (17). This suggests i) that the CGTCA motif is not the only element confering the cAMP effect to the IGFBP-1 promoter, a possibility supported by the deletion and mutation studies outlined above, and ii) that the other elements may be as important as the CGTCA motif in confering multihormonal regulation of IGFBP-1 expression.

Effect of glucocorticoids on IGFBP-1 Expression

Glucocorticoids appear able to increase serum IGFBP-1 levels (11), and this effect may well be mediated at the level of IGFBP-1 gene transcription (12,15,18). Initial studies using HEP G2 cells found that 1 μM dexamethasone failed to stimulate IGFBP-1 protein or mRNA levels (unpublished observations). However, 1 μM dexamethasone increased IGFBP-1 promoter activity by 190% over control values (Figure 4). In addition, previous studies showed that 1 μM dexamethasone enhanced the stimulatory effects of the theophylline/cAMP combination on IGFBP-1 protein levels, mRNA levels and promoter activity; in these studies, the theophylline/cAMP combination increased p1205CAT activity ~6-fold whereas the same combination plus 1 μM dexamethasone increased p1205CAT activity by up to ~50-fold (18). Dexamethasone was also able to potentiate the effect of the theophylline/cAMP combination on p357CAT, but not on p103CAT, suggesting that the effects of these hormones are confered between -357 and -103 bp of the IGFBP-1 promoter (18). Since the effects of theophylline may not be mediated exclusively through its' effects on cAMP, studies were performed which looked at the ability of 1 μM dexamethasone to potentiate the stimulatory effect of 0.5 mM cAMP on p357CAT. As shown in Figure 4, cAMP increased p357CAT activity by 123%, dexamethasone by 190%, and the combination by 500%; thus, dexamethasone acts synergistically with cAMP alone to increase IGFBP-1 promoter activity. In contrast, this combination had no effect on p103CAT activity, consistent with the published data noted above. Studies are in progress to characterize *cis* elements which confer the dexamethasone effect to the IGFBP-1 promoter, and to determine whether the effect of dexamethasone is direct (i.e. involves binding of glucocorticoid receptors (GRs) to the IGFBP-1 promoter).

Multihormonal Regulation of IGFBP-1 and PEPCK Promoter Activities

Recent studies demonstrate that PEPCK promoter activity i) is stimulated by cAMP, an effect confered through multiple CREs which bind CREB and C/EBP (9,21); ii) is stimulated by glucocorticoids through a complex glucocorticoid response unit (GRU) spanning 110 bp of the PEPCK promoter and containing 2 GR binding sites in addition to 2 sites which bind required accessory factors (20,40); and iii) is inhibited by insulin, an effect confered in part by the ability of insulin to in some way alter the nature of the protein complex found at one of the GRU accessory factor sites (10). Recent studies demonstrating binding of CREB to GR proteins gives an early glimpse of one possible mechanism by which insulin can affect one GRU accessory factor binding site to simultaneously inhibit the stimulatory effects of glucocorticoids and cAMP on PEPCK promoter activity (20).

As shown in Figure 4, insulin profoundly inhibits the stimulatory effect of the cAMP/dexamethasone combination on p357CAT activity; insulin had a similar effect on theophylline/cAMP/dexamethasone-stimulated increases in IGFBP-1 protein levels, mRNA levels and promoter activity (18). As stated above, the element confering the dexamethasone effect to the IGFBP-1 promoter has not been mapped, and the CRE identified in the human IGFBP-1 promoter is not conserved in the rat and mouse promoters, suggesting that other *cis* elements are important in confering the cAMP effect. The ability of insulin to inhibit basal IGFBP-1 promoter activity is confered through the -120 to -96 bp element, which has similarity to the IRE identified in the PEPCK promoter (10,35). The possibility that this highly conserved element also plays an important role in confering the effect of dexamethasone to the IGFBP-1 promoter, perhaps as part of a complex GRU, seems likely and is currently under investigation. The role of this element in confering the effect of cAMP is also being studied. If confirmed, investigation of the proteins responsible for this simple functional organization should yield valuable insight into how these three hormone pathways converge to regulate transcription of IGFBP-1 and other genes.

REFERENCES

1. Lee, P.D.K., Conover, C.A., and Powell, D.R., Regulation and function of insulin-like growth factor binding protein-1, *Proc. Soc. Exp. Biol. Med.*, in press.
2. Brinkman, A., Groffen, C., Kortleve, D.J., Geurts van Kessel, A., and Drop, S.L.S., Isolation and characterization of a cDNA encoding the low MW IGFBP (IBP-1). *EMBO J* 7:2417-2423 (1988).
3. Mohn, K., Melby, A., Tewari, D., Laz, T., and Taub, R. The gene encoding IGFBP1 is rapidly and highly induced in regenerating liver. *Mol. Cell. Biol.* 11:1393-1401 (1991).
4. Drop, S.L.S., Kortleve, D.J., and Guyda, H.J., Immunoassay of a somatomedin BP from human amniotic fluid: levels in fetal, neonatal and adult sera. *J. Clin. Endocrinol. Metab.* 59:908-915 (1984).
5. Conover, C.A., Butler, P.C., Wang, M., Rizza, R.A., and Lee, P.D.K., Lack of growth hormone effect on insulin-associated suppression of IGFBP-1 in humans. *Diabetes* 39:1251-1256 (1990).
6. Lewitt, M.S., Denyer, G.S., Cooney, G.J., and Baxter, R.C., Insulin-like growth factor binding protein-1 modulates blood glucose levels. *Endocrinol.* 129:2254-2256 (1991).
7. Sasaki, K., Cripe, T.P., Koch, S.R., Andreone, T.L., Peterson, D.D., and Granner, D.K., Multihormonal regulation of PEPCK gene transcription. *J. Biol. Chem.* 259:15242-15251 (1984).
8. Magnuson, M.A., Quinn, P.G., and Granner, D.K., Multihormonal regulation of phosphoenolpyruvate carboxykinase-CAT fusion gene. *J. Biol. Chem.* 262:14917-14920 (1987).
9. Liu, J., Park, E.A., Gurney, A.L., Roesler, W.J., and Hanson, R.W., cAMP induction of PEPCK gene transcription is mediated by multiple promoter elements. *J. Biol. Chem.* 266:19095-19102 (1991).
10. O'Brien, R.M., Lucas, P.C., Forest, C.D., Magnuson, M.A., and Granner, D.K., Identification of a sequence in the PEPCK gene that mediates a negative effect of insulin on transcription. *Science* 249:533-537 (1990).
11. Conover, C.A., Divertie, G.D., and Lee, P.D.K., Cortisol increases plasma insulin-like growth factor binding protein-1 in humans. *Acta Endocrinol.*, in press.
12. Luo, J., Reid, R.E., and Murphy, L.J., Dexamethasone increases hepatic IGFBP-1 mRNA and serum IGFBP-1 concentrations in the rat. *Endocrinol.* 127:1456-1462 (1990).
13. Ooi, G.T., Tseng, L.Y.-H., Tran, M.Q., and Rechler, M.M., Insulin rapidly decreases IGFBP-1 gene transcription in streptozotocin-diabetic rats. *Mol. Endocrinol.* 6:2219-2228 (1992).
14. Lewitt, M.S., and Baxter, R.C., Regulation of growth hormone-independent IGFBP (BP-28) in cultured human fetal liver explants. *J. Clin. Endocrinol. Metab.* 69:246-252 (1989).
15. Orlowski, C.C., Ooi, G.T., and Rechler, M.M., Dexamethasone stimulates transcription of the IGFBP-1 gene in H4-II-E rat hepatoma cells. *Mol. Endocrinol.* 4:1592-1599 (1990).
16. Powell, D.R., Suwanichkul, A., Cubbage, M.L., DePaolis, L.A., Snuggs, M.B., and Lee, P.D.K., Insulin inhibits transcription of the human gene for IGFBP-1. *J. Biol. Chem.* 266:18868-18876 (1991).
17. Unterman, T.G., Oehler, D.T., Murphy, L.J., and Lacson, R.G., Multihormonal regulation of IGFBP-1 in rat H4IIE hepatoma cells: The dominant role of insulin. *Endocrinol.* 128:2693-2701 (1991).
18. Powell, D.R., Lee, P.D.K., DePaolis, L.A., Morris, S.L., and Suwanichkul, A., Dexamethasone stimulates expression of IGFBP-1 in HEP G2 human hepatoma cells. *Growth Reg.* 3:11-13 (1993).
19. Suwanichkul, A., DePaolis, L.A., Lee, P.D.K., and Powell, D.R., Identification of a promoter element which participates in cAMP-stimulated expression of human IGFBP-1. *J. Biol. Chem.*, in press.

20. Imai, E., Miner, J.N., Mitchell, J.A., Yamamoto, K.R., and Granner, D.K., Glucocorticoid receptor-cAMP response element binding protein interaction and the response of the phosphoenolpyruvate carboxykinase gene to glucocorticoids. *J. Biol. Chem.* 268:5353-5356 (1993).

21. Park, E.A., Roesler, W.J., Liu, J., Klemm, D.L., Gurney, A.L., Thatcher, J.D., Shuman, J., Friedman, A., and Hanson, R.W., The role of the CCAAT/Enhancer binding-protein in the transcriptional regulation of the gene for phosphoenolpyruvate carboxykinase (GTP). *Mol. Cell. Biol.* 10:6264-6272 (1990).

22. Suwanichkul, A., Cubbage, M.L., and Powell, D.R., The promoter of the human gene for insulin-like growth factor binding protein-1. *J. Biol. Chem.* 265:21185-21193 (1990).

23. deWet, J., Wood, K., De Luca, M., Helinski, D., and Subramani, S., Firefly luciferase gene: Structure and expression in mammalian cells. *Mol. Cell. Biol.* 7:725-737 (1987).

24. Borman, C., Moffat, L., and Howard, B., Recombinant genomes which express chloramphenicol acetyltransferase in mammalian cells. *Mol. Cell. Biol.* 2:1044-1051 (1982).

25. Needleman, S.B., and Wunsch, C.D., A general method applicable to the search for similarities in the amino acid sequence of two proteins. *J. Mol. Biol.* 48:443-453 (1970).

26. Unterman, T.G., Lacson, R.G., McGary, E., Whalen, C., Purple, C., and Goswami, R.G., Cloning of the rat IGFBP-1 gene and analysis of its' 5' promoter region. Biochem. Biophys. Res. Comm. 185:993-999 (1992).

27. Mendel, D.B., and Crabtree, G.R., HNF-1, a member of a novel class of dimerizing homeodomain proteins. *J. Biol. Chem.* 266:677-680 (1991).

28. Powell, D.R., and Suwanichkul, A., HNF1 activates transcription of the human gene for IGFBP-1. *DNA Cell Biol.* 12:283-289 (1993).

29. Blumenfeld, M., Maury, M., Chouard, T., Yaniv, M., and Condamine, H., HNF1 shows a wider distribution than products of its known target genes in developing mouse. *Development* 113:589-599 (1991).

30. De Simone, V., DeMagistris, L., Lazzaro, D., Gerstner, J., Monaci, P., Nicosia, N., and Cortese, R., LFB3, a heterodimer-forming homeoprotein of the LFB1 family, is expressed in specialized epithelia. *EMBO J* 10:1435-1444 (1991).

31. Chin, E., Zhou, J., and Bondy, C., Anatomical relationships in the patterns of IGF-I, IGFBP-1, and IGF-I receptor gene expression in the rat kidney. *Endocrinol* 130:3237-3245 (1992).

32. Suikkari, A.-M., Koivisto, V.A., Koistinen, R., Seppala, M., and Yki-Jarvinen, H., Dose-response characteristics for suppression of low MW IGFBP by insulin. *J. Clin. Endocrinol. Metab.* 68:135-140 (1989).

33. Conover, C.A., and Lee, P.D.K., Insulin regulation of IGFBP production in cultured HEP G2 cells. *J. Clin. Endocrinol. Metab.* 70:1062-1067 (1990).

34. Lee, P.D.K., Suwanichkul, A., DePaolis, L.A., Snuggs, M.B., Morris, S.L., and Powell, D.R., IGF suppression of IGFBP-1 production: Evidence for mediation by the type I IGF receptor.. *Reg. Peptides, in press.*

35. Suwanichkul, A., Morris, S.L., and Powell, D.R., Identification of an insulin responsive element in the promoter of the human gene for IGFBP-1. *J. Biol. Chem.*, in press.

36. Johnson, T.M., Rosenberg, M.P., and Meisler, M.H., An insulin-responsive element in the pancreatic enhancer of the amylase gene. *J. Biol. Chem.* 268:464-468, 1993.

37. Hilding, A., Thoren, M., and Hall, K., The effect of glucagon on serum IGFBP-1 levels in GH-deficient patients. In: Binoux, M., Zapf, J., eds. *2nd International Workshop on IGFBPs*, Opio, France, 111 (abstract 98B) (1992).

38. Montminy, M.R., Gonzalez, G.A., and Yamamoto, K., Characteristics of the cAMP response unit. *Rec. Prog. Hormone Res.* 46:219-230 (1990).

39. Habener, J.F., Cyclic AMP response element binding proteins: A cornucopia of transcription factors. *Mol. Endocrinol.* 4:1087-1094 (1990).

40. Imai, E., Stromstedt, P.-E., Quinn, P.G., Carlstedt-Duke, J., Gustafsson, J.-A., and Granner, D.K., Characterization of a complex glucocorticoid response unit in the PEPCK gene. *Mol. Cell. Biol.* 10:4712-4719 (1990).

INSULIN-LIKE GROWTH FACTOR BINDING PROTEIN-1: IDENTIFICATION, PURIFICATION, AND REGULATION IN FETAL AND ADULT LIFE

Terry G. Unterman

Department of Medicine, University of Illinois College Medicine at Chicago and VA West Side Medical Center, Chicago, IL

INTRODUCTION

Insulin-like growth factors (IGFs) are mitogenic peptides which resemble proinsulin in structure and are thought to regulate cellular proliferation and other anabolic processes in many tissues (1). Early bioassay studies demonstrated that circulating IGF bioactivity is reduced in a variety of conditions where anabolism is impaired, including fasting, uremia and streptozotocin (STZ)-induced diabetes (2). Phillips and co-workers demonstrated that this reduction in IGF bioactivity reflected changes in circulating levels of both IGFs (reduced) and IGF inhibitors (increased), factors which inhibit the effects of IGFs in bioassay (2). Recent studies have shown that IGFs circulate in association with specific binding proteins (IGFBPs) which are thought to modulate the availability and the biological effects of IGFs on target tissues (3). To date, 6 distinct IGFBPs have been identified and their cDNAS cloned (4,5). We considered the possibility that alterations in circulating levels of specific IGFBPs might contribute to the modulation of IGF bioactivity in metabolic disease. The purpose of this paper is to review our major findings related to this issue.

IGF BINDING ACTIVITY AND IGFBP LEVELS IN STZ-DIABETES

We first examined changes in circulating IGFBPs in acutely ketotic STZ-diabetic rats, since circulating levels of bioassayable IGF inhibitors are markedly elevated in these animals (6). Early studies relying on charcoal absorption techniques to separate bound and free $[^{125}I]$IGF-I failed to detect an increase in IGF binding activity in diabetic serum (7,8). However, we considered the possibility that the availability of low mol wt IGFBPs might be selectively increased in STZ-diabetic serum and not detected in assays of total serum IGF binding activity. Thus, we incubated $[^{125}I]$IGF-I with serum from diabetic and non-diabetic rats, and separated bound and free tracer by sizing chromatography (9).

Current Directions in Insulin-Like Growth Factor Research,
Edited by D. LeRoith and M.K. Raizada, Plenum Press, New York, 1994

215

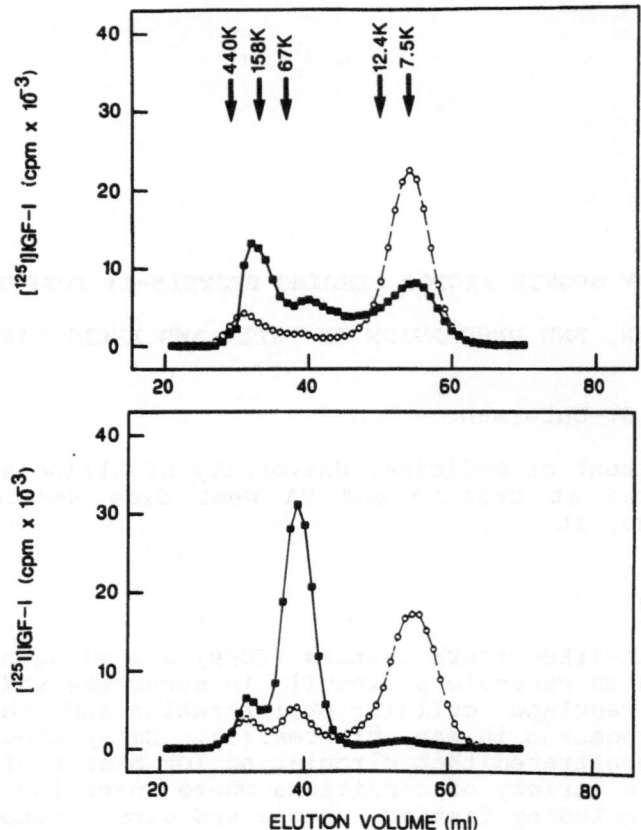

Figure 1. Circulating [^{125}I]IGF-I binding activity. [^{125}I]IGF-I was incubated overnight at 4 C with 500 μl/ml serum from control (Upper Panel) or STZ-diabetic serum (Lower Panel) with (Open Circles) or without (Solid Squares) 500 ng/ml unlabeled IGF-I, then filtered on a 1.5x50 cm Sephacryl S-200 column. Radioactivity in eluate was measured and normalized to 1x10^6 cpm per experiment. The mean value for 7 experiments is shown. Arrows indicate mol wt. From reference (9), with permission.

As shown in Figure 1, [^{125}I]IGF-I binding activity was markedly increased in diabetic serum, due to increased binding to low mol wt IGFBPs (30-40 K); less tracer eluted with high mol wt IGF-IGFBP complexes (>150 K), consistent with reduced levels of IGFBP-3 (10). Studies with excess unlabeled IGF-I confirmed that this binding was competitive (Figure 1). Binding also was specific since it was not inhibited by excess insulin (not shown). Subsequent studies with lower concentrations of serum revealed that IGF binding activity was increased ~100-fold in diabetic rats. Further, IGF binding activity was lowered by treatment with insulin and rose again after insulin was discontinued (9), demonstrating that these differences reflected changes in insulin status and were not due to non-specific effects of STZ. Of note, activated charcoal stripped tracer off of these IGFBPs (not shown), explaining why previous studies failed to detect increased IGF binding activity in diabetic serum.

216

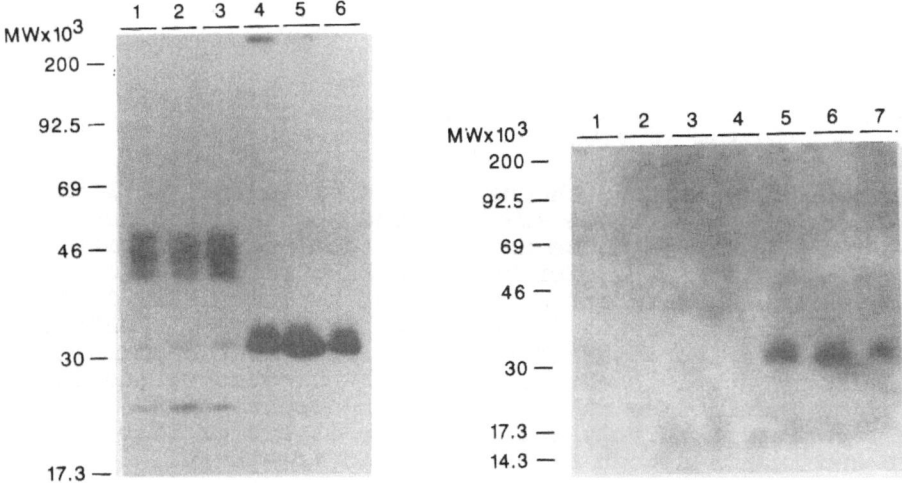

Figure 2. Western ligand and immunoblotting of serum IGFBPs.
Left panel. Serum from control (1-3), STZ-diabetic
(4-6) rats was loaded for 13% SDS/PAGE and proteins
transferred to nitrocellulose. Membranes were
probed with [^{125}I]IGF-I prior to autoradiography.
Right panel. Fetal (1), control (2-4) and STZ-
diabetic (5-7) adult rat serum was loaded for
SDS/PAGE and transfer. Membranes were probed with
antiserum against human IGFBP-1 (by D. Clemmons)
and then [^{125}I]protein A. From reference (11), with
permission.

        To determine whether differences in circulating IGF
binding activity reflected changes in the levels of specific
IGFBPs, we next performed western ligand and immunoblotting
studies. As shown in Figure 2, ligand blotting studies
revealed that 32-34 K IGFBPs were markedly increased while
40-50 K IGFBPs (representing IGFBP-3) were reduced in STZ-
diabetic serum. Subsequent ligand blots also showed that the
predominant IGFBP in diabetic serum (32 K) was slightly
smaller than the major IGFBP present in fetal rat serum (34
K) (IGFBP-2) (not shown), the only other low mol wt IGFBP
that had been identified in the rat at that time. Further,
immunoblotting with antiserum developed against human IGFBP-1
(by D. Clemmons) recognized the 32 K IGFBP present in STZ-
diabetic rat serum, but not IGFBPs in adult or fetal rat
serum (Figure 2), providing further confirmation that this
IGFBP is distinct from rat IGFBP-2.
        Subsequent studies confirmed that levels of immunore-
active IGFBP-1 were reduced in insulin-treated STZ-diabetic
animals and rose again after insulin was discontinued (9). Of
note, immunoblotting with antiserum against rat IGFBP-2
showed that serum levels of IGFBP-2 also were increased in
STZ-diabetic rats: however, levels of IGFBP-2 remained high
after 2 days of insulin treatment (9). Further, affinity
labeling and immunoprecipitation studies confirmed that
IGFBP-1 is the major available IGFBP in diabetic serum (9)

and accounts for the increase in circulating IGF binding activity in STZ-diabetes.

HEPATIC IGFBP mRNA IN STZ-DIABETES: EFFECTS OF INSULIN AND GLUCOCORTICOIDS IN VIVO

To better understand the regulation of low mol wt IGFBPs in STZ-diabetic animals, we examined the abundance of hepatic IGFBP-1 and -2 mRNA in these animals in collaboration with G. Ooi and M. Rechler (12). Northern blotting (Figure 3) demonstrated that changes in the abundance of hepatic IGFBP-1 and -2 mRNA paralleled changes in serum levels (above). IGFBP-1 mRNA levels were increased 100-fold in STZ-diabetic animals, decreased to control levels after insulin treatment, and rose again after insulin was withdrawn. In contrast, IGFBP-2 mRNA levels were increased only 8-fold in the livers of diabetic animals and remained high despite insulin treatment.

Recently, we asked whether the increase in circulating levels of IGFBP-1 and hepatic IGFBP-1 mRNA in STZ-diabetic animals reflects an effect of insulin deficiency _per se_, or whether insulin deficiency represents a permissive state

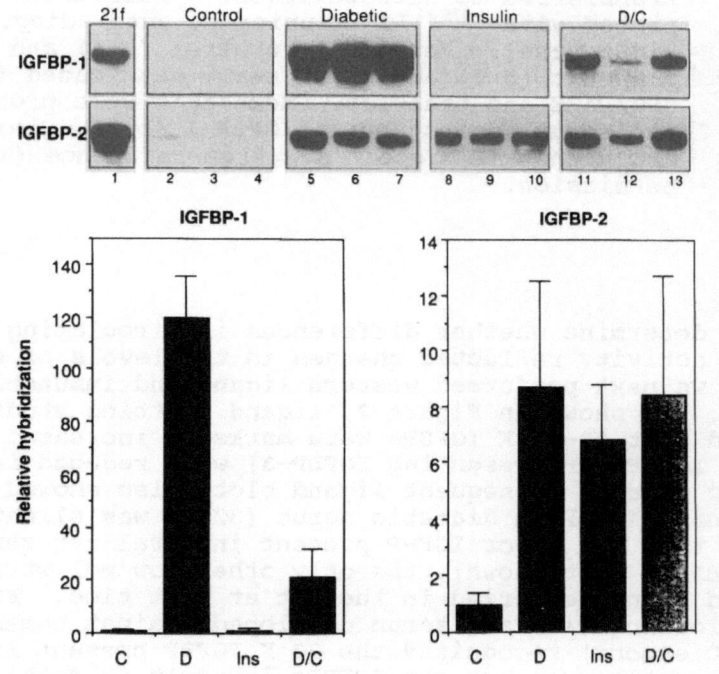

Figure 3. Northern blotting of total cellular RNA prepared from the livers of fetal (d 21 gestation), control, STZ-diabetic, and insulin-treated rats, and diabetic animals after insulin was discontinued. **Upper Panel.** 20 μg of RNA was loaded for northern blotting and probed with ^{32}P-labeled cDNA probes for IGFBP-1 and -2. **Lower Panel.** Blots were quantified by direct counting of radioactivity and hybridization intensity was expressed relative to control. From reference (12), with permission.

where increased levels of counter-regulatory factors, including glucocorticoids, more readily promote IGFBP-1 expression. To address this question, we examined serum IGF binding activity, serum levels of immunoreactive IGFBP-1, and levels of hepatic IGFBP-1 mRNA in intact and adrenalectomized STZ-diabetic and non-diabetic animals with and without corticosterone treatment. As shown in Figure 4, adrenalectomy prevented much of the increase in circulating IGF binding activity was restored to intact diabetic levels by treatment with 50 mg/kg/d corticosterone (Figure 4). Further, adrenalectomy and corticosterone treatment had similar effects on serum levels of immunoreactive IGFBP-1 and the abundance of hepatic IGFBP-1 mRNA in STZ-diabetic animals (13).

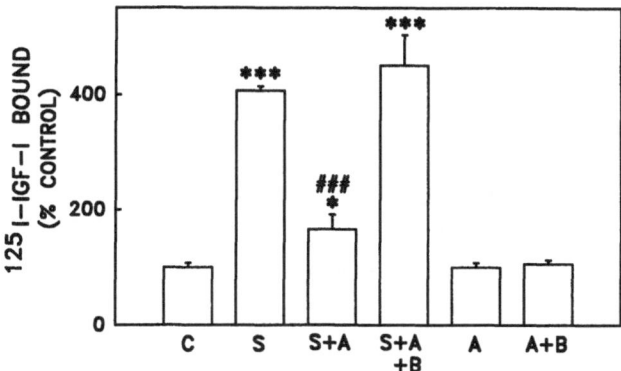

Figure 4. Serum [^{125}I]IGF-I binding activity in control (C) and STZ-diabetic (S) rats, with/without adrenalectomy (A) treatment with corticosterone 50 mg/kg/d. IGF binding activity was measured by 1 h incubation of tracer with 2.5 µl/ml serum and precipitation with polyethylene glycol. ***, P<0.001 vs C. ###, P<0.001 vs S. (Copyright, Endocrine Society, 1993).

Taken together, these findings support the concepts that 1) circulating IGF binding activity is markedly increased and may limit the availability of IGFs to target tissues when insulin levels are low in acutely diabetic animals; 2) this increase in IGF binding activity reflects increased availability of IGFBP-1; 3) serum IGF binding activity and levels of IGFBP-1 are regulated according to changes in insulin and glucocorticoid status in experimental diabetes; and 4) insulin and glucocorticoids regulate hepatic production of IGFBP-1 at the level of mRNA abundance.

IGFBPs AND IGFs IN THE SMALL FOR GESTATIONAL AGE FETAL RAT

Early studies suggested that the IGF system functions differently <u>in utero</u> compared to post-natal life, since 1)

IGF-II levels are high in the fetal rat but low in adult animals (14); 2) serum IGF-I levels are relatively low in the fetus and rise during neonatal life (15); and 3) in adult life, most IGFs circulate as part of a high mol wt (150 K) complex with IGFBP-3 which has limited access to the extra-vascular space, while low molecular weight IGFBPs predominate in the fetal circulation and may traverse the vascular barrier (5,16). We asked whether alterations in circulating levels of low molecular weight IGFBPs may have important consequences for fetal growth. Since insulin is an important regulator of IGFBP-1 expression and levels in adult animals (above), and since insulin appears to be an important determinant of fetal growth (17), we examined circulating levels and hepatic expression of IGFBPs 24 hr after uterine artery ligation (18,19), when the delivery of maternal nutrients and fetal insulin levels are reduced and fetal growth is impaired.

Serum [^{125}I]IGF-I binding activity was increased 6-fold (P<0.001) and ligand blotting revealed that circulating levels of 32-34 K IGFBPs were increased in SGA fetuses compared to controls (19). Densitometric analysis of western ligand blots revealed a 4-fold increase in 32 K IGFBP levels and immunoblotting confirmed that levels of IGFBP-1 were increased in SGA serum (19). Similarly, northern blotting showed that the abundance of IGFBP-1 mRNA was increased in the livers of SGA litters relative to controls (18).

Of note, densitometric analysis showed that levels of 34 K IGFBPs also were increased 2-fold in SGA animals, while no

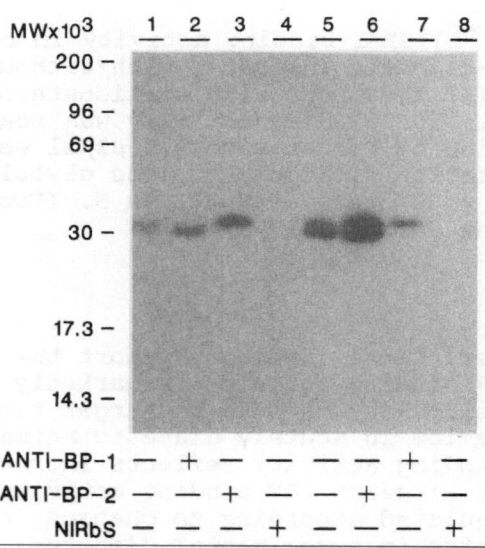

Figure 5. 2 μl of control (lane 1) or SGA (lane 2) fetal rat serum was prepared for SDS/PAGE and ligand blot-ting, as in Figure 2. Alternatively, IGFBPs in 10 μl of control (lanes 3,5,7) or SGA (lanes 4,6,8) serum were first precipitated with antiserum against rat IGFBP-1 or -2 or with non-immune rabbit serum (NIRbS) prior to loading for ligand blotting. From reference (19), with permission.

change in serum levels of 34 K IGFBP-2 was detected by immunoblotting and the abundance of hepatic IGFBP-2 mRNA was not increased in SGA fetuses. Immunoprecipitation studies provided an explanation for this apparent discrepancy.

As shown in Figure 5, circulating forms of IGFBP-1 (32 K) and IGFBP-2 (34 K) were clearly separable after immunoprecipitation from control fetal serum, and serum levels of immunoprecipitable IGFBP-2 were not altered in SGA litters. In contrast, antiserum against IGFBP-1 precipitated increased amounts of IGFBPs from SGA serum, and these IGFBPs appeared to contain both 32 and 34 K components. More recent studies suggest that this difference in the apparent mol wt of IGFBP-1 reflects the presence of both phosphorylated and nonphosphorylated forms of IGFBP-1 in SGA serum (R. McCusker and T. Unterman, unpublished observations). Since phosphorylated forms of IGFBP-1 bind IGF-I with a higher affinity than nonphosphorylated IGFBP-1 (20), the presence of phosphorylated IGFBP-1 may further increase IGF binding activity and reduce the availability of IGFs to target tissues in the SGA fetus.

Correlation analysis revealed significant inverse relationships between serum levels of IGFBP-1 (estimated by densitometric analysis of 32 K IGFBPs) and fetal body (r=-0.499, P<0.004) and liver weight (r=-0.512; P<0.001), consistent with the concept that IGFBP-1 contributes to the regulation of fetal somatic growth. Serum levels of 32 K IGFBPs also correlated with levels of both insulin and IGF-I (P<0.002 for each). Since hepatic IGF-I receptors are abundant in the fetal hepatocyte (21), and since IGF-I has potent effects on hepatocellular metabolism in the fetal rat (22), one may speculate that IGF-I, together with insulin, may contribute to the regulation of hepatic production of IGFBP-1 in fetal life.

We also examined levels of insulin, IGF-I and -II and their relationships to fetal weights in SGA and control litters. Levels of insulin were reduced in SGA fetuses compared to controls (70±5 vs 134±12 μU/ml, respectively, P<0.001) and insulin levels correlated with levels of IGF-I (r=0.480, P<0.02), but not with IGF-II. Levels of insulin and IGF-I also correlated with fetal body (r=0.531, P<0.001 and r=0.501, P<0.004, respectively) and liver weight (r=0.480, P<0.05 and r=0.578, P<0.005, respectively). Of note, relationships between insulin and fetal body and liver weight were no longer significant after controlling for the effects of IGF-I. In contrast, relationships with IGF-I remained significant after controlling for the effects of insulin (P<0.03), supporting the concept that IGF-I may be important in the regulation of fetal somatic growth, and that changes in the availability of IGF-I may mediate some effects of insulin on fetal development.

These findings, and similar results we have obtained in recent studies of fetal growth during maternal starvation (22), indicate that circulating levels of IGF-I (and not IGF-II) are reduced and that levels of IGFBP-1 (and not IGFBP-2) are increased in the fetal rat when serum insulin levels are low and somatic growth is impaired. It is interesting to note that this pattern of adaptation is similar to changes observed in adult animals and patients when nutrition is restricted and insulin levels are low (2,23). This observation is even more striking when one recognizes that growth hormone, which is known to be important in the regulation of IGF-I expression and levels in childhood and adult life, is not thought to be important in the regulation of intrauterine growth. Thus, in the setting of nutritional restriction, similar growth hormone-independent mechanisms may reduce the availability of IGFs, both by lowering IGF levels and by increasing levels of IGFBP-1, and thereby limit somatic growth in both fetal and post-natal life.

PURIFICATION AND REGULATION OF IGFBP-1 EXPRESSION IN VITRO

Since the liver appears to be an important source of IGFBP-1 (24), we examined the direct effects of insulin, glucocorticoids, and cyclic AMP analogues on hepatocellular expression of IGFBP-1 utilizing rat H4IIE hepatoma cells, since they are well-differentiated and hormone responsive (25). Our initial studies revealed that H4IIE cells produce a 32 K IGFBP that is recognized by antiserum against human IGFBP-1, like the major IGFBP present in diabetic rat serum (11,26). We purified this protein to homogeneity by ammonium sulfate precipitation, G-75 chromatography, and neutral and acid reverse phase HPLC. Amino acid sequencing confirmed that this protein is the rat form of IGFBP-1 (26).

IGF binding assays revealed that the availability of IGFBPs in H4IIE conditioned medium was hormonally regulated, and ligand and immunoblotting studies confirmed that insulin inhibits, while glucocorticoids and cAMP increase the production of IGFBP-1 by H4IIE cells (27). Further, northern blotting revealed that 100 nM insulin rapidly lowers IGFBP-1 mRNA levels below control (< 10% of control at 4 h), while glucocorticoids and cAMP increase the abundance of IGFBP-1 mRNA ~9-fold and ~3-fold, respectively (27). Of note, insulin exerted a dominant effect on the expression of IGFBP-1 mRNA levels, and both prevented and reversed the effects of gluco-

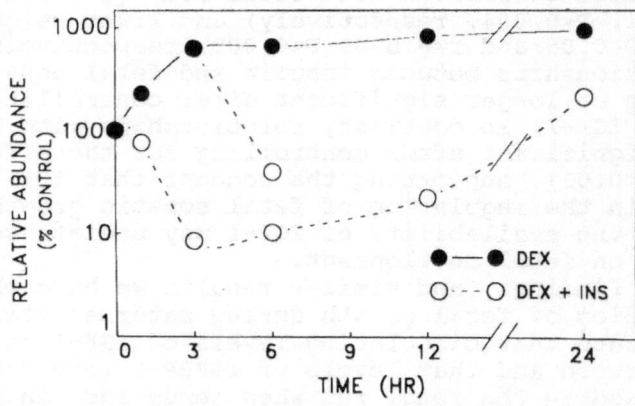

Figure 6. Effect of 1 μM dexamethasone (DEX) and/or 100 nM insulin (INS) on IGFBP-1 mRNA levels in rat H4IIE hepatoma cells. IGFBP-1 mRNA levels were measured by northern blotting of total cellular RNA and results expressed relative to control cells. Reprinted from reference (27), with permission.

corticoids (Figure 6) and cAMP (not shown). Subsequent studies confirmed that similar effects were achieved with lower concentrations of insulin (ED_{50} = 40 pM).

To determine whether insulin regulates the abundance of IGFBP-1 mRNA at the level of IGFBP-1 gene transcription and/or mRNA stability, we next examined the effects of insulin on the survival of IGFBP-1 mRNA during incubation with actinomycin D, a potent inhibitor of transcription. We estimated that the half-life for IGFBP-1 mRNA in H4IIE cells is ~80 minutes and found that the stability of IGFBP-1 mRNA is not altered by the addition of insulin. Based on these results, we inferred that insulin must regulate the level of IGFBP-1 mRNA at the level of gene transcription. Subsequent studies have demonstrated by run-on assay that insulin rapidly inhibits IGFBP-1 gene transcription in vivo (28). Taken together, these findings support the view that insulin plays a major role in regulating circulating levels of IGFBP-1, and that this regulation occurs at the level of hepatic IGFBP-1 gene transcription.

Based on these observations, we have sought to investigate specific mechanisms by which insulin and other factors regulate hepatic production of IGFBP-1 at the level of gene transcription. Towards this end, we have cloned the rat IGFBP-1 gene and its 5' promoter region (29). Dideoxy sequencing of 1100 bases of the rat IGFBP-1 promoter upstream form the transcription initiation site reveals that the human and rat IGFBP-1 promoters are highly homologous for the first 480 bases upstream and computer analysis indicates the presence of several putative response elements within this region (30).

We recently have used reporter gene constructs with the firefly luciferase gene to confirm that this region of the IGFBP-1 promoter initiates luciferase transcription in transient transfection studies with H4IIE cells, and have found that specific sequences within this proximal promoter region mediate the effects of insulin and glucocorticoids on IGFBP-1 promoter activity. Studies with 5'-nested deletions indicate that the effects of glucocorticoids are greatly reduced when constructs are truncated within 90 bases from the transcription initiation site, and DNAse I footprinting with purified glucocorticoid receptor protein (provided by Dr. L. Friedman) and site-directed mutagenesis of the corresponding sequences have identified a specific glucocorticoid response sequence in this region (31). We used internal deletions to identify a 23 base pair sequence which is required for insulin to exert its full effects on IGFBP-1 promoter activity (31). Further analysis of these components of the IGFBP-1 promoter and trans-acting factors which interact with these DNA sequences should help to clarify specific mechanisms by which insulin and glucocorticoids regulate hepatic expression of IGFBP-1, and thereby modulate the biological effects of IGFs on target tissues.

SUMMARY

Circulating IGF bioactivity is reduced under a variety of conditions where anabolism is impaired. Serum IGF binding activity is increased in acutely diabetic animals and is regulated according to insulin status. Alterations in serum

IGF binding activity reflects, at least in part, changes in circulating levels of IGFBP-1, suggesting that IGFBP-1 is an important modulator of IGF availability in post-natal life. Serum IGF binding activity and levels of IGFBP-1 also are high in the hypoinsulinemic SGA fetal rat and levels of IGFBP-1 correlate with fetal liver and body weight, indicating that IGFBP-1 contributes to the regulation of somatic growth _in utero_. Hepatic expression of IGFBP-1 is regulated at the level of gene transcription by insulin in a dominant negative fashion, while glucocorticoids and cAMP analogues exert positive effects on hepatocellular IGFBP-1 mRNA. Glucocorticoids exert important effects on circulating levels and hepatic expression of IGFBP-1 _in vivo_ under conditions where insulin levels are low. Regulation of hepatic production of IGFBP-1 may provide a mechanism by which insulin and counter-regulatory factors may modulate the availability of IGFs and the biological effects of IGFs in both fetal and adult life.

REFERENCES

1. Zapf, J, Schmid, C and Froesch, ER 1984 Biological and immunological properties of insulin-like growth factors (IGF) I and II. Clin Endocrinol Metab 13:3-30.

2. Phillips, LS and Unterman, TG 1984 Somatomedin activity in disorders of nutrition and metabolism. Clin Endocrinol Metab 13:145-189.

3. Baxter, RC and Martin, JL 1989 Binding proteins for the insulin-like growth factors: structure, regulation and function. Progr Growth Factor Res 1:49-68.

4. Shimasaki, S and Ling, N 1991 Identification and molecular characterization of insulin-like growth factor binding proteins (IGFBP-1, -2, -3, -4, -5, and -6). Progr Growth Factor Res 3:243-266.

5. Rechler, MM and Brown, AL 1992 Insulin-like growth factor binding proteins: gene structure and expression. Growth Regulation 2:55-68.

6. Phillips, LS, Fusco, AC and Unterman, TG 1985 Nutrition and somatomedin. XIV. Altered levels of somatomedins and somatomedin inhibitors in rats with streptozotocin-diabetes. Metabolism 34:765-770.

7. Rieu, M and Binoux, M 1985 Serum levels of insulin-like growth factor (IGF) binding protein in insulin-dependent diabetics during an episode of severe metabolic decompensation and the recovery phase. J Clin Endocrinol Metab 60:781-789.

8. Glaser, EW, Goldstein, S and Phillips, LS 1987 Nutrition and Somatomedin. XVII. Circulating somatomedin C during treatment of diabetic ketoacidosis. Diabetes 36:1152-1160.

9. Unterman, TG, Patel, K, Mahathre, VK, Rajamohan, G, Oehler, DT and Becker, RE 1990 Regulation of low

molecular weight insulin-like growth factor binding
proteins (IGF BPs) in experimental diabetes mellitus.
Endocrinology 126:2614-2624.

10. Zapf, J, Hauri, C, Waldvogel, M, Futo, E, Hasler, H,
    Binz, K, Guler, HP, Schmid, C and Froesch, ER 1989
    Recombinant human insulin-like growth factor I induces
    its own specific carrier protein in hypophysectomized and
    diabetic rats. Proc Natl Acad Sci USA 86:3813-3817.

11. Unterman, TG, Oehler, DT and Becker, RE 1989
    Identification of a type 1 insulin-like growth factor
    binding protein (IGF BP) in serum from rats with diabetes
    mellitus. Biochem Biophys Res Commun 163:882-887.

12. Ooi, GT, Orlowski, CC, Brown, AL, Becker, RE, Unterman,
    TG and Rechler, MM 1990 Tissue distribution and hormonal
    regulation of mRNAs encoding rat insulin-like growth
    factor binding proteins rIGFBP-1 and rIGFBP-2. Mol
    Endocrinol 4:321-328.

13. Unterman, TG, Hofert, J, Oehler, D and Lacson, R 1993
    Effects of glucocorticoids on circulating levels and
    hepatic expression of insulin-like growth factor binding
    proteins and IGF-I in the adrenalectomized
    streptozotocin-diabetic rat. (Manuscript Submitted).

14. Moses, AC, Nissley, SP, Short, PA, Rechler, MM, White,
    RM, Knight, AB and Higa, OZ 1980 Increased levels of
    multiplication-stimulating activity, an insulin-like
    growth factor, in fetal rat serum. Proc Natl Acad Sci USA
    77:3649-3653.

15. Donovan, SM, Oh, Y, Pham, H and Rosenfeld, RG 1989
    Ontogeny of serum insulin-like growth factor binding
    protein in the rat. Endocrinology 125:2621-2627.

16. Bar, RS, Clemmons, DR, Boes, M, Busky, WH, Booth, BA,
    Dake, BL and Sandra, A 1990 Transcapillary permeability
    and subendothelial distribution of endothelial and
    amniotic fluid insulin-like growth factor binding
    proteins in the rat heart. Endocrinology 127:1078-1086.

17. Hill, DJ and Milner, RDG 1985 Insulin as a growth factor.
    Pediatr Res 19:879-886.

18. Unterman, TG, Lacson, R, Gotway, MB, Oehler, DT, Gounis,
    A, Simmons, RA and Ogata, ES 1990 Circulating levels of
    insulin-like growth factor binding protein-1 (IGFBP-1)
    and hepatic mRNA are increased in the small for
    gestational age (SGA) fetal rat. Endocrinology
    127:2035-2037.

19. Unterman, TG, Simmons, RA, Glick, RP and Ogata, ES 1993
    Circulating levels of insulin, insulin-like growth
    factor-I (IGF-I), IGF-II, and IGF-binding proteins in the
    small for gestational age fetal rat. Endocrinology
    132:327-336.

20. Jones, JI, Busby, WH,Jr., Wright, G, Smith, CE,

Kimacki,NM and Clemmons, DR 1993 Identification of the
sites of phosphorylation in insulin-like growth factor
binding protein-1. Regulation of its affinity by
phosphorylation of serine 101. J Biol Chem 268:1125-1131.

21. Caro, JF, Poulos, J, Ittoop, O, Pories, WJ, Flickinger,
EG and Sinha, MK 1988 Insulin-like growth factor I
binding in hepatocytes from human liver, human hepatoma,
and normal, regenerating, and fetal rat liver. J Clin
Invest 81:976-981.

22. Shambaugh, GE,III, Radosevich, JA, Glick, RP, Gu, DS,
Metzger, BE and Unterman, TG 1993 Insulin-like growth
factors and binding proteins in the fetal rat:
alterations during maternal starvation and effects in
fetal brain cell culture. Neurochem Res. (In Press).

23. Clemmons, DR and Underwood, LE 1991 Nutritional
regulation of IGF-I and IGF binding proteins. Annu Rev
Nutr 11:393-412.

24. Brinkman, A, Groffen, C, Kortleve, DJ, Guerts van Kessel,
A and Drop, SL 1988 Isolation and characterization of a
cDNA encoding the low molecular weight insulin-like
growth factor binding protein (IBP-1). EMBO J
7:2417-2423.

25. Sasaki, K, Cripe, TP, Koch, SR, Andreone, TL, Peterson,
DD, Beale, EG and Granner, DK 1984 Multihormonal
regulation of phosphoenolpyruvate carboxykinase gene
transcription. The dominant role of insulin. J Biol Chem
259:15242-15250.

26. Unterman, TG, Oehler, DT, Gotway, MB and Morris, PW 1990
Production of the rat type 1 insulin-like growth
factor-binding protein by well differentiated H_4EIIC_3
hepatoma cells: Identification, purification, and
N-terminal amino acid analysis. Endocrinology
127:789-797.

27. Unterman, TG, Oehler, DT, Murphy, LJ and Lacson, RG 1991
Multihormonal regulation of insulin-like growth factor-
binding protein-1 in rat H4IIE hepatoma cells: the
dominant role of insulin. Endocrinology 128:2693-2701.

28. Ooi, GT, Tseng, LY-H, Tran, MQ and Rechler, MM 1992
Insulin rapidly decreases insulin-like growth factor-
binding protein-1 gene transcription in streptozotocin-diabetic rats. Mol Endocrinol 6:2219-2235.

29. Unterman, TG, Lacson, RG, Goswami, R, Whalen, C and
McGarry, E 1992 Cloning of the rat insulin-like growth
factor binding protein-1 (IGFBP-1) gene and analysis of
its 5' promoter region. Biochem Biophys Res Commun
185:993-999.

30. Goswami, R, Lacson, R and Unterman, T 1993 Identification
of insulin and glucocorticoid response sequences in the
rat IGF binding protein-1 (IGFBP-1) promoter. Proc
Endocrine Soc 75th Annual Meeting (Abstract).

226

RAPID REGULATION OF INSULIN-LIKE GROWTH FACTOR BINDING PROTEIN-1 TRANSCRIPTION BY INSULIN IN VIVO AND IN VITRO

Matthew M. Rechler, Guck T. Ooi, Dae-shik Suh and Lucy Tseng

Growth and Development Section
Molecular and Cellular Endocrinology Branch
National Institute of Diabetes and Digestive and Kidney Diseases
National Institutes of Health
Bethesda, MD 20892

INTRODUCTION

Insulin-like growth factor binding protein-1 (IGFBP-1) is distinctive among the IGFBPs in human plasma in its dynamic regulation by metabolic changes [reviewed in [1]]. IGFBP-1 levels are increased by acute fasting [2,3] and in diabetes [4,5], and normalized by refeeding and insulin treatment, respectively. IGFBP-1 also is increased by prolonged exercise, growth hormone deficiency and pregnancy [1], and following insulin-induced hypoglycemia [6].

We have used the streptozotocin-induced diabetic rat as a convenient animal model to elucidate the molecular basis of IGFBP-1 regulation. IGFBP-1 mRNA abundance in diabetic rat liver is regulated similarly to IGFBP-1 in human plasma: levels are increased ~100-fold in untreated diabetic animals and normalized by insulin treatment [7]. This paper summarizes our recent studies demonstrating that this regulation results from changes in IGFBP-1 transcription, and describes similar effects of insulin on IGFBP-1 transcription and IGFBP-1 mRNA abundance in H4-II-E rat hepatoma cells.

RESULTS

Transcriptional regulation of IGFBP-1 in diabetic and insulin-treated diabetic rats

We used a stable, non-ketotic model of insulin-deficient diabetes in order to be able to see rapid effects of insulin. Rats were made diabetic by intraperitoneal injection of moderate doses of streptozotocin (100 mg/kg) and studied 7 d after injection [8]. Diabetic animals were glycosuric, hyperglycemic, and had not gained weight, but were not ketotic and did not require insulin. As seen in Fig. 1, 1.55 kb IGFBP-1 mRNA (determined by Northern blotting) and IGFBP-1 transcription (determined by nuclear run-on transcription) were increased in 6 of 6 diabetic rats, compared to 5 non-diabetic control rats. In this experiment, the magnitude of increase varied for individual animals.

Current Directions in Insulin-Like Growth Factor Research,
Edited by D. LeRoith and M.K. Raizada, Plenum Press, New York, 1994

227

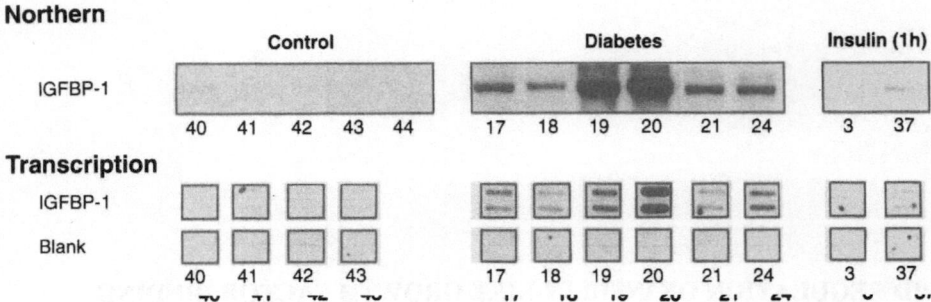

Northern

	Control	Diabetes	Insulin (1h)

IGFBP-1

40 41 42 43 44   17 18 19 20 21 24   3 37

Transcription

IGFBP-1

Blank

40 41 42 43   17 18 19 20 21 24   3 37

Figure 1. Comparison of IGFBP-1 mRNA levels and IGFBP-1 transcription in the livers of individual diabetic rats. Seven days after intraperitoneal injection of streptozotocin (100 mg/kg), diabetic animals were sacrificed (animals 17-21, 24) or injected subcutaneously with 2.4 U regular insulin / 100 g (animals 3, 37). Control rats that had not been injected with streptozotocin were sacrificed at the same time (animals 40-44). Total liver RNA from each animal was examined by Northern blotting (top) using a rat IGFBP-1 cDNA probe (nucleotides -116 to +557 relative to ATG, +1; random-primed using $[\alpha$-^{32}P]dCTP). Radiolabeled elongated nuclear transcripts were prepared from liver nuclei from the same animals using $[\alpha$-^{32}P]CTP and $[\alpha$-^{32}P]GTP. The transcripts were hybridized to nitrocellulose filter segments containing either an IGFBP-1 DNA target [a PCR-amplified 1374 bp fragment (nt -116 to 1258 of the cDNA)] or to blank filters without DNA, and autoradiographed (bottom). [Modified from reference 9].

The levels of IGFBP-1 mRNA and IGFBP-1 transcription in individual rats were compared after quantitation of the hybridized radioactivity by ß-scanning. As seen in Fig. 2, IGFBP-1 transcription increased in proportion to mRNA levels in individual animals (r=0.88, p<.0001). From these and other experiments [8,10], we conclude that IGFBP-1 transcription is increased in diabetic rat liver, and that this increase is sufficient to account for the increase in steady state IGFBP-1 mRNA.

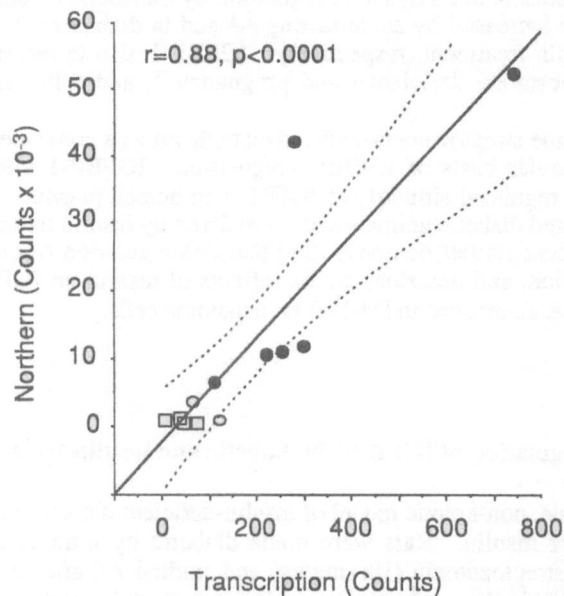

Figure 2. Quantitative relationship between hepatic IGFBP-1 mRNA abundance and IGFBP-1 gene transcription in individual control and diabetic rats. The hybridization signals in the autoradiographs shown in Fig. 1 were quantitated by computer-driven scanning of ß-radioactivity. Hepatic IGFBP-1 mRNA levels are plotted against IGFBP-1 transcription for individual rats. Non-diabetic control, open squares; untreated diabetic, closed circles; insulin-treated diabetic, open circles. The dashed lines show the 95% confidence limits of the regression line. [Modified from reference 9].

Insulin treatment of diabetic rats for 1 h (the shortest time examined) decreased both IGFBP-1 mRNA and IGFBP-1 transcription to the levels seen in non-diabetic control rats (Fig. 1). Similar results have been seen in other experiments [Fig. 3 and 8,10]. Although euglycemia was restored in the insulin-treated animals shown in Figs. 1 and 3, IGFBP-1 mRNA also was normalized in animals that remained hyperglycemic [8]. Thus, insulin decreases hepatic IGFBP-1 transcription in diabetic rats within 1 h. Growth hormone also rapidly inhibits IGFBP-1 transcription [11], and stimulates transcription of the IGF-I [12] and Spi-2.1 [13,14] genes. Since the decrease in IGFBP-1 transcription 1 h after insulin treatment resulted in a decrease in IGFBP-1 mRNA, our results further suggest that IGFBP-1 mRNA has a rapid turnover in vivo.

Insulin rapidly inhibits IGFBP-1 transcription and decreases IGFBP-1 mRNA abundance in H4-II-E rat hepatoma cells

Insulin also inhibits IGFBP-1 gene expression in H4-II-E cells, a cell line established from a well-differentiated rat hepatoma that expresses IGFBP-1 as its predominant IGFBP [15]. Cells were grown to confluence and equilibrated in serum-free medium. Total RNA was prepared at different times from cells treated with insulin and from control cells, and examined by Northern blotting. The results are shown in Fig. 4. IGFBP-1 mRNA decreased ~50% by 1 h after insulin addition, and 90% after 2-6 h.

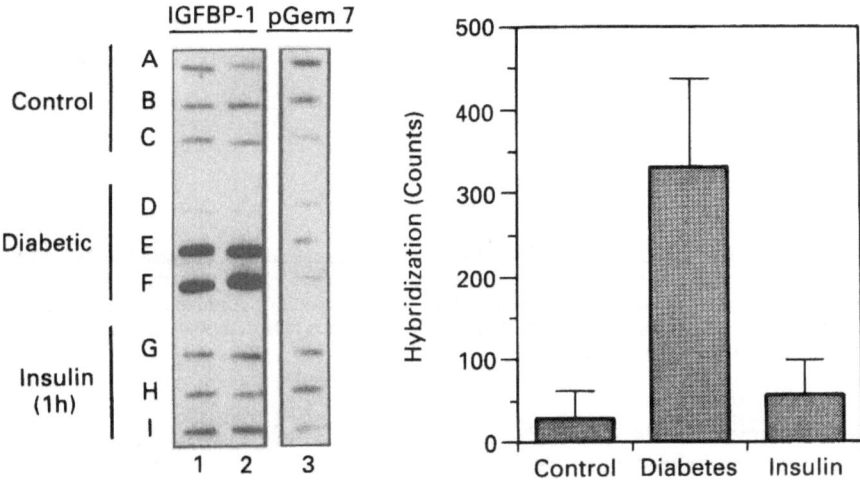

Figure 3. Insulin rapidly inhibits IGFBP-1 transcription in diabetic rat liver. Rats were made diabetic with streptozotocin, after which some animals were killed and others injected subcutaneously with a mixture of 2 U regular and 3.2 U intermediate-acting insulin, and killed 1 h later when they were euglycemic. Radio-labeled nuclear transcripts were prepared from the livers of 3 control (A, B, C), 3 diabetic (D, E, F) and 3 insulin-treated diabetic rats (G, H, I). Transcripts (2.5-5.5 million cpm per incubation) were hybridized to duplicate nitrocellulose filters containing an IGFBP-1 cDNA target (lanes 1 and 2) and to a single filter containing pGem7 plasmid DNA (lane 3). An autoradiograph of the hybridization is shown in the left panel. Radioactivity is increased in diabetic livers E and F, and decreased in insulin-treated diabetic rats G-I to the levels seen in non-diabetic control This can be appreciated in samples hybridized with comparable amounts of input radioactivity: for example, lanes C (5.5 million cpm), F (4.4 million cpm), and I (4.4 million cpm); lane E (2.7 million cpm), G (3.3 million cpm), and H (2.7 million cpm). Diabetic rat D was unusual in that hepatic IGFBP-1 mRNA abundance was not increased although it was hyperglycemic; animal D lost more weight than its cohorts, suggesting that an unrelated process may have blunted the IGFBP-1 response.

In the right panel, hybridized radioactivity was quantitated by ß-scanning, normalized to 5.5 million cpm input radioactivity, and corrected for hybridization of the same transcript to pGem7 plasmid. The corrected hybridized counts (Mean ± SD) are plotted for 3 control, 2 diabetic, and 3 insulin-treated diabetic rats. Results from non-responding animal D were excluded. [Reprinted from reference 8 with permission].

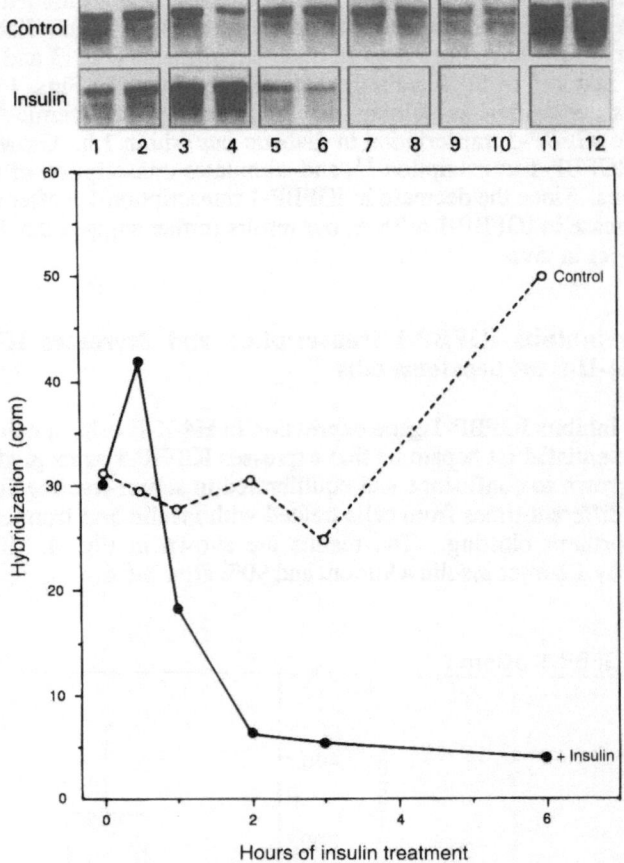

Figure 4. Insulin rapidly inhibits IGFBP-1 mRNA in H4-II-E rat hepatoma cells. Confluent H4-II-E cells, equilibrated in serum-free medium, were incubated with or without insulin (1 μg/ml)) for the indicated times. Total RNA was prepared, and IGFBP-1 mRNA examined by Northern blotting. Autoradiographs of the 1.55 kb IGFBP-1 mRNA is shown above; ß-scan quantitation of the hybridized radioactivity is shown below. [Reprinted from reference 16 with permission].

Insulin also rapidly inhibits IGFBP-1 transcription in H4-II-E cells (Fig. 5). In this experiment, cells were preincubated with dexamethasone for 3 h to stimulate IGFBP-1 transcription. [We previously have shown that insulin inhibits IGFBP-1 transcription in dexamethasone-treated cells as in cells that had not received dexamethasone [16]]. In the absence of insulin, IGFBP-1 transcription continued to increase during the 80 min incubation with dexamethasone, indicating that the peak induction of transcription by dexamethasone had not been reached. In cultures treated with insulin, IGFBP-1 transcription decreased by ~75% after 20 min (the earliest time examined), and by ~95% after 80 min. This is one of the most rapid effects of insulin on gene transcription [16]; for example, inhibition of PEPCK transcription in H4-II-E cells was detectable 5-10 min after insulin addition [17]. Thus, the effects of insulin on IGFBP-1 transcription in H4-II-E cells mimic its effects in diabetic rat liver, suggesting that they are direct effects rather than being secondary to in vivo metabolic alterations. The fact that transcription is inhibited faster than the decrease in IGFBP-1 mRNA suggests that the primary regulatory effect of insulin is at the level of transcription. As with diabetic liver, there is no need to postulate a post-transcriptional component of IGFBP-1 regulation. These results validate the use of the H4-II-E cell line as a model system to study IGFBP-1 gene regulation.

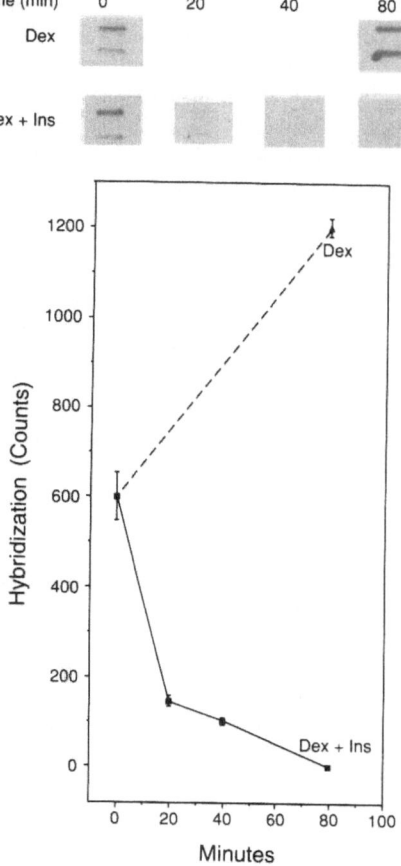

Figure 5. Insulin rapidly inhibits IGFBP-1 transcription in H4-II-E cells. Confluent H4-II-E cells were incubated with 1 µM dexamethasone for 3 h to induce IGFBP-1 transcription. Then, insulin (1 µg/ml) was added to some cultures, and nuclei prepared for run-on transcription assays 20, 40 and 80 min later. Control samples that did not receive insulin were examined at 0 and 80 min. Two preparations of nuclei (six 100-mm plates per preparation) were examined for each experimental treatment at each time point, and nuclear transcripts hybridized to nitrocellulose filters containing immobilized rat IGFBP-1 cDNA or pGem7 DNA (background) as targets. The top panel shows an autoradiograph of the radioactivity hybridized to the IGFBP-1 cDNA target. In the bottom panel, hybridized radioactivity, determined by ß-scanning and corrected for background hybridization, is plotted at different times after insulin addition. Cultures that were incubated with dexamethasone plus insulin are indicated by the solid line; cultures that received dexamethasone alone are shown by the dashed line. [Reprinted from reference 18 with permission].

Regulation of rat IGFBP-1 promoter activity

We have isolated genomic fragments of the rat IGFBP-1 gene containing the promoter, and coupled them to a promoterless firefly luciferase reporter plasmid. Constructs containing a promoter fragment extending from -925 to + 77 (with respect to the transcription initiation site, +1) were transiently transfected into H4-II-E cells using lipofectin [19]. The cultures were changed to serum-free media and treated with dexamethasone (1 µM), dibutyryl cyclic AMP (0.5 mM) or phorbol 12-myristate 13-acetate (PMA; 1 µM) for 24 h, after which the cells were lysed. Luciferase activity in the cell lysates was increased 20-fold by dexamethasone treatment, and ~3-fold by incubation with cyclic AMP or PMA [19]. These three agents previously have been reported to stimulate

231

IGFBP-1 transcription: dexamethasone in H4-II-E cells [16], cyclic AMP analogues in HepG2 human hepatocarcinoma cells [20], and phorbol esters in HEC-50 human endometrial carcinoma cells [21]. Coincubation with 1 μg/ml insulin abolished the stimulation of luciferase activity in transfected H4-II-E cells by all 3 agents [19]. These results indicate that the regulatory elements for insulin, as well as for dexamethasone, cyclic AMP and phorbol esters, are contained within this promoter fragment, and confirm the dominant negative regulation by insulin. Preliminary deletion mapping of the rat IGFBP-1 promoter suggests that a promoter fragment whose 5' end is at nt -327 is sufficient to confer responsiveness to all 4 agents. Responsiveness to cyclic AMP and phorbol esters is lost when the region from nt -327 to nt -235 is deleted. Most of the response to dexamethasone is lost when nt -235 to nt -135 are deleted, although the fragment whose 5' end is at nt -135 retains some dexamethasone responsiveness. Residual dexamethasone-stimulated promoter activity in the nt -135 construct still is inhibited by insulin [19], suggesting that the insulin response element is located within this proximal fragment. Finer deletion mapping within this region shows that responsiveness to dexamethasone is lost when nt -108 to -87 are mutated. Further work is required to define the relationship between the cis-elements and trans-acting factors involved in the regulation of IGFBP-1 transcription by dexamethasone and insulin.

Protein synthesis is required for rapid turnover of IGFBP-1 mRNA in H4-II-E rat hepatoma cells

Another approach to studying the pathways of transcriptional regulation has been through the use of protein synthesis inhibitors such as cycloheximide. Transcription of some genes involves labile, rapidly-turning over proteins, and is inhibited by cycloheximide [22]. Transcription of other genes, known as immediate early genes, involves activation of stable existing proteins rather than de novo protein synthesis, and hence is insensitive to cycloheximide [23]. We have examined the effects of the protein synthesis inhibitor cycloheximide on the regulation of IGFBP-1 gene expression by insulin [18].

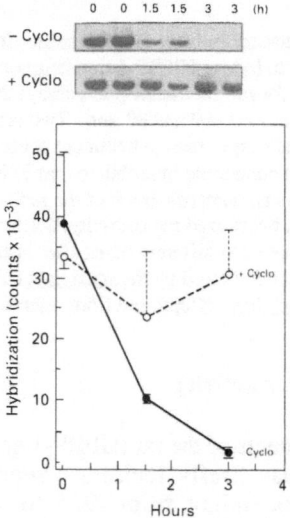

Figure 6. Cycloheximide abolishes the decrease in IGFBP-1 mRNA following insulin treatment of H4-II-E cells. H4-II-E cells were grown to confluence, then equilibrated with serum-free medium. Cycloheximide was added to half of the cultures to give a final concentration of 10.7 μM (3 μg/ml). After 1.5 h, insulin (1 μg/ml) was added to all of the cultures without media change, and the incubation continued. Total RNA was prepared from duplicate cultures 0, 1.5 and 3 h after insulin addition, and examined by Northern blot hybridization using a rat IGFBP-1 cDNA probe. An autoradiograph of the hybridization is shown in the top panel. The hybridized radioactivity was quantitated by ß-scanning; the Mean ± SD are plotted in the bottom panel at different times after insulin addition. - cycloheximide, solid circles; + cycloheximide, open circles. Reprinted with permission from reference 18.

First, we studied the effect of cycloheximide-treatment on the insulin-induced decrease in IGFBP-1 mRNA (Fig. 6). Some cultures were incubated with cycloheximide for 1.5 h, at concentrations that inhibited protein synthesis by 95%. Insulin was added to cycloheximide-treated and untreated cultures for an additional 1.5 or 3 h. In cultures that had not been treated with cycloheximide, IGFBP-1 mRNA decreased rapidly after insulin treatment, reaching undetectable levels by 3 h. These results are similar to those presented in Fig. 4. In cultures that had been incubated with cycloheximide, however, IGFBP-1 mRNA did not decrease following insulin addition.

Nuclear run-on transcription experiments were performed to determine whether cycloheximide abolished the insulin-induced decrease in IGFBP-1 mRNA by blocking the inhibition of IGFBP-1 transcription caused by insulin (Fig. 7). Cultures were incubated overnight with dexamethasone to induce IGFBP-1 transcription. Then, half of the cultures were preincubated with cycloheximide for 1.5 h, after which insulin was added to some cycloheximide-treated and untreated cultures for another 1.5 h. In the absence of cycloheximide, insulin inhibited IGFBP-1 transcription by ~90%, similar to the results presented in Fig. 5. However, in cycloheximide-treated cells, IGFBP-1 transcription was decreased ~64% by treatment with cycloheximide alone, and a further 90% by the addition of insulin. Both effects of cycloheximide, partial inhibition of basal IGFBP-1 transcription without abolishing the inhibition of transcription by insulin, would be expected to decrease IGFBP-1 mRNA, and so could not account for the higher levels of IGFBP-1 mRNA seen in cells treated with cycloheximide and insulin.

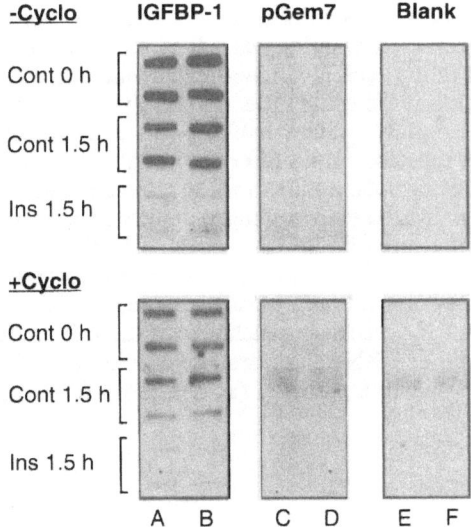

Figure 7. Insulin inhibits IGFBP-1 transcription in cycloheximide-treated H4-II-E cells. Confluent cultures equilibrated in serum-free medium were incubated overnight with 1 µM dexamethasone. Cycloheximide was then added to one-half of the cultures for 1.5 h, after which insulin (1 µg/ml) was added to some of the cycloheximide-treated and untreated cultures. Nuclei were isolated 0 and 1.5 h after insulin addition, and elongated radiolabeled RNA transcripts prepared from duplicate preparations of nuclei for each time and experimental condition. Transcripts from a single preparation of nuclei were hybridized to duplicate nitrocellulose filters containing immobilized IGFBP-1 cDNA (A, B), pGem7 plasmid DNA (C, D), or blank filters containing no DNA (E, F), and are shown as horizontal rows. Each hybridization received 2.1×10^6 cpm of radioactive transcripts. An autoradiograph of the hybridization is shown. [Modified from reference 18.]

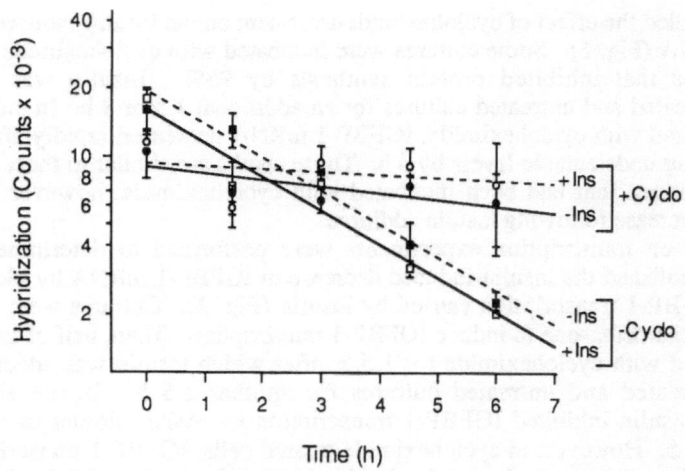

Figure 8. Cycloheximide stabilizes IGFBP-1 mRNA in actinomycin-D-treated H4-II-E cells. Confluent H4-II-E cells in serum-free medium were incubated with dexamethasone overnight. One-half of the cultures were treated with cycloheximide for 1.5 h, after which insulin was added to half of the cycloheximide-treated and untreated cultures (time zero). Actinomycin-D (5 µg/ml) was added at the same time as insulin, and total RNA prepared at the indicated times thereafter. After Northern blot analysis, the IGFBP-1 hybridization signal was quantitated by ß-scanning, corrected for background radioactivity, and plotted at different times after actinomycin-D addition. + cycloheximide, circles; − cycloheximide, squares; + insulin, open symbols, dashed lines; −insulin, solid symbols and solid lines. [Reprinted with permission from reference 18.]

This suggested an alternative explanation of these results, namely, that the predominant effect of cycloheximide was to stabilize IGFBP-1 mRNA. To demonstrate this directly, IGFBP-1 mRNA turnover was measured in cells treated with actinomycin-D to block the initiation of new transcripts (Fig. 8). Cultures were treated overnight with dexamethasone, after which cylohexamide was added to half of the cultures. At time zero, insulin was added to half of the cycloheximide-treated and untreated cultures; actinomycin-D was added to all cultures at the same time. In the absence of cycloheximide, IGFBP-1 mRNA levels decreased rapidly, with a half-life of ~2 h, agreeing well with previous results [24]. As previously reported [25], insulin does not affect the turnover rate of IGFBP-1 mRNA. In the presence of cycloheximide, however, the turnover of IGFBP-1 mRNA was considerably slower, the half-life being prolonged approximately 10-fold. Insulin again did not significantly affect the rate of IGFBP-1 mRNA disappearance. We conclude that IGFBP-1 mRNA persists in insulin-treated cells in the presence of cycloheximide despite decreased transcription because the mRNA is stabilized. Inhibition of protein synthesis may stabilize IGFBP-1 mRNA because translation of the mRNA is inhibited (as reported for histone [26] and ß-tubulin [27] mRNAs), or because of inhibition of the synthesis of a labile protein in the mRNA degradative pathway (as described for human transferrin receptor mRNA [28]).

CONCLUSIONS

IGFBP-1 inhibits or potentiates the actions of IGF-I and IGF-II in vitro under different conditions [29]. The biological activity of IGFBP-1 is determined by many factors. In addition to the abundance of the protein, these include transcapillary permeability [30], post-translational modifications such as phosphorylation [31], proteolysis [32], and association with the extracellular matrix or cell surface [29]. We have focussed on the regulation of IGFBP-1 gene expression since abundance of the mRNA correlates with the abundance of the protein. IGFBP-1 mRNA abundance, in turn, reflects the regulation of gene transcription.

We have shown that IGFBP-1 gene transcription is regulated in vivo (in diabetic rat liver) and in vitro (in a rat hepatoma cell line). Insulin rapidly decreases IGFBP-1 transcription in both systems. Combined with the rapid turnover of IGFBP-1 mRNA, the transcriptional regulation of IGFBP-1 by insulin is sufficient to account for the rapid decrease in IGFBP-1 mRNA abundance. Transcriptional regulation also may contribute to the rapid regulation of IGFBP-1 in human plasma. Ongoing studies of the IGFBP-1 promoter using the rat hepatoma cell model system should help us understand in precise detail the molecular basis of this transcriptional regulation.

REFERENCES

1. M. M. Rechler, Insulin-like growth factor binding proteins, *Vitamins & Hormones* 47:1 (1993).

2. A. M. Cotterill, C. T. Cowell, R. C. Baxter, et al, Regulation of the growth hormone-independent growth factor-binding protein in children, *J. Clin. Endocrinol. Metab.* 67:882 (1988).

3. W. H. Busby, D. K. Snyder, and D. R. Clemmons, Radioimmunoassay of a 26,000-dalton plasma insulin-like growth factor-binding protein: control by nutritional variables, *J. Clin. Endocrinol. Metab.* 67:1225 (1988).

4. K. Brismar, M. Gutniak, G. Povoa, et al, Insulin regulates the 35 kDa IGF binding protein in patients with diabetes mellitus, *J. Endocrinol. Invest.* 11:599 (1988).

5. A. M. Suikkari, V. A. Koivisto, E. M. Rutanen, et al, Insulin regulates the serum levels of low molecular weight insulin-like growth factor-binding protein, *J. Clin. Endocrinol. Metab.* 66:266 (1988).

6. S. I. Yeoh, and R. C. Baxter, Metabolic regulation of the growth hormone independent insulin-like growth factor binding protein in human plasma, *Acta Endocinol. (Copenhagen).* 119:463 (1988).

7. G. T. Ooi, C. C. Orlowski, A. L. Brown, et al, Different tissue distribution and hormonal regulation of mRNAs encoding rat insulin-like growth factor binding proteins rIGFBP-1 and rIGFBP-2, *Mol. Endocrinol.* 4:321 (1990).

8. G. T. Ooi, L. Y. -H. Tseng, M. Q. Tran, et al, Insulin rapidly decreases IGFBP-1 gene transcription in streptozotocin-diabetic rats, *Mol. Endocrinol.* 6:2219 (1992).

9. G. T. Ooi, L. Y. -H. Tseng, and M. M. Rechler, Post-transcriptional regulation of insulin-like growth factor binding protein-2 in diabetic rat liver, *Biochem. Biophys. Res. Commun.* 189:1031 (1992).

10. G. T. Ooi, L. Y. -H. Tseng, and M. M. Rechler, Transcriptional regulation of the rat IGFBP-1 and IGFBP-2 genes, *Growth Regulation* 3:12 (1993).

11. C. Seneviratne, L. Jiangming, and L. J. Murphy, Transcriptional regulation of rat insulin-like growth factor-binding protein-1 expression by growth hormone, *Mol. Endocrinol.* 4:1199 (1990).

12. D. P. Bichell, K. Kikuchi, and P. Rotwein, Growth hormone rapidly activates insulin-like growth factor I gene transcription *in vivo, Mol. Endocrinol.* 6:1899 (1992).

13. J. B. Yoon, H. C. Towle, and S. Seelig, Growth hormone induces two mRNA species of the serine protease inhibitor gene family in rat liver, *J. Biol. Chem.* 262:4284 (1987).

14. J. B. Yoon, S. A. Berry, S. Seelig, et al, An inducible nuclear factor binds to a growth hormone-regulated gene, *J. Biol. Chem.* 265:19947 (1990).

15. Y. W-H. Yang, A. L. Brown, C. C. Orlowski, et al, Identification of rat cell lines that preferentially express insulin-like growth factor binding proteins rIGFBP-1, 2, or 3, *Mol. Endocrinol.* 4:29 (1990).

16. C. C. Orlowski, G. T. Ooi, D. R. Brown, et al, Insulin rapidly inhibits insulin-like growth factor binding protein-1 gene expression in H4-II-E rat hepatoma cells, *Mol. Endocrinol.* 5:1180 (1991).

17. D. T. Chu, C. M. Davis, N. B. Chrapkiewicz, et al, Reciprocal regulation of gene transcription by insulin. Inhibition of the phosphoenolpyruvate carboxykinase gene and stimulation of gene 33 in a single cell type, *J. Biol. Chem.* 263:13007 (1988).

18. G. T. Ooi, D. R. Brown, D. Suh, et al, Cycloheximide stabilized insulin-like growth factor binding protein-1 mRNA and inhibits IGFBP-1 transcription in H4-II-E rat hepatoma cells, *J. Biol. Chem.* (1993). In press.

19. D. Suh and G. T. Ooi, Inhibition of IGFBP-1 gene expression by insulin and stimulation by dexamethasone, cyclic AMP, and phorbol esters are mediated by different cis-acting elements in the rat IGFBP-1 promoter, *Prog. of the 75th Ann. Mtg. of the Endoc. Soc.* (1993). Abstract

20. D. R. Powell, A. Suwanichkul, Cubbage,M.L., et al, Insulin inhibits transcription of the human gene for insulin-like growth factor-binding protein-1, *J. Biol. Chem.* 266:18868 (1991).

21. Y. Gong, G. Ballejo, B. Alkhalaf, et al, Phorbol esters differentially regulate the expression of insulin-like growth factor-binding proteins in endometrial carcinoma cells, *Endocrinology* 131:2747 (1992).

22. K. R. Yamamoto and B. M. Alberts, Steroid receptors: elements for modulation of eukaryotic transcription, *Annu. Rev. Biochem.* 45:721 (1976).

23. H. R. Herschman, Primary response genes induced by growth factors and tumor promoters, *Ann. Rev. Biochem.* 60:281 (1991).

24. C. C. Orlowski, G. T. Ooi, and M. M. Rechler, Dexamethasone stimulates transcription of the insulin-like growth factor binding protein-1 (IGFBP-1) gene in H4-II-E rat hepatoma cells, *Mol. Endocrinol.* 4:1592 (1990).

25. T. G. Unterman, D. T. Oehler, L. J. Murphy, et al, Multihormonal regulation of insulin-like growth factor binding protein-1 in rat H4IIE hepatoma cells: the dominant role of insulin , *Endocrinology* 128:2693 (1991).

26. R. A. Graves, N. B. Pandey, N. Chodchoy, et al, Translation is required for regulation of histone mRNA degradation, *Cell* 48:615 (1987).

27. D. A. Gay, S. S. Sisodia, and D. W. Cleveland, Autoregulatory control of b-tubulin mRNA stability is linked to translation elongation, *Proc. Natl. Acad. Sci. USA* 86:5763 (1989).

28. D. M. Koeller, J. A. Horowitz, J. L. Casey, et al, Translation and the stability of mRNAs encoding the transferrin receptor and c-*fos*, *Proc. Natl. Acad. Sci. USA* 88:7778 (1991).

29. D. R. Clemmons, IGF Binding Proteins: Regulation of Cellular Actions, *Growth Regulation* 2:80 (1992)

30. R. S. Bar, D. R. Clemmons, M. Boes, et al, Transcapillary permeability and subendothelial distribution of endothelial and amniotic fluid insulin-like growth factor binding proteins in the rat heart, *Endocrinology* 127:1078 (1990).

31. J. I. Jones, A. J. D'Ercole, C. Camacho-Hubner, et al, Phosphorylation of insulin-like growth factor (IGF)-binding protein 1 in cell culture and in vivo: Effects on affinity for IGF-1, *Proc. Natl. Acad. Sci. USA* 88:7481 (1991).

32. P. G. Campbell, J. F. Novak, T. B. Yanosick, et al, Involvement of the plasmin system in dissociation of the insulin-like growth factor-binding protein complex, *Endocrinology* 130:1401 (1992).

IGF BINDING PROTEIN-3 AND THE ACID-LABILE SUBUNIT: FORMATION OF THE TERNARY COMPLEX *IN VITRO* AND *IN VIVO*

Robert C. Baxter

Department of Endocrinology
Royal Prince Alfred Hospital
Sydney, NSW 2050, Australia

INTRODUCTION

In the adult circulation, IGF-I and IGF-II are found predominantly in association with the growth hormone-dependent IGF binding protein, IGFBP-3. The occupied binding protein combines with a third protein, the acid-labile subunit (ALS or α-subunit), to form a ternary complex of ~140 kDa which acts as a stable reservoir of IGFs[1,2]. Recent studies *in vitro* and *in vivo* have shed new light on the structure and regulation of the components of the complex, and on the kinetics of ternary complex formation. This paper will discuss some of these new observations, which are relevant both to the regulation of endogenous IGF bioavailability and to the pharmacokinetics of administered IGFs.

COMPLEX FORMATION *in vitro*

Structural Features of the Acid-labile Subunit

The cloning of cDNAs for human and rat ALS has recently been reported[3,4]. The mature proteins have a predicted molecular mass of 63-64 kDa, similar to the value of 66 kDa estimated by SDS-PAGE of the purified human serum protein following enzymatic deglycosylation[5]. Six potential N-glycosylation sites are conserved between the human and rat proteins, and while it is not known if all sites are used, the active protein isolated from human serum appears on SDS-PAGE as a doublet of 84-86 kDa[5], indicative of at least two glyco-forms. IGFBP-3 in serum also contains about 15 kDa of carbohydrate[6]; thus up to 25% (i.e. 35 kDa) of the 140 kDa complex with IGFBP-3 consists of carbohydrate. Since there is ample evidence that the carbohydrate moiety of glycoproteins affects their secretion, biological activity and clearance[7], it can be assumed that the functions of the ternary

Current Directions in Insulin-Like Growth Factor Research,
Edited by D. LeRoith and M.K. Raizada, Plenum Press, New York, 1994

237

complexes will be dependent on their carbohydrate composition — however, little direct evidence is currently available to support this.

An unusual feature of ALS is its high leucine content — over 20% of total amino acids. Most of the leucine residues are found within repeating units of 24 amino acids, which comprise the central 80% of the protein[3,4]. Proteins containing similar leucine-rich repeats form a diverse group, among which the structure of the repeating sequence is highly conserved, although the length varies from 23-26 residues[3,8-10]. Consensus repeating sequences of three proteins in this group are shown in Fig. 1; ALS contains 18-20 such units. An asparagine residue is also highly conserved in the repeating units, but never (at least in the case of ALS) forms part of potential N-glycosylation sites. All members of the leucine-rich family share the property of forming protein-protein interactions, which may result in soluble protein complexes (as in the case of ALS), receptor-ligand complexes, or cell interactions.

```
ALS (human)          L-x-L-S-x-N-x-L-x-x-L-x-x-x-A-F-x-G-L-x-x-L-x-x
Chaoptin
  (drosophila)       L-x-L-S-x-N-x-α-x-x-L-x-x-   α-F-x-x-L-x-x-L-x-x
Adenylyl
cyclase (yeast)      L-x-L-x-x-N-x-α-x-x-α-x-x-x-α-   x-x-L-x-x-L-x-x
```

Figure 1. Consensus leucine-rich repeating units of human acid-labile subunit[3], *Drosophila* chaoptin (a photoreceptor-specific cell-adhesion molecule)[8], and *Saccharomyces* adenylyl cyclase[9]. Spaces are included to improve alignment. α = aliphatic residue (L, I, A, or V).

Structural Determinants of Ternary Complex Formation

Neither IGFBP-3 nor IGFs alone bind to ALS[1,5]. There may, however, be some interaction between IGF and ALS once the ternary complex has formed. This is suggested by two lines of evidence. First, there appears to be some cross-linking of IGF-I to ALS when reconstituted ternary complexes are treated with the bifunctional reagent disuccinimidyl suberate, suggesting that the IGF and ALS are within 10Å of each other in the complex[1]. Second, the affinity of ALS binding can be influenced by the nature of the IGF bound to IGFBP-3. For example, although IGF-II binds to IGFBP-3 with slightly higher affinity than IGF-I[12], the resulting IGF-II•IGFBP-3 complex has a lower affinity for ALS than IGF-I•IGFBP-3[5]. In a more striking example of the same phenomenon[11], IGF-I analogs lacking the C- and D-domains bind with increased potency to IGFBP-3 (Fig. 2a), but the resulting binary complexes bind ALS with much lower affinity than the normal IGF-I•IGFBP-3 complex (Fig. 2b). In contrast, IGF-I analogs mutated in the first 16 residues of the B-domain bind to IGFBP-3 with much lower affinity than normal IGF-I, and thus form decreased amounts of binary complexes; but when complexes do form, their affinity for ALS is normal[11]. Since the residues of IGF-I involved in the putative interaction with ALS include those thought to interact with the type I IGF receptor[13], it can be predicted that receptor binding of IGF in the ternary complex would be impaired.

238

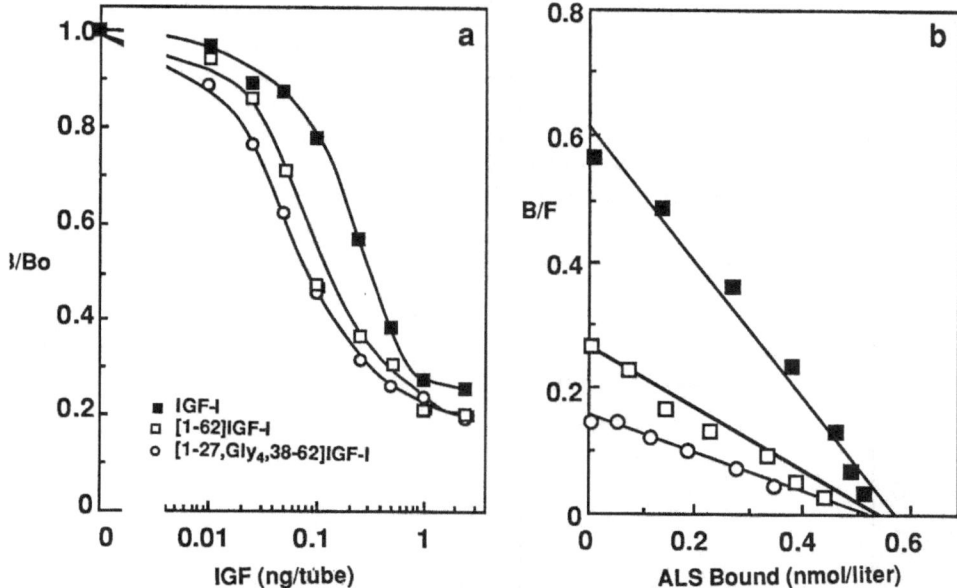

Figure 2. (a) Competition by IGF-I (■) and the analogs [1-62]IGF-I (D-domain truncation, □) and [1-27,Gly₄,38-62]IGF-I (D-domain truncation plus C-domain deletion, O), for the binding of iodinated IGF-I to human IGFBP-3 (0.5 ng/tube). (b) Scatchard plots of ALS binding to human IGFBP-3 (10 ng/tube) in the presence of IGF-I and analogs (10 ng/tube). Reproduced from data in Ref. 11, with permission.

An interesting alteration in IGF specificity is seen in IGFBP-3 from human pregnancy serum. During pregnancy, cation-dependent proteolytic activity appears in serum, rendering IGFBP-3 undetectable by ligand blotting[14]. In investigating this phenomenon, it was observed that pregnancy IGFBP-3 is unable to form a ternary complex with iodo-IGF-I, while complexing almost normally in the presence of native IGF-I[15]. Residues in the vicinity of Tyr[60] of IGF-I appear to be involved, since the analogue [Leu[60]]IGF-I, like iodo-IGF-I, is entirely inactive in ternary complex formation from pregnancy IGFBP-3, while complexing with greater than normal potency to normal IGFBP-3[16].

Immunoblotting after SDS-PAGE suggests that proteolysed IGFBP-3 is reduced in size to about 30 kDa[14]; however, the "truncation" only becomes apparent after SDS-PAGE, since FPLC of serum samples after transient acidification to destroy endogenous ALS reveals no difference in size — about 50 kDa — between IGFBP-3 from pregnancy and nonpregnancy samples[17]. In contrast, serum from patients with end-stage renal failure (ESRF) frequently contains immunoreactive IGFBP-3 that appears about 30 kDa even by gel chromatography[18]. Complex formation studies with this IGFBP-3 indicate a greatly reduced, though still measurable, ability to bind ALS. Fig. 3a compares immunoreactive IGFBP-3 profiles in sera from a normal adult and one with ESRF. In Fig. 3b, ALS binding to IGFBP-3 from the 140 kDa, 50 kDa and ~30 kDa peaks are compared. In the ESRF sample, IGFBP-3 in the 30 kDa and 50 kDa peaks shows considerably decreased complex formation activity, although the small amount of IGFBP-3 found at 140 kDa complexes normally. Interestingly, IGFBP-3 from the 50 kDa fraction of normal serum also shows less ALS binding activity than that from the 140 kDa peak (not shown), suggesting that even in normal serum, 50 kDa IGFBP-3 may not be in simple equilibrium with ternary complexed IGFBP-3, but may have impaired complex-forming activity. Whether this impairment results from proteolysis or some other post-translational modification is not known.

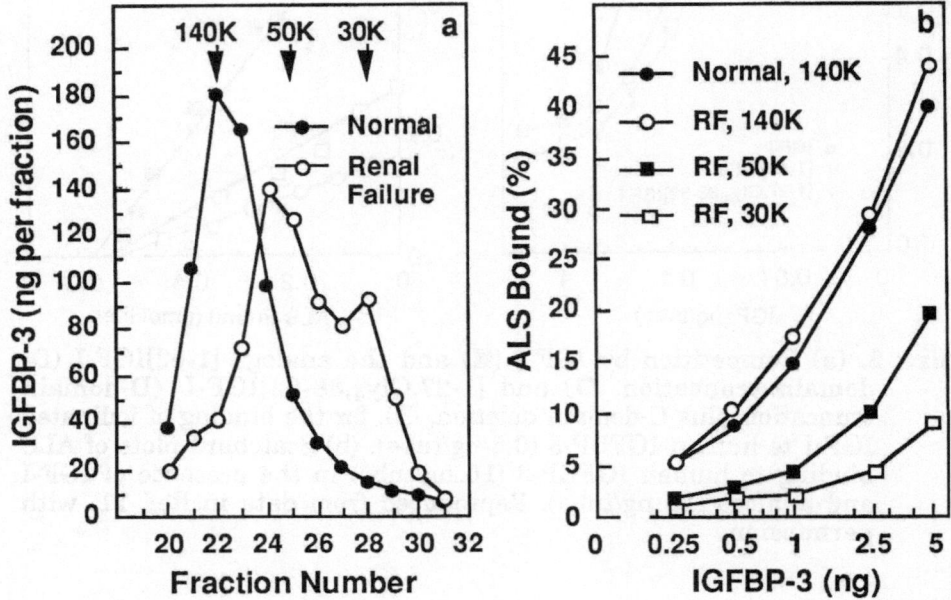

Figure 3. (a) Superose chromatography of serum from a healthy adult (●) and an adult with end-stage renal failure (O). Samples of 0.2 ml were fractionated at 1 ml/min, and IGFBP-3 measured by RIA in each 0.5 ml fraction. Arrows indicate approximate elution positions of proteins of 140, 50 and 30 kDa. (b) Binding of iodinated ALS to increasing concentrations of immunoreactive IGFBP-3 in 140, 50 and 30 kDa peak fractions from normal or renal failure (RF) serum, as indicated. Incubations also contained 5 ng IGF-I. Before assay, fractions were transiently acidified and neutralized to destroy endogenous ALS. Ternary complexes were immunoprecipitated with IGFBP-3 antiserum[11].

COMPLEX FORMATION *in vivo*

Studies with Exogenous IGFBP-3

Recent studies of hIGFBP-3 administration in the rat indicate that it forms the ternary complex extremely rapidly[19]. Two minutes after injecting a 100 μg bolus, the majority of hIGFBP-3, measured with a primate-specific RIA in serum fractionated by gel chromatography, is found in the 140 kDa peak, where much of it remains for hours. In contrast, IGFBP-3 detectable after 2 min in the 50 kDa peak has virtually disappeared by 15 min[19]. Since ALS is known to circulate in 2- to 3-fold excess of the other components of the complex[20],there should be sufficient available ALS to complex even the high serum concentrations of hIGFBP-3 achieved (5-10 mg/l). However, estimated "free" IGF levels do not exceed 10-20 µg/l, raising the question of the origin of the IGFs which complex with the administered hIGFBP-3.

Co-injection of IGF-I with hIGFBP-3 increases the percentage found in the ternary complex, and significantly prolongs retention of hIGFBP-3 in the circulation[19]. Conversely, hIGFBP-3 administration to streptozotocin-diabetic rats, in which hepatic IGF-I production and serum IGF-I levels are markedly reduced[21], results in less complexing of hIGFBP-3[19]. Thus IGF-I appears limiting for complex formation in this model. It is clear that "steady-state" free IGF in the circulation is insufficient to complex the hIGFBP-3, since little hIGFBP-3 shifts to the 140 kDa form when incubated with rat serum *in vitro*. One potential source of IGFs for complexing with exogenous hIGFBP-3 is the endogenous ternary complex, from which IGFs might dissociate. However, measurement of the distribution of rIGFBP-3 in these experiments, using a rat-specific RIA, indicates that little exchange of IGFs between endogenous and exogenous IGFBP-3 occurs, since the majority of rIGFBP-3 remains in the 140 kDa form (Lewitt et al., manuscript submitted).

It thus appears that there is a large pool of available IGFs, not measurable as "free" peptide in steady-state experiments, which fluxes through the circulation and can complex with IGFBP-3 as it enters the circulation. Some or all of this IGF pool might be bound to IGFBPs with lower affinity than IGFBP-3, while uncomplexed IGFs (which have a circulating half-life of only a few minutes) might also contribute. If this interpretation is correct, the mitogenic and metabolic potential of "endocrine" IGFs may be much greater than previously recognized.

Studies with Exogenous IGF-I

In several rat models of IGF-deficiency — hypophysectomy, diabetes, and malnourishment — where IGFBP-3-deficiency is also observed, administration of IGF-I significantly restores IGFBP-3 levels, but GH administration is required for the IGFBP-3 to re-appear in the ternary complex[22,23]. Taken together with observations that IGF-I increases the appearance of IGFBP-3 in cell culture medium, either by new synthesis or by release from the cell surface or matrix[24,25], these studies suggest that IGF-I is the primary regulator of IGFBP-3, whereas GH regulates ALS.

Studies of IGF-I administration to humans, however, lead to a different conclusion. In healthy subjects treated with IGF-I by daily sc injection for 7 days, a significant decline in steady-state IGFBP-3 levels, of approximately 20%, has been observed[26]. Concomitantly, a fall in steady-state ALS of almost 30% was seen. A possible explanation of this decline in the ternary complex is the suppression of pituitary GH secretion by IGF-I; however this would imply that GH, rather than IGF-I, is the primary regulator of IGFBP-3 as well as ALS.

The concept of regulation of both IGFBP-3 and ALS by GH is supported by a recent study of Kupfer et al.[27], in which calorically restricted adults were treated with IGF-I alone or a combination of IGF-I and GH. As seen in the study of healthy subjects, IGF-I alone, administered daily for 5 days, caused significant decreases in both IGFBP-3 and ALS. When IGF-I was co-administered with GH, however, a significant increase in both IGFBP-3 and ALS levels was seen, concomitantly with a marked improvement in nitrogen retention[27]. Thus, in contrast to the rat studies, it appears that in humans, IGF-I treatment alone is insufficient to increase IGFBP-3 levels, and in fact leads to a fall. Whether the contrasting results are due to a species difference, or to methodological differences between the human and rat studies, remains an important question.

CONCLUSION

Considerable advances have been made in recent years in understanding the way in which IGFBP-3 forms a stable complex with IGFs in the circulation. Although many tissues produce IGFs which appear important in mediating paracrine or autocrine activities, the original concept of IGFs as endocrine mediators of GH action, and perhaps of other forms of metabolic regulation, is increasingly supported by recent studies. The regulation of ternary complex formation is essentially the regulation of IGF delivery to the tissues, and thus is of key importance in understanding endocrine IGF actions.

Although initial studies suggest widespread tissue expression of ALS mRNA[4], little is yet known of the major locations of ALS synthesis, and the factors that can influence it at the cellular level. Current investigations of these topics will provide new information about the role and regulation of ALS, and may yet point to unrecognized activities of this protein in modulating IGF and IGFBP-3 activities at the cellular level, in addition to its established role in the circulation.

ACKNOWLEDGEMENTS

These studies were supported by the National Health and Medical Research Council, Australia. IGF-I analogs were generously provided by Dr. M.A. Cascieri of Merck, Sharp and Dohme.

REFERENCES

1. R.C. Baxter and J.L. Martin, Structure of the M_r 140,000 growth hormone-dependent insulin-like growth factor binding protein complex: Determination by reconstitution and affinity-labeling. *Proc. Natl. Acad. Sci. USA* 86:6898 (1989).

2. J.L. Martin and R.C. Baxter, Insulin-like growth factor binding protein-3: Biochemistry and physiology, *Growth Regulation* 2:88 (1992).

3. S.R. Leong, R.C. Baxter, T. Camerato, J. Dai and W.I. Wood, Structure and functional expression of the acid-labile subunit of the insulin-like growth factor-binding complex. *Mol. Endocrinol.* 6:870 (1992).

4. J. Dai and R.C. Baxter, Molecular cloning of the acid-labile subunit of the rat insulin-like growth factor binding protein complex. *Biochem. Biophys. Res. Commun.* 188:304 (1992).

5. R.C. Baxter, J.L. Martin and V.A. Beniac, High molecular weight insulin-like growth factor binding protein complex. Purification and properties of the acid-labile subunit from human serum. *J. Biol. Chem.* 264:11843 (1989).

6. W.I. Wood, G. Cachianes, W.J. Henzel, G.A. Winslow, S.A. Spencer, R. Hellmiss, J.L. Martin and R.C. Baxter, Cloning and expression of the growth hormone-dependent insulin-like growth factor-binding protein. *Mol. Endocrinol.* 2:1176 (1988).

7. T.W. Rademacher, R.B. Parekh and R.A. Dwek, Glycobiology. *Ann. Rev. Biochem.* 57:785 (1988).

8. D.E. Krantz and S.L. Kipursky, *Drosophila* chaoptin, a member of the leucine-rich repeat family, is a photoreceptor cell-specific adhesion molecule. *EMBO J.* 9:1969 (1990).

9. J. Field, H.-P. Xu, T. Michaeli, R. Ballester, P. Sass, M. Wigler and J. Colicelli, Mutations of the adenylyl cyclase gene that block RAS function in *Saccaromyces cerevisiae. Science* 247:464 (1990).

10. F.S. Lee and B.L. Vallee, Modular mutagenesis of human placental ribonuclease inhibitor, a protein with leucine-rich repeats. *Proc. Natl. Acad. Sci. USA* 87:1879 (1990).

11. R.C. Baxter, M.L. Bayne and M.A. Cascieri, Structural determinants for binary and ternary complex formation between insulin-like growth factor (IGF) I and IGF binding protein-3. *J. Biol. Chem.* 267:60 (1992).

12. J.L. Martin and R.C. Baxter, Insulin-like growth factor-binding protein from human plasma: Purification and characterization. *J. Biol. Chem.* 261:8754 (1986).

13. M.L. Bayne, J. Applebaum, D. Underwood, G.G. Chicchi, B.G. Green, N.S. Hayes and M.A. Cascieri, The C region of human insulin-like growth factor (IGF) I is required for high affinity binding to the type I IGF receptor. *J. Biol. Chem.* 264:11004 (1989).

14. P. Hossenlopp, B. Segovia, C. Lassarre, M. Roghani, M. Bredon and M. Binoux, Evidence of enzymatic degradation of insulin-like growth factor binding proteins in the 150K complex during pregnancy. *J. Clin. Endocrinol. Metab.* 71:797 (1991).

15. A.-M. Suikkari and R.C. Baxter, Insulin-like growth factor (IGF) binding protein-3 in pregnancy serum binds native IGF-I but not iodo-IGF-I. *J. Clin. Endocrinol. Metab.* 73:1377 (1991).

16. A.-M. Suikkari and R.C. Baxter, Structural regions of IGF-I which determine differential binding to IGFBP-3 from nonpregnancy and pregnancy serum. Abstracts., 9th Int. Congr. Endocrinol., Nice, p. 348 (1992).

17. A.-M. Suikkari and R.C. Baxter, Insulin-like growth factor binding protein-3 is functionally normal in pregnancy serum. *J. Clin. Endocrinol. Metab.* 74:177 (1992).

18. W.F. Blum, M.B. Ranke, K. Kietzmann, B. Tönshoff and O. Mehls, Excess of IGF-binding proteins in chronic renal failure: Evidence for relative GH resistance and inhibition of somatomedin activity, *in*: "Insulin-like Growth Factor Binding Proteins", S.L.S. Drop and R.L. Hintz, eds., Excerpta Medica, Amsterdam, p. 93 (1989).

19. M.S. Lewitt, H. Saunders, A.J. Lennon, S.R. Holman and R.C. Baxter, Distribution and actions of human IGFBP-1 and IGFBP-3 in the rat. *Growth Regulation* 3:42 (1993).

20. R.C. Baxter, Circulating levels and molecular distribution of the acid-labile (α) subunit of the high molecular weight insulin-like growth factor-binding protein complex. *J. Clin. Endocrinol. Metab.* 70:1347 (1990).

21. C.D. Scott and R.C. Baxter, Production of insulin-like growth factor I and its binding protein in rat hepatocytes cultured from diabetic and insulin-treated diabetic rats. *Endocrinology* 119:2346 (1986).

22. J. Zapf, C. Hauri, M. Waldvogel, E. Futo, H. Häsler, K. Binz, H.P. Guler, C. Schmid and E.R. Froesch, Recombinant human insulin-like growth factor I induces its own specific carrier protein in hypophysectomized and diabetic rats. *Proc. Natl. Acad. Sci. USA* 86:3813 (1989).

23. D.R. Clemmons, J.P. Thissen, M. Maes, J.M. Ketelslegers and L.E. Underwood, Insulin-like growth factor-I (IGF-I) infusion into hypophysectomized or protein-deprived rats induces specific IGF-binding proteins in serum. *Endocrinology* 125:2967 (1989).

24. L.K. Bale and C.A. Conover, Regulation of insulin-like growth factor binding protein-3 messenger ribonucleic acid expression by insulin-like growth factor I. *Endocrinology* 131:608 (1992).

25. J.L. Martin, M. Ballesteros and R.C. Baxter, Insulin-like growth factor-I (IGF-I) and transforming growth factor-β1 release IGF-binding protein-3 from human fibroblasts by different mechanisms. *Endocrinology* 131:1703 (1992).

26. R.C. Baxter, N. Hizuka, K. Takano, S.R. Holman and K. Asakawa, Responses of insulin-like growth factor binding protein-1 (IGFBP-1) and the IGFBP-3 complex to administration of insulin-like growth factor-I. *Acta Endocrinol.* 128:101 (1993).

27. S.R. Kupfer, L.E. Underwood, R.C. Baxter and D.R. Clemmons, Enhancement of the anabolic effects of growth hormone and insulin-like growth factor I by use of both agents simultaneously. *J. Clin. Invest.* 91:391 (1993).

ROLE OF POST TRANSLATIONAL MODIFICATIONS IN MODIFYING THE BIOLOGIC ACTIVITY OF INSULIN LIKE GROWTH FACTOR BINDING PROTEINS

David R. Clemmons
Division of Endocrinology
Department of Medicine
CB# 7170, MacNider
University of North Carolina
Chapel Hill, NC 27599-7170

INTRODUCTION

The six insulin-like growth factor binding proteins (IGFBP's) have been shown to be produced by a variety of cell types. When present in the local pericellular microenvironment these proteins have very high affinity for insulin like factors (IGF's) and can regulate the amount of each IGF that is available to bind to receptors. Estimates of the affinity constants for each protein show that they are at least equal to the Type 1 IGF receptor and in many cases significantly greater. Studies of the biologic actions of IGF-I and II in vitro in the presence of IGFBP's have shown that the BP's consistently inhibit the acute metabolic effects of IGF-I such as glucose transport and lipid synthesis. However, analysis of longer term effects such as stimulation of protein synthesis and DNA synthesis have shown that binding proteins can either enhance or inhibit these IGF-I actions. The major focus of research in our laboratory has been to determine if post-translational modifications of these binding proteins result in alterations in their ability to either inhibit or stimulate IGF-I actions. Studies in our lab have linked changes in several of these post-translational modifications to changes in the cellular responsiveness to IGF-I.

PROTEIN PHOSPHORYLATION

Insulin like growth factor binding protein-1 is phosphorylated on serine residues[1,2] Serine phosphorylation results in a substantial increase in the affinity of IGFBP-1, for and IGF-I and II (Figure 1). Specifically dephosphorylated IGFBP-1 obtained either by alkaline phosphatase treatment of IGFBP-1 or expression in E-Coli has an affinity of $1.0 \times 10^9 L/mg$ for IGF-I. Phosphorylated IGFBP-1 that has been purified from either Chinese Hamster Ovary (CHO) cells Hep G-2 (human hepatoma) cells has an affinity that is six-fold greater. The purified dephosphorylated protein has been shown to markedly potentiate the action of

Current Directions in Insulin-Like Growth Factor Research,
Edited by D. LeRoith and M.K. Raizada, Plenum Press, New York, 1994

245

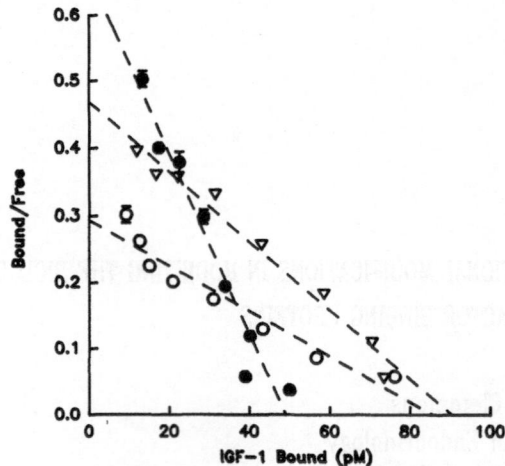

Figure 1. Scatchard plots showing the affinity of different phosphorylated forms of IGFBP-1. IIGFBP-1 purified from Hep G2 cell conditioned media or E. Coli supernatants were used as sources of phosphorylated (●—●) and dephosphorylated (▽-▽) IGFBP-1 respectively. Dephosphorylated IGFBP-1 (○-○) was also prepared by exposing the phosphorylated form to alkaline phosphatase then repurifying the protein. The data are plotted according to the method of scatchard. The affinity constant of each form of protein is shown.

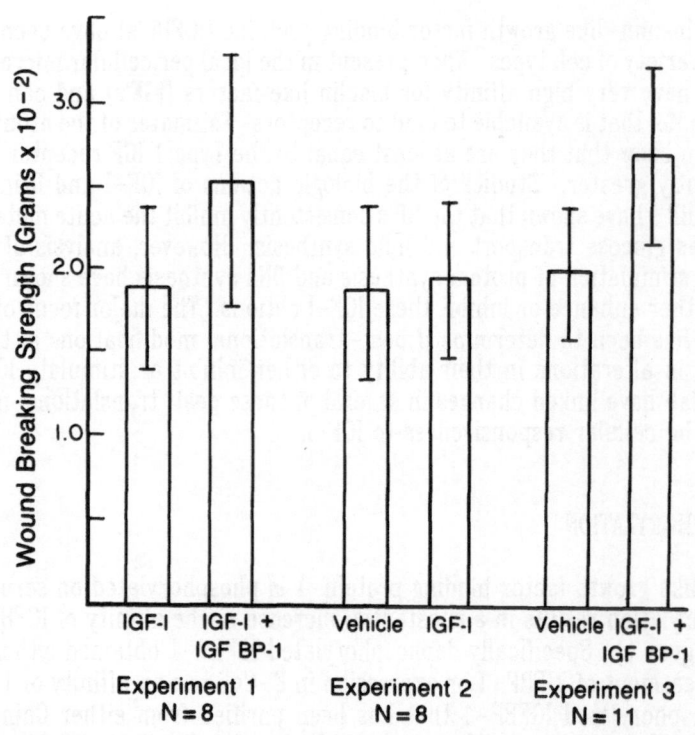

Figure 2. Potentiation of the wound healing response by IGFBP-1. Pure dephosphorylated IGFBP-1 was applied to linear incision wound in a methyl cellulose vehicle with or without IGF-I. Seven days later the wounds were excised and breaking strength measured. Each bar represents the mean ± one SD of 24-33 wounds.

IGF-1 in stimulating DNA synthesis in porcine aortic smooth muscle cells and in human fibroblasts. In contrast the purified phosphoprotein is consistently inhibitory and inhibits IGF-1 actions. Presumably this affinity shift allows more growth factor to equilibrate with cell surface receptors and thus changes the distribution of IGF-1 among the binding protein in the interstitial fluid and cell surface IGF-1 receptors.

In order to more accurately determine the role of phosphorylation in normal physiology we have recently extended these studies to an in vivo test model. In this model of wound healing six parallel linear incisions are made in the abdominal wall of rats[3]. Each pair of parallel incisions is made approximately 1-1/2 cm apart. Eight to 11 rats were used in each treatment group. A single application of growth factor was applied to the wounds using a methylcellulose vehicle containing either IGF-1 alone, IGFBP-1 alone or IGF-1 plus IGFBP-1. The parallel wound received the same vehicle with the paired treatment. Only two treatments (e.g. IGF-1 vs. IGF-1 plus IGFBP-1) could be compared in a single experiment. A minimum of 21 pairs of wounds were analyzed per experiment and in some experiments the wound number was as high as 33. After placement of the material it was allowed to remain in the wound bed for eight days at which time the entire wound was excised and breaking strength was assessed by tensiometry. Tensiometry measures the force in Newtons that is necessary to break the wound. This is a function of the wound collagen and glucosaminoglycan content and the degree of collagen crosslinking. Wound cellularity was also analyzed by histologic analysis. A further parameter of wound healing hydroxyproline accumulation which is an index of collagen synthesis and cross linking was also determined. Application of the combination of IGFBP-1 plus IGF-1 as compared to control wounds receiving vehicle alone showed that this combination caused a 38% increase in wound breaking strength (Figure 2). Likewise increased cellularity was noted in these wounds as compared to control wounds. When the combination was compared to IGF-1 alone an approximately 28% increase in wound breaking strength was noted. This was repeated in two separate experiments with two groups of animals. To further determine the significance of phosphorylation, dephosphorylated IGFBP-1 plus IGF-1 was compared to phosphorylated IGFBP-1 plus IGF-1. While dephosphorylated IGFBP-1 caused a 31% increase in breaking strength in the presence of IGF-1 the combination of phosphorylated IGFBP-1 and IGF-1 did not cause a significantly greater response than IGF-1 alone. Cross comparisons between groups of animals showed that the groups of animals were comparable as were the numbers and qualities of wounds. Therefore it appears that the combination of dephosphorylated IGFBP-1 plus IGF-1 is a potent potentiator of wound healing in rats. The enhanced response to phosphorylated IGFBP-1 on wound healing validates our previous in vitro observations in an in vivo animal model. Whether this change is due to a simulation of mitogenesis or stimulation of collagens synthesis or crosslinking remains to be ascertained.

Because of the importance of phosphorylation of IGFBP-1 in regulating IGF actions we attempted to determine its sites of phosphorylation. Radiolabelling of IGFBP-1 with purification and subsequent proteolytic cleavage and radiosequencing showed that serines 101, 119 and 169 were the sites of phosphorylation (Figure 3). Counting each of these peptides revealed that serine 101 accounted for approximately 60% of the radiolabelled protein whereas serine 169 was approximately 25% of the remainder was accounted for by serine 119[4]. Direct radiosequencing and with counting showed that serine 101 and 169 were definitely detectible as labelled residues. More importantly serines 98 and 96 which were also favorable phosphorylation sites were definitely not phosphorylated indicating that the pattern of phosphorylation is quite specific. In order to validate the importance of serine 101 in IGFBP-1 function site-directed mutagenesis was used to convert this serine to an alanine. Radiolabelling and immunoprecipitation studies showed that mutagenesis resulted in a 58% reduction in total phosphoprotein accumulation in the IGFBP-1 immunoprecipitable band indicating that mutagenesis had been successful and that our estimate determining by counting purified tryptic fragments was accurate. More importantly we estimated that the affinity constant of ala[101] IGFBP-1 for IGF-1 was reduced approximately 3.5 fold as compared to the wild type protein. However the affinity of ala[101] was still approximately two fold greater than dephosphorylated wild type protein indicating that phosphorylation at positions 119 and 169 contribute to the net affinity constant. Therefore it appears that phosphorylation of discrete serine residues in IGFBP-1 can markedly alter its affinity of IGFBP-1 for IGF-1 and presumably through this mechanism will alter its bioactivity. Biologic studies are underway at this time to determine if the mutants have enhanced bioactivity compared to the wild type protein.

ADHERENCE TO EXTRACELLULAR MATRIX

IGF synthesis and secretion by connective tissue cells has raised the possibility that locally synthesized IGF's may accumulate in connective tissues for use during remodeling on in response to injury. Recently we have shown that specific forms of IGFBP's are adherent to the extracellular matrix of connective tissue cells. Specifically when extracellular matrix is prepared from normal diploid skin fibroblasts it can be shown to bind IGFBP-3 and IGFBP-5[5]. Since fibroblasts synthesize much more IGFBP-3 than IGFBP-5, yet IGFBP-5 is the predominant form of IGFBP in ECM, it appears that it is preferentially localized in the ECM. Proof that the 30Kda IGFBP band detected in the ECM was IGFBP-5 was obtained with immunoblotting using specific antisera for this protein. Interestingly although fibroblast conditioned media contains predominately a 22KDa fragment of IGFBP-5 the ECM contains only the intact protein indicating that some substance in the ECM may be protecting the protein from proteolysis (Figure 4). The function of the matrix associated IGFBP-5 was examined in two ways. First the affinity of the ECM associated protein was estimated and was shown to be approximately twelve fold lower than IGFBP-5 in fibroblast conditioned medium (Figure 5). Therefore it appears that immobilization of IGFBP-5 in ECM results in a reduction in its affinity for both IGF-1 and II. This is similar to a finding reported previously from our laboratory that cell surface associated IGFBP-3 had a much reduced affinity as compared to IGFBP-3 in solution[6]. Localization of IGFBP-5 in the EMC provides not only a storage compartment for IGF's but also alters cellular responsiveness to this IGF-1. Preparation of fibroblast ECM and layering the IGFBP-5 onto the ECM followed by plating of a fibroblast cell line that secretes very little IGFBP-5 on top of this ECM, showed that ECM associated IGFBP-5 in the presence of IGF-1 was capable of potentiating the fibroblast growth in response to this growth factor. Specifically cell number increases were potentiated by approximately 55% (Figure 6). Analysis of ECM at the end of the experiment showed significant amounts of IGFBP-5 were retained in the matrix although the amount was much less than at the initiation of the experiment indicating that IGFBP-5 was diffusing off the matrix while the experiment was in progress. To prevent diffusion pure IGFBP-5 was embedded in a collagen synthetic matrix which was then allowed to dry on the tissue culture plates. Plating of cells on top of this matrix and analysis of cell growth showed that ECM associated IGFBP-5 was capable of potentiating the cell growth response to IGF-1 by approximately 40%. This ability of ECM associated IGFBP-5 to potentiate the cell growth response to IGF-1 may have important ramifications for understanding how IGF-1 stimulates cell growth. Since IGF-1 is an important stimulant of ECM synthesis and since IGFBP-5 is widely distributed throughout connective tissue cell ECM, the findings suggest that this could be an important means for sequestering large amounts of IGF's near sites of IGF action in mesenchymal tissues. This could also be an important mechanism in the repair response that follows tissue injury since it would allow for release of large amounts of IGF when needed.

248

₁ APWQCAPCSAEKLALCPPVSASCSEVTRSAGCGCCPM

₃₈ CALPLGAACGVATARCARGLSCRALPGEQQPLHALTR

₇₅ GQGACVQESDASAPHAAEAGSPESPES*TEITEEELL

₁₁₁ DNFHLMAPS*EEDHSILWDAISTYDGSK   75% ³²P

₁₃₆ ALHVTNIKKWKEPCRIELYRVVESLAK

₁₆₅ AQETS*GEEISK 25% ³²P

₁₇₆ FYLPNCNKNGFYHSRQCETSMDGEAGLCWCVYPWNG

₂₁₂ KRIPGSPEIRGDPNCQIYFNVQN

Figure 3. Sites of phosphorylation in IGFBP-1. ^{32}P labelled IGFBP-1 was prepared and purified to homogeneity. The pure protein was cleaved with trypsin and the purified peptide fragments sequenced. A phosphoserine at position 169 was identified. A large peptide containing several potential phosphorylation sites was further digested with V-8 protease. The purified peptides were sequenced and serines at positions 101 and 119 were shown to be labelled.

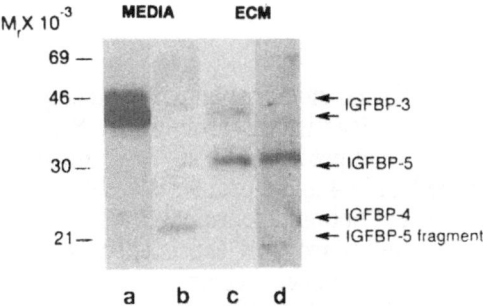

Figure 4. Forms of IGFBP's that are secreted by human fibroblasts and are present in ECM. Conditioned media and fibroblast ECM were prepared using confluent monolayers. Both ligand (lanes a and c) and immunoblots (lanes b and d) are shown. The 30 KDa band that is present in ECM is identified as IGFBP-5. A corresponding intact IGFBP-5 band is not present in conditioned media and only a 22 KDa fragment can be detected.

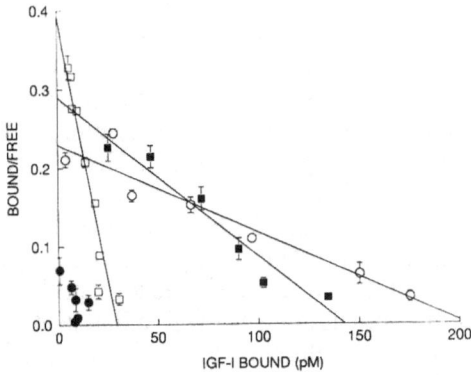

Figure 5. Scatchard plot showing the affinity of IGFBP-5 in media and ECM. IGFBP-5 was purified and a solution binding assay (□ - □) used to estimate its affinity for IGF-I. It was also layered onto ECM (○ - ○) and its binding affinity determined.

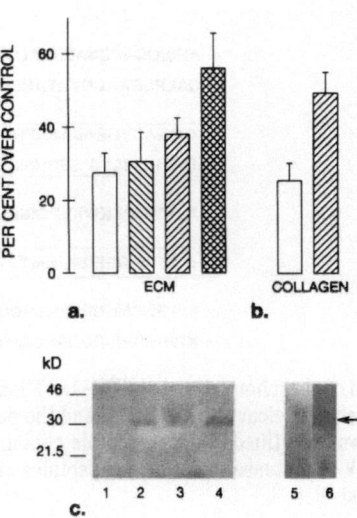

Figure 6. Potentiation of the fibroblast growth response to IGF-I by ECM associated IGFBP-5. ECM was prepared from confluent fibroblast monolayers then exposed to pure IGFBP-5. Fresh human fibroblasts were plated onto the ECM and cell number determined after 48 hours. The control cultures were exposed to IGF-I alone and had no IGFBP-5 added to their ECM. Additional cultures had IGFBP-5 added to media, to the ECM or to media and ECM. Additional cultures were plated on collagen (Type I) with or without IGFBP-5.

PROTEOLYSIS OF IGFBP'S

An additional post translational modification of IGF binding proteins that appears to be important in modulating their bioactivity is proteolysis. Specific proteases have been identified for IGFBP-2, 3, 4 and 5 and proteases for IGFBP-1 and 6 have not been excluded. Our laboratory has studied the IGFBP-4 and 5 proteases[7,8]. These are serine proteases that are calcium dependent but they are not metalloproteases since they cannot be activated by zinc. Their protease inhibitor profiles are most consistent with the serine proteases since they are easily inhibited by aprotinin and benzamidine. The biologic relevance of proteolysis to IGF-I action has been shown for both IGFBP-4 and 5. Addition of intact IGFBP-4 to fibroblasts that do not release IGFBP-4 protease results in a dose dependent inhibition of IGF-I stimulated DNA synthesis (Figure 7). Specifically concentrations between 50 and 250 ng/ml of IGFBP-4 completely eliminate the IGF-I stimulated response. In contrast the addition of pure IGFBP-4 to confluent smooth muscle cell cultures resulted in no inhibition of IGF-I stimulated DNA synthesis. Collection of conditioned media after 24 hours shows that a 24 to 48 hour incubation results in almost total cleavage of IGFBP-4 to 14 and 16 KDa non-IGF binding fragments by the smooth muscle cell cultures whereas the fibroblasts showed no degradation (Figure 8). These fragments are clearly non-NGF binding and even ligand blotting using radiolabelled IGF-II and long exposure times failed to reveal significant binding of radiolabelled IGF. However, we have not conducted solution binding assays with purified fragments to prove that they have absolutely no binding affinity. It is apparent that cleavage of intact IGFBP-4 results in at least a marked reduced of affinity of this protein for its ligand. Thus it appears that proteolytic cleavage of IGFBP-4 results in complete inactivation its ability to inhibit IGF-I action.

In contrast proteolytic cleavage of IGFBP-5 has been associated with a different set of responses. Specifically Andress, et al. reported that proteolytic cleavage of IGFBP-5 to a 22 KDa fragment occurred in osteoblast conditioned media[9]. Purification of this 22 KDA fragment to homogeneity resulted in a preparation of IGFBP-5 that modestly enhanced IGF-I activity when added to human osteoblast cultures, whereas intact IGFBP-5 had no effect. This suggests that proteolysized IGFBP-5 is an important potentiator of IGF action and that proteolytic cleavage of IGFBP-5 may have very different affects on IGF action as compared to proteolytic cleavage of IGFBP-4. Furthermore this finding suggests that if proteolysis of

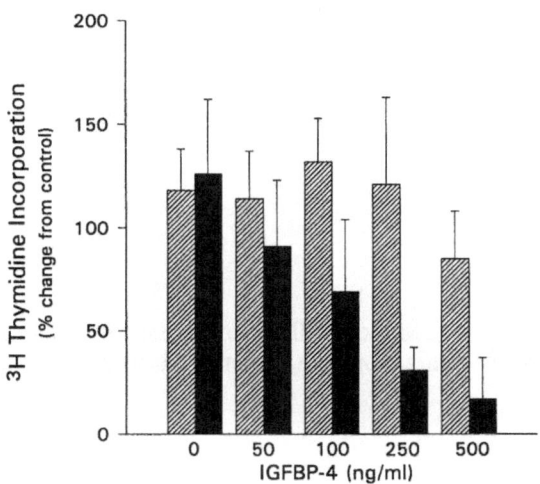

Figure 7. DNA synthesis response of cultured smooth muscle cells and human fibroblasts to IGF–I and IGFBP–4. Quiescent smooth muscle cells (hatched bars) or human fibroblasts (solid bars) were exposed to IGF–I and increasing concentrations of IGFBP–4. The DNA synthesis response was measured after 36 hours.

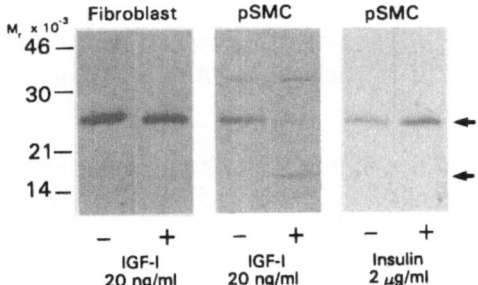

Figure 8. Proteolysis of IGFBP–4 by smooth muscle cells. The cell culture supernatants from the experiment shown in Figure 7 were immunoblotted using anti IGFBP–4 antiserum. In the presence of IGF–I the smooth muscle cell cultures degrade IGFBP–4 into a 16 KDa non–IGF binding fragment whereas the fibroblast cultures have minimal proteolytic activity.

an IGF binding protein causes a reduction in affinity but the fragment still retains some binding activity, this may result in potentiation of IGF action whereas if proteolysis occurs it results in complete loss and capacity to bind the growth factor this may result in reversal the IGFBP's inhibitory effects.

In summary, IGF binding proteins can act to modulate significantly cellular responses to the IGF's. Reduction in the affinity of IGF binding proteins for the IGF's will result in marked changes in IGF action. Future studies using site directed mutagenesis should be very useful in determining the regions of IGFBP's that account for both matrix binding, cell association and proteolysis. Identification of the proteases and the factors that control their activity opens up another level of control for understanding the role of factors that modulate IGF action.

ACKNOWLEDGEMENTS

The author wishes to thank Ms. Leigh Elliott for her help in preparing the manuscript. This work was supported by grants from the National Institutes of Health AG-02331 and HL26309.

REFERENCES

1. Jones JI, D'Ercole AJ, Camacho-Hubner C, Clemmons DR. Phosphorylation of insulin-like growth factor binding protein in cell culture and in vivo: effects on affinity for IGF-I. Proc Natl Acad Sci USA. 88:7481-7485 (1991).

2. Frost JP, Tseng LT. Insulin-like growth factor binding protein-1 is phosphorylated by cultured human endometrial cells and multiple protein kinases in vitro. J Biol Chem. 266:18082-18088 (1991).

3. Jyung J, Mustoe T, Busby WH, Clemmons DR. Increased wound breaking strength induced by insulin-like growth factor-1 in combination with IGF binding protein-1. Surg. In press (1993).

4. Jones JI, Busby WH, Wright G, Smith CE, Kimack NM, Clemmons DR. Identification of the sites of phosphorylation in insulin like growth factor binding protein-1: Regulation of its affinity by phosphorylation of serine 101. J Biol Chem. 268:1125-1131 (1993).

5. Jones JI, Busby WH, Gockerman A, Camacho-Hubner C, Clemmons DR. Extracellular matrix contains insulin-like growth factor for IGF binding protein 5: potentiation of the effects of IGF-I. J. Cell Biol. 121:679-687 (1993).

6. McCusker RH, Camacho-Hubner C, Bayne ML, Cascieri MA, Clemmons DR. Insulin-like growth factor (IGF) binding to human fibroblast and glioblastoma cells: The modulating effect of cell released IGF binding proteins (IGFBPs). J Cell Physiol. 144:244-253 (1990).

7. Cohick WS, Gockerman A, Clemmons DR. Vascular smooth muscle cells synthesize two forms of insulin like growth factor binding proteins (IGFBP) which are regulated differently by the IGF's. J. Cell Physiol; in press (1993).

8. Camacho-Hubner C , Busby WH , McCusker RH , Wright G , Clemmons DR . Identification of the forms of -insulin-like growth factor binding proteins produced by human fibroblasts and the mechanisms that regulate their secretion. J Biol Chem. 267:11949-11956 (1992).

9. Andress DL, Birnbaum RS. A novel human insulin-like growth factor binding protein secreted by osteoblast cells. Biochem Biophys Res Commun. 176:213-218 (1991).

CELLULAR ACTIONS OF INSULIN-LIKE GROWTH FACTOR BINDING PROTEIN-3

Cheryl A. Conover, Jay T. Clarkson, Susan K. Durham, and Laurie K. Bale

Endocrine Research Unit
Mayo Clinic and Mayo Foundation
5-164 West Joseph
Rochester, MN 55905

INTRODUCTION

Insulin-like growth factor I (IGF-I) plays a major role in regulating growth of cells in vivo and in vitro (1). IGF-I initiates metabolic and mitogenic processes in a wide variety of cell types by binding to specific type I IGF receptors in the plasma membrane (2). In addition, recent evidence suggests that association of IGF-I with distinct binding proteins present in extracellular fluids plays a pivotal role in determining IGF-I availability and bioactivity (3,4).

IGF binding protein-3 (IGFBP-3) is one of six IGFBPs identified to date (5,6). Although structurally related, these IGFBPs differ in molecular size, biochemical characteristics, IGF preference, tissue-specificity, hormonal regulation, and, presumably, physiological function. IGFBP-3 is the major circulating form in adults. As part of a large 150-kD ternary complex with an 85-kD acid-labile subunit and IGF peptide, plasma IGFBP-3 prolongs the half-life of IGFs in the circulation, serves as a reservoir for IGFs, and restricts extravascular IGF transit (3,7,8). In addition, IGFBP-3 mRNA is expressed in most adult tissues and a variety of cells in culture synthesize and secrete IGFBP-3 (9-16). Thus, IGFBP-3 present in the cellular microenvironment has the potential to directly alter local cell response to IGF-I.

We have developed and characterized two model systems for studying the regulation and biological actions of IGFBP-3 at the cellular level: human fibroblasts and bovine fibroblasts. The salient features of these cell models are summarized in Table 1. The human fibroblast monolayer

Table 1. Fibroblast model systems for studying IGF-I and IGFBP-3 cellular physiology.

	Human	Bovine
Type I IGF receptors	++	+++
IGFBP-3 secretion	constitutive	inducible
Membrane-associated IGFBP-3	+++	0
IGF peptide secretion	+	0

system has been utilized extensively in our laboratory as a model for studying various aspects of IGF physiology. Receptor binding for IGF-I, IGF-II, and insulin, and peptide effects on IGFBP expression and cell metabolism and proliferation have been compared and detailed (11,17-22). In particular, human fibroblasts secrete copious IGFBP-3, and have been useful for studying the cellular effects of endogenous IGFBP-3. Bovine fibroblasts are a complementary model system, with distinct advantages for investigating the effects of exogenous IGFBP-3 on IGF cellular action. Bovine fibroblasts in monolayer culture have abundant type I IGF receptors, which account for 80-90% of cell surface [^{125}I]IGF-I binding in the basal state (23,24). In comparison, human fibroblasts have high-affinity cell-associated IGFBP-3 which can represent greater than 50% of the surface binding capacity. In accord with the large number of type I IGF receptors, bovine fibroblasts are exquisitely sensitive to the metabolic and mitogenic effects of low nanomolar concentrations of IGF-I (23-25). Unlike human fibroblasts, these cells do not secrete IGF peptides or IGFBP-3 under basal conditions; however, IGFBP-3 expression is inducible with IGF-I or insulin (13). Thus, with cultured bovine fibroblasts, IGFBP-3 action can be evaluated in a physiologically relevant system with minimal interference by endogenous IGFBP-3.

Inhibitory and Potentiating Effects of IGFBP-3

In our initial studies (23), we used cultured bovine fibroblasts and IGFBP-3 purified from bovine serum to determine the effect of IGFBP-3 on IGF-I binding and cellular action (Fig. 1). [^{3}H]Aminoisobutyric acid (AIB) uptake was chosen to monitor cell response because there is little IGF binding activity and no IGFBP-3 detectable during this 6-h bioassay. The experimental design was patterned after the IGFBP-3 coincubation/preincubation scheme described by DeMellow and Baxter (26). Coincubation of bovine fibroblasts with IGF-I and increasing concentrations of IGFBP-3 produced a dose-dependent inhibition of IGF-I-stimulated [^{3}H]AIB uptake. This inhibition paralleled the ability of IGFBP-3 to prevent IGF-I cell surface binding. In contrast, preincubation of bovine fibroblasts with IGFBP-3, followed by extensive washing, resulted in a dose-dependent enhancement of subsequent IGF-I-stimulated [^{3}H]AIB uptake; a 20-86% increase was seen after a 24h preexposure to IGFBP-3 and a 2- to 6-fold potentiation was seen after a 72h preincubation. The potentiating effect of IGFBP-3 correlated with increased [^{125}I]IGF-I binding to cultured bovine fibroblasts. Affinity cross-linking experiments indicated that the increase in IGF-I binding was due to the appearance of cell-associated IGFBP-3. These data suggested that soluble IGFBP-3 inhibits IGF-I action by sequestering and preventing IGF-I receptor binding whereas cell-associated IGFBP-3 somehow enhances the growth-promoting effects of IGF-I.

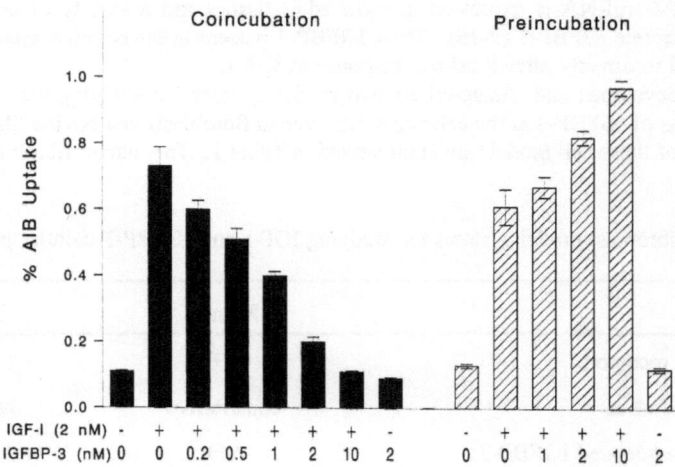

Figure 1. Effect of IGFBP-3 on IGF-I stimulation of AIB uptake in bovine fibroblasts. IGF-I (2 nM) stimulation of [^{3}H]AIB uptake was measured in the presence of IGFBP-3 (Coincubation) or following a 24h preincubation period with IGFBP-3. Adapted from reference 23.

Physiological Implications of Soluble IGFBP-3 as an Inhibitor

Many peptide hormones, including IGF-I, insulin, and growth hormone (GH) are able to induce a decrease in the number of their cell-surface receptors as a result of receptor binding and internalization (down-regulation). Insulin and GH are secreted in a pulsatile fashion in vivo and generally circulate at low levels. Total circulating IGF-I concentrations, on the other hand, are high and relatively constant. The 150-kD IGFBP-3 complex in plasma acts to restrict IGF-I access to target tissues, thereby preventing extensive receptor down-regulation. We propose a similar role for locally secreted IGFBP-3.

Preincubation of bovine fibroblasts with IGF-I results in a striking dose-dependent decrease in maximum specific [^{125}I]IGF-I binding; IGF-I at 1 nM significantly decreases binding, and a half-maximal effect is seen with approximately 4 nM IGF-I (25). Affinity cross-linking experiments indicated a specific decrease in the number of type I IGF receptors on bovine fibroblasts after treatment with IGF-I. In the experiments depicted in Figure 2, we preincubated bovine fibroblasts

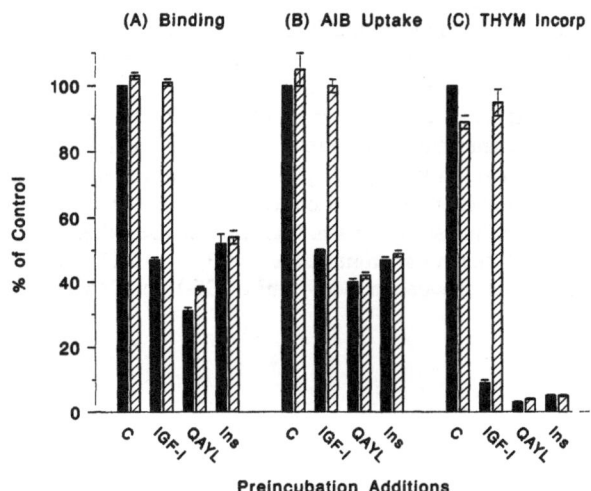

Figure 2. Effect of IGFBP-3 on ligand-induced receptor down-regulation and cell desensitization. Bovine fibroblasts were incubated for 24h at 37°C in serum-free medium (SFM) with no peptide [control C] or containing IGF-I (4 nM), [QAYL]IGF-I (QAYL; 4 nm), or insulin (Ins; 100 nM) in the absence (solid bars) or presence (hatched bars) of 6 nM IGFBP-3. Cells were washed to remove reversibly bound peptides, and [^{125}I]IGF-I specific binding (A), IGF-I-stimulated [^{3}H]AIB uptake (B), and IGF-I-stimulated [^{3}H]thymidine incorporation (C) were measured. Results are the mean ± SEM of three determinations, expressed as a percentage of the control value (no additions during the serum-free preincubation period). From reference 25, with permission.

with IGF-I (4 nM), [QAYL]IGF-I (4 nM), or insulin (100 nM) in the absence or presence of IGFBP-3 (6 nM) for 24h. [QAYL]IGF-I is an IGF-I analog with normal type I IGF receptor binding and activation, but with 600-fold reduced affinity for IGFBP-3 (27). Similarly, insulin at high concentrations can cross-react with the type I IGF receptor in these cells, but does not bind to IGFBPs (2-4,23). After the preincubation period, cells were washed to remove reversibly bound peptides, and [^{125}I]IGF-I binding and IGF-I stimulation of [^{3}H]AIB uptake and [^{3}H]thymidine incorporation were measured. Preincubation with IGF-I resulted in a 50% decrease in IGF-I-stimulated [^{3}H]AIB uptake commensurate with a 53% decrease in IGF-I cell binding; IGF-I-stimulated [^{3}H]thymidine incorporation was inhibited 90% under these conditions. Preincubation with [QAYL]IGF-I or insulin had effects comparable to IGF-I, decreasing both [^{125}I]IGF-I binding and action 50-95%. The addition of purified bovine IGFBP-3 with IGF-I during the preincubation period blocked the decrease in binding and prevented the cells from becoming desensitized to IGF-I stimulation. IGFBP-3 did not prevent the [QAYL]IGF-I- or insulin-induced decreases in IGF-I binding and cell responsiveness. IGFBP-3 alone had no effect on IGF-I binding and action in these cells. These data indicated that IGFBP-3 can prevent IGF-I-induced receptor down-regulation, a process that renders cells refractory to further stimulation by IGF-I.

Cultured bovine fibroblasts seem to be particularly susceptible to IGF-I-induced receptor down-regulation and cellular desensitization, possibly because these cells are not normally shielded by IGFBP-3 (13). Cells that secrete IGFBP-3 constitutively and/or are able to respond quickly to IGF-I with increased IGFBP-3 might maintain growth-promoting processes better during excursions in local IGF-I levels. We explored this idea using cultured human fibroblasts which express IGFBP-3 under basal conditions and increase soluble IGFBP-3 as a direct response to IGF-I (21). Normal human fibroblasts were incubated for 24h with IGF-I (which binds to and increases soluble IGFBP-3) or [QAYL]IGF-I (which does not). Cells were washed to remove reversibly bound ligand, and [^{125}I-QAYL]IGF-I binding performed. The use of this radiolabeled IGF permits determination of type I receptor binding on cells independent of the presence of cell-associated IGFBPs (28). As shown in Figure 3A, a 24-h preincubation with [QAYL]IGF-I caused a dose-dependent decrease in [^{125}I-QAYL]IGF-I binding to human fibroblasts. With 1 nM [QAYL]IGF-I, receptor binding was 55-60% of maximum. In comparison, preincubation with 1 nM IGF-I did not affect [^{125}I-QAYL]IGF-I binding to normal human fibroblasts. Changes in receptor availability were reflected in functional responsiveness to IGF-I (Figure 3B). Preincubation with [QAYL]IGF-I for 24h caused a dose-dependent decrease in IGF-I-stimulated [^{3}H]thymidine incorporation in human fibroblasts; after exposure to 1 nM [QAYL]IGF-I, cell responsiveness was only 35% of control. This relatively resistant state was also seen after a 48-h preincubation period and persisted for more than 24h after removal of the [QAYL]IGF-I. In contrast, human fibroblasts were fully responsive to IGF-I stimulation after preincubation with equivalent concentrations of IGF-I. IGF-I-stimulated [^{3}H]AIB uptake in human fibroblasts was also impaired after preincubation with 1 nM [QAYL]IGF-I (62 ± 5% of maximum stimulation, n = 3, P < 0.05), whereas cells maintained responsiveness to IGF-I after incubation with 1 nM IGF-I for 24h (91 ± 5% of maximum stimulation, P = NS).

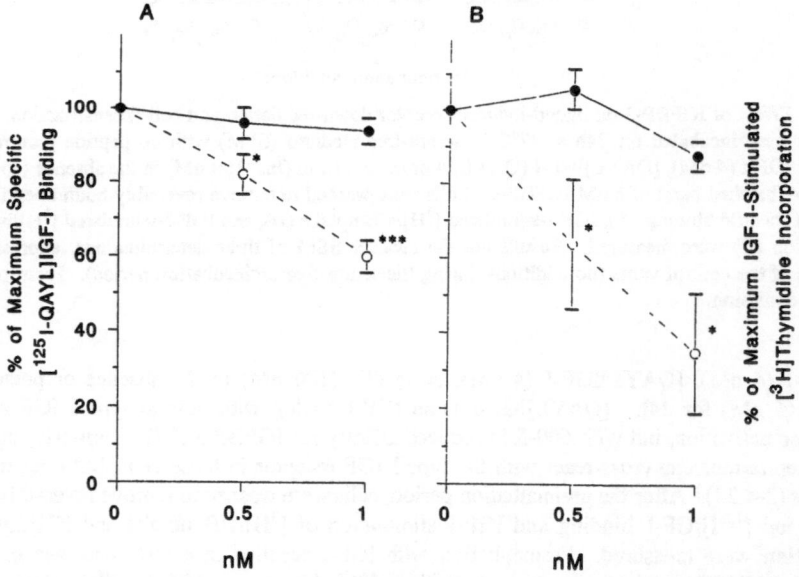

Figure 3. Human fibroblast receptor binding and cell responsiveness to IGF-I stimulation: preincubation with IGF-I and [QAYL]IGF-I. Human fibroblasts were preincubated for 24h in SFM with IGF-I (•) or [QAYL]IGF-I (o) at the indicated concentrations. Cultures were washed to remove reversibly bound peptide and then (A) [^{125}I-QAYL]IGF-I binding or (B) IGF-I stimulated [^{3}H]thymidine incorporation was measured. Results are mean ± SE of 3 experiments expressed as % of maximum (no additions during preincubations = 100%). Adapted from reference 21.

We propose that IGFBP-3 in its free soluble form functions in a buffering capacity to limit IGF-I and target cell interaction, thereby modulating the biological response to changes in local IGF-I levels. Under normal circumstances, IGFBP-3 secreted into the microenvironment could play a key role in maintaining IGF-I receptor availability and ensuring continued cell response to an essential growth factor. From this perspective, an acute inhibitory effect of IGFBP-3 could be regarded as growth-promoting. Relative overexpression or oversecretion of IGFBP-3 may result in it being a negative regulatory factor for cell growth, however.

Glycosylated and Nonglycosylated IGFBP-3

IGFBP-3 is unique among the IGFBPs in its extensive glycosylation in the native state. Sequence analysis of IGFBP-3 cDNA predicts a core protein size of ~29-kD (9,29,30). However, IGFBP-3 from mammalian sources migrates on SDS-PAGE as two or more glycoprotein forms with mol wt of 37- to 48-kD (3,4,7,9-16). To determine the functional significance of carbohydrate moieties on IGFBP-3, we examined the effects of nonglycosylated *Escherichia coli*-derived recombinant human IGFBP-3 (hIGFBP-3$^{E.\ coli}$) and glycosylated Chinese hamster ovary cell-derived hIGFBP-3 (hIGFBP-3^{CHO}) on IGF-I action in cultured bovine fibroblasts (24,30,31). Both IGFBP-3 preparations bound IGF-I with high affinity and were 5- to 10-fold more potent than unlabeled IGF-I in inhibiting [^{125}I]IGF-I binding to bovine fibroblasts. These results suggested that soluble hIGFBP-3, in glycosylated and nonglycosylated forms, would compete effectively with membrane receptor for IGF-I peptide. Indeed, addition of hIGFBP-3$^{E.\ coli}$ or hIGFBP-3^{CHO} produced a dose-dependent inhibition of IGF-I- stimulated [^{3}H]AIB uptake in bovine fibroblasts, whereas the two hIGFBP-3 preparations had no effect on basal or insulin-stimulated [^{3}H]AIB uptake under these conditions (Fig. 4).

In contrast to the inhibitory effect of IGFBP-3 when coincubated with IGF-I, preincubation of bovine fibroblasts for 72h with hIGFBP-3$^{E.\ coli}$ or hIGFBP-3^{CHO} potentiated subsequent IGF-I-stimulated [^{3}H]AIB uptake. As shown in Figure 5, preexposure of bovine fibroblasts to hIGFBP-3$^{E.\ coli}$ resulted in a dose-dependent increase in IGF-I-stimulated [^{3}H]AIB uptake. A maximal potentiating effect (2.5-fold) was obtained with 50 nM hIGFBP-3$^{E.\ coli}$. This was equivalent to the

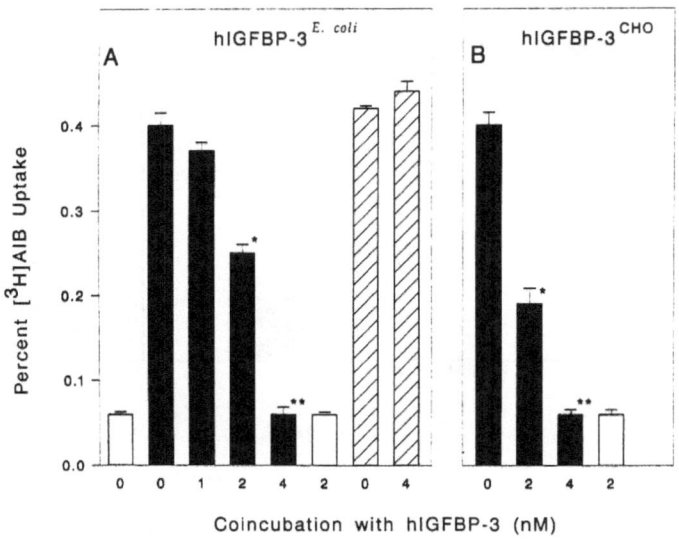

Figure 4. IGF-I-stimulated [^{3}H]AIB uptake in bovine fibroblasts: effect of coincubation with hIGFBP-3. [^{3}H]AIB uptake was measured after stimulation with 2 nM IGF-I (solid bars), 100 nM insulin (hatched bars), or serum-free medium alone (open bars) with or without coincidental addition of (A) hIGFBP-3$^{E.\ coli}$ or (B) hIGFBP-3^{CHO} at the indicated concentrations. From reference 24, with permission.

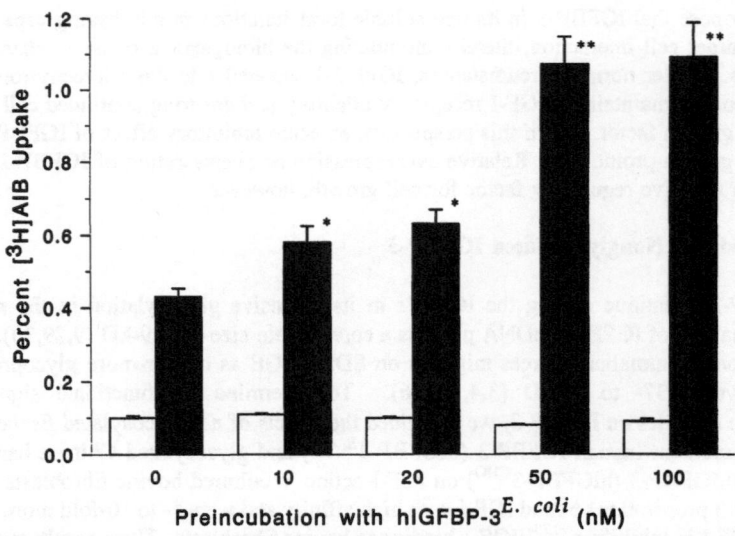

Figure 5. IGF-I-stimulated [³H]AIB uptake in bovine fibroblasts: dose-response effect of preincubation with hIGFBP-3$^{E.\ coli}$. Bovine fibroblasts were preincubated with or without the indicated concentrations of hIGFBP-3$^{E.\ coli}$ for 72h. Cells were washed and stimulation of [³H]AIB uptake by 2 nM IGF-I (solid bars) or serum-free medium alone (open bars) was measured. From reference 24, with permission.

2.2- and 2.5-fold stimulatory effect of bovine serum-purified IGFBP-3 and hIGFBP-3^{CHO} under the same experimental conditions. Even at the highest concentration tested, preincubation with hIGFBP-3$^{E.\ coli}$ had no effect on basal [³H]AIB uptake in these cells. In the experiment shown in Figure 6, we measured [¹²⁵I]IGF-I binding and IGF-I-stimulated [³H]AIB uptake following preincubation with 50 nM hIGFBP-3$^{E.\ coli}$ for 1, 24, and 72h. A 1h preincubation with

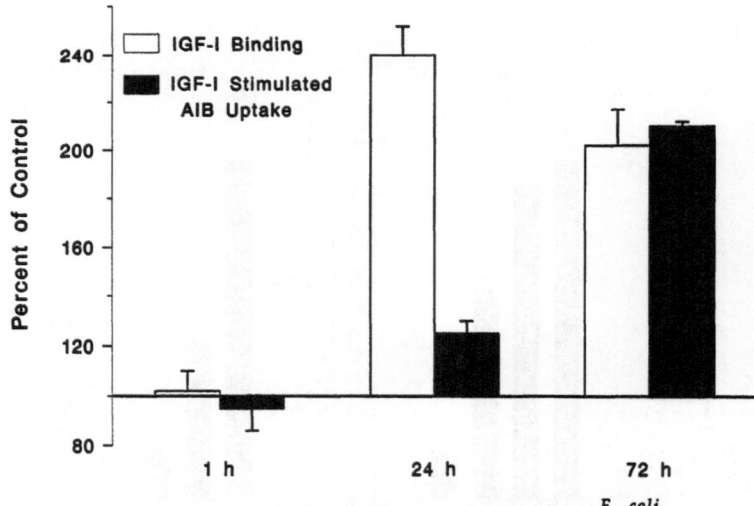

Figure 6. IGF-I binding and stimulated [³H]AIB uptake in bovine fibroblasts: time-dependent effect of preincubation with hIGFBP-3. Bovine fibroblasts were preincubated with or without 50 nM hIGFBP-3$^{E.\ coli}$ for 1, 24, or 72h. Cells were washed and [¹²⁵I]IGF-I binding (open bars) and IGF-I-stimulated [³H]AIB uptake (solid bars) were determined. Results are means ± SE of three separate experiments, expressed as percent of control values for each time period (i.e., IGF-I binding and stimulated [³H]AIB uptake without hIGFBP-3 preincubation. From reference 24, with permission.

hIGFBP-3$^{E. coli}$ had no significant effect on IGF-I binding to bovine fibroblasts or cell response to IGF-I stimulation. When cells were preincubated with hIGFBP-3$^{E. coli}$ for 24h, [^{125}I]IGF-I binding to bovine fibroblasts increased 2.4-fold, whereas responsiveness to IGF-I was increased only 25%. IGF-I cell binding was still 2-fold increased after a 72h preincubation with corresponding enhancement of IGF-I-stimulated [^{3}H]AIB uptake. The increase in [^{125}I]IGF-I binding to bovine fibroblast monolayers appeared due to association of hIGFBP-3$^{E. coli}$ with the cell surface rather than to an increase in type I IGF receptor number or affinity. Affinity cross-linking experiments indicated exogenous IGFBP-3 adhered to the fibroblast surface and exhibited time-dependent processing to lower molecular weight forms that retained the ability to bind radiolabeled IGF-I (Fig. 7). Intense binding of [^{125}I]IGF-I to cell-associated 29-kD hIGFBP-3$^{E. coli}$ seen after 24h of incubation was reduced approximately 70% after 72h, concomitant with the appearance of smaller bands indicating hIGFBP-3$^{E. coli}$ forms of 12- to 27-kD. These data demonstrate that glycosylation is not obligatory for biologically functional IGFBP-3. Furthermore, they indicate that enhancement of IGF-I action correlates with the apparent processing of surface-bound IGFBP-3.

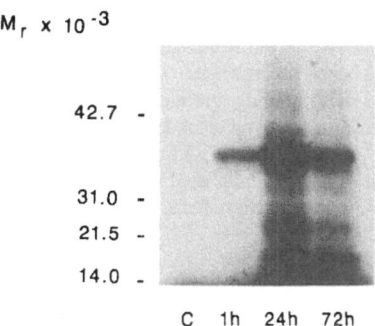

M$_r$ x 10^{-3}

42.7 -

31.0 -
21.5 -

14.0 -

C   1h   24h   72h

Figure 7. Autoradiogram of [^{125}I]IGF-I cross-linked to monolayer cultures of bovine fibroblasts: effect of preincubation with hIGFBP-3. Bovine fibroblasts were either untreated (C, Control) or preincubated with 50 nM hIGFBP-3$^{E. coli}$ for 1, 24, or 72h before [^{125}I]IGF-I cell binding and cross-linking in the presence of 10 μM insulin. Solubilized complexes (reducing conditions) were separated by 5-15% gradient SDS-PAGE. Adapted from reference 24.

Underlying Mechanisms of IGFBP-3 Potentiation

We evaluated the functional significance of this processing of cell-bound IGFBP-3 by comparing the relative affinities of soluble IGFBP-3, 72h cell-associated IGFBP-3, and the type I IGF receptor for IGF-I peptide. As shown in Figure 8, IGFBP-3 in solution had a higher affinity than the type I IGF receptor for IGF-I. However, there was a marked reduction in affinity for ligand following IGFBP-3 cell-association and processing (i.e., after a 72-h preincubation) such that the apparent affinity was less than that of the receptor.

Although the biological impact of such a change in affinity is unknown, high affinity IGFBP-3 could capture IGF-I peptide and, anchored in the microenvironment, gradually release IGF-I in the vicinity of its receptor through a progressive reduction in affinity. In this way, cell-associated IGFBP-3 could act not only to concentrate IGF-I locally, but also to deliver it in a controlled manner, such that ligand-induced receptor down-regulation and attendant cell desensitization would be prevented (25).

Regulated IGFBP-3 delivery of IGF peptide to receptor may be an important mechanism underlying IGFBP-3's potentiating effect. However, if direct interaction of IGF-I with cell-bound IGFBP-3 is a prerequisite for enhanced cell responsiveness, then the stimulatory effects of IGF-like peptides that bind the type I IGF receptor but do not bind IGFBP-3 should not be amplified -- peptides such as [QAYL]IGF-I and insulin (27). Nonetheless, we found significant enhancement

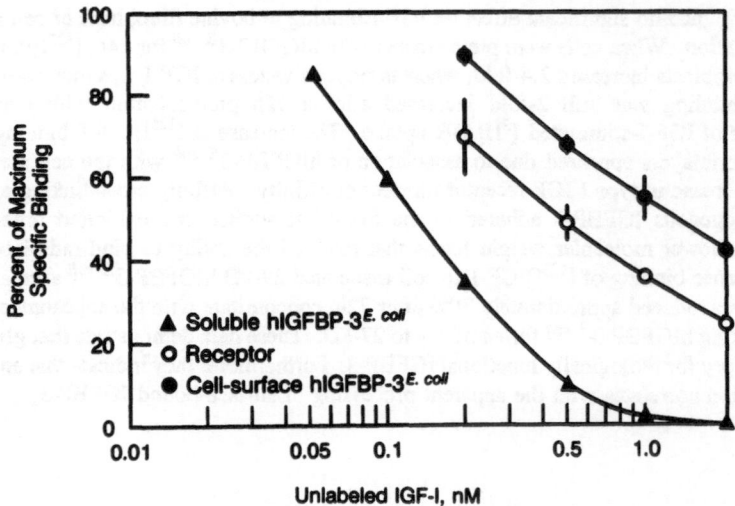

Figure 8. IGF-I binding to soluble IGFBP-3, cell surface-associated IGFBP-3, and type I IGF receptors on bovine fibroblasts. Bovine fibroblasts were preincubated with 50 nM hIGFBP-3[E. coli] for 72h. For measurement of cell surface-associated IGF binding proteins (•), [^{125}I]IGF-I monolayer binding was performed in the presence of 10 μM insulin. For receptor binding (○), [QAYL]IGF-I was used as the radioligand (28). Soluble hIGFBP-3[E. coli] binding affinity (▲) was assessed by [^{125}I]IGF binding in a solution assay. In each experiment, various concentrations of unlabeled IGF-I were added to compete for radioligand binding. Results are means ± SE of triplicate determinations, expressed as percent of maximum specific binding. From reference 24, with permission.

of [QAYL]IGF-I- and insulin-stimulated [^{3}H]AIB uptake after a 72h preincubation with IGFBP-3. In experiments summarized in Figure 9, 2 nM IGF-I, 2 nM [QAYL]IGF-I, and 100 nM insulin had equivalent bioactivity under control conditions, stimulating [^{3}H]AIB uptake approximately 6-fold. A 72-h preincubation with IGFBP-3 alone had no effect on [^{3}H]AIB uptake, but significantly increased the biological effectiveness of IGF-I, [QAYL]IGF-I, and insulin (2.1-, 1.7-, and 1.6-fold, respectively).

On the other hand, we saw little or no stimulation of [^{3}H]AIB uptake under basal conditions or following IGFBP-3 incubation when using 2 nM concentrations of IGF-II, [Ser[24]]IGF-I, and [1-27,Gly$_4$,38-70]IGF-I, IGFs that have 10- to 30-fold reduced affinity for the type I IGF receptor but normal IGFBP-3 binding (32,33) (Fig. 10). At 10 nM, these IGFs activate type I IGF receptor signalling and significant potentiation by IGFBP-3 was observed. IGF-I and [QAYL]IGF-I stimulated [^{3}H]AIB uptake at 2 and 10 nM, and at both peptide concentrations bioactivity was enhanced by pretreatment of cells with IGFBP-3. In other experiments, preincubation with IGFBP-3 had no effect on bovine fibroblast responsiveness to growth factors structurally unrelated to IGF-I: epidermal growth factor, basic fibroblast growth factor or platelet-derived growth factor.

These experiments suggest that there is a modification of type I IGF receptor mediated-signal transduction resulting from prolonged incubation of bovine fibroblasts with IGFBP-3. The exact mechanism by which cell-associated IGFBP-3 alters type I IGF reactivity is unclear. However, employing [QAYL]IGF-I as the radioligand to measure specific receptor binding or by monitoring insulin-inhibitable binding, we could show no significant increase in type I IGF receptor number or affinity as a result of IGFBP-3 cell treatment.

Taken together, the data suggest that IGFBP-3 potentiation of IGF-I action in bovine fibroblasts may involve changes in IGFBP-3 form and function and in type I IGF receptor responsiveness. Cell-associated IGFBP-3 may provide a mechanism for optimal presentation of IGF-I to its receptor as well as a means to heighten receptor reactivity to IGF-I and related peptides. Cell types that secrete and bind IGFBP-3 might respond more effectively to IGFs, providing a regional growth advantage.

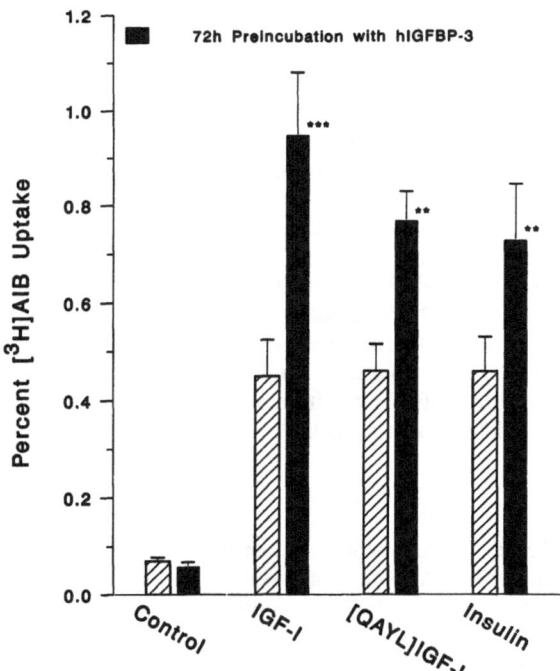

Figure 9. Stimulation of [³H]AIB uptake in bovine fibroblasts by IGF-I, [QAYL]IGF-I, and insulin: preincubation with IGFBP-3. Bovine fibroblasts were preincubated without (hatched bars) or with (solid bars) 50 nM hIGFBP-3[E. coli] for 72h. Cultures were washed and stimulation of [³H]AIB uptake by medium alone (Control), 2 nM IGF-I, 2 nM [QAYL]IGF-I, or 100 nM insulin was measured. Results are the mean ± SE of five experiments. From reference 31, with permission.

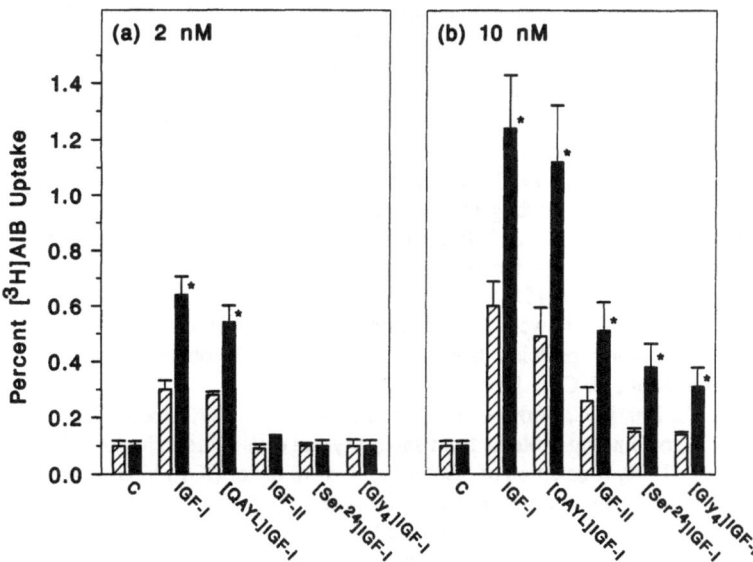

Figure 10. Stimulation of [³H]AIB uptake in bovine fibroblasts by IGFs and IGF-I analogs: preincubation with IGFBP-3. Bovine fibroblasts were preincubated without (hatched bars) or with (solid bars) 50 nM hIGFBP-3[E. coli] for 72h. Cells were washed and [³H]AIB uptake stimulated by 2 or 10 nM IGF was measured. [Gly₄]IGF-I, [1-27,Gly₄,38-70] IGF-I. From reference 31, with permission.

CONCLUSION

These observations are in keeping with an overall concept of IGFBP-3 acting to assist IGF-I in promoting growth

1) The 150-kD plasma IGFBP-3 complex appears to function as a large IGF reservoir, ensuring a steady and ready supply of an essential growth factor.

2) Intact, soluble IGFBP-3 may serve as a buffer to maintain type I IGF receptor availability and allow continued cell responsiveness to IGF-I stimulation.

3) Cell-associated IGFBP-3 may provide a mechanism for enhanced type I IGF receptor reaction to IGF-I and related compounds.

At the cellular level, the biological effect IGFBP-3 will depend upon the balance between the inhibitory effect of soluble IGFBP-3 and the potentiating effects of cell-associated IGFBP-3. Of course, the total picture of cell response to IGF-I includes receptor concentration, local IGF concentration and other IGFBPs. Further studies are necessary to understand fully the structure/function relationship of IGFBP-3 and its interactive role with other IGFBPs in modulating IGF-I receptor binding and intracellular signalling.

REFERENCES

1. Froesch, E.R., Schmid, C., Schwander, Z., and Zapf, J.A., 1985, Actions of insulin-like growth factors. Annu. Rev. Physiol. 47:443-467.
2. Nissley, P., and Lopaczynski, W., 1991, Insulin-like growth factor receptors. Growth Factors 5:29-43.
3. Baxter, R.C., and Martin, J.L., 1989, Binding proteins for the insulin-like growth factors: structure, regulation and function. Prog. Growth Factor Res. 1:49-68.
4. Rosenfeld, R.G., Lamson, G., Pham, H., Oh, Y., Conover, C.A., DeLeon, D.D., Donovan, S.M., Ocrant, I., and Giudice, L., 1990, Insulin-like growth factor binding proteins. Rec. Prog. Horm. Res. 46:99-163.
5. 1990, Report on the nomenclature of the IGF binding proteins. J. Clin. Endocrinol. Metab. 70:817.
6. 1992, Report on the nomenclature of the IGF binding proteins. Endocrinology 130:1736.
7. Hardouin, S., Hossenlopp, P., Segovia, B., Seurin, D., Portolan, G., Lassare, C., and Binoux, M., 1987, Heterogeneity of insulin-like growth factor binding proteins and relationships between structure and affinity. 1. Circulating forms in man. Eur. J. Biochem. 170:121-132.
8. Hintz, R.L., 1990, Role of growth hormone and insulin-like growth-factor-binding proteins. Horm. Res. 33:105-110.
9. Wood, W.I., Cachianes, G., Henzel, W.J., Winslow, G.A., Spencer, S.A., Hellmiss, R., Martin, J., and Baxter, R.C., 1988, Cloning and expression of the growth hormone-dependent insulin-like growth factor-binding protein. Mol. Endocrinol. 2:1176-1185.
10. Martin, J.L., and Baxter, R.C., 1988, Insulin-like growth factor-binding proteins (IGFBPs) produced by human skin fibroblasts: immunological relationship to other human IGF-BPs. Endocrinology 123:1907-1915.
11. Conover, C.A., Liu, F., Powell, D., Rosenfeld, R.G., and Hintz, R.L., 1989, Insulin-like growth factor binding proteins from cultured human fibroblasts: characterization and hormonal regulation. J. Clin. Invest. 83:852-859.
12. Bachrach, L.K., Liu, F.R., Borrow, G.N., and Eggo, M.C., 1989, Characterization of insulin-like growth factor-binding proteins from sheep thyroid cells. Endocrinology 125:2831-2838.
13. Conover, C.A., 1990, Regulation of insulin-like growth factor (IGF) binding protein synthesis by insulin and IGF-I in cultured bovine fibroblasts. Endocrinology 126:3139-3145.
14. Mondschein, J.S., Smith, S.A., and Hammond, J.M., 1990, Production of insulin-like growth factor binding proteins (IGFBPs) by porcine granulosa cells: identification of IGFBP-2 and -3 and regulation by hormones and growth factors. Endocrinology 127:2298-2306.
15. Smith, E.P., Dickson, B.A., and Chernausek, S.D., 1990, Insulin-like growth factor binding protein-3 secretion from cultured rat sertoli cells: dual regulation by follicle stimulating hormone and insulin-like growth factor-I. Endocrinology 127:2744-2751.

16. Ernst, M., and Rodan, G.A., 1990, Increased activity of insulin-like growth factor (IGF) in osteoblastic cells in the presence of growth hormone (GH): positive correlation with the presence of the GH-induced IGF-binding protein BP-3. Endocrinology 127:807-814.

17. Conover, C.A., Dollar, L.A., Hintz, R.L., and Rosenfeld, R.G., 1983, Insulin-like growth factor I/somatomedin C (IGF-I/SM-C) and glucocorticoids synergistically regulate mitosis in competent human fibroblasts. J. Cell. Physiol. 116:502-509.

18. Conover, C.A., Hintz, R.L., and Rosenfeld, R.G., 1985, Comparative effects of somatomedin C and insulin on the metabolism and growth of cultured human fibroblasts. J. Cell. Physiol. 122:133-141.

19. Conover, C.A., Misra, P., Hintz, R.L., and Rosenfeld, R.G., 1986, Effect of an anti-insulin-like growth factor I receptor antibody on insulin-like growth factor II stimulation of DNA synthesis in human fibroblasts. Biochem. Biophys. Res. Commun. 139:501-508.

20. Conover, C.A., Rosenfeld, R.G., and Hintz, R.L., 1987, Insulin-like growth factor II binding and action in human fetal fibroblasts. J. Cell. Physiol. 133:560-566.

21. Conover, C.A., 1991, A unique receptor-independent mechanism by which insulin-like growth factor-I regulates the availability of insulin-like growth factor binding proteins in normal and transformed human fibroblasts. J. Clin. Invest. 88:1354-1361.

22. Bale, L.K., and Conover, C.A., 1992, Regulation of insulin-like growth factor binding protein-3 messenger ribonucleic acid expression by insulin-like growth factor I. Endocrinology 131:608-614.

23. Conover, C.A., Lombana, F., Ronk, M., and Powell, D.R., 1990, Structural and biological characterization of bovine insulin-like growth factor binding protein-3. Endocrinology 127:2795-2803.

24. Conover, C.A., 1991, Glycosylation of insulin-like growth factor binding protein-3 (IGFBP-3) is not required for potentiation of IGF-I action: evidence for processing of cell-bound IGFBP-3. Endocrinology 129:3259-3268.

25. Conover, C.A., and Powell, D.R., 1991, Insulin-like growth factor (IGF) binding protein-3 blocks IGF-I-induced receptor down-regulation and cell desensitization in cultured bovine fibroblasts. Endocrinology 129:710-716.

26. DeMellow, J.S.M., and Baxter, R.C., 1988, Growth hormone-dependent insulin-like growth factor (IGF) binding protein both inhibits and potentiates IGF-I-stimulated DNA synthesis in human skin fibroblasts. Biochem. Biophys. Res. Commun. 156:199-204.

27. Bayne, M.L., Applebaum, J., Chicchi, G.G., Hayes, N.S., Green, B.G., and Cascieri, M.A., 1988, Structural analogs of human insulin-like growth factor I with reduced affinity for serum binding proteins and the type 2 insulin-like growth factor receptor. J. Biol. Chem. 263:6233-6239.

28. McCusker, R.H., Camacho-Hubner, C., Bayne, M.L., Cascieri, M.A., and Clemmons, D.R., 1990, Insulin-like growth factor (IGF) binding to human fibroblast and glioblastoma cells: the modulating effect of cell released IGF binding proteins (IGFBPs). J. Cell. Physiol. 144:244-253.

29. Shimasaki, S., Koba, A., Mercado, M., Shimonaka, M., and Ling, N., 1989, Complementary DNA structure of the high molecular weight rat insulin-like growth factor binding protein (IGFBP-3) and tissue distribution of its mRNA. Biochem. Biophys. Res. Commun. 165:907-912.

30. Sommer, A., Maack, C.A., Spratt, S.K., Mascarenhas, D., Tressel, T.J., Rhodes, E.T., Lee, R., Roumas, M., Tatsuno, G.P., Flynn, J.A., Gerber, N., Taylor, J., Cudney, H., Nanney, L., Hunt, T.K., and Spencer, E.M., 1991, Molecular genetics and actions of recombinant insulin-like growth factor binding protein-3. in "Modern Concepts of Insulin-Like Growth Factors", E.M. Spencer, ed, Elsevier, Amsterdam, pp. 715-728.

31. Conover, C.A., 1992, Potentiation of insulin-like growth factor (IGF) action by IGF-binding protein-3: studies of underlying mechanism. Endocrinology 130:3191-3199.

32. Cascieri, M.A., Chicchi, G.G., Applebaum, J., Hayes, N.S., Green, B.G., and Bayne, M.L., 1988, Mutants of human insulin-like growth factor I with reduced affinity for the type I insulin-like growth factor receptor. Biochemistry 27:3229-3233.

33. Bayne, M.L., Applebaum, J., Underwood, D., Chicchi, G.G., Green, B.G., Hayes, N.S., and Cascieri, M.A., 1988, The C region of human insulin-like growth factor (IGF) I is required for high affinity binding to the type 1 IGF receptor. J. Biol. Chem. 264:11004-11008.

GENE EXPRESSION OF THE IGF BINDING PROTEINS DURING POST-IMPLANTATION EMBRYOGENESIS OF THE MOUSE;
COMPARISON WITH THE EXPRESSION OF IGF-I AND -II AND THEIR RECEPTORS IN RODENT AND HUMAN

Alwin G.P. Schuller,[1] Johan W. van Neck,[1] Dicky J. Lindenbergh-Kortleve,[1] Cora Groffen,[1] Ilona de Jong,[1] Ellen C. Zwarthoff[2] and Stenvert L.S. Drop[1]

[1]Department of Pediatrics, Division of Endocrinology, Sophia Children's Hospital, Rotterdam
and
[2]Department of Pathology, Erasmus University, Rotterdam
P.O.Box 70029, 3000 LL Rotterdam, The Netherlands

Summary

The IGF binding proteins (IGFBPs) comprise at least six distinct species which may modulate the action of IGFs. IGFs are important regulators of fetal growth and differentiation. We have studied the mRNA expression of the six IGFBPs during post-implantation embryogenesis (day 11-18) by in situ hybridization techniques.

Expression of IGFBP-1 was detected in mouse conceptuses after day 12 of gestation and seemed restricted to the liver. Transcripts for IGFBP-2, -4 and -5 were detected in various tissues and were found in all stages tested. In contrast, expression of IGFBP-3 and -6 could be detected only weakly in late gestational embryos. Comparison of the expression pattern of IGFBP-2, -4 and -5, which were found widely distributed in mouse conceptuses, revealed that IGFBP-2 was expressed mainly in the ectodermal layer and also in the mesoderm derived part of the tongue (day 13.5). Transcripts for IGFBP-4 however, only were detected in the mesoderm derived tissues, whereas expression of IGFBP-5 was restricted to the ectodermal layer. A similar distribution pattern was observed in the lung. In general, expression of IGFBP-2 and -5 was detected in the same cells, whereas IGFBP-4 and -5 were expressed mainly in different cell types.

In rodents as in the human there is widespread expression of the genes coding IGFs, the IGFBPs and the receptors during pre- and postimplantation embryogenesis.
These data support the assumption that the IGFs play an important role during embryogenesis.

Introduction

The Insulin-like growth factors IGF-I and -II are small molecular weight peptides with both mitogenic and metabolic properties. Both IGFs are thought to be involved prenatally in the regulation of growth and differentiation, IGF-II being even more prominent than IGF-I. Direct evidence for a physiological role of the IGFs in embryonic growth was obtained by embryonic transplantation studies and in the gene disruption mouse model. Lui et al tranplanted 10d old rat embryos under the capsule of both kidneys of syngeneic hosts, where they grew rapidly and differentiated normally. Infusion of rabbit IGF-I antiserum into the renal artery for 9 days resulted in a marked reduction of growth. Infusion of IGF-II did promote growth even more so in hypophysectomized hosts, whereas infusion of IGF-I did not.[1] In order to create a gene deletion mouse model a gene mutation disrupting one of the IGF-II alleles was introduced in mouse embryonic stem cells and chimaeric animals were constructed. Heterozygous progeny had 60% reduction of body weight

Current Directions in Insulin-Like Growth Factor Research,
Edited by D. LeRoith and M.K. Raizada, Plenum Press, New York, 1994

267

when compared to their ES cell-derived wild-type littermates but were otherwise normal and fertile.[2] Homozygous mutants were in appearance as compared to normals due to parental imprinting.[5]

The IGF-II gene deletion studies and other studies have indicated that gene expression of both IGFs and the IGF receptors occur very early in the pre-implantation period of rodent embryogenesis.[2,5,6,7,8,9] IGF-I added to the culture medium of two cell mouse embryos stimulated the number of cells in the resultant blastocysts.[3]

IGF-I and -II exert their metabolic and growth promoting actions via high-affinity binding to specific IGF receptors[4] Paracrine and autocrine regulatory mechanisms are presumed to play an important role during the early stages of embryonic development.

Pre- and postnatally, IGFs occur in plasma and other biological fluids bound to IGF binding proteins (IGFBPs). These IGFBPs are thought not only to function as carrier proteins but probably more importantly to modulate paracrine (and autocrine) actions of the IGFs at a pericellular level.[10,11]

In view of the predominantly local activity of the various proteins, the distribution of the expression of the IGFs, their binding proteins and receptors may have a particular functional significance.

In this review we will describe the sites of mRNA expression of all six IGFBPs in the mouse during mid- and late gestation and we relate our findings to recent published reports on the mRNA expression of IGF-I and -II and their specific receptors in rodents.

Additionally we will compare the mRNA expression of the six IGFBPs in the mouse with published data on expression in the human fetus.

Table 1. Major tissues in which IGFBP-1, -2, -3, -4, -5 and -6 mRNA expression in the mid gestational mouse embryo (day 12) was detected.

	IGFBP-1	IGFBP-2	IGFBP-3	IGFBP-4	IGFBP-5	IGFBP-6
Mesencephalon	-	+	-	+	-	-
Telencephalon	-	+	-	+	-	-
Tongue	-	+	-	+	+	-
Liver	+	+	-	+	-	-
Sclerotomes	-	+	-	+	-	-
Snout	-	-	-	+	-	-
Nasal Placode	-	+	-	-	+	-

+ = detected
- = not detected

IGFBP mRNA expression

The in situ hybridization patterns of the six mIGFBPs on autoradiogram of tissue sections revealed that IGFBP -2, -4 and -5 are highly expressed between gestational day 11-18 and widely distributed in the mouse conceptuses[12] (see figure 1). The major tissues in which mRNA expression of the six mIGFBPs was detected in mid- and late gestational mouse conceptuses are summarized in table 1 and -2 respectively.

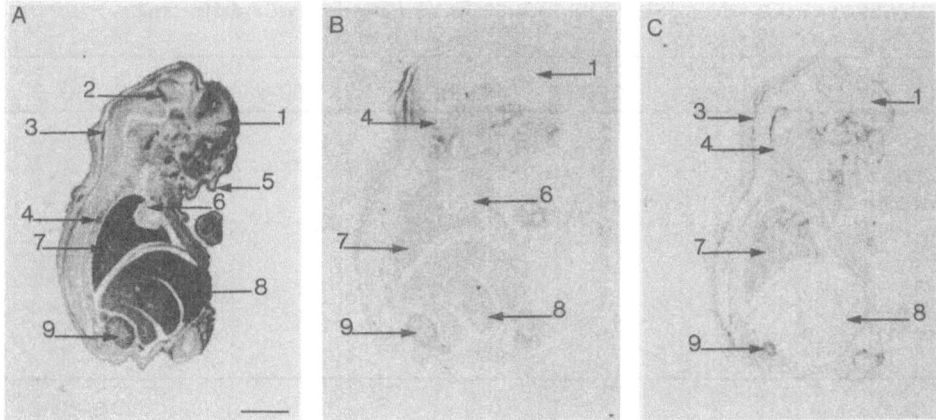

Figure 1.   *In situ* hybridization to sagittal sections of 18 day old mouse fetuses.
5 μm thick paraffin sections were hybridized with ^{35}S-labelled RNA probes as described[12]. Figure
A, B and C show auto-radiographic images of paraffin sections of an 18 day mouse fetus hybridized
with probes for IGFBP-2 (A), IGFBP-4 (B) and IGFBP-5 (C). The auto-radiogram of IGFBP-2 was
exposed for 1 day, the others for 3 days.
1=Brain, 2=Choroid plexus, 3=Meninges, 4=Vertebrae, 5=Skin, 6=Heart, 7=Lung, 8=Liver,
9=Kidney. Scale bar=2 mm.

Expression of IGFBP-1 mRNA was restricted to the liver only. Transcripts of IGFBP-3 and -6 were very weakly expressed. As early as day 11 of gestation IGFBP-2 expression was detected in telencephalon, mesencephalon and tongue. From day 13, the IGFBP-2 transcript was found in differentiating sclerotomes, tongue, oesophagus, nasal placode, lung and liver. After day 14 the expression of IGFBP-2 was also found in the choroid plexus, meninges, cornea, sclera, sub-maxillary gland, thymus, vertebrae, kidney, intestine, bladder and hair follicles.

Transcripts of IGFBP-4 were detected as early as day 11 in telencephalon, mesencephalon, snout, tongue, and differentiating sclerotomes. After day 14 IGFBP-4 expression was undetectable in the brain areas. In contrast, IGFBP-4 transcripts were clearly detectable in lung, liver, kidney, intestine, vertebrae, ribs and incissivus.

Table 2. Major tissues in which IGFBP-1, -2, -3, -4, -5 and -6 mRNA expression in the late gestational mouse fetus (day 18) was detected.

	IGFBP-1	IGFBP-2	IGFBP-3	IGFBP-4	IGFBP-5	IGFBP-6
Choroid Plexus	-	+	-	-	-	-
Meninges	-	+	-	-	+	-
Vertebrae	-	+	+	+	+	+
Heart	-	+	-	-	-	-
Lung	-	+	-	+	+	+
Liver	+	+	+	+	-	+
Kidney	-	+	-	+	+	-
Intestine	-	+	-	+	+	-

+ = detected
- = not detected

Expression of IGFBP-5 was also detectable as early as day 11 of gestation in differentiating sclerotomes and the ectodermal layer of the tongue. Furthermore, expression of IGFBP-5 in 14 day old embryos was found in nasal placodes, pharynx and oesophagus. After day 14 of gestation expression of IGFBP-5 was found in the cornea and sclera of the eye, meninges, lung, kidney, intestine, bladder, vertebrae and ribs. Expression of IGFBP-5 was not above background level in the liver.

Comparison of the hybridization patterns revealed that although several tissues were found to express more than one IGFBP the expression pattern of these proteins was distinct. For instance, in the tongue of a mouse embryo of 13 days gestation highest expression of IGFBP-2 was found in the ectodermal layer, with moderate expression in the mesodermal derived part of the tongue. Transcripts for IGFBP-4 were only expressed in the mesoderm derived part (figure 2A, B), whereas IGFBP-5 was expressed solely in the ectodermal layer of the developing tongue (figure 2C, D).

A similar distribution between IGFBP-2, -4 and -5 was seen in the lung. Here, expression of IGFBP-2 was highest in or around respiratory epithelium, but also detectable in the interstitium. Transcripts for mIGFBP-4 however, were found mainly in the interstitium (figure 3A, B), whereas highest expression of mIGFBP-5 was located in or around the respiratory epithelium (figure 3C, D).

IGFBP expression in the human fetus in the early second trimester has been described partly using immunohistochemical methods (IGFBP-1/-3) partly using mRNA expression by in situ

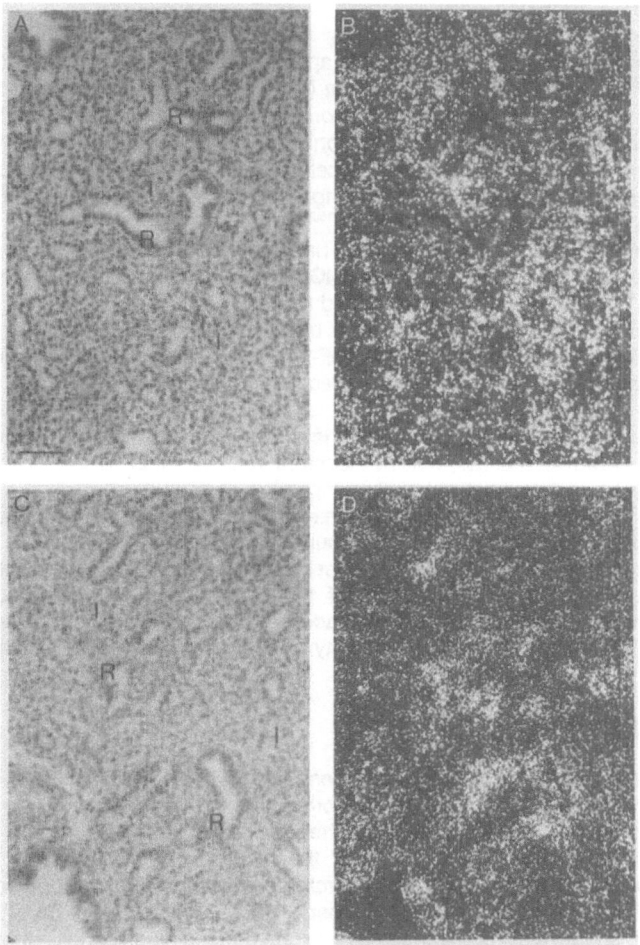

Figure 2. Light field (A and C) and dark field (B and D) images of a paraffin section of the lung of a 18 day mouse fetus. Sections were hybridized with probes for IGFBP-4 (A and B) and IGFBP-5 (C and D). I=interstitium, R=Respiratory epithelium. Scale bar=70 μm.

hybridization techniques (IGFBP-4/-6).[13,14] Strong immunological staining was found for IGFBP-1/-3 in epithelial lining of gut, lung, kidney and skin; in endoderm derived cell types (pancreatic, hepatic, adrenal cortical tissue) and in mesoderm derived tissues such as skeletal muscle, cardiac and smooth muscle. mRNA for IGFBP-4/-6 were found in many tissues and were particularly localized in populations of proliferating cells, such as epidermis of the skin, crypt epithelia of small intestine and epithelia of developing glomeruli and ureteric buds.

IGF-I and -II mRNA expression

IGF-I and -II gene expression in the rodent is apparent already early in gestation (day 11). The distribution of IGF-I & -II mRNA in multiple rodent (rat, mouse) and human tissues is similar.[15]

IGF-I mRNA is not produced exclusively by connective tissue cells of mesenchymal origin but also parenchym. In the early period of rat embryonic development (ie embryonic days 14-15) IGF-I and -II have been found in distinctly different cell types. IGF-I mRNA expression was found in undifferentiated mesenchymal tissue of sprouting nerves and spinal ganglia, in areas of active tissue remodeling such as cardiac outflow tract.[16,22]

In contrast, abundant IGF-II gene expression has been found in developing muscle, cartilage, vascular tissue, liver and pituitary. In addition IGF-II mRNA expression was found in areas of vascular interface with the brain, such as choroid plexus.[18,19,20]

In rats IGF-I mRNA levels increase markedly between day 11-13 of gestation. At this period organogenesis takes place and contrast with IGF-II mRNA levels which remain constantly high.[16] It is of interest that the IGF-I gene expression precedes that of growth hormone.

IGF-II mRNA levels remain at a constant high level during fetal development and decline in the postnatal period. However the IGF-II mRNA content remains high in the brain into adult life, expressed by the choroid plexus and leptomeninges.[17]

Also in human fetuses of IGF-I and -II mRNA expression has been documented.[15] Using in situ hybridization a wide variety of cell types within human fetal tissues showed IGF-I and IGF-II gene expression from day 18 to 14 wk of gestation. Localization of both mRNAs was found in connective tissue of mesenchymal origin. The pattern of IGF-II expresion showed specific age-related differences in different tissues such as kidneys, adrenal glands and liver.[21] The sites of expression did not correlate with areas of high mitotic activity or specific types of differentiation.

IGF receptor mRNA expression

Type 1 IGF receptor gene expression has been found widespread during the early stage of development of the rat (day 14-15) being most prominent in the developing nervous system and muscle and being relatively underexpressed in the liver. Most intense expression was determined in the neuroepithelial cells of the floorplate of the hindbrain but additionally expression was determined in tongue, myotomes, vertebral sclerotomes, mesonephros and bowel walls.[22]

There are two types of the IGF type 2 Mannose-6-Phosphate receptors (MPR): MPR 300 (MW 300.000) mediates endocytosis of mannose-6-phosphate containing ligands, whereas MPR 46 (M W46.000) mediates the secretion of parts of its ligands. There is an almost complementary and non-overlapping expression of both receptors during mouse embryogenesis.[23] Up to embryonic day 15 MPR 46 is highly expressed at sites of hematopoiesis and in the thymus. MPR 300 is expressed in the cardiovascular system. Later on at day 17.5 there is a wide variety of expression without overlap.

It is of specific interest that highest gene expression for both type 1 and type 2 IGF receptors was found in fetal tissues with levels rapidly decreasing after birth.[22,24]

Discussion

IGFBP-1, -2, -4 and -5 are highly expressed during embryogenesis of the mouse. Transcripts for IGFBP-1 seemed to be restricted to the liver, whereas expression of IGFBP-2, -4 and -5 was found in various tissues in all stages tested. Expression of IGFBP-3 and -6 could only be detected after a long exposure time.

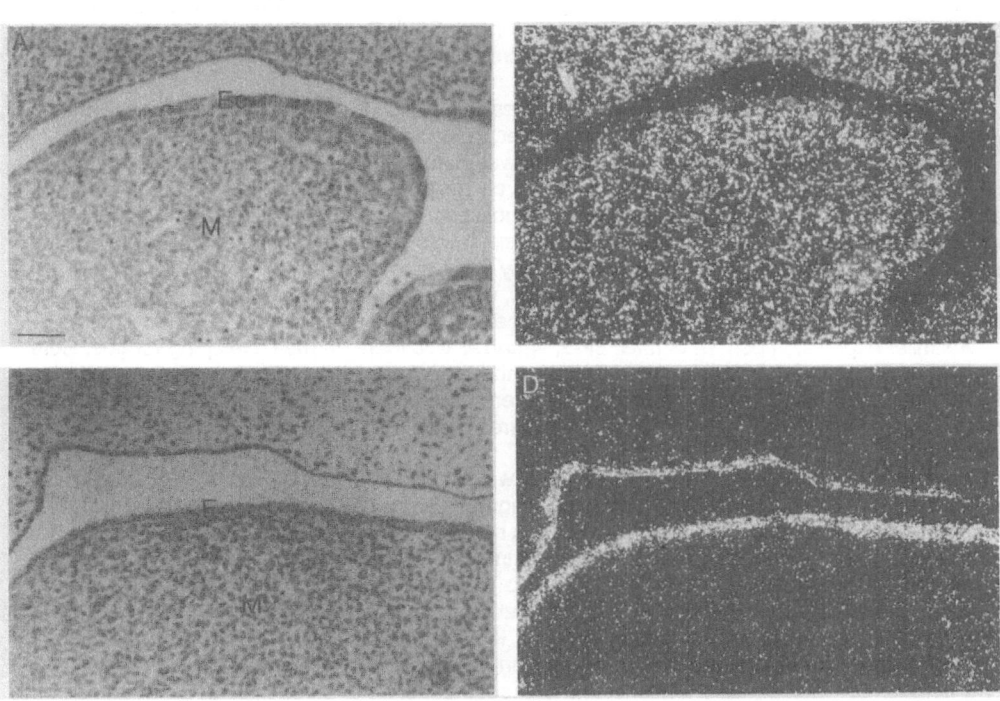

Figure 3.　　Light field (A and C) and dark field (B and D) images of a paraffin section of the tongue of a 13.5 day mouse embryo. Sections were hybridized with probes for IGFBP-4 (A and B) and IGFBP-5 (C and D). Ec=Ectodermal layer, M=mesodermal derived part. Scale bar=70 μm.

Comparison of the expression patterns of the IGFBPs revealed that they were clearly distinct, although several tissues were found to express more than one IGFBP. For instance, mRNA expression of IGFBP-1, -2, and -4 was high in fetal liver as early as day 11 of embryogenesis. Also, in rat fetal serum high levels of these IGFBPs were found.[25] Although synthesis of IGF-II has been demonstrated in a variety of fetal rat tissues,[26] the major source of IGF-II in fetal serum is thought to be the liver. Taken together these findings suggest that IGFBP-1, -2 and -4 are synthesized and secreted by the liver and that they may play a role in transporting IGFs in fetal serum. Furthermore, the finding that these IGFBPs are all expressed in the liver raises the question whether these IGFBPs might have distinct effects.

Also in the epithelium of the choroid plexus high levels of IGFBP-2 transcripts were found. Recently, it was shown that in mid-gestational rat brain IGFBP-2 mRNA was expressed in epithelium of the choroid plexus, whereas IGF-II was expressed in adjacent mesenchymal cells. It was suggested that epithelial derived IGFBP-2 might mediate delivery of mesenchymal synthesized IGF-II to the cerebrospinal fluid.[27]

Expression of several IGFBPs in the same tissue was observed in the developing lung and tongue. However, in the lung, highest expression of IGFBP-2 and -5 was found in or around respiratory epithelium, but also detectable in interstitium. In contrast, expression of IGFBP-4 was located mainly in interstitium.

Earlier studies have shown that mRNA expression of IGF-II was present in lung of human conceptuses, but that IGF-II could not be detected in respiratory epithelium.[28]

In contrast, immunohistochemical studies showed that the IGFs were present in the epithelium of the airways, with only slight immunostaining of the pulmonary interstitium.[29,30] Since mRNA expression was limited to fibroblasts and mesenchymal cells, these immunoreactive cells were thought not to be the primary sites of IGF synthesis. It was suggested that these cells may define sites of IGF action.[29] Altogether, these findings may suggest that in the lung expression of mIGFBP-4 and IGF-II may take place in the same cell types, whereas mIGFBP-5 might be expressed in cells that do not synthesize IGF-II, but may be target cells of IGFs.

A similar expression pattern of mIGFBP-2, -4 and -5 was found in the developing tongue. Here highest expression of mIGFBP-2 was found in the ectodermal layer, but expression was also detectable in the mesoderm derived part. However, mIGFBP-4 transcripts could only be detected in the mesoderm derived part of the tongue, whereas mIGFBP-5 expression was restricted to the ectodermal layer.

Recently, in mid gestational rat embryos it was shown that expression of IGF-II was also restricted to the mesoderm derived part of the tongue.[27] Furthermore, transcripts for the type 1 IGF receptor were found in the tongue. Although it is not known whether the expression includes the ectodermal layer of the tongue, these results indicated that IGFs may act locally in tongue development.[22] These findings suggest that also in the developing tongue IGF-II, IGFBP-4 and IGFBP-5 mRNA may be expressed compartmentalized and that expression of IGF-II and IGFBP-4 may occur in the same compartment, whereas IGFBP-5 may be expressed mainly in another compartment. Furthermore, the distinct expression patterns of IGFBP-2, -4 and -5 in both the lung and tongue may indicate different functions or modes of action of these mIGFBPs.

From the various studies in rodents and in the human it has become quite evident that their is widespread expression of the genes encoding IGFs, the IGFBP and the receptors during pre- and postimplantation embryogene-sis.[6,12,22,23,27,36] That the various mRNA are also translated early and in many fetal tissues is suggested by several immunohistochemical studies.[15,18,31]

In addition these data support the assumption that the IGFs play an important role during embryogenesis. It is of specific interest that IGF-II is paternally imprinted[5] whereas the type 2 IGF receptor is maternally imprinted.[32] Thus embryos that receive a normal functioning ligand allele from the father and a normally functioning receptor from the mother will inherit a unit essential for paracrine/autocrine function.[33]

A highly relevant but difficult question to answer is whether the expression of the various components are complementary expressed in time and localization representing specific developmental functions and whether this would be developmentally regulated. There are several examples that would suggest such regulatory mechanisms. For instance, embryonic and fetal expression of type 2 receptor mRNA is seen at sites where there are also high levels of IGF-II expression.[7] The genes of IGF-I and of the type 1 receptor are expressed in a diametrically opposed fashion,[34] necessitating tranportation of the IGFs for instance by binding proteins. During outgrowth and differentiation of the limbs IGF-I and -II mRNA expression changes dramatically in mesoderm derived cells. IGFBP-2 is present abundantly in the apical ectodermal ridge and disappears as the underlying mesoderm begins to differentiate.[35]

Complementary patterns of IGF-II and IGFBP-2 expression during outgrowth of facial processes and limbs has been suggested by Wood et al.[36] Our data suggest that mRNA expression of IGFBP-2 and -5 mainly in the same cell types, wheras complementary expression of IGFBP-4 and -5 was found.

The expression of the IGFs, the IGFBPs and the receptors is often distinctly localized in actively proliferating tissues suggesting a crucial role during growth and organogenesis. However, it should be stressed that additionally there are several pre- and posttranslational regulatory mechanisms that ultimately determine the very delicate organ or tissue specific interrelationship of the IGFs, their receptors and binding proteins.

Further detailed organ- or tissue specific in situ hybridization and immunohisto-chemical studies and ultimately in vivo gene disruption studies will be required to resolve the question as to these role of the various proteins.

Acknowledgements

The authors would like to thank Drs. L.G. Wilming and Dr. C. Meyers for their contribution in setting up the technique of *in situ* hybridization, Dr. C. Vermey-Keers for her kind help and knowledge on mouse embryogenesis and F.L. van der Panne for photography. This work was supported by grants from the Sophia Foundation for Medical Research and Novo-Nordisk Insulin Laboratories, Denmark.

References

1. L.L. Liu, S. Greenberg, S.M. Russell, and C.S. Nicoll, Effects of IGF I and II on growth and differentiation of transplanted rat embryos and fetal tissues, Endocrinology 124:3077-3082 (1989).

2. T.M. DeChiara, A. Efstratiadis, and E.J. Robertson, A growth deficiency phenotype in heterozygous mice carrying an insulin-like growth factor II gene disrupted by targeting, Nature 345:78-80 (1990).

3. M.B. Harvey, P.L. Kaye, IGF I stimulates growth of mouse preemplantation embryos in vitro, Mol Reprod Dev 31:195-199 (1992).

4. C. Moxham, and S. Jacobs, Insulin-like growth factor receptors, In: P.N. Schofield ed. The Insulin-like growth factors, Oxford University press, Oxford (1992).

5. T.M. DeChiara, E.J. Robertson, and A. Efstratiadis, Parental imprinting of the mouse insulin-like growth factor II gene, Cell 64:849-859 (1991).

6. J.E. Lee, J. Pintar, and A. Efstradiadis, Pattern of the insulin-like growth factor II gene expression during early mouse embryogenesis, Development 110:151-159 (1990).

7. P.V. Senior, S. Byrne, W.J. Brammar, and F. Beck, Expression of the IGF-II/mannose-6-phosphate receptor mRNA and protein in the developing rat, Development 109:67-73 (1990).

8. S. Kapur, H. Tamada, S.K. Dey, and G.K. Andrews, Expression of insulin-like growth factor-I (IGF-I) and its receptor in the peri-implantation mouse uterus, and cell-specific regulation of IGF-I gene expression by estradiol and progesterone, Biol Rep 46:208-219 (1992).

9. N.A. Telford, A. Hogan, C.R. Franz, G.A. Schultz, Expression of genes for insulin and insulin-like growth factors and receptors in early postimplantation mouse embryos and embryonal carcinoma cells, Mol Reprod Dev 27:81-92 (1990).

10. S.L.S. Drop, A.G.P. Schuller, D.J. Lindenbergh-Kortleve, C. Groffen, A. Brinkman, and E.C. Zwarthoff, Structural aspects of the IGFBP family, Growth Regulation 2: 80-87 (1992).

11. D.R. Clemmons, IGF binding proteins: regulation of cellular actions, Growth Regulation 2:80-87 (1992).

12. A.G.P. Schuller, E.C. Zwarthoff, and S.L.S. Drop, Gene expression of the six IGF binding proteins in the mouse conceptus during mid- and late gestation, Endocrinology in press (1993).

13. D.J. Hill, and D.R. Clemmons, Similar distribution of insulin-like growth factor binding proteins-1, -2, -3 in human fetal tissues, Growth Factors 6:315-326 (1992).

14. P.J.D. Delhanty, D.J. Hill, S. Shimasaki and V.K.M. Han, Insulin-like growth factor binding protein-4, -5 and -6 mRNAs in the human fetus: localization to sites of growth and differentiation?, Growth Regulation 3:8-11 (1993).

15. V.K.M. Han, and D.J. Hill, The involvement of IGFs in embryonic and fetal development, In P.N. Schofield ed., The insulin-like growth factors, Oxford University press, Oxford (1992).

16. P. Rotwein, K.M. Pollo, M. Watson, and J.D. Milbrandt, IGF gene expression during rat embryonic development, Endocrinology 121:2141-2144 (1987).

17. P. Rotwein, S.K. Burgess, J.D. Mibrandt, and J.E. Krause, Differential expression of IGF genes in rat central nervous system, Proc Natl Acad Sci USA 85:265-269 (1988).

18. F. Beck, N.J. Samani, J.D. Penschow, B. Thorley, G.W. Tregear, and J.P. Coghlan, Histochemical localization of IGF-I and -II mRNA in the developing rat embryo, Development 101:175-184 (1987).

19. F. Stylianopoulou, A. Efstratiadis, J. Herbert, and J. Pintar, Pattern of the insulin-like growth factor II gene expression during rat embryogenesis, Development 103: 497-506 (1988).

20. R.S.K. Florance, P.V. Senior, S. Byrne, and F. Beck, The expression of IGF-II in the early post-implantation rat conceptus, J Anat 175:169-179 (1991).

21. A.L. Brice, J.E. Cheetman, V.N. Bolton, N.C.W. Hill, and P.N. Schofield, Temporal changes in the expression of the insulin-like growth factor II gene associated with tissue maturation in the human fetus, Development 106:543-554 (1989).

22. C.A. Bondy, H. Werner, C.T. Roberts Jr., and D. LeRoith, Cellular pattern of IGF-I, and type I IGF receptor gene expression in early organogenesis: comparison with IGF-II gene expression, Mol Endocrinol 4:1386-1398 (1990).

23. U. Matzner, K. Figura von, and R. Pohlmann, Expression of the two mannose 6-phosphate receptors is spatially and temporally different during mouse embryogenesis, Development 114:965-972 (1992).

24. M. Ballesteros, C.D. Scott, and R.C. Baxter, Developmental regulation of IGF-mannose 6-phosphate receptor mRNA in the rat, Biochem Biophys Res Commun 172:775-779 (1990).

25. P.J. Fielder, G. Thordarson, F. Talamantes, and R.G. Rosenfeld, Characterization of IGF binding proteins during gestation in mice: effects of hypophysectomy and a specific serum protease activity, Endocrinology 127:2270-2280 (1990).

26. J.A. Romanus, Y.W.H. Yang, S.O. Adams, A.N. Sofair, L.Y.H. Tseng, P. Nissley, and M.M. Rechler, Synthesis of insulin-like growth factor II (IGF-II) in fetal rat tissues: translation of IGF-II ribonucleic acid and processing of pre-pro-IGF-II, Endocrinology 122:709-716 (1988).

27. T.L. Wood, A.L. Brown, M.M. Rechler, and J.E. Pintar, The expression pattern of an insulin-like growth factor (IGF)-binding protein gene is distinct from IGF-II in the midgestational rat embryo, Mol Endocrinol 4:1257-1263 (1990).

28. V.K.M. Han, A.J. D'Ercole, and P.K. Lund, Cellular localization of somatomedin (insulin-like growth factor) messenger RNA in the human fetus, Science 236:193-197 (1987a).

29. V.K.M. Han, D.J. Hill, A.J. Strain, A.C. Towle, J.M. Lauder, L.E. Underwood, and A.J. D'Ercole, Identification of somatomedin/insulin-like growth factor immunoreactive cells in the human fetus, Pediatr Res 22:245-249 (1987b).

30. D.J. Hill, D.R. Clemmons, S. Wilson, V.K.M. Han, A.J. Strain, and R.D.G. Milner, Immunological distribution of one form of insulin-like growth factor (IGF)-binding protein and IGF peptides in human fetal tissues, J Mol Endocrinol 2:31-38 (1989).

31. F. Beck, N.J. Samani, S. Byrne, K. Morgan, R. Gebhard, and W.J. Brammar, Histochemical localization of IGF-I and IGF-II mRNA in the rat between birth and adulthood, Development 104:29-38. (1988).

32. D.P. Barlow, R. Stoger, B.G. Hermann, K. Saito, and N. Schweifer, The mouse insulin-like growth factor type-2 receptor is imprinted and closely linked to the Tme locus, Nature 349;84-87 (1992).

33. P.L. Kaye, K.L. Bell, L.F.S. Beebe, G.F. Dunglison, H.G. Gardner, and M.B. Harvey, Insulin and the insulin-like growth factors (IGFs) in preimplantation development, Reprod Fertil Dev 4:373-386 (1992).

34. H. Werner, M. Woloschak, M. Adamo, Z. Shen-Orr, C.T. Roberts Jr., and D. Leroith, Developmental regulation of the rat insulin-like growth factor I receptor gene, Proc Natl Acad Sci 86:7451-7455 (1989).

35. R.D. Streck, T.L. Wood, M-S. Hsu, and J.E. Pintar, Insulin-like growth factor I and II and insulin-like growth factor binding protein-2 RNAs are expressed in adjacent tissues within rat embryonic and fetal limbs, Development 151:586-596 (1992).

36. T.L. Wood, R.D. Streck, and J.E. Pintar, Expression of the IGFBP-2 gene in post-implantation rat embryos, Development 114:59-66 (1992).

HORMONAL REGULATION OF INSULIN-LIKE GROWTH FACTOR BINDING PROTEIN-1 EXPRESSION AND THE DEVELOPMENT OF TRANSGENIC MOUSE MODELS TO STUDY IGFBP-1 FUNCTION

Liam J. Murphy, Douglas Barron and Charita Seneviratne

Departments of Internal Medicine and Physiology
Faculty of Medicine, University of Manitoba
Winnipeg, Canada, R3E 0W3

INTRODUCTION

The insulin-like growth factors (IGF) are present in the serum, other biological fluids and tissue extracts in association with high affinity binding proteins. Six members of this family of binding proteins have been identified and both rodent and human cDNAs encoding these proteins have been isolated. In serum from rodents and humans the majority of IGF is present as a complex of approximately 150-200 kDa. This complex is composed of IGF-I or IGF-II, a 100 kDa acid-labile protein and IGFBP-3. Under normal conditions only a small fraction of IGF-I or IGF-II in the plasma is free and there appears to be an excess of available binding sites for IGF. Although IGFBP-3 is responsible for the majority of the IGF binding capacity in the plasma, other binding proteins are also detectable in plasma and there is evidence that the different binding proteins serve to compartmentalize the IGFs into pools with different functional half-lifes and capacities[1].

The major IGF binding protein present in serum, IGFBP-3 is growth hormone dependent and this growth hormone dependence appears to be mediated via IGF-I, since treatment of hypophysectomized rats with either growth hormone or IGF-I is able to restore the 150-200 kDa IGF-I complex[2].

In serum from fetal and neonatal animals the majority of IGF-I and II is associated with a 40-50 kD complex rather than the 150 kD complex. In the rodent the predominant IGF binding protein is IGFBP-2. In tissues other than

Current Directions in Insulin-Like Growth Factor Research,
Edited by D. LeRoith and M.K. Raizada, Plenum Press, New York, 1994

279

the brain the expression of this binding protein is reduced after the neonatal period and little IGFBP-2 is detected in adult rat plasma[3,4].

The first binding protein to be completely purified, and the binding protein which has been the most intensely studied to date is IGFBP-1. This binding protein was first purified from human amniotic fluid but is also produced by a variety of cell lines including human hepatoma and endometrial cancer cells[5,6]. It accounts for a small percentage of the IGF binding capacity in human serum and probably an even a smaller percentage of IGF binding capacity in the rodent serum. However, the regulation of this binding protein appears to quite different from IGFBP-3 and suggests a functionally important role in both growth and metabolism.

In this chapter I will review the current understanding or the role of the IGFBPs, regulation of IGFBP-1 in the rodent and discuss our attempts at developing transgenic mouse strains as tools to investigate the physiological role of the IGFBP-1.

THE FUNCTIONAL ROLE OF THE IGF BINDING PROTEINS

Although there is now a considerable body of literature on the expression of the various IGFBPs, the functional role of these proteins remains unclear. A simplistic view is that they serve to block the insulin-like activity of the relatively large concentrations of the IGFs present in the circulation[7]. However a number of other functions for these binding proteins have been proposed. These include, prolongation of the half-life of IGF, facilitation of delivery of IGFs to cell surface receptor and facilitation of transcapillary transport of IGF to tissues. Although it has been possible to demonstrate that the IGFBPs inhibit IGF action *in vitro*, the results obtained using in these assays are variable and synergism rather than inhibition can also be demonstrated under certain circumstances. Both IGFBP-1 and IGFBP-3 are able to inhibit IGF-I binding and IGF-I action in a variety of assay systems including, DNA synthesis in fibroblasts[8], IGF-I stimulated α-aminoisobutyric acid transport in choriocarcinoma cell[9], lipogenesis and glucose oxidation in adipocytes[10]. However, both IGFBP-1 and IGFBP-3 can also enhance the effects of IGF-I on DNA synthesis in fibroblasts[8,11]. The variable effects of the IGFBPs in these assays systems appears to be dependent upon the relative concentrations and timing of addition of IGF-I and the IGFBP. Post-translational modification of IGFBP-1 is thought to be important in modulation of the affinity of this binding protein for IGF-I. Isoforms of IGFBP-1 which differ in the degree of phosphorylation have been detected in amniotic fluid and culture medium of endometrial cells[12]. The phosphorylated isoform of IGFBP-1 has a 5 fold higher affinity for IGF-I than does the non-phosphorylated form and would be expected to have a greater capacity to inhibit IGF-I action. These *in vitro* studies have not as yet provided a definitive understanding of the physiological role of any of the IGFBPs however they do suggest that the effects of the binding proteins on IGF-I action may not be simple inhibition. Since the ontogeny, regulation, cellular and tissue specific expression of the known IGF binding proteins are unique, it is likely that each of the binding proteins subserves slightly different functions.

REGULATION OF IGFBP-1 EXPRESSION

The regulation of IGFBP-1 expression is complex and it is possible to demonstrate that a variety of hormonal and non-hormonal factors have effects on expression of this binding protein. In man, serum IGFBP-1 concentrations appear to be inversely correlated with serum insulin levels[13,14] and elevated IGFBP-1 concentrations are apparent in normal individuals after an overnight fast[15,16]. The major source of circulating IGFBP-1 in man is presumed to be the liver although this has not been clearly established. In human fetal liver explants and human hepatoma cell lines synthesis of IGFBP-1 is suppressed by insulin[17]. In the rat, IGFBP-1 mRNA is most abundant in the liver, however there is a considerable amount of IGFBP-1 mRNA in the kidney. In the mouse, IGFBP-1 mRNA appears to be equally abundant or even more abundant in the kidney than the liver (Fig 1). In both the rat and mouse, food deprivation results in a dramatic increase in both hepatic and renal IGFBP-1 mRNA abundance and an increase in circulating IGFBP-1 levels[18]. Refeeding with standard rodent chow or glucose infusion results in a prompt decline in hepatic

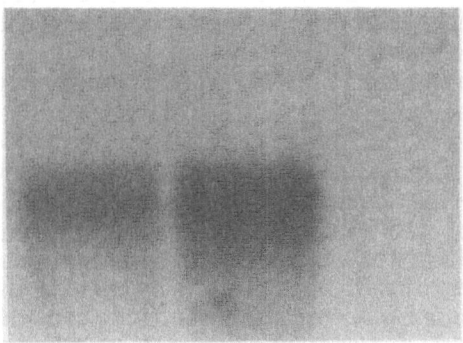

Figure 1. Northern blot of hepatic and renal RNA from normal mice. Fifteen micrograms of RNA from the liver (lane 1), kidney (lane 2) and brain (lane 3) were analyzed per lane. A mouse IGFBP-1 cDNA was used as the hybridization probe.

IGFBP-1 mRNA in the fasted rat[18]. Surprisingly, administration of insulin to fasted rats did not decrease hepatic IGFBP-1 expression suggesting that although insulin may be a major regulator of IGFBP-1 expression other factors are also important. When insulin was administered to overnight fasted rats it either had no significant effect on hepatic IGFBP-1 abundance or at some doses resulted in an increase in IGFBP-1 mRNA levels[18]. Under these circumstances, it is possible that the insulin induced hypoglycemia or some secondary response such as elevated corticosterone levels were able to mask the effects of insulin on IGFBP-1 expression. In culture rat hepatocytes and rat hepatoma cells, insulin suppresses IGFBP-1 expression even in the presence of low concentrations of dexamethazone[19,20]. Insulin suppression of rat IGFBP-1/CAT- reporter constructs can be demonstrated in gene transfer experiments using primary hepatocyte cultures[21].

IGFBP-1 expression is also increased in diabetic animals and this increase is reversed by insulin administration. Unterman and colleagues were the first to demonstrate increased concentrations of a low molecular weight IGF binding protein in sera from diabetic rats[22]. This binding protein ultimately proved to be IGFBP-1. In our studies an increase in IGFBP-1 mRNA was observed in both the liver and the kidney of streptozotocin diabetic rats[23]. The increase in message abundance was more marked in the liver than the kidney. IGFBP-1 mRNA abundance correlated with duration of the diabetes and also with the degree of hyperglycemia[23]. The increase in IGFBP-1 mRNA in diabetic rats can be explained at least in part by enhanced transcription [24]. Sequence analysis of the 5' flanking sequence of the rat IGFBP-1 reveal a number of potential insulin response elements (IRE). The most likely candidate region for insulin responsiveness appears to be in the region between -98 and -110. An IRE of the amylase/proximal phosphoenolpyruvate carboxykinase type is located between -111 to -102 upstream of the transcription start site in the rat IGFBP-1 gene (Table 1). This region confers insulin-responsiveness in experiments where CAT-reporter gene are transfected into hepatocytes [21]. An additional potential IRE of this type is located between -1286 and -1293. The core sequence of the glucagon type IRE is found in the rat IGFBP-1 5' flanking sequence at -289 to -294 and -364 to -369 however deletional analysis and transfection experiments with primary hepatocytes suggests that these regions are less important the -98 to -110 region[21,25].

Early reports referred to IGFBP-1 as a growth hormone independent binding protein[17]. However plasma IGFBP-1 levels have been demonstrated to be elevated in growth hormone deficient individuals by a number of investigators [26,27] and more careful investigation has revealed that hepatic expression of IGFBP-1 is downregulated by growth hormone in the rat. Since circulating levels of IGFBP-1 correlate inversely with plasma insulin concentrations and *in vitro* experiments with hepatic tissue have demonstrated that secretion of this protein is inhibited by insulin the effects of growth hormone deficiency and replacement were initially considered to be secondary to effects of growth hormone on insulin secretion [18]. This does not appear to be the case.

In the hypophysectomized rat hepatic and renal IGFBP-1 mRNA is increased and an increased in IGFBP-1 concentrations can be demonstrated in hepatic extracts and serum[6]. Growth hormone administration to hypophysectomized rats rapidly lowers hepatic IGFBP-1 mRNA levels. Furthermore, it is possible to demonstrate that the increase in hepatic IGFBP-1 mRNA is predominantly due to an increase in transcription rate and that growth hormone rapidly suppresses transcription [28]. A single injection of growth hormone to hypophysectomized rats reduced IGFBP-1 transcription rate to levels seen in the sham-operated control rats within 30 minutes [28].

Using mobility shift assays and DNAseI footprinting and nuclear extracts from hypophysectomized rats, pituitary intact rats and growth hormone treated hypophysectomized rats we have scanned the 1 kb 5' flanking region of the rat IGFBP-1 gene in search of growth hormone response elements. We have now identified a number of regions where the retardation pattern appears to be different between hypophysectomized and intact rats. The region -83 to -278 upstream from the transcription start site looks particularly promising. This region binds a factor or factors whose abundance is enhanced in hypophysectomized rats. The gel retardation pattern rapidly reverts to normal after growth hormone administration. These observations suggest that growth hormone is able to acutely downregulate expression of this gene and that downregulation of IGFBP-1 expression may involve removal of a DNA binding protein in the proximal 5' flanking region. Since few growth hormone responsive genes have been studied to date there is little information on the types of cis acting elements which may mediate growth hormone effects on transcription. Growth hormone does induce *c-fos* and *c-jun*[29] and thus interaction with the Ap1 response element TGGGTCA at -1101 to -1094 may be involved. The hepatic serine protease inhibitor, Spi 2.1 is induced by growth hormone and comparison of the sequence thought to be involved in the growth hormone responsiveness of this gene with the 5' flanking region of IGFBP-1 did not reveal any striking similarities [30].

Table 1. Potential cis-acting sequences in the rat IGFBP-1 gene which may confer hormone responsiveness.

```
Amylase/Proximal PEPCK
     IRE                          T A G T C A A A C A
          rIGFBP-1 (-111to -102)  a A G c a A A A C A
                   (-1293 to -1286)  T A acg C A A A ttA

Glucagon type IRE                  G/C G C C T G
          rIGFBP-1 (-294 to -289)    a G C C T G
                   (-369 to -364)  G   G C C T G

Consensus GRE   T/G G T A C A n n n T G T T C T
rIGFBP-1 (-353 to - 339)  G G T c t A A T C T G T T C T
         (-1078 to -1064)  T  G T c C A G G G T G T T C T

Consensus ERE        G G T C A n n n T G A C C
rIGFBP-1 (-825 to -812)  G c T C A G A G c G A C C
         (-837 to -824)  G G T C g C T C T G A g C
         (-1509 to -1496)  G G T C A G C C T G c a C
```

As discussed above, hepatic and renal expression of IGFBP-1 is also enhanced in the fasted rat where pituitary growth hormone secretion is suppressed. Furthermore, the increase in IGFBP-1 mRNA abundance is not seen if growth hormone is administered during the period of food-deprivation and growth hormone administration but not insulin administration to fasted rats results in a rapid decrease in IGFBP-1 mRNA levels prior to any increase in serum insulin concentration [18]. It is possible that the enhanced IGFBP-1 expression seen in the fasted and diabetic rat may result, not only from insulin deficiency but also from the functional growth hormone deficiency that occurs in these conditions in the rat.

This effect of growth hormone on IGFBP-1 expression does not appear to be mediated via IGF-I since administration of IGF-I at a dose of 75ug/100g body weight to hypophysectomized rats did not reduce hepatic IGFBP-1 mRNA levels. It has been difficult to demonstrate effects of growth hormone on IGFBP-1 expression in culture hepatocytes however there is one report that demonstrated that bovine growth hormone is able to suppress IGFBP-1 production by isolated rat hepatocytes[31]. However, other investigators have not been able to demonstrate a growth hormone effect. This may be due to the rapid loss of growth hormone receptor expression which occurs in primary culture. In addition, dexamethasone is routinely added to primary hepatocytes in order to enhance IGFBP-1 expression it may inhibit the effects of growth hormone. The growth hormone effect we have demonstrated *in vivo* does not appear to be due to secondary hyperinsulinism since there was no significant increase in insulin levels over the first few hours after growth hormone administration. Furthermore, as discussed above, insulin administration to fasted animals did not reduce hepatic IGFBP-1 mRNA levels.

Glucocorticoid excess in the rat is associated with increased serum IGF binding capacity. Using specific molecular probes we have demonstrated that hepatic expression of IGFBP-1 is enhanced by dexamethasone administration. A significant increase in hepatic IGFBP-1 mRNA levels were apparent as early as 1 hour following dexamethasone administration and the response was dose-dependent[32]. An accompanying increase in serum IGFBP-1 after dexamethasone was demonstrated by immuno-blotting. The mechanism whereby glucocorticoid excess increase IGFBP-1 mRNA levels may involve both stability of messenger RNA and increased transcription. In our in vivo experiments we were unable to detect an effect of dexamethasone on IGFBP-1 transcription rate in hepatic nuclei from rats which had been pretreated for 1 hour with dexamethasone whereas in H4-II-E rat hepatoma cells dexamethasone enhanced IGFBP-1 transcription[33]. There are potential glucocorticoid response elements with only 1 or 2 base mismatchs from the consensus glucocorticoid response element at -353 to -339 and at -1078 to -1064 upstream of the transcription start-site (Table 1). There may be differences in regulation between the rodent and human gene since dexamethasone has been reported to inhibited production of IGFBP-1 in human fetal liver explants [17].

In the uterus but not in hepatic or renal tissue estrogen downregulates IGFBP-1 expression[34]. A cyclical variation in uterine IGFBP-1 expression is also seen throughout the estrus cycle in the rat suggesting that the effects of estrogen on IGFBP-1 expression are physiologically important. Three potential estrogen response elements are present in the IGFBP-1 gene between 0.8 and 1.5 kb upstream of the transcription start site (Table 1). The functional importance of these sequences in estrogen responsiveness remains to be determined.

These observation support the notion that multiple hormonal factors are involved in the regulation of IGFBP-1 expression in the rat. Expression of IGFBP-1 is enhanced in many situation associated with growth retardation and hormonal regulation of IGFBP-1 expression appears to be the inverse of that of IGF-I expression. In the uterus, the major growth stimulus is estrogen which enhances expression of IGF-I but inhibits expression of IGFBP-1[34]. Thus, although IGFBP-1 appears to be a relatively minor component of the total serum IGF binding capacity it may well be an important component of the overall growth response.

DEVELOPMENT OF TRANSGENIC MOUSE MODELS

We have started to develop transgenic mouse models in an attempt to understand the physiological role of IGFBP-1 in growth and development. Our initial experiments involve the over-expression of IGFBP-1 using a rat IGFBP-1 cDNA driven by the SV-40 promoter. This construct produces high levels of IGFBP-1 when transfected into COS-1 cells [35]. Out of a total of 10 transfers of 25-30 micro-injected embryos four male and two female transgenic founders were obtained. The average litter size was 5 pups (Table 2). Three of these mice were breed with non-transgenic mice. Approximately 50% of the offspring of each of the matings carried the transgene. Some of the F1 transgenic demonstrated a modest elevation IGFBP-1 mRNA in the kidney (5.1±1.5 fold above control) but no significant increase in IGFBP-1 mRNA levels in other tissues examined such as the liver, brain and testes (Fig. 2).

TABLE 2. Constructs used to generate IGFBP-1 transgenic mice.

Construct[a]	Transfers[b]	Litter size[c]	No litter[d]	Transgenics
PSVCSL	10	5.0 ±1.9	2	2F/4M
PGKK1	9	1.7 ± 1.2	6	nil (2 dead)
MTK1	5	4.0 ± 3.6	3	not tested (1 dead)
BPCAT	12	6.9 ± 2.7	3	5F/2M
BPGAL	10	6.3 ± 2.6	4	2M

a. The construct used were; PSVCSL, a full-length rat IGFBP-1 cDNA driven by the SV-40 promoter; PGKK1, a 5 kb BamH1-EcoR1 fragment of the rat IGFBP-1 gene driven by the mouse phosphoglycerate kinase promoter; MTK1, a 5 kb BamH1-EcoR1 fragment of the rat IGFBP-1 gene driven by the mouse metalothionine-1 promoter; BPCAT, a 1.7 kb Pst1-BamH1 fragment of the 5' flanking DNA of the rat IGFBP-1 gene upstream of the bacterial chloramphenicol transferase reporter gene; BPGAL, a 1.7 kb Pst1-BamH1 fragment of the 5' flanking DNA of the rat IGFBP-1 gene upstream of the b galactosidase reporter gene. b. Between 25-35 micro-injected embryo were transferred into foster mothers. c. The mean ± SEM number of pups per litter excluding those embryo transfers which did not yield any pups.. d. Technically satisfactory transfers which yielded no offspring.

When sera from these animals were analyzed by ligand blotting the pattern obtained was similar to the non-transgenic wild-type mice and there appeared to be no consistent increase the IGFBPs which migrated in the 30 kD range (Fig. 3). In these F1 litters, litters derived from crossing two founders and litters from transgenic F1 mice backcrossed with founders there was no marked difference in litter size, sex ratio, birth weight or weight gain (Fig. 4). Thus although there appeared to be slightly higher levels of IGFBP-1 mRNA in kidneys from these transgenic animals we were not able to confirm an increase in synthesis of IGFBP-1 protein and there appeared to be no gross difference in the phenotype of the mice carrying the transgene. The failure to obtain significant over-expression of IGFBP-1 could be attributable to the choice of the SV-40 promoter which has subsequently shown to direct expression of transgenes in very limit tissues or the fact that a cDNA rather than a genomic fragment was used.

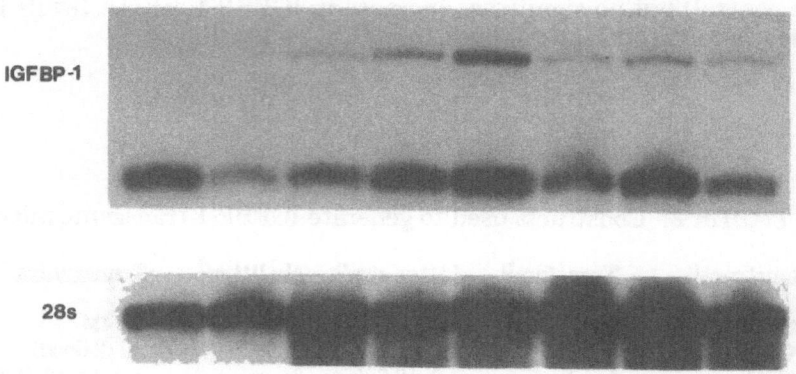

Figure 2. Northern blot of renal RNA from normal wild type mice (lane 1 and 2) and transgenic mice (lanes 3 to 8). Fifteen micrograms of RNA was analyzed per lane. A rat IGFBP-1 cDNA was used as the hybridization probe.

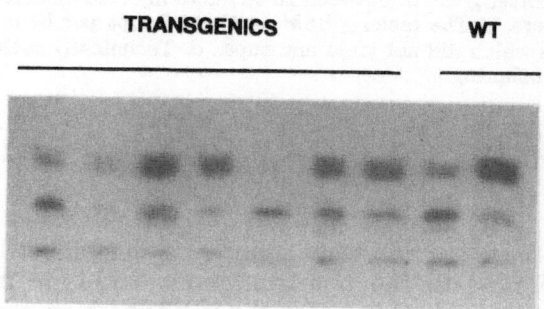

Figure 3. Ligand blot of serum from normal wild type mice and transgenic mice. Three microliters of serum was analyzed per lane.

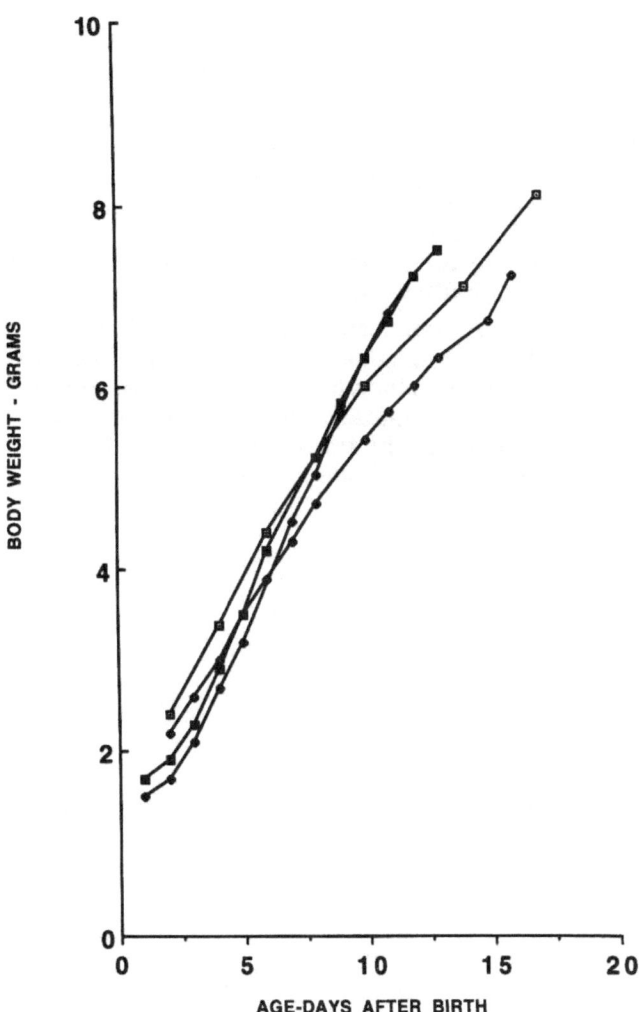

Figure 4. Body weight gain in mice from three transgenic litters and a non transgenic control litter (◇).

Transgenes were constructed using genomic fragments containing the coding region of rat IGFBP-1. A 5kb BamH1-EcoR1 fragment containing the entire coding region of the rat IGFBP-1 gene, the 3' untranslated sequence, the 5' untranslated region and 78 bp of 5' flanking DNA was inserted downstream of the mouse metalothionine promoter (MTK1) or the mouse phosphoglycerate kinase promoter (PGKK1). When embryos which had been micro-injected with these constructs were transferred into foster mothers there was a high number of apparently successful transfers which yielded no litters and litter size was markedly reduced (Table 2). Approximately 60 percent of successful transfers yielded no litters compared with a rate of 10-25% for a variety of other transgenes studied in this laboratory. In addition, litter size was small and 3 phenotypically normal mice died at or about the time of birth. These mice were not genotyped. To date we have not been successful in obtaining transgenic mice with the PGKK1 construct. We have yet to determine the genotype of the 12 mice obtained from embryos micro-injected with the MTK1 construct.

To study developmental regulation of IGFBP-1 expression constructs were made where 1.7 kb of IGFBP-1 5' flanking DNA was cloned upstream of the chloramphenicol acetyl transferase or β-galactosidase gene. Embryos micro-injected with these transgenes resulted in higher yields viable offspring and to date 9 transgenic founder mice with these reporter genes have been obtained (Fig. 5). These animals are currently being breed to obtain mouse strains and if they express the transgene they should prove to be useful models to examine tissue and cell specific expression and developmental regulation.

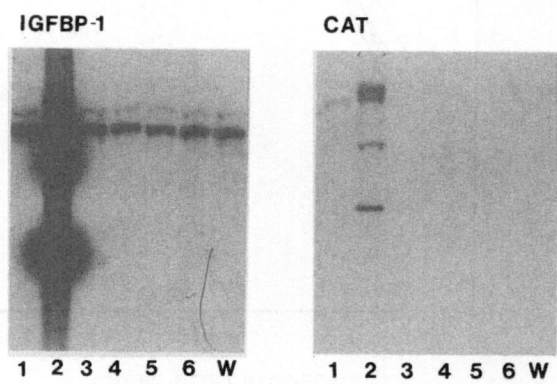

Figure 5. Southern blot of mouse tail genomic DNA from two groups of transgenic mice. The panel on the right is from a litter of mice where the PSVCSL construct was micro-injected whereas the panel on the left is a litter where the BPCAT construct was used. In each case DNA from a non transgenic mouse (W) has been included. The probes used were a rat IGFBP-1 cDNA and the pBPCAT plasmid respectively.

CONCLUSIONS AND FUTURE DIRECTIONS.

Expression of IGFBP-1 correlates inversely with IGF-I expression in may conditions such as diabetes, starvation, growth hormone deficiency and glucocorticoid excess. Since IGFBP-1 can at least under in vitro assay conditions inhibit IGF-I action, these observations provide circumstantial evidence that IGFBP-1 is intimately involved in the growth process. To provide definitive proof for this hypothesis we have started to develop transgenic mouse strains to examine the effects of exogenous IGFBP-1 on weight gain and skeletal growth. To date we have had limited success in obtaining transgenic mice which overexpress IGFBP-1 at significant levels. In the transgenic mice so far generated the level of expression of the transgene has been quite low and there appears to be no gross disturbance in body growth. Using IGFBP-1 genomic constructs and strong promoters which are expressed in a wide variety of tissues we have noted a marked reduction in litter size and it is possible that over-expression of IGFBP-1 may have deleterious effects on early fetal development. In this regard it is important bear in mind that elevated fetal expression of IGFBP-1 is seen in a variety of different models of intrauterine growth retardation[35,36]. We are now examining the transgenic animals containing other IGFBP-1 constructs. We believe this approach of over expressing IGFBP-1 together with homologous recombination gene knock out experiments will provide insights into the physiological role of the IGFBP-1 in normal growth and development.

ACKNOWLEDGMENTS

The work described in this review was supported by grants from the Medical Research Council of Canada and the National Cancer Institute of Canada. L.J.M is the recipient of an endowed research professorship in metabolic diseases.

REFERENCES

1. J.L. Martin and R.C. Baxter, Insulin-like growth factor-binding proteins from human plasma. *J. Biol. Chem.* 261:8754 (1986).
2. J. Zapf, C. Hauri, M. Waldvogel, E. Futo, H. Hasler, K. Benz, H.P. Guler, C. Schmid and E.R. Froesch, Recombinant human insulin-like growth factor-I induces its own specific carrier protein in hypophysectomized and diabetic rats.
    Proc. Natl. Acad. Sci. U.S.A. 86:3813 (1989).
3. S.M. Donovan, Y. Oh, H. Pham and R.G. Rosenfeld, Ontogeny of serum insulin-like growth factor binding proteins in the rat.
    Endocrinology 125:2621 (1989).
4. J.A. Romanus, J.E. Terrell, Y.W.H. Yang, S.P. Nissley and M.M. Rechler, Insulin-like growth factor carrier proteins in neonatal and adult rat serum are immunologically different: demonstration using a new radioimmunoassay for the carrier protein from BRL-3A rat liver cells. *Endocrinology* 118:1743 (1986).
5. S.L.S. Drop, G. Valiquette, H.J. Guyda, M.T. Corvol, and B.I. Posner, Partial purification and characterization of a binding protein for insulin-like activity (ILAs) in human amniotic fluid: a possible inhibitor of insulin-like activity. *Acta Endocrinol. (Copenh.)* 90:505 (1979).
6. Y. Gong, G. Ballejo, B. Alkhalaf, P. Molnar, L.C. Murphy and L.J. Murphy, Phorbol esters differentially regulate expression of insulin-like growth factor binding proteins in human endometrial carcinoma cells. *Endocrinology* 131:2747 (1992).
7. J. Zapf, M. Waldvogel and E.R. Froesch, Binding of nonsuppressible insulinlike activity to human serum. *Arch. Biochem. Biophys.* 168:638 (1975).

8.   J.S.M. De Mellow and R.C. Baxter, Growth hormone dependent insulin-like growth factor binding protein both inhibits and potentiates IGF-I stimulated DNA synthesis in human skin fibroblasts. *Biochem. Biophys. Res. Commun.* 156:199 (1988).

9.   O. Ritvos, T. Ranta, J. Jalkanen, A-M. Suikkari, R. Voutilainen, H. Bohn, and E-M Rutanen, Insulin-like growth factor (IGF) binding protein from human decidua inhibits the binding and biological action of IGF-I in cultured choriocarcinoma cells. *Endocrinology* 122:2150 (1988).

10.   P.E. Walton, R. Gopinath and T.D. Etherton, Porcine insulin-like growth (IGF) binding protein blocks IGF-I action on porcine adipose tissue. *Proc. Soc. Exp. Biol.* 190:315 (1989).

11.   R.G. Elgin, W.H. Busby, and D.R. Clemmons, An insulin-like growth factor (IGF) binding protein enhances the biological response to IGF-I. *Proc. Natl. Acad. Sci. U.S.A.* 84:3254 (1987).

12.   R.A. Frost and L. Tseng, Insulin-like growth factor-binding protein-1 is phosphorylated by cultured human endometrial stromal cells and multiple protein kinases in vitro, *J. Biol. Chem.* 266:18082 (1991).

13.   A-M. Suikkari, V.A. Koivisto, R. Koistinen, M. Seppala and H. Yki-Jarvivnen, Dose-response characteristics for suppression of low molecular weight plasma insulin-like growth factor-binding protein by insulin. *J. Clin. Endocrinol. Metab.* 68:135 (1989).

14.   A-M. Suikkari, V.A. Koivisto, E-M. Rutanen, H. Yki-Jarvinen, S-L. Karonen and M. Seppala, Insulin regulates the serum levels of low molecular weight insulin-like growth factor-binding protein. *J. Clin. Endocrinol. Metab.* 66:266 (1988).

15.   W.H. Busby, D.K. Snyder and D.R. Clemmons, Radioimmunoassay of a 26,000-dalton plasma insulin-like growth factor-binding protein: control by nutritional variables. *J. Clin. Endocrinol. Metab.* 67:1225 (1988).

16.   R.C. Baxter and C.T. Cowell, Diurnal rhythm of growth hormone-independent binding protein for insulin-like growth factors in human plasma. *J. Clin. Endocrinol. Metab.* 65:432 (1987).

17.   M.S. Lewitt and R.C. Baxter, Regulation of growth hormone-independent insulin-like growth factor-binding protein (BP-28) in cultured human fetal liver explants. *J. Clin. Endocrinol. Metab.* 69:246 (1989).

18.   L.J. Murphy, C. Seneviratne, P. Moreira and R. Reid, Enhanced expression of insulin-like growth factor binding protein-I in the fasted rat: The effects of insulin and growth hormone administration. *Endocrinology* 128:689 (1991).

19.   B.C. Villafuerte, S. Goldstein, L.J. Murphy and L. S. Phillips, Nutrition and somatomedin XXV. Regulation of insulin-like growth factor binding protein-1 in primary cultures of normal rat hepatocytes. *Diabetes* 40:837 (1991).

20.   T.G. Unterman, D.T. Oehler, L.J. Murphy and R.S. Lacson, Multihormonal regulation of insulin-like growth factor binding protein-1 in rat H4IIE hepatoma cells: the dominant role of insulin. *Endocrinology* 128:2693 (1991).

21.   P.M. Thule, C.K. Seneviratne, C.I. Pao, E.M. Marino-Rodriguez, C.S. Noel, L.J. Murphy and D.G. Robertson, Characterization of the rat IGFBP-1 promoter: Regulation by insulin. *Proc. 75th Annual Meeting of the Endocrine Society*, Las Vagas, (1993).

22.   T.G. Unterman, D.T. Oehler, R.E. Becker, Identification of a type-I insulin-like growth factor binding proetin (IGFBP) in serum from rats with diabetes mellitus. *Biochem. Biophys. Res. Commun.* 163:882 (1989).

23. J-M. Luo and L.J. Murphy, Differential expression of insulin-like growth factor-I and insulin-like growth factor binding protein-1 in the diabetic rat. *Mole. Cell. Biochem.* 103:41 (1991).

24. G.T. Ooi, L.Y-H. Tseng and M.M. Rechler, Post-transcriptional regulation of insulin-like growth factor binding protein-2 mRNA in diabetic rat liver.*Biochem. Biophys. Res. Commun*.189:1031 (1992).

25. T.G. Uterman, R.G. Lacson, E. McGary, C. Whalen, C. Purple and R.G. Goswami, Cloning of the rat insulin-like growth factor binding protein-1 gene and analysis of its 5' promoter region. *Biochem. Biophys. Res. Commun*. 185:993 (1992).

26. S.L.S. Drop, D.J. Kortleve, H.J. Guyda and B.I. Posner, Immunoassay of a somatomedin-binding protein from human amniotic fluid: levels in fetal, neonatal and adult sera. *J. Clin. Endocrinol. Metab.* 59:908 (1984).

27. G. Povoa, A. Roovete and K. Hall, Cross-reaction of serum somatomedin-binding protein in a radioimmunoassay developed for somatomedin-binding protein isolated from amniotic fluid. *Acta Endocrinol. (Copenh.)* 107:56 (1984).

28. C. Seneviratne, J. Luo and L. J. Murphy, Transcriptional regulation of insulin-like growth factor binding protein-1 expression by growth hormone. *Molecular Endocrinology* 4:1199 (1990).

29. M.C. Slootweg, R.P. de Groot, M.P. Herrman-Erlee, I. Koornneef, W. Kruijer and Y.M. Kramer, Growth hormone induces expression of c-jun and jun B oncogenes and employs a protein kinase C signal transduction pathway for induction of c-fos oncogene expression. *J. Mol. Endocrinol.* 6:179 (1991).

30. J-B. Yoon, S.A. Berry, S. Seelig and H.C. Towle, An inducible nuclear factor binds to a growth hormone-regulated gene. *J. Biol. Chem.* 265:19947 (1990).

31. Z. Kacchra, C. Yannopoulos, I. Barash, H.J. Guyda, L.J. Murphy and B.I. Posner, The differential regulation by glucagon and growth hormone of IGF-I and IGF binding proteins (IGF-BPs) in cultured rat hepatocytes. *Proc. 72th Annual Meeting of the Endocrine Society*, Abstract 1133, 1990.

32. J-M. Luo, R.E. Reid and L.J. Murphy, Dexamethasone increases hepatic insulin-like growth factor binding protein-1 (IGFBP-1) mRNA and serum IGFBP-1 concentrations in the rat. *Endocrinology* 127:1456 (1990).

33. C.C. Orlowski, G.T. Ooi and M.M. Rechler, Dexamethasone stimulates transcription of the insulin-like growth factor-binding protein-1 gene in H4-II-E rat hepatoma cells. *Mole. Endocrinol.* 4:1592 (1990).

34. L. J. Murphy. The uterine insulin-like growth factor system, in: "Modern concepts of insulin-like growth factors," E.M. Spencer, ed., Elsevier, New York pp 275-285, (1991).

35. L.J. Murphy, C. Seneviratne, G. Ballejo, F. Croze, T.G. Kennedy, Identification and characterization of a rat decidual insulin-like growth factor binding protein cDNA. *Mole. Endocrinol.* 4:329 (1990).

36. 37. intrauterine growth

LIMITED PROTEOLYSIS OF INSULIN-LIKE GROWTH FACTOR BINDING PROTEIN-3 (IGFBP-3) : A PHYSIOLOGICAL MECHANISM IN THE REGULATION OF IGF BIOAVAILABILITY

Michel Binoux, Claude Lalou, Claudine Lassarre,
Christiane Blat, and Paul Hossenlopp

Unité de Recherches sur la Régulation de la Croissance, Inserm
U.142, Hôpital Saint Antoine, 75571 Paris Cédex 12, France

INTRODUCTION

Enzymes are known to play a role in activating or releasing growth factors like the FGFs and TFGß from the extra-cellular matrix (1, 2). When first discovered, *in vivo* proteolysis of serum IGFBPs, and particularly IGFBP-3, by pregnancy-associated protease(s) was naturally suspected to be physiologically significant (3, 4). Although the structural alteration of IGFBP-3 and its loss of affinity for the IGFs were clearly demonstrated by Western ligand blotting, immunoblotting and competitive binding studies (3 - 5), some doubt was expressed as to its physiological nature in view of the harsh experimental conditions used (SDS, acidification). Both IGFBP-3 and the acid-labile (α) subunit were found in the 150-kDa material and ternary complex formation was found to be possible with acidified pregnancy serum in the presence of α-subunit (6). Actually, in our initial work on pregnancy serum, we showed that the 150-kDa complex was not disrupted in pregnancy serum analysed by gel filtration at neutral pH, although there were also small IGFBP-3 fragments in the fractions containing low molecular weight proteins (3). It therefore seemed possible that the stability of the 150-kDa complex would be diminished and the IGFBP-3 functionally altered, even if proteolysis remained limited.

Our new data indicate that during pregnancy the ternary complex is functionally altered, with repercussions on the bioavailability of the IGFs. In addition, IGFBP-3 proteolysis has been found to occur in the normal state, its extent varying with GH/IGF status, which further supports the physiological significance of the phenomenon.

Current Directions in Insulin-Like Growth Factor Research,
Edited by D. LeRoith and M.K. Raizada, Plenum Press, New York, 1994

293

STRUCTURAL AND FUNCTIONAL ALTERATIONS OF SERUM IGFBP-3 DURING PREGNANCY

1.   In pregnancy serum, ligand blotting reveals a strong reduction or near disappearance of the characteristic IGFBP-3 42-39-kDa doublet, which is confirmed by immunoblotting using an antibody raised against recombinant IGFBP-3. The major proteolytic fragment of IGFBP-3 appears as a broad, dense band at 30 kDa (Figure 1). This fragment fails to be detected by ligand blotting in native serum, although it does appear in gel filtration eluates where it is concentrated in the 150-kDa material (3). The structural alteration of IGFBP-3 therefore causes a significant loss of binding activity.

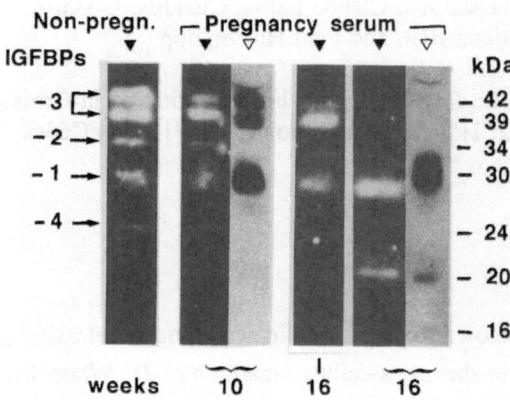

Figure 1 : Western-ligand blot (▼) and immunoblot (anti-IGFBP-3) (▽) analysis of sera from pregnant women

2.   Competitive binding experiments performed with IGFBP-3 isolated from pregnancy serum have shown that, compared with IGFBP-3 from normal serum, there is a loss of affinity which is 2-fold for IGF-II and 15-fold for IGF-I (unpublished). The markedly larger loss of affinity for IGF-I appears to be related to the greater propensity of the 42-kDa form to be proteolysed. In ligand blot studies of large numbers of pregnancy serum samples, a band corresponding to the 39-kDa form was sometimes still visible, with greater or lesser intensity, whereas the 42-kDa form was barely or not detectable (Figure 1). Furthermore, earlier studies with normal serum had indicated that the 39-kDa form has a preferential affinity for IGF-II, and the 42-kDa form, a better affinity for IGF-I (7). These findings would explain the selective loss of affinity for IGF-I of proteolysed IGFBP-3.

3.   The outcome of the loss of affinity for IGFs, and particularly IGF-I, is an increased rate of dissociation. In kinetics of dissociation studies at 37°C, more than 50%

of IGF-I is dissociated from proteolysed IGFBP-3 after only 20 minutes, whereas with intact IGFBP-3 this level of dissociation is not yet reached after 4 hours (Figure 2).

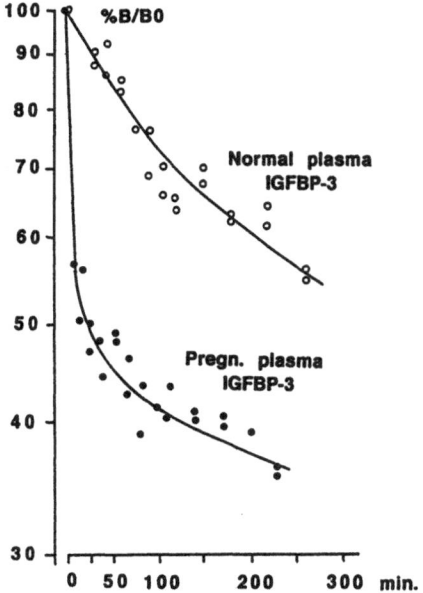

Figure 2 : Kinetics of dissociation at 37°C of ^{125}I-IGF-I bound to IGFBP-3 from normal and pregnancy serum

4.   The structural alteration of IGFBP-3 in native serum was also revealed by competitive binding studies using the same antibody as that used for immunoblotting. IGFBP-3's affinity for the antibody in normal serum proved to be 10 times that of the IGFBP-3 in pregnancy serum, which means that the protease-induced alteration of IGFBP-3 is detectable by the antibody within the ternary complex in native serum. The apparent serum concentrations of IGFBP-3 as measured by RIA would therefore differ depending upon the affinity with which the antibody used recognizes the intact and proteolysed forms of the protein (8).

5.   IGFBP-3's differential losses of affinity for IGF-I and IGF-II would result in redistribution of serum IGFs among the pools of 150-kDa and (binary) 40-kDa complexes and free IGFs. When pregnancy and non-pregnancy serum samples are incubated at 37°C, then gel filtered at neutral pH at room temperature, IGF-I and IGF-II assays of the three pools reveal significant changes in their relative proportions (Table I). In pregnancy serum, the 150-kDa complex has a lower percentage of IGF-I, but a higher percentage of IGF-II. The latter is however lower in the 40-kDa complex. The percentage of free IGF-I more than doubles, whereas that of free IGF-II either remain stable or tend to drop (Table I). From data on the half-lives of IGFs in man (9), we have calculated that the production rate of free IGF-I would be almost quadrupled in pregnant women. It is therefore evident that the bioavailability of IGF-I is increased during pregnancy, in response to the enhanced metabolic and growth requirements accompanying placental and foetal development. This increased availability of IGF-I would also account for certain acromegaly-like features observed during pregnancy.

Table I : Changes in the distribution of serum IGFs during pregnancy and estimation of production rates of free IGFs. p ≤ *0.02, **0.01, ***0.005. Normal (n = 5) ; Pregnant women (n = 7).

	IGF-I		IGF-II				Production rates mg/day	
	Normal	Preg.	Normal	Preg.	Free	Half-life	Normal	Preg.
% 150-kDa	78	69**	77	84.5*	IGF-I	12 min	2.95	11.4***
% 40-kDa	18	20	21	14*				
% Free	4	11***	2	1.5	IGF-II		6.95	5.26

6.   Figure 3 illustrates this pregnancy-associated increase in IGF-I bioavailability at the cellular level. Comparisons were made of the effects on DNA synthesis in chick embryo fibroblasts of two pools of normal and pregnancy serum mixed in such a way that each would contain identical concentrations of IGF-I and IGF-II. Pregnancy serum proved to have a stronger stimulatory effect than normal serum. This suggests that the enhancement of DNA synthesis is the result of increased dissociation of bound IGFs and would agree with C. Conover's recent observations that in bovine fibroblasts pretreated with IGFBP-3 the potentiation of IGF activity is related to proteolysis of cell-surface-associated IGFBP-3 (10).

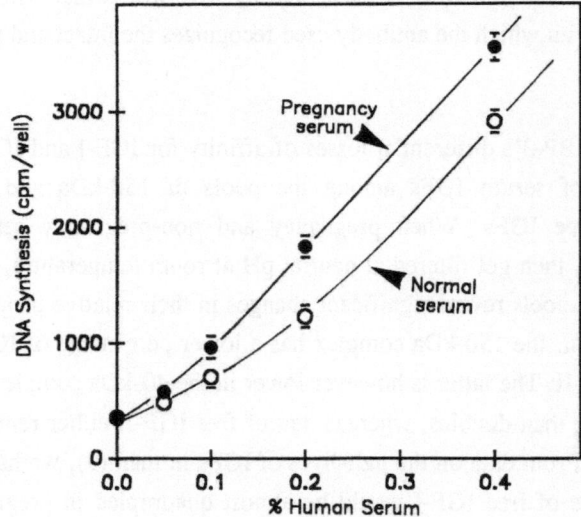

Figure 3 : Comparative effects of pregnancy and normal serum pools with identical concentrations of IGF-I and IGF-II on ¹⁴C-thymidine incorporation in chick embryo fibroblasts

IGFBP-3 PROTEOLYSIS IN THE NORMAL STATE AND AS RELATED TO GH/IGF-I STATUS

1.   With the development of the ultrasensitive technique of revealing immunoblots by chemiluminescence (ECL Western blotting detection system, Amersham, U.K.), we were able to demonstrate the occurrence of proteolysis of serum IGFBP-3 in the normal state and its detectability in all serum samples investigated. The SDS-PAGE profiles were similar to that of pregnancy serum, with a major, and usually the only visible, fragment migrating at 30 kDa. Estimated by densitometric scanning, which reflects the extent of *in vivo* proteolysis, the proportion of proteolysed IGFBP-3 was 30-40% in normal sera, and nearly 100% in third-trimester pregnancy sera.

Comparisons of lymph (which reflects the interstitium) and serum revealed greater proportions of 30-kDa proteolysed IGFBP-3 in the former than in the latter (Figure 4). In proteolysis assays using ^{125}I-IGFBP-3 as substrate, which give a notion of residual proteolytic activity at any given time *in vivo*, lymph had weak, and serum, negligible activity compared with that in pregnancy serum. Nevertheless, that in lymph was almost 8 times that in the corresponding serum. Moreover, the protease activity was suppressed

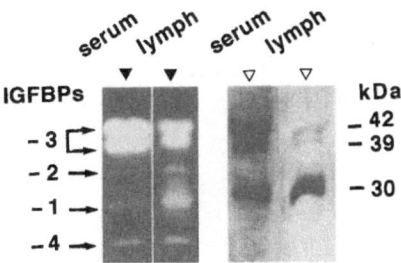

Figure 4 : Western ligand blot (▼) and immunoblot (anti-IGFBP-3) (▽) analysis of serum (3 μl) and lymph (15 μl) taken from the same normal subject.

by the same protease inhibitors as the pregnancy-associated protease(s). These findings confirm that there is a strong similarity between the IGFBP-3 proteases active in the normal state and during pregnancy. They also suggest that the initial sites of proteolysis are outside of the bloodstream and probably in the tissues (11).

2.   From comparisons by ligand- and immunoblotting of serum samples from GH-deficient, normal and acromegalic subjects, a parallel increase emerged in the density

of the 39-42-kDa doublet corresponding to intact IGFBP-3. Conversely, the proportions of proteolysed IGFBP-3 revealed by immunoblotting were seen to increase from acromegalic to normal to GH-deficient subjects (Figure 5 and Table II). This suggests that IGF-I may be involved in the regulation of either IGFBP proteases or their inhibitors, as has recently been reported for the plasmin system (12 - 14).

Functionally, this means that the regulation of protease activity would be adapted to enhance (in GH deficiency) or depress (in acromegaly) IGF-I availability as needed. However, recent observations in our laboratory indicate excessive proteolysis of IGFBP-3 in constitutionally tall children, which would suggest dysregulation of protease/antiprotease equilibrium (unpublished).

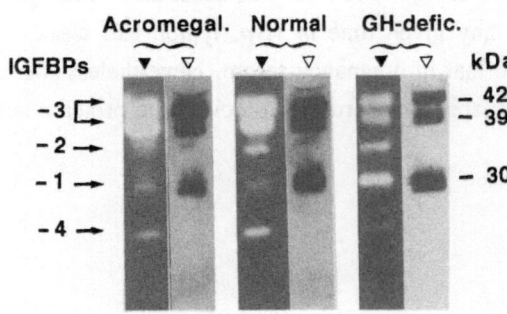

Figure 5 : Western ligand blot (▼) and immunoblot (anti-IGFBP-3) (▽) analysis of sera from normal, acromegalic and GH-deficient subjects.

3.   A further question raised by these data concerns interpretation of IGF-I and IGF-II levels in serum. To date, these have generally been considered as a reflection of liver production in view of the long half-life (12 to 15 hours) of the 150-kDa complexes, the major carriers of serum IGFs in healthy adults and adolescents. Our studies of pregnant women indicate that proteolysis of IGFBP-3 promotes dissociation of IGF-I (but not IGF-II) from the 150-kDa complexes, thereby reducing its half-life, although serum levels do not drop, which would suggest increased synthesis. Gestating rats, by contrast, exhibit a decrease in serum IGF-I levels in the latter half of pregnancy (15), although no change is observed in liver IGFBP-3 mRNA (16). It is therefore possible that the increase in IGF-II/IGF-I ratio in GH-deficient patients and the corresponding decrease in acromegalics (Table II), classically interpreted as meaning that IGF-II is less GH-dependent, in part reflect changes in the half-lives of IGF-I and IGF-II as a result of their redistribution

among the circulating pools, which itself is dependent on the synthesis of the different forms of IGFBP and the extent of their (particularly IGFBP-3's) proteolysis. Further research on metabolic clearance of the two IGFs and regulation of their synthesis in the liver as related to GH status will provide the answers to these questions.

Table II : Serum levels of IGF-I and IGF-II and percentage of proteolysed IGFBP-3 (mean ± SEM) in acromegalic, normal and GH-deficient subjects. The percentages were calculated from densitometry scanning of Western immunoblots.

	Acromegalic (n = 22)	Normal (n = 22)	GH-deficient (n = 18)
IGF-I ng/ml	1220 ± 52	300 ± 12	58 ± 7
IGF-II ng/ml	1346 ± 55	1318 ± 56	471 ± 56
IGF-II/IGF-I	1.16 ± 0.09	4.5 ± 0.3	8.9 ± 0.6
% proteolysed IGFBP-3	15 ± 2.9	37 ± 3.6	53 ± 3.3

CONCLUSION

The 150-kDa complexes, initially seen as a buffering system for IGFs, should essentially be considered as a reservoir, from which the flow of IGFs is controlled by serine protease activity. Both in the bloodstream and at the cellular level, proteolysis of the IGFBPs and consequently the changes in their affinities for the IGFs provide additional complexity in the modulation of IGF activity by which fine regulation of cell growth and metabolism is achieved.

ACKNOWLEDGEMENTS

This work was supported by the Institut National de la Santé et de la Recherche Médicale. We thank Berta Segovia for her technical assistance.

REFERENCES

1. Bashkin P, Doctrow S, Klagsbrun M, Svahn CM, Folkman J, Vlodavsky I. 1989. Basic Fibroblast growth factor binds to subendothelial extracellular matrix and is released by heparitinase and heparin-like molecules. Biochemistry 28:1737-1743.

2. Sato Y, Rifkin DB. 1989. Inhibition of endothelial cell movement by pericytes and smooth muscle cells : activation of a latent Transforming growth factor-ß1-like molecule by plasmin during co-culture. J Cell Biol 109:309-315.

3. Hossenlopp P, Segovia B, Lassarre C, Roghani M, Bredon M, Binoux M. 1990. Evidence of enzymatic degradation of insulin-like growth factor binding proteins in the "150 K" complex during pregnancy. J Clin Endocrinol Metab 71:797-805.

4. Giudice LC, Farrell BM, Pham H, Lamson G, Rosenfeld RG. 1990. Insulin-like growth factor binding proteins in maternal serum throughout gestation and in the puerperium : effects of a pregnancy-associated serum protease activity. J Clin Endocrinol Metab 71:806-816.

5. Binoux M, Hossenlopp P, Lassarre C, Segovia B. 1991. Degradation of IGF binding protein-3 by proteases: physiological implications. In : Modern Concepts of Insulin-Like Growth Factors. Proc. 2nd. Int. Symp. Insulin-Like Growth Factors/Somatomedins, Spencer EM (ed), Elsevier, pp 329-336.

6. Suikkari AM, Baxter RC. 1992. Insulin-like growth factor-binding protein-3 is functionally normal in pregnancy serum. J Clin Endocrinol Metab 74:177-183.

7. Hardouin S, Hossenlopp P, Segovia B, Seurin D, Portolan G, Lassarre C, Binoux M. 1987. Heterogeneity of insulin-like growth factor binding proteins and relationships between structure and affinity. 1. Circulating forms in man. Eur J Biochem 170:121-132.

8. Lassarre C, Lalou C, Perin L, Binoux M. Further evidence for structural alteration of insulin-like growth factor binding protein-3 as a result of limited proteolysis *in vivo*. Agreement between radioimmunoassay and ligand blotting data (submitted).

9. Guler HP, Zapf J, Schmid C, Froesch ER. 1989. Insulin-like growth factors I and II in healthy man. Estimations of half-lives and production rates. Acta Endocrinol (Copenh) 121:753-758.

10. Conover CA. 1992. Potentiation of Insulin-like Growth Factor (IGF) action by IGF-Binding Protein-3 : studies of underlying mechanism. Endocrinology 130:3191-3199.

11. Lalou C, Binoux M. 1993. Evidence that limited proteolysis of insulin-like growth factor binding protein-3 (IGFBP-3) occurs in the normal state outside of the bloodstream. Regulatory Peptides (in press).

12. Campbell PG, Novak JF, Wines K, Walton PE. 1993. Localization of plasmin activity on osteosarcoma cells : cell surface proteolysis of Insulin-like Growth Factor Binding Proteins. Growth Regulation 3:95-98.

13. Lalou C, Silve C, Segovia B, Binoux M. 1992. The role of plasminogen activator/plasmin system in regulated proteolysis of insulin-like Growth Factor Binding Proteins (IGFBPs) produced by human osteosarcoma cells in culture. 2nd Int Workshop on IGF Binding Proteins, Opio (France), Aug 27-30, 1992 (Abstract).

14. Schneider DJ, Sobel BE. 1991. Augmentation of synthesis of plasminogen activator inhibitor type 1 by insulin and insulin-like growth factor type 1 : implications for vascular disease in hyperinsulinemic states. Proc Natl Acad Sci USA 88:9959-9963.

15. Davenport ML, Clemmons DR, Miles MV, Camacho-Hubner C, D'Ercole J, Underwood LE. 1990. Regulation of serum insulin-like growth factor-I (IGF-I) and IGF binding protein during pregnancy. Endocrinology 127:1278-1286.

16. Donovan SM, Giudice LC, Murphy LJ, Hintz RL, Rosenfeld RG. 1991. Maternal insulin-like growth factor-binding protein messenger ribonucleic acid during rat pregnancy. Endocrinology 129:3359-3366.

EFFECTS OF INSULIN-LIKE GROWTH FACTOR I (IGF-I) ADMINISTRATIONS ON SERUM IGF BINDING PROTEINS (IGFBPS) IN PATIENTS WITH GROWTH HORMONE DEFICIENCY

Naomi Hizuka[1], Kazue Takano[1], Kumiko Asakawa-Yasumoto[1]
Izumi Fukuda[1], Tomoko Suzuki[1], Hiroshi Demura[1], Chika Shimojoh[2]
and Kazuo Shizume[2]

[1] Department of Medicine, Institute of Clinical Endocrinology, Tokyo
Women's Medical College, Tokyo, 162, Japan
[2] Research Laboratory, The Foundation for Growth Science, Tokyo, 162,
Japan

INTRODUCTION

Recently, with an availability of recombinant human insulin-like growth factor I (IGF-I), IGF-I has been tried for various clinical therapeutic applications. IGFs are found in blood complexed to specific binding proteins (IGFBPs), and these IGFBPs might influence the biological effect of administered IGF-I. Therefore, it is interesting to study the responses of serum IGFBPs to IGF-I administrations. We have previously reported that serum IGFBP-1 levels increased after single sc injection of IGF-I, and serum IGFBP-2 levels increased and decreased serum IGFBP-3 and acid labile subunit levels after repetitive IGF-I injections in normal subjects[1]. We report here the responses of IGFBPs to single and repetitive IGF-I sc injections in patients with growth hormone (GH) deficiency.

MATERIALS AND METHODS

IGF-I Preparation

Recombinant human IGF-I was kindly provided by Fujisawa Pharmaceutical Co. Ltd. Osaka. The preparation of IGF-I was synthesized by DNA technology as described[2]. IGF-I was dissolved in physiological saline at a concentration of 0.6% just before use.

Subjects

Five male patients with GH deficiency (20 - 31yrs) participatied in this study (Table 1).

Current Directions in Insulin-Like Growth Factor Research,
Edited by D. LeRoith and M.K. Raizada, Plenum Press, New York, 1994

301

Table 1. Clinical findings of five patients with GH deficiency.

Patient No.	Sex	Age	Etiology	Hormone Deficiency
1	M	31	Idiopathic	GH, LH, FSH
2	M	27	Idiopathic	GH, LH, FSH
3	M	20	Sept-opticdysplasia	GH, LH, FSH, TSH
4	M	21	Germinoma	GH, LH, FSH, TSH, ACTH, AVP
5	M	22	Idiopathic	GH, LH, FSH, TSH, ACTH

These patients had ceased GH therapy more than three years ago and they have had replacement therapy for other pituitary hormone deficiency. Informed consent was obtained from each subject and the experimental protocol was approved by Human Subjects Investigation Committee of Tokyo Women's Medical College.

Administration of IGF-I

Four patients with GH deficiency (GHD: #1, 2, 3 and 4) received sc administration of IGF-I at a dose of 0.05 - 0.01 mg/kg after overnight fasting (Table 1). A patient (#1) underwent three experiments (0.05, 0.075 and 0.1 mg/kg IGF-I administration, respectively). Blood samples were taken before and after the sc administration. The subjects for 0.05 or 0.1 mg/kg administration and for 0.075 mg/kg administration had lunch at 6 h and 4h after IGF-I administration, respectively. The subjects had dinner at 10 h after IGF-I administration.

Two patients with GHD (#3 and 5) received sc injection of IGF-I (0.1 mg/kg) once a day at 30 min after breakfast for consecutive 7 days[3]. Lunch, snack and dinner were taken at 4, 6 and 10 h after IGF-I administration, respectively. Blood samples were taken before and after the first and the seventh IGF-I administration. The results for serum IGF-I, blood glucose and biochemical data in this study have been reported[3].

Assays

IGFBPs Analysis

Serum IGFBPs were analyzed by Western ligand blotting using the method of Hossenlopp et al[4]. Briefly, serum (2 μl) were electrophoresed on 12% SDS polyacrylamide gel under nonreducing conditions. The size fractionated proteins were electroblotted onto nitrocellulose sheet. The nitrocellulose sheet was treated with Nonidet P-40, BSA and Tween 20, and incubated with a mixture of ^{125}I-IGF-I and ^{125}I-IGF-II (1x10[6] cpm each) for 2 days. After extensive washing of the nitrocellulose, the IGFBPs were detected by autoradiography. The autoradiographic intensity of the bands of IGFBPs were determined using densitometer.

Serum IGFBP-1 was measured by enzyme immunoassay (Medix Biochemica Ab, Kauniainen, Finland).

Serum IGFBP-3 were analyzed by Western immunoblot. Serum samples were electrophoresed on 12% SDS-acrylamide gel under non-reducing condition. The size-fractionated proteins were electroblotted onto nitrocellulose sheet. The sheet was blocked with 5% (w/v) skim milk, and then incubated with anti-hIGFBP-3 antibody

(α-IGFBP-3-3g1)[5], kindly provided by Dr. R G Rosenfeld. After extensive washing, the sheet was incubated with HRP-conjugated anti-rabbit IgG, and then IGFBP-3-anti-IGFBP-3 anitbody complexes were detected with ECL (Enhanced Chemiluminescence) system (Amersham). To validate this method, sera from normal subject , patients with acromegaly, GHD, liver cirrhosis, chronic renal failure, anorexia nervosa, pregnant woman, and short child with malnutrition were analyzed.

Measurements for IGF-I, Insulin, C-peptide and Glucose

Plasma IGF-I level was measured using acid-ethanol extracted plasma by radioimmunoassay as described previously[6].

Serum C-peptide levels were measured by commercially available RIA kit: minimal detectable level was 0.1 ng/ml. Blood glucose levels were measured with Autoanalyzer.

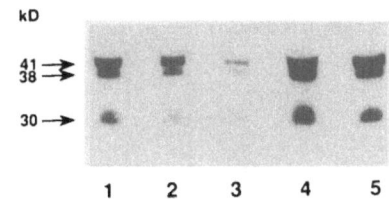

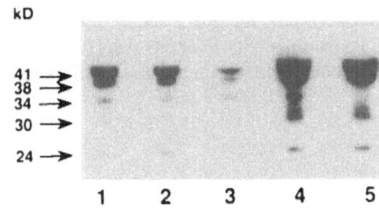

Figure 1. Western immunoblot of IGFBP-3 and Western ligand blot in serum from normal subject (1), patients with GHD (2, 3), and patients with acromegaly (4, 5).

RESULTS

Serum IGFBP-3 by Western Immunoblot

Serum IGFBP-3 from normal subjects, patients with acromegaly, GHD, liver cirrhosis, chronic renal failure, anorexia nervosa, pregnant woman, and short child with malnutrition were analyzed and compaired with Western ligand blot (Fig.1 and 2). IGFBP-3 immunoreactive bands were recognized at 41/38 and 30kDa. Serum IGFBP-3 at 41/38kDa increased in patients with acromagaly and decreased in patients with GHD, liver cirrhosis, anorexia nervosa and short child with malnutrition. In pregnant woman, serum IGFBP-3 by Western ligand blot was not found, however, 30kDa immunoreactive IGFBP-3 but not 41/38kDa was found.

Western Immunoblot

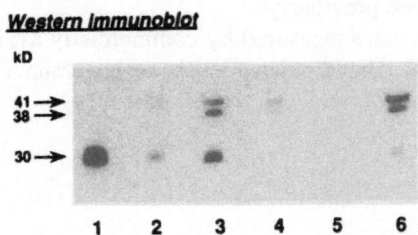

Western Ligand Blot

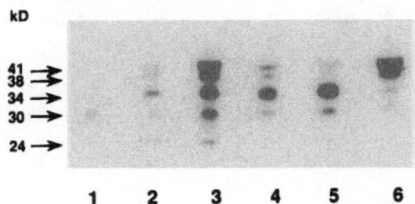

Figure 2. Western immunoblot of IGFBP-3 and Western ligand blot in serum from pregnant woman (1), patients with liver cirrhosis (2), chronic renal failure (3), and anorexia nervosa (4), short child with malnutrition (5), and normal subject (6).

Serum IGFBP Responses to Single sc Administration of IGF-I

Basal serum IGFBP-1 levels in patients with GHD were greater than those in normal subjects (9.1 ± 3.9 vs 3.5 ± 1.5 ng/ml, $p<0.05$). Serum IGFBP-1, blood glucose, serum C-peptide and plasma IGF-I levels in four patients with GHD after single sc administration of IGF-I were shown in Fig. 3. The blood glucose levels decreased after IGF-I administration as a function of dosage of IGF-I. Serum IGFBP-1 increased and serum C-peptide levels decreased after IGF-I administration. The increase of serum IGFBP-1 levels did not correlate with the nadir values for blood glucose and C-peptide although the number of subjects investegated was small.

Increased responses of IGFBP-1 (30kDa) to IGF-I administration were also found by Western ligand blot (Fig. 4). By Western ligand blot, serum IGFBP-2 (34kDa), -3 (41/38kDa) , and -4 (24kDa) levels did not change.

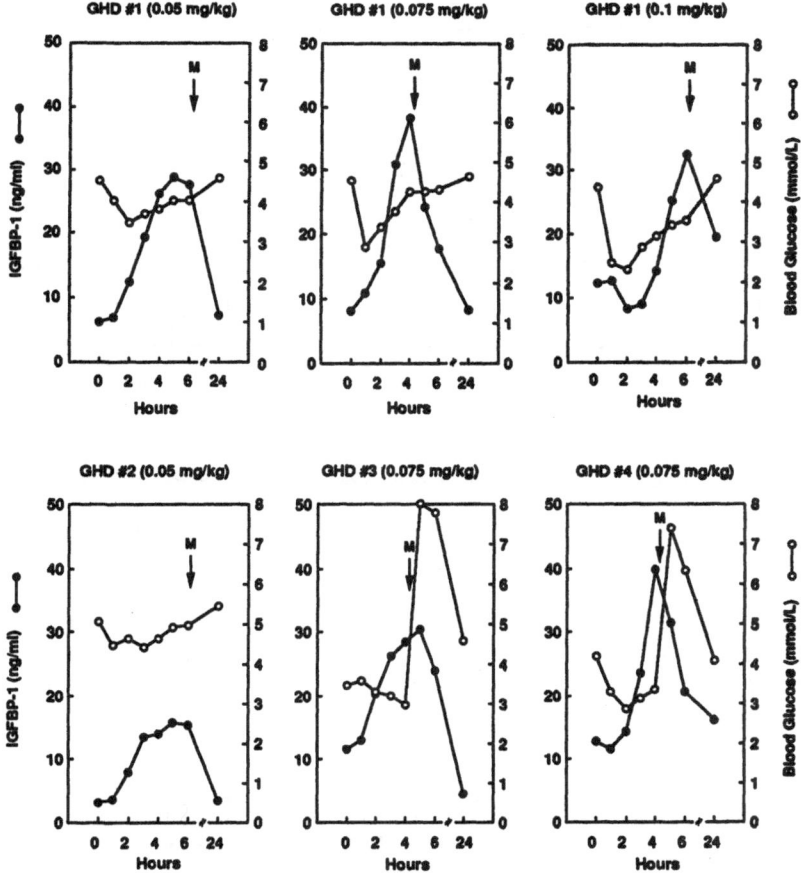

Figure 3A. Serum IGFBP-1 and blood glucose levels after single sc administration of IGF-I in four patients with GHD. M indicates lunch.

Serum IGFBP Responses to Repetitive sc Administrations of IGF-I

Serum IGFBP-3 by both Western immunoblot (Fig. 5) and Western ligand blot (Fig. 6) did not change after repetitive administrations of IGF-I for seven days in patients with GHD. Serum IGFBP-2 levels increased after the repetitive IGF-I administrations by Western ligand blot (Fig. 6). Serum IGFBP-2 after the repetitive administration increased to 154% and 147% in two patients with GHD (#3 and #5), respectively. By Western ligand blot, serum IGFBP-1 levels increased at 4 and 6 h after either first or seventh IGF-I injections (Fig. 6).

DISCUSSION

In the present study, we have investigated serum IGFBP responses to either single or repetitive sc administrations of IGF-I in patients with GH deficiency. As reported previously[7], serum IGFBP-1 levels were greater than those those in normal subjects. We have previouly reported that serum IGFBP-1 levels increased after IGF-I administration in

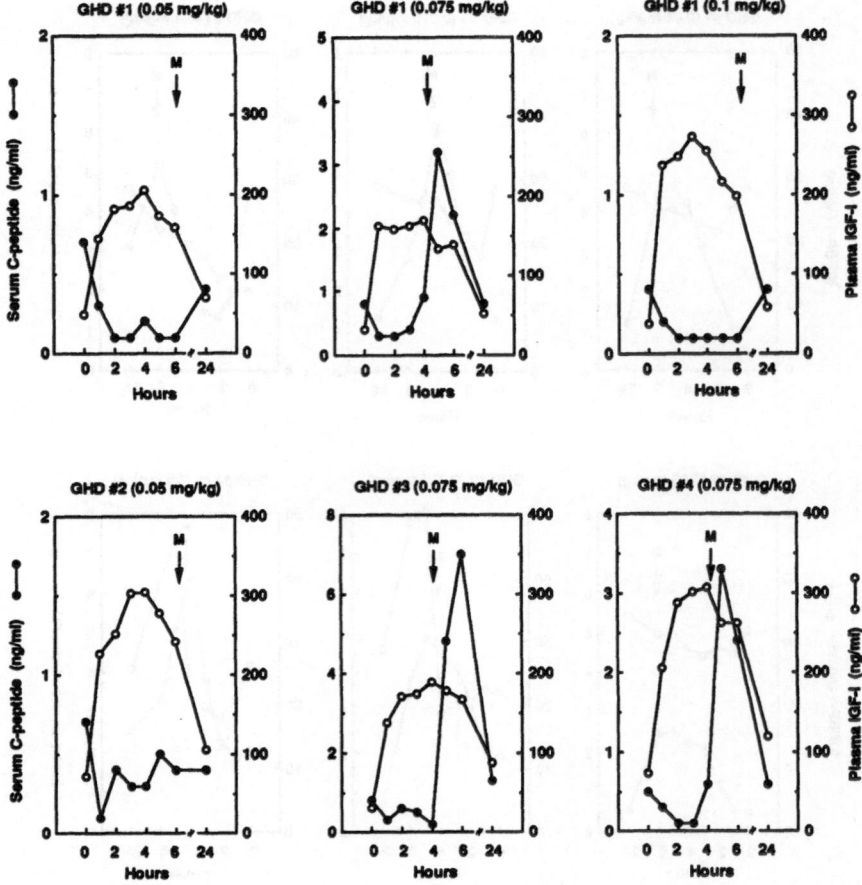

Figure 3B. Serum C-peptide and plasma IGF-I levels after single sc administration of IGF-I in four patients with GHD. M indicates lunch.

normal subjects[1]. In this study, we found that serum IGFBP-1 levels increased,and serum C-peptide and blood glucose levels decreased in patients with GH deficiency after IGF-I administration. The increased IGFBP-1 responses to IGF-I might be partly related to decreased blood glucose and decreased insulin seecretion. However, the peak values did not correlate with the nadir values for blood glucose and C-peptide levels although number of patients investigated in this study was small. Thus, the mechanism for increased responses to IGF-I administration remains unclear.

Serum IGFBP-2 values in patients with GH deficiency after repetitive IGF-I administrations increased as same as normal subjects[1,8]. Serum IGFBP-3 values by Western ligand blot and Western immunoblot did not increase after the administrations. In normal subjects, we have reported that serum IGFBP-3 levels and acid labile subunit levels by RIA decreased after repetitive administrations of IGF-I in normal subjects. In rats, IGF-I induced serum IGFBP-3[9], however, the results in human subjects were controversial[1,10,11]. Further studies will be required.

GHD #1: IGF-I 0.1 mg/kg sc

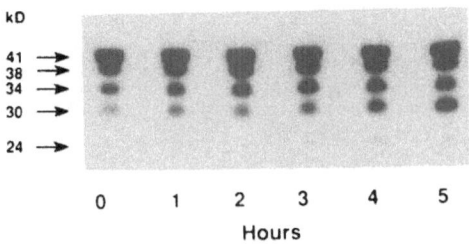

GHD #2: IGF-I 0.05 mg/kg sc

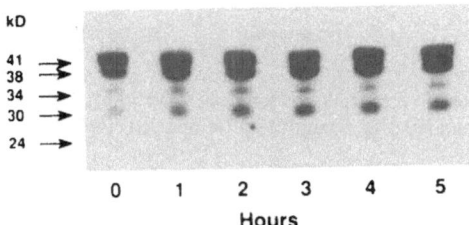

Figure 4. Western ligand blot analysis of serum IGFBPs in patients with GHD after single sc IGF-I injection.

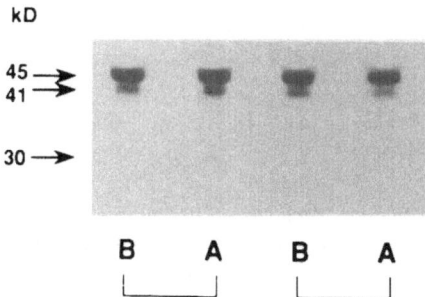

Figure 5. Western immunoblot of serum IGFBP-3 in two patients with GHD (left: #5; right: #3) after repetitive IGF-I for 7 days.

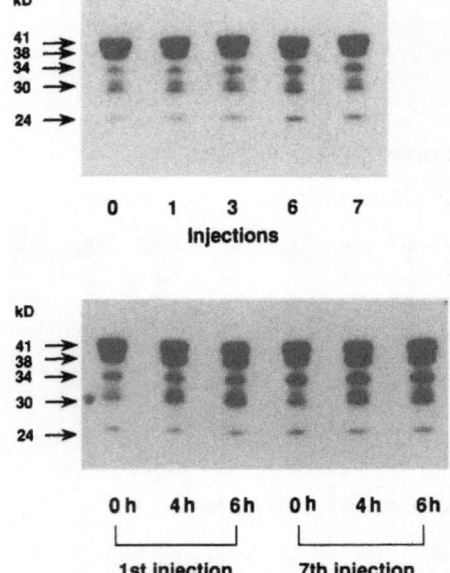

Figure 6. Western ligand blot of serum IGFBPs in two patient with GHD
(top: #5; bottom: #3) after repetitive IGF-I injections for 7 days.

ACKNOWLEDGMENTS

We are greatly indebted to Dr. Ron G. Rosenfeld, Stanford University, CA, and
Fujisawa Pharmaceutical Co. Ltd. (Osaka, Japan) for supplying anti-hIGFBP-3 antibody,
and recombinant human IGF-I, respectively. This work was supported in part by Grant-in-
Aid for General Scientific Research (No. 03671168 and No. 04671499) from the Ministry of
Education, Science, and Culture and a research grant from the Intractable Disease Division,
Public Health Bureau, Ministry of Health and Welfare.

REFERENCES

1. R.C. Baxter, N. Hizuka, K. Takano, et al., Responses of insulin-like growth factor binding protein-1
   (IGFBP-1) and the IGFBP-3 complex to administration of insulin-like growth factor-I. Acta Endocrinol
   128:101 (1993).
2. M. Niwa, S. Sato, Y. Saito, et al., Chemical synthesis, cloning, and expression of genes for human
   somatomedin C (insulin-like growth factor I) and 59Val-somatomedin C. Ann NY Acad Sci 469:31
   (1986).
3. K. Takano, N. Hizuka, K. Shizume, et al., Repeated sc administration of recombinant human insulin-like
   growth factor I (IGF-I) to human subjects for 7 days. Growth Regulation 1:23 (1991).
4. P. Hossenlopp, D. Seurin, B. Segovia-Quinson, et al., Analysis of serum insulin-like growth factor
   binding proteins using Western blotting: Using the method for titration of binding proteins and
   competitive binding studies. Anal Biochem 154:138 (1986).

5. P.C. Cohen, H.C.B. Graves, D.M. Peehl, et al., Prostate-specific antigen (PSA) is an IGF binding protein-3 (IGFBP-3) protease found in seminal plasma. J Clin Endocrinol Metab 75:1046 (1992).

6. M. Miyakawa, N. Hizuka, K. Takano, et al., Radioimmunoassay for insulin-like growth factor I (IGF-I) using biosynthetic IGF-I. Endocrinol Japon 33:795 (1986).

7. R.C. Baxter, Physiological roles of IGF binding proteins. in: Modern Concepts of Insulin-like Growth Factors, E.M. Spencer ed., Elsevier, New York p. 371 (1991).

8. J. Zapf, C. Schmid, H.-P. Guler, et al., Regulation of binding proteins for insulin-like growth factors (IGF) in humans. Increased expression of IGF binding protein 2 during IGF treatment of healthy adults and in patients with extrapancreatic tumor hypoglycemia. J Clin Invest 86:952 (1990).

9. J. Zapf, M. Hauri, M. Waldvogel, et al., Recombinant human insulin-like growth factor I induces its own specific carrier protein in hypophysectomized and diabetic rats. Proc Natl Acad Sci U.S.A. 86:3813 (1989).

10. Z. Laron, B. Klinger, W.F. Blum, et al., IGF binding protein-3 in patients with Laron-type dwarfism: Effect of exogenous rIGF-I. Clin Endocrinol 36:301 (1992).

11. H. Kanety, A. Karasik, B. Klinger, et al., Long-term treatment of Laron type dwarfs with insulin-like growth factor-I increases serum insulin-like growth factor-binding protein-3 in the absence of growth hormone activity. Acta Endocrinol 128:144 (1993).

METABOLIC EFFECTS OF rhIGF-1 IN NORMAL HUMAN SUBJECTS

N.J. Rennert,[1] S.D. Boulware,[2] D. Kerr,[1] S. Caprio,[2]
W.V. Tamborlane[2] and R.S. Sherwin[1]

Departments of Internal Medicine[1] and Pediatrics[2]
Yale University School of Medicine
New Haven, CT

INTRODUCTION

The availability of human IGF-1 synthesized by recombinant DNA techniques has provided the impetus for a systematic assessment of rhIGF-1's metabolic actions *in vivo*. Classic studies of Froesch and coworkers[1,2] have established IGF-1 as the only other naturally occurring hypoglycemic hormone, besides insulin. In order to more precisely define the mechanism of IGF-1's hypoglycemic effects, studies were performed in our laboratory in which rhIGF-1 was administered to awake, 24 hour fasted rats using the euglycemic clamp technique.[3] The effects of rhIGF-1 were compared to those of insulin using doses of each hormone that produced a comparable stimulation of glucose uptake. Although rhIGF-1 and insulin increased glucose uptake and suppressed protein breakdown to a similar extent, their effects on hepatic glucose production (as measured by 3-^{3}H-glucose) and circulating free fatty acids (FFA) were strikingly different (Fig. 1). In contrast to insulin, rhIGF-1 did not suppress hepatic glucose production in the rat during sustained euglycemia. Additionally, in the rat no effect on FFA levels was detected, a feature that further distinguished rhIGF-1 from insulin.[3] Remarkably, infusion of rhIGF-1 also produced a profound reduction in circulating insulin in spite of maintenance of euglycemia. The concomitant portal hypoinsulinemia thus might have obscured effects of rhIGF-1 on the liver and fat cell.

Much less is known about the tissues involved in the metabolic response to rhIGF-1 in human subjects. Guler *et al.* demonstrated rhIGF-1's hypoglycemic effect in healthy human volunteers using both bolus intravenous[4] and chronic subcutaneous[5] infusions. A transient fall in circulating free fatty acid (FFA) levels was also seen during bolus intravenous rhIGF's administration, but was relatively small .[4]

To further explore the spectrum of rhIGF-1's metabolic effects in normal human subjects and to examine the mechanisms underlying these effects, we infused rhIGF-1 (provided by Genentech, Inc., So. San Francisco, CA) under conditions of euglycemia, hyperglycemia and hypoglycemia, using the glucose clamp technique. The scope of this work includes rhIGF-1's measured effects on glucose and protein metabolism, circulating FFA, and islet hormones.

Current Directions in Insulin-Like Growth Factor Research,
Edited by D. LeRoith and M.K. Raizada, Plenum Press, New York, 1994

311

Euglycemia

To examine the influence of rhIGF-1 on islet hormones and substrate metabolism, we studied nine healthy non-obese adult volunteers (two females and seven males, age 25±3 yrs) after a 10 h overnight fast .[6] Each subject received a primed continuous infusion of [6-³H] glucose (25 μCi bolus, 0.25 μCi/min) for 6 h. After a 3 h isotope equilibration period, a primed (20 μg kg, over 10 min) continuous (0.4 μg kg^{-1}·min^{-1}) rhIGF-1 infusion was

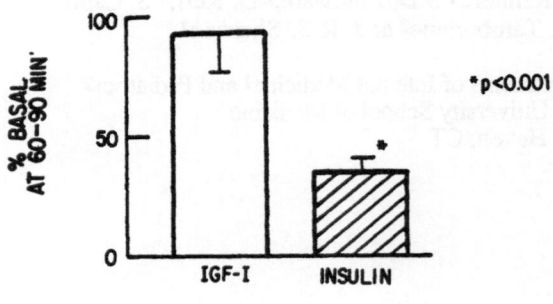

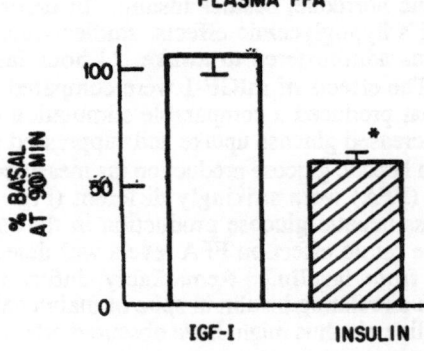

Figure 1. Effects of rhIGF-1 and insulin on hepatic glucose production and circulating FFA in the awake, fasted rat. Based on the data of Jacob, et. al. (3)

administered. Using the glucose clamp technique,[7] a variable rate of glucose was infused to maintain euglycemia (~15 mM).

During rhIGF-1 infusion, total IGF-1 levels increased 2 to 3-fold (p<0.05) and free IGF-1 rose from undetectable to 72±3 ng/ml (p<0.05). Despite maintenance of euglycemia, plasma insulin and C-peptide declined by 31% and 56% (p<0.01), respectively (Fig. 2). Glucagon also fell from 72±9 to 42±4 pg/ml (p<0.005). As shown in Figure 2, hepatic glucose production decreased by 68% in the final hour of the study (p<0.01). At the same time, there was a two- to three-fold increase in glucose uptake. Concomitantly, carbohydrate oxidation increased by 47%, while fat oxidation decreased by 37% (p<0.05) and plasma FFA declined by 60% (p<0.001). rhIGF-1 infusion also produced a consistent reduction in circulating essential amino acids, including total branched chain amino acids which fell from 406±23 to 219±14 μM (p<0.05). Plasma alanine, on the other hand, remained unchanged.

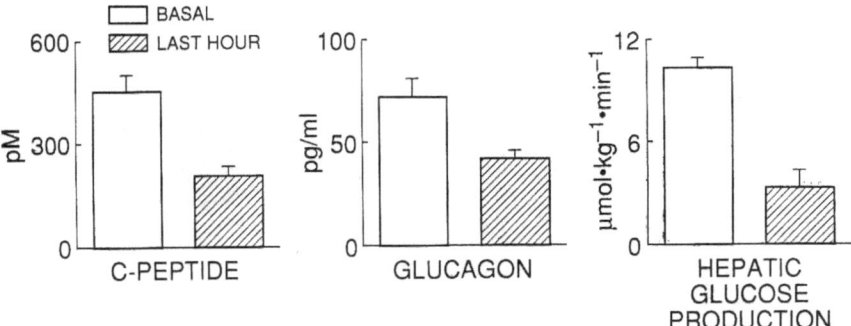

Figure 2. Effects of rhIGF-1 on C-peptide, glucagon and hepatic glucose production during euglycemic clamp. Based on the data of Boulware *et al.* (6)

Hyperglycemia

To address the issue of whether rhIGF's insulin lowering effect may be overcome when insulin secretion is stimulated by glucose, we used the glucose clamp technique to produce two standardized levels of hyperglycemia (+2.8 mmol/l and +7 mmol/l).[8] Fourteen healthy, nonobese adult volunteers (six male, eight female) were studied on two separate days after a 12 hr overnight fast. The subjects received, in random order, either rhIGF-1 or saline prior to and during a hyperglycemic clamp study. After a 30 min basal rest period, a primed-continuous infusion of rhIGF-1 was administered (the same dose as was used in the euglycemia studies) while maintaining euglycemia for 60 min. Then plasma glucose was raised either 50 mg/dl (n=6) or 125 mg/dl (n=8) above baseline using a variable glucose infusion for the remaining 2 h.

During the rhIGF-1 infusion, total IGF-1 increased rapidly from 196±37 to 449±71 ng/ml. IGF-1 concentrations were not significantly different in the two hyperglycemic studies and did not change throughout the saline control studies. The effect of rhIGF-1 on C-peptide concentrations at the different levels of hyperglycemia is shown in Figure 3. When plasma glucose was raised by 2.8 mmol/l (+50 mg/dl), first (2-10 min) and second (20-120)min phase C-peptide levels were suppressed during rhIGF-1 infusion by 39% (p< 0.05) and 40% (p< 0.05), respectively. Insulin levels were also suppressed, but only during the second phase (215±42 vs 151±28 pmol/l, p<0.05). Nevertheless, the rate of glucose metabolism was two-fold higher in the rhIGF-1 infused group (8.0±0.5 vs 3.5±0.1 mg kg^{-1}·min^{-1}, p<0.001). At the higher +7 mmol/l (or +125 mg/dl) glucose stimulus, the suppression appeared to be less remarkable. Second phase C-peptide (by 24%) but not insulin concentrations were significantly reduced. Again rates of glucose metabolism were significantly higher during rhIGF-1 infusion (11.8±1.2 vs 8.9±0.8 mg kg-1·min-1, p<0.01).

Hypoglycemia

To compare the kinetic mechanisms underlying IGF-1 induced hypoglycemic counterregulation with that of insulin, 18 healthy, nonobese volunteers (age range 18-33 yrs) received infusions of rhIGF-1 using either a free fall or hypoglycemic clamp protocol.[9] In the free fall hypoglycemia protocol, ten subjects (eight male) first received a primed-continuous infusion of 3-[3H]-glucose over a 120 min tracer equilibration period and then

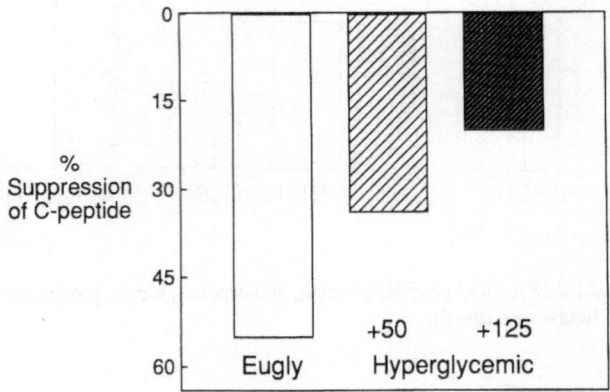

Figure 3. Effect of rhIGF-1 on C-peptide levels during euglycemia and the two levels of hyperglycemia. Based on the data of Rennert, et. al. (8).

either a primed-continuous infusion of human regular insulin (0.8 mU kg-1 min-1) or rhIGF-1 (0.7 μg kg-1 min-1 for 120 min. Plasma glucose was allowed to fall and measured every 5 min at the bedside. After 120 min, the infusion of rhIGF-1 or insulin was discontinued and recovery from hypoglycemia was monitored over a 120 min period. In the hypoglycemic clamp protocol, eight men received either a primed-continuous infusion of human regular insulin (6.5 mU/kg over 10 min, 1.3 mU kg-1·min-1) or rhIGF-1 (40 μg/kg prime, 0.8 μg·kg-1·min-1 for 240 min. Plasma glucose levels were "clamped" at 5.0 mM for 120 min (euglycemic period), thereafter lowered over 30 min to 2.8 mM and maintained at this level for 2 h (hypoglycemic period). Additionally, assessments of hypoglycemic symptoms and cortical evoked potentials (P300) were recorded.[10]

In the free fall protocol, the fasting, rate of fall, and nadir plasma glucose values were similar in all studies, as were the plateau values over the final 60 min of the infusion periods. However, recovery from hypoglycemia was significantly delayed after infusion of rhIGF-1 vs insulin(p<0.001). In the rhIGF-1-1 study, IGF-1 levels reached a plateau 3-fold above baseline values by 40 min, whereas in the insulin study, levels of insulin rose 5-fold above baseline (p<0.001). The rate of fall and magnitude of suppression of C-peptide levels were similar in both studies. However, C-peptide levels promptly increased after stopping the insulin infusion, but remained suppressed for 100 min after discontinuing rhIGF-1. Remarkably, the rise in plasma glucagon seen during insulin induced hypoglycemia was totally suppressed by rhIGF-1 infusion (glucagon levels rose from 111±18 to 158±25 ng/l in the insulin study but were unchanged form baseline, 109±17 vs 105±21 ng/l with rhIGF-1, p<0.005). Similarly, the rise in growth hormone was also attenuated. In contrast, norepinephrine levels were 44% higher at the end of the rhIGF-1 infusion and remained higher in the rhIGF-1 study (vs insulin) after the hormone infusions were stopped. Epinephrine levels were similar during the hypoglycemic phases of both studies, however they returned to baseline more rapidly after insulin as compared to rhIGF-1. Cortisol responses (data not shown) were similar in the two studies.

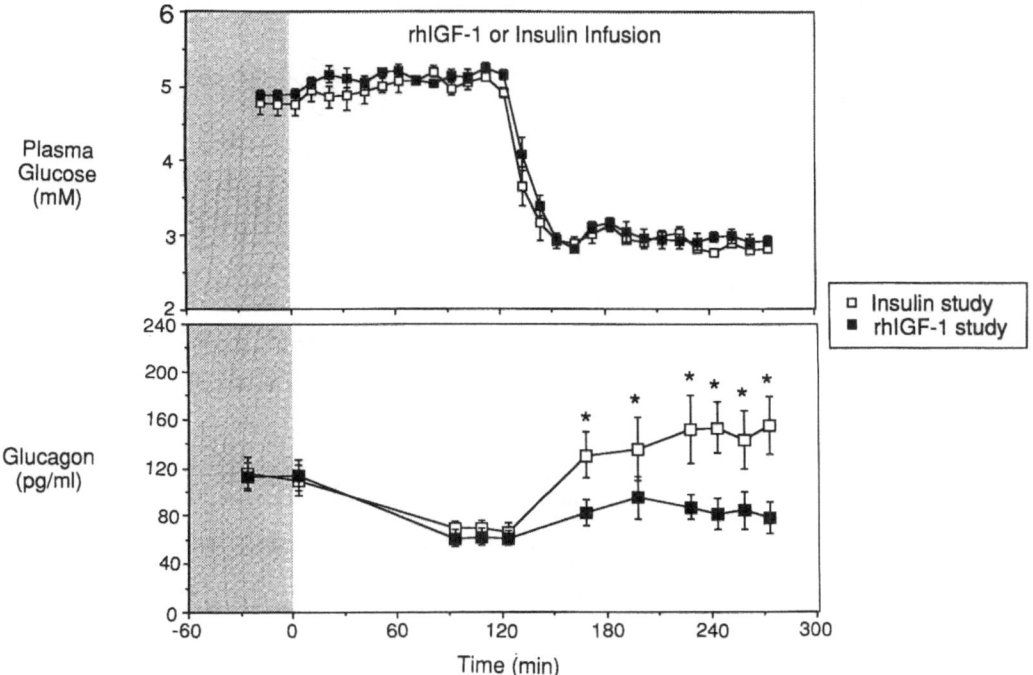

Figure 4. Effects of rhIGF-1 vs insulin on plasma glucagon during hypoglycemia. Based on the data of Kerr *et al.* (10).

The plasma glucagon response to the hypoglycemic clamp protocol is shown in Figure 4. During the euglycemic phase of both studies, plasma glucagon was suppressed by approximately 50%. However, when blood glucose was lowered to 2.8 mM by insulin, plasma glucagon promptly rose. In contrast, rhIGF-1 induced hypoglycemia did not cause a significant change in glucagon from basal values. The initial growth hormone response to hypoglycemia was delayed in the rhIGF-1 study. However, after 60 min growth hormone levels were similar in both studies. Once again, the pattern of the norepinephrine response to hypoglycemia was strikingly different. The rise in norepinephrine was significantly greater during rhIGF-1 induced hypoglycemia when compared with insulin (2.7±0.3 vs 2.2±0.3 nM, p<0.05 (Fig. 5). In keeping with the norepinephrine results, heart rate was higher and symptomatic awareness of hypoglycemia was strikingly enhanced by rhIGF-1 infusion (Fig. 5). On the other hand, the effect of hypoglycemia to increase latency and decrease the amplitude of the P300 potential was similar in both studies.

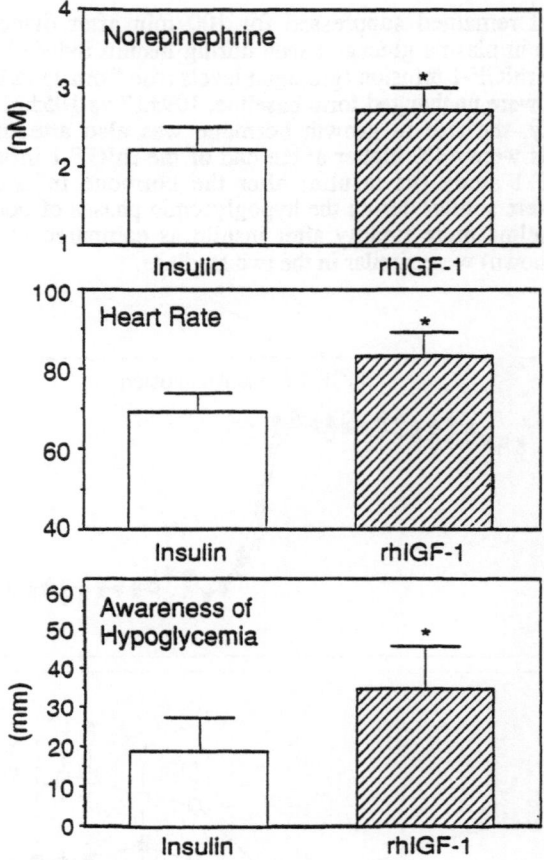

Figure 5. Effect of rhIGF-1 on norepinephrine, heart rate and awareness of hypoglycemia during hypoglycemic clamp protocol. Based on the data of Kerr *et al.* (10).

DISCUSSION

The studies we have described explore the metabolic effects of intravenously administered rhIGF-1 in normal humans during euglycemia, hypoglycemia and hyperglycemia. rhIGF-1's diverse metabolic effects are strikingly similar to those seen with insulin, however remarkably they occur despite a fall in measured insulin and C-peptide levels. The stimulation of glucose uptake and suppression of amino acids achieved by rhIGF-1 infusion in the human studies resembled those seen in the rat.[3] However, the effects on hepatic glucose production and fat metabolism were quite different in humans vs. rats.

In human subjects, rhIGF-1 markedly suppressed hepatic glucose production and circulating FFA levels, effects not seen in the rats.[3] It is possible that the suppression of hepatic glucose production may be explained by either rhIGF-1 crossreacting with hepatic insulin receptors (since few rhIGF-1 receptors are present in normal adult liver)[11] or by direct suppression of glucagon production by pancreatic alpha cells. The fall in FFA seen during euglycemia is especially remarkable since it occurred when insulin concentrations were reduced. *In vitro* studies have suggested that adipose tissue is not as sensitive to IGF-1 when compared with insulin.[12] In concordance with these data, studies in normal and diabetic rats did not show a significant effect on FFA.[3] The marked inhibitory effect seen in our studies may reflect spillover binding of rhIGF-1 to the insulin receptor.

Based on earlier work [13], we expected that rhIGF-1 infusion would suppress growth hormone levels and our results corroborate this. The remarkable effect of rhIGF-1 to suppress islet hormones (insulin and glucagon), however, was not expected. During sustained hyperglycemia (a potent stimulus for insulin secretion), the inhibitory effect of rhIGF-1 on insulin secretion seemed to be partially overcome by increasing the hyperglycemic stimulus. Thus, it appears that rhIGF-1's influence on glucose-stimulated insulin secretion may be dependent on a balance between the dose of rhIGF-1 administered and the intensity of the insulin secretory stimulus generated by exogenous glucose. Importantly however, rhIGF-1's direct stimulatory effects on glucose metabolism appear to be sufficient to override the limitations imposed by its capacity to inhibit insulin secretion. The marked inhibition of insulin and glucagon secretion noted in our studies might be explained by a direct effect of rhIGF-1 on the pancreatic beta and alpha cell. Direct effects of rhIGF-1 on both pancreatic alpha and beta cells have recently been demonstrated *in vitro*. These effects appear to be modulated by specific pancreatic IGF-1 receptors.[14,15] Alternatively, delta cell somatostatin secretion might be involved and might account for the coincident suppression of many hormones including insulin, growth hormone and glucagon.

The effect of rhIGF-1 to suppress glucagon is particularly important during hypoglycemia. This effect undoubtedly contributes to the delayed recovery from hypoglycemia seen with rhIGF-1 and could potentially limit its therapeutic value. Another unexpected finding during hypoglycemia was the stimulation of catecholamines and the increase in symptoms noted with rhIGF-1 compared with insulin. This sympathetic stimulation might to some extent offset the delayed recovery from hypoglycemia due to glucagon suppression.

In summary, rhIGF-1 has widespread acute metabolic effects in healthy human volunteers which closely resemble those of insulin , but occur despite inhibition of insulin secretion. The metabolic actions of rhIGF-1 are likely the result of both its direct insulin-like effects and its capacity to modulate the secretion of other glucoregulatory hormones.

REFERENCES

1. J. Zapf, C. Haun, M. Waldvogel, and G.R. Froesch, Acute metabolic effects and half-lives of intravenously administered insulin-like growth factors I and II in normal and hypophysectomized rats, *J. Clin. Invest.* 77:1768 (1986).

2. G.R. Froesch, C. Schmidt, J. Schwander, and J. Zapf, Actions of insulin-like growth factors, *Annu. Rev. Physiol.* 47:443 (1985).

3. R. Jacob, E. Barrett, G. Plewe, K.D. Fagin, and R.S. Sherwin, Acute effects of insulin-like growth factor-I on glucose and amino acid metabolism in the awake, fasted rat, *J. Clin. Invest.* 83:1717 (1989).

4. H. Guler, J. Zapf, and G.R. Froesch, Short-term metabolic effects of recombinant human insulin-like growth factor I in healthy adults, *N. Engl. J. Med.* 317:137 (1987).

5. H. Guler, S. Schmid, J. Zapf, and E.R. Froesch, Effects of recombinant insulin-like growth factor I on insulin secretion and renal function in normal human subjects, *Proc. Natl.Acad. Sci. USA* 86:2868 (1989).

6. S.D. Boulware, W.V. Tamborlane, L.S. Matthews, and R.S. Sherwin, Diverse effects of insulin-like growth factor I on glucose, lipid and amino acid metabolism, *Am. J. Physiol.* 262:E130 (1992).

7. R.A. DeFronzo, J.D. Tobin, and R. Andres, Glucose clamp technique: a method for quantifying insulin secretion and resistance, *Am. J. Physiol. (Endocrinol. Metab.Gastrointest. Physiol.* 6):E214 (1979).

8.   N.J. Rennert, S. Caprio, and R.S. Sherwin, Insulin-like growth factor I inhibits glucose-stimulated insulin secretion but does not impair glucose metabolism in normal humans,. *Clin. Endo. Metab.* 76:804 (1993).

9.   D. Kerr, W.V. Tamborlane, F. Rife, and R.S. Sherwin, Effect of insulin-like growth factor-I on the responses to and recognition of hypoglycemia in humans: a comparison with insulin, *J. Clin. Invest.* 91:141 (1993).

10.   T.W. Jones, G. McCarthy, W.V. Tamborlane, S. Caprio, E. Roessler, D. Kraemer, K. Starich-Zych, T. Avison, S.D. Boulware, and R.S. Sherwin, Mild hypoglycemia and impairment of brain stem ad cortical evoked potentials in healthy subjects, *Diabetes* 39:1550 (1990).

11.   J.F. Caro, J. Poulos, O Itteup, W.J. Purie, E.G. Flickinger, and M.K. Sinha. Insulin-like-growth factor I binding from human liver, human hepatoma, and normal regenerating and fetal rat liver. *J. Clin. Invest.* 81:976 (1988).

12.   J. Zapf, E. Schoenle, M. Waldvogel, I. Sand, and E.R. Froesch. Effect of trypsin treatment of rat adipocytes on biological effects and binding of insulin and insulin-like growth factors. Further evidence for the action of insulin-like growth factors through the insulin receptor. *Eur. J. Biochem.* 1123:605 (1981).

13.   M. Berelowitiz, M. Szabo, L.A. Frohman, S. Firestone and L. Chu, Somatomedin-C mediates growth hormone negative feedback by effects on both the hypothalamus and the pituitary. *Science,* 212:1279 (1981).

14.   J.L., Leahy, and K.M. Vandekerkhove, Insulin-like growth factor-I at physiological concentration is a potent inhibitor of insulin secretion. *Endocrinology* 126:1593 (1990).

15.   C.H.F. Van Schravendijk, L. Leyleu, J.L. Vanden-Brande, and D.G. Pipeteers. Direct effect of high and insulin-like growth factor-I on the secretory activity of rat pancreatic beta cells, *Diabetologia* 33:649 (1990).

IGFS AND MUSCLE DIFFERENTIATION

J. R. Florini, D. Z. Ewton, K. A. Magri, and F. J. Mangiacapra

Biology Department
Syracuse University
Syracuse, NY 13244

INTRODUCTION

We have studied the stimulation of differentiation by the IGFs for more than a decade, and have now reached a reasonably comprehensive view of this important action of these major hormones. Our recent work makes it clear that a primary mechanism involves increased expression of one of the recently discovered myogenesis genes, myogenin (Florini et al., 1991a). Research on the control of muscle differentiation has been significantly advanced by the discovery of MyoD and three closely related genes, which have been shown to be capable of converting non-muscle cells to the myogenic line, and which are normally expressed only in skeletal muscle cells or their precursors. Myogenin is the member of the family most closely correlated with terminal differentiation, which involves the fusion of myoblasts to form postmitotic myotubes and the concomitant expression of some 20 muscle-specific genes. The IGFs control this process even when they are not added to myoblasts, as incubation of these cells in low-serum medium (the technique most often used to induce differentiation) induces the expression of IGF-II, which then gives autocrine stimulation of myogenesis (Florini et al., 1991b). Indeed, one cell line (Sol 8) secretes so much IGF-II into the medium that we have often found relatively little effect of exogenous IGFs on these cells. The essential role of myogenin in IGF-induced myogenesis, and of autocrine IGF-II in the absence of exogenous IGFs, have been demonstrated by the use of antisense oligodeoxynucleotides complementary to portions of their mRNAs (Florini et al., 1991a, 1991b).

As is always the case, questions remain. Among the most significant of these are: (1) What is the role of the other three members of the MyoD family in IGF-induced myogenesis, (2) Why is IGF-II "better" than IGF-I at inducing differentiation (in spite of the fact that both act via the type I IGF receptor), and (3) What suppresses IGF-II expression in myoblasts incubated under "growth" conditions in high serum? This report presents some results of our ongoing attempts to answer these questions.

EFFECTS OF IGFs ON INDIVIDUAL MyoD FAMILY MEMBERS

After the discovery of MyoD (Davis et al., 1987), three other genes also capable of converting fibroblasts to the myogenic cell line were cloned: myogenin, myf-5, and MRF4/herculin/myf-6 (one gene named by three different laboratories) were all shown to share many of the structural and biological properties of MyoD (for reviews see Olson, 1990; Weintraub et al., 1991; Li and Olson, 1992; Edmondson and Olson, 1993). Initially, it seemed that the four genes functioned in similar ways, but differences in their patterns of expression in cultured cells and in their developmental expression (Ott et al., 1991; Hinterberger et al., 1991) indicated that there might be differences in their function. The observation that BC3H1 cells (which exhibit biochemical differentiation but do not express MyoD or fuse to form myotubes) could be induced to fuse upon transfection with a MyoD expression vector suggested that MyoD plays a crucial role in fusion. This conclusion is

Current Directions in Insulin-Like Growth Factor Research,
Edited by D. LeRoith and M.K. Raizada, Plenum Press, New York, 1994

319

considerably weakened by the repeated observations that L6 myoblasts, which fuse quite well, do not express MyoD to a detectable extent (Wright et al., 1989; Mangiacapra, 1992).

Comparisons of Effects on Myogenin and myf-5 mRNA Levels

We now believe that all myogenic differentiation is, to at least some extent, controlled by IGFs, whether autocrine or added to the medium. We feel that our approach (adding exogenous IGFs at zero time) offers significant advantages because we can thus control the timing and amount of the triggering hormone the cells encounter. Because L6 cells express no detectable MyoD, and because MRF4 is expressed only quite late in differentiation (Rhodes and Konieczny, 1989), we can initially restrict our attention to only two of the four myogenesis genes, myogenin and myf-5. The observations we made on the expression of these two genes were strikingly different, as is shown in Figure 1.

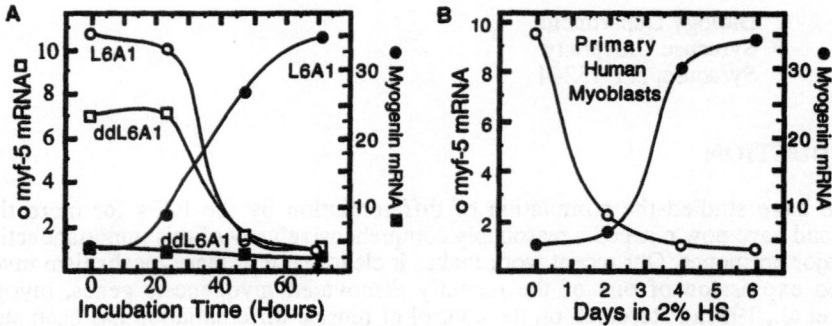

Figure 1. Steady State Levels of Myogenin and myf-5 mRNAs during Myogenic Differentiation.
A. Parental L6A1 myoblasts and a differentiation defective subclone of this cell line were incubated with IGF-II (160 ng/ml) for the indicated times. Northern blot analyses of mRNA content of the preparations were done as detailed by Mangiacapra et al. (1992). B. Primary human myoblasts (a gift of Helen Blau) were plated and grown, then treated with 2% horse serum for the indicated times. Reprinted with permission from Mangiacapra et al. (1992). Results are in arbitrary densitometer units.

Contrary to our expectations, the levels of myf-5 mRNA decreased sharply as those of myogenin mRNA increased during differentiation. As shown in panel B, the phenomenon is not an idiosyncrasy of a rodent cell line; very similar results were obtained with primary cultures of human myoblasts. The amounts of myf-5 and myogenin are strikingly reciprocal in normal myoblasts, but the relationship is not direct. In differentiation defective cells (ddL6A1, derived by repeated passaging of cells), the decrease in myf-5 occurred even when there was no elevation of myogenin in response to IGF treatment. It appears that myogenin mRNA is accumulated at a higher level than myf-5 mRNA; substantially longer exposure times were required for the myf-5 blots than for myogenin, although this interpretation is limited by the fact that the absolute specific activities of the probes were not rigorously established. Because myoblasts express IGF-II when incubated in low serum medium, these observations do not distinguish between the possibilities that the decrease in myf-5 expression is an effect of IGFs or a response to incubation in low serum.

Necessity for Early myf-5 Expression

In spite of its relatively early disappearance from the cells, myf-5 appears to be essential for at least some early aspects of myogenic differentiation. Treatment of the cells with an antisense oligodeoxyribonucleotide complementary to the first five codons of myf-5 gave substantial inhibition of differentiation and suppressed induction of myogenin expression (Mangiacapra et al., 1992). It appears that myf-5 is essential for initial induction of myogenin expression, but the subsequent rise in myogenin expression may involve autoinduction, as has been reported for several members of the MyoD family (Olson, 1990).

Effects on MyoD Expression in C2 Cells

Obviously, any effects of IGFs on MyoD cannot be studied using L6 myoblasts because these cells do not express the MyoD gene under any known conditions. We investigated this point using cells of the C2 line, although the relatively high rate of IGF-II production by these cells (Florini et al., 1991b) poses some difficulties for this approach. Although MyoD (unlike myogenin) is expressed to a substantial extent in growing C2 myoblasts; treatment with IGFs gave a two- to four-fold increase in levels of the mRNA with a time course similar to that shown by myogenin (Mangiacapra, 1992). We have not yet examined effects of the IGFs on MRF4.

Anomalous Elevation of Myogenin in Some Control Samples

In some of our experiments, we have detected a mystifying elevation of myogenin mRNA levels in "control" cells incubated in low serum. Normally, myogenin mRNA levels are relatively low during the first 48 hours of such incubation, but occasionally we find controls that appear as high as those from cells treated with IGFs. This is never accompanied by an increased rate of differentiation, measured either as fusion or as elevation of creatine kinase. We have tried a large number of variations in experimental conditions, but have been unable to define situations that consistently give high myogenin mRNA in controls. In addition to the considerable frustration these observations have caused, they also warn that a simple elevation of myogenin gene expression by the IGFs may not be sufficient to explain completely their effects in stimulating myogenesis.

RELATIONSHIP OF MITOGENIC AND MYOGENIC EFFECTS OF IGF-I

One aspect of the stimulation of muscle differentiation by the IGFs has puzzled us for some time. This is the greater absolute stimulation of differentiation by IGF-II and insulin compared to IGF-I, in spite of strong evidence that the stimulation is mediated by the type I IGF receptor (Ewton et al., 1987). When plotted as % of maximal stimulation, the concentration dependency curves are essentially parallel for the four anabolic actions of the IGFs we have measured routinely — stimulation of amino acid uptake, cell proliferation, and differentiation, and inhibition of proteolysis. In all cases the relative potency is IGF-I> IGF-II > insulin — i.e., it parallels the binding properties of the type I receptor. Furthermore, IGF-I analogs that exhibited little or no binding to the type II receptor were as active as IGF-I in stimulating all four processes. Recent unpublished experiments in our laboratory (Ewton et al., manuscript in preparation) with a larger series of IGF analogs have added strong additional evidence to support this conclusion; overall, effectiveness of the analogs in stimulating myogenesis closely paralleled their affinity binding to the type I receptor and was not dependent on their cross-reaction with the type II receptor.

The greater quantitative activity of IGF-II and insulin in stimulating L6A1 differentiation is illustrated in Figure 2.

Note that the curves in Figure 2 show the order of affinities mentioned above; the concentration of hormone giving the half-maximal effect (an approximation of the K_D in most situations) shows that IGF-I is more potent but less quantitatively active than IGF-II or insulin. Thus it has been puzzling to us that IGF-II and insulin could elicit such high CK levels and large myotubes, while apparently exerting their effects via the type I IGF receptor.

A Possible Explanation for the Relatively Low Response to IGF-I

One hint toward a possible explanation comes from a long-held view on the negative control of myogenesis by mitogens. Reasoning from a series of experiments using medium conditioning and other methods to remove "mitogens" from the medium, Konigsberg (1971) suggested that myoblasts make a choice during G1 phase of the cell cycle. If they are stimulated by mitogens present in the medium, they reenter the cell cycle, and thus areprevented from differentiation (all myotube nuclei are 2N, so presumably fusion occurs after M phase and prior to S). This view was strongly supported by a series of studies from Hauschka's laboratory (Linhkart et al., 1982) which showed that a purified mitogen, FGF, was a potent inhibitor of myogenic differentiation, and similar results have been obtained more recently with PDGF and EGF. Our observations that IGFs, which are mitogenic and *also* stimulate differentiation, appear to be in disagreement with this view, or at least to limit the validity of a generalization that mitogens inhibit myogenesis.

It has been widely observed that IGF-I is more mitogenic than IGF-II in many kinds of cells, and we find this is also true of L6 myoblasts (Ewton et al., 1987). So we reasoned that IGF-I might be delaying differentiation by forcing the cells to undergo a round of cell division before allowing the onset of differentiation. To investigate this possibility, we have recently used several parallel approaches: (1) Incubation with an inhibitor of cell proliferation, cytosine arabinoside, (2) Initial plating of the cells at high density to suppress proliferation, (3) Use of analogs of IGF-I that are less active mitogens, and (4) Isolation of a subclone of L6A1 cells that exhibits a substantially lower response to the mitogenic effects of IGF-I.

All of these approaches have given similar results; under these conditions, the quantitative effect of added IGF-I was substantially increased relative to that of IGF-II and insulin. This was observed both in measurements of biochemical (creatine kinase levels) and morphological (fusion) differentiation. Thus it appears that the lesser stimulation of myogenesis by IGF-I results from its mitogenic activity, as we suspected. As is so often the case, this "solution" of one problem now raises another question; why is IGF-I significantly more mitogenic that IGF-II if both act through the type I receptor?

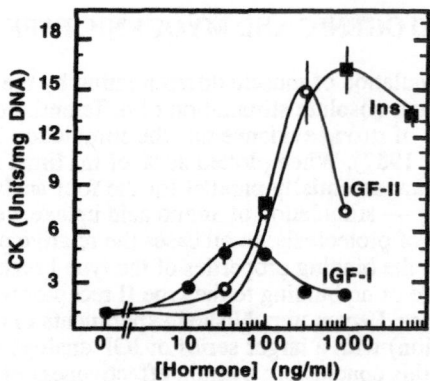

Figure 2. Effects of the IGFs on Differentiation of L6A1 Myoblasts. Cells were plated, treated, and analyzed by our standard techniques (Florini et al., 1991a), using the final concentrations of the hormones as specified. Creatine kinase and DNA were measured after 3 days on triplicate plates for each concentration. The extent of fusion observed microscopically closely paralleled the creatine kinase levels, as is always the case in our experiments on IGF effects on myogenesis.

Secretion of IGF Binding Proteins by L6 Cells

Our studies with the IGF analogs also revealed another aspect of IGFs stimulation of myogenesis. Uniformly, IGF-I analogs that bound weakly or not at all to the IGF binding proteins were substantially more active (up to 100-fold) in stimulating myogenesis than was native IGF-I. Thus it appeared that binding proteins were significantly inhibiting the effects of exogenous IGF-I, and in the serum-free medium we employed, this meant that the cells were secreting substantial amounts of IGF binding proteins. Accordingly, we confirmed the reports by McCusker and Clemmons (1988) and McCusker et al. (1988, 1989) that L6 cells secrete two binding proteins subsequently shown to be IGFBP-4 and IGFBP-5 (McCusker, personal communication). As has been reported by others, we found that IGF-I and IGF-II were more active than insulin or the longR3 IGF analog (which is poorly bound by IGFBPs) in increasing the amount of binding proteins in the medium. This

322

has been attributed to a stabilizing effect of the IGF-I ligand that makes the binding proteins more resistant to proteolysis (Camacho-Hubner et al., 1992; Conover, 1991;). We have no direct evidence on this point, but our observations are quite consistent with this conclusion.

AUTOCRINE EFFECTS OF IGF-II ON MYOBLASTS

The reports by Tollefsen et al. (1989a, 1989b) of expression of the IGFs and their binding proteins and receptors in differentiating C2 cells prompted us to initiate a collaborative investigation of the possibility that this might account for the differentiation of myoblasts incubated in the absence of exogenous IGFs. We and others had previously noted that various cell lines exhibited substantial differences in their rates of differentiation in the absence of exogenous IGFs, as well as in their responsiveness to exogenous IGFs. Accordingly, we conducted a study in which we compared rates of "spontaneous" (i.e., not stimulated by exogenous IGFs) differentiation in several myogenic lines and sublines with their expression of IGF-II; under the conditions of these experiments, there was little or no detectable expression of IGF-I. As shown in Figure 3, there was a striking correlation between the rate of differentiation and the amount of autocrine expression of IGF-II, both at the mRNA and protein levels (Florini et al., 1991b).

The amounts of IGF-II secreted into the medium by Sol 8 and C2-DZE cells approach those that give maximal stimulation of differentiation, and it seems likely that the concentration of hormone in the immediate vicinity of the cells is substantially greater than the average throughout the unstirred medium. A comparison of the response of these myogenic lines to exogenous IGF-II gave results consistent with our observations on autocrine secretion of the hormone; lines such as Sol 8 that are very active secretors of IGF-II show little or no response to exogenous hormone, presumably because they are producing a saturating quantity of IGF-II (Figure 4 in Florini et al., 1991b). In contrast, the L6A1 subline, which was initially cloned on the basis of a high degree of sensitivity to exogenous insulin, secretes relatively little IGF-II, and continues to exhibit such sensitivity (Figure 3B).

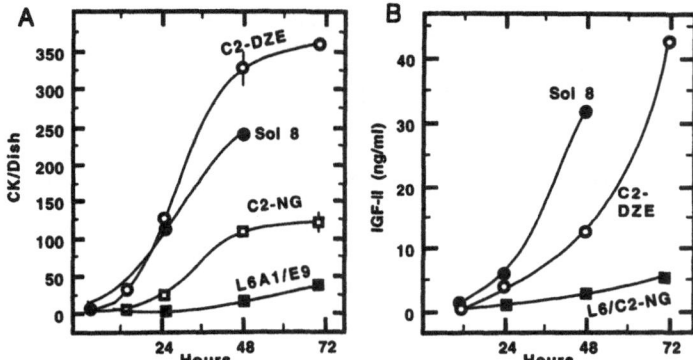

Figure 3. Comparison of Rate of "Spontaneous" Differentiation and Secretion of IGF-II Into the Medium by Muscle Cells. A. Differentiation of Several Cell Lines in the Absence of Added IGFs. For these incubations, cells were plated, grown to approximately 80% confluence, and then shifted to 2% horse serum at zero time. Differentiation was quantitated by our usual CK assay (Florini, 1989).
B. Secretion of IGF-II into the Medium by Myoblasts. Conditioned medium from the cells was collected as the cells were harvested for part A and analyzed for IGF-II by radioimmunoassay. Reprinted with permission from Florini et al (1991b).

Effects Of Antisense Oligos Complementary To IGF-II

These observations illustrate a good correlation between autocrine secretion of the IGFs and the differentiation that occurs in the absence of added IGFs, but they do not establish a causal relationship. To attempt to do this, we again turned to antisense oligos, this time using sequences complementary to the first five codons on IGF-I and IGF-II. The results, summarized in Figure 4, demonstrate that autocrine IGF-II secretion plays an essential role in myogenic differentiation, even when it is not added to the medium by the experimenter.

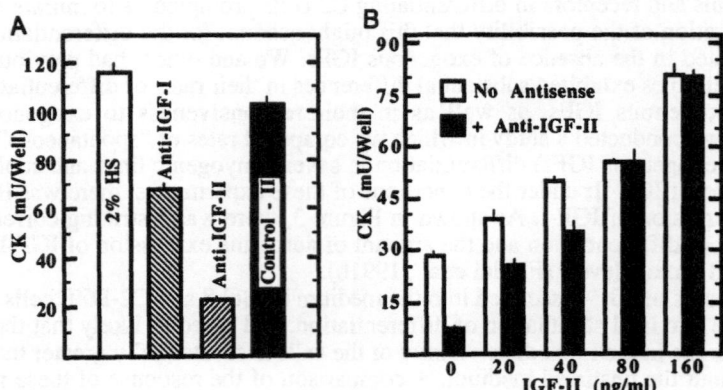

Figure 4. Inhibition of Myogenesis by Antisense Oligodeoxyribonucleotides. A. Specificity of Inhibition by Anti-IGF-II. Cells were plated, grown for 24 hours, and then switched to 2% serum with the indicated antisense oligos at 150 μg/ml. CK analyses were done 72 hours later. B. Stimulation of Differentiation by Exogenous IGF-II in the Presence of Anti-IGF-II. Cells were plated and analyzed as indicated for A, except that in this case parallel wells also received IGF-II at the indicated concentrations. Reprinted with permission from Florini et al. (1991b).

These observations suggest that there may be some contribution by autocrine IGF-I in spite of its apparently low expression, but clearly the primary stimulation is by IGF-II. Figure 4B provides a persuasive control observation that is particularly useful in experiments with the necessarily high levels of added oligos. If the oligo blocks differentiation solely by inhibiting expression of IGF-II by the cells, it should have no effect on stimulation by exogenous IGF-II. This is the case, as shown at exogenous IGF-II levels above those (20 and 40 ng/ml) at which autocrine secretion makes a detectable contribution to stimulation. We have also shown (Figure 8 in Florini et al., 1991b) that elevation of myogenin expression is essential for myogenesis stimulated by autocrine IGF-II as it is for exogenous IGFs — i.e., the mechanism is the same whether exogenous or autocrine hormone is the stimulus.

Serum Components that Suppress IGF-II Expression

Thus we believe that we have clearly established a role for autocrine expression of the IGF-II gene in myogenesis. This raises additional questions, as is always the case. For us, the most interesting one is what component of serum suppresses autocrine expression of IGF-II. Initial experiments in our laboratory show that TGF-ß is a potent inhibitor of IGF-II expression (Magri et al., manuscript in preparation), and Rosenthal et al. (1992) have demonstrated that FGF inhibits expression of the IGF-II gene in BC3H1 cells. For us, the most interesting inhibitors are the IGFs themselves.

Feedback Inhibition by the IGFs

Research currently in progress in our laboratory has demonstrated that IGF-I, IGF-II, and insulin are all potent inhibitors of IGF-II gene expression in L6A1 myoblasts incubated in low serum, and their relative potency is that associated with the IGF-I receptor; i.e., IGF-I > IGF-II >= insulin. Additional evidence that the type I receptor mediates the suppression of IGF-II expression is now being obtained with IGF analogs. Those that bind poorly to the type II receptor show only slightly diminished suppression of IGF-II expression, but reduced binding to the type I receptor is associated with less activity in suppressing IGF-II. IGF-II analogs that bind poorly to the type I receptor have only modest activity. It is interesting that very actively secreting cells such as those of the Sol 8 do not exhibit such feedback inhibition by exogenous IGFs; possibly their lack of this control may account for their high levels of IGF-II expression.

Lack of "Escape" from IGF Feedback

The induction of myogenesis by the IGFs involves early responses (most probably the induction of myogenin gene expression) that initiate a set of later processes insensitive to the administration of exogenous inhibitors such as TGF-ß (Florini et al., 1986). It seemed possible that cells such as the Sol 8 myoblasts might similarly have progressed beyond a sensitive stage and thus exhibited little or no feedback by exogenous IGFs. Accordingly we investigated the possibility that autocrine expression of IGF-II exhibits a "commitment" like that of myogenesis; i.e., that there might be a time after which exogenous IGFs no longer feed back to suppress IGF-II mRNA elevation? The answer given by a series of such experiments is an unequivocal "no." At all times we have tried, addition of exogenous IGFs always gave a subsequent decrease in the IGF-II mRNA content of myoblasts.

SUMMARY AND CONCLUSIONS

The Role of IGFs in Myogenesis

Thus we are now convinced that the control of myogenesis by IGFs is a general phenomenon that occurs in all skeletal muscle cells, whether or not IGFs are added to the "differentiation" medium. We believe that several medium components contribute to the suppression of IGF-II expression in myoblasts incubated in high serum "growth" medium, and conclude that the IGF-I receptor mediates the feedback inhibition of IGF-II gene expression in muscle cells.

Mechanism of Induction of Myogenesis by IGFs

The observations summarized here now permit a reasonably coherent overview of the stimulation of myogenic differentiation by the IGFs. It seems clear that all IGFs act by binding to the Type I IGF receptor, and that this process is inhibited to a significant extent by IGF binding proteins secreted by the target myoblasts. A major, but possibly not the only relevant effect of this binding is the induction of expression of the myogenin gene; this induction appears to require the presence of myf-5 protein, at least during the early part of the response. Cells capable of a mitogenic response undergo a round of division in response to IGF-I, thus delaying their entry into the final processes of postmitotic terminal differentiation. Other laboratories have shown that myogenin complexes with one or more widely occurring proteins such as E12 or E47 to form an active complex that interacts with CAnnTG elements in muscle specific genes, turning on expression of those genes and thus initiating the phenotype associated with terminally differentiated skeletal muscle.

ACKNOWLEDGMENTS

The work summarized here was supported by grants HL11551 and AG05557 from the National Institutes of Health, and grant 9000716 from the United States Department of Agriculture. Essential technical assistance was provided by Suzette Roof and Melleny Hale. We thank Drs. Mark Benedict for ^{125}I-IGFs, and Margaret Cascieri, Ron Rosenfeld, John Ballard and Paul Walton for IGF analogs that were very helpful to us. We are very grateful to Dr. Helen Blau for the primary human muscle cells, to Eric N. Olson for the myogenin constructs and important suggestions concerning our research.

REFERENCES

Camacho-Hubener, C., Busby, W. H., McCusker, R. H., Wright, G., and Clemmons, D. R., 1992, Identification of the forms of Insulin-like Growth Factor-binding proteins produced by human fibroblasts and the mechanisms that regulate their secretion, *J. Biol. Chem.* 267:11949.

Conover, C. A., 1991, A unique receptor-independent mechanism by which Insulin-like Growth Factor-I regulates the availability of Insulin-like Growth Factor binding in normal and transformed human fibroblasts, *J. Clin. Invest.* 88:1354.

Davis, R. L., Weintraub, H., and Lassar, A. B., 1987, Expression of a single transfected cDNA converts fibroblasts to myoblasts, *Cell* 51:987.

Edmondson, D. G., and Olson, E. N., 1993, Helix-loop-helix proteins as regulators of muscle-specific transcription, *J. Biol. Chem.* 268:755.

Ewton, D. Z., Falen, S. L., and Florini, J. R., 1987, The type II IGF receptor has low affinity for IGF-I analogs: pleiotypic actions of IGFs on myoblasts are apparently mediated by the type I receptor, *Endocrinol.* 120:115.

Florini, J. R., 1989, Assay of creatine kinase in microtiter plates using thio-NAD to allow monitoring at 405 nM, *Anal. Biochem.* 182:399.

Florini, J. R., Ewton, D. Z., Falen, S. L., and Van Wyk, J. J., 1986, Biphasic concentration dependency of the stimulation of myoblast differentiation by somatomedins, *Am. J. Physiol. (Cell. Physiol).* 250:771.

Florini, J. R., Ewton, D. Z., and Roof, S. L., 1991a, IGF-I stimulates terminal myogenic differentiation by induction of myogenin gene expression, *Mol. Endocrinol.* 5:718.

Florini, J., Magri, K., Ewton, D., James, P., Grindstaff, K., and Rotwein, P., 1991b, "Spontaneous" differentiation of skeletal myoblasts is controlled by autocrine secretion of insulin-like growth factor-II., *J. Biol. Chem.* 266:15917.

Hinterberger, T. J., Sassoon, D., A., Rhodes, S. J., and Konieczny, S. F., 1991, Expression of the muscle regulatory factor MRF4 during somite and skeletal myofiber development, *Develop. Biol.* 147:144.

Konigsberg, I. R., 1971, Diffusion-mediated control of myoblast fusion, *Develop. Biol.* 26:133.

Li, L., and Olson, E. N., 1992, Regulation of muscle cell growth and differentiation by the MyoD family of helix-loop-helix proteins, *Adv. in. Cancer. Res.* 58:95.

Linkhart, T A., Clegg, C. H., Lim, R. W., Merrill, G. F., Chamberlain, J. S., and Hauschka, S. D., 1982, Control of mouse myoblast commitment to terminal differentiation by mitogens. (in Pearson, M. L., and Epstein, H. F. (Eds): Molecular and Cellular Control of Muscle Development., Cold Springs Harbor Press. p 877.

Mangiacapra, F., 1992, Molecular characterization of myogenic differentiation: Expression of myf-5 and myogenin, M. S. Thesis, Syracuse University.

Mangiacapra, F. J., Roof, S. L., Ewton, D. Z., and Florini, J. R., 1992, Paradoxical decrease in myf-5 mRNA levels during induction of myogenic differentiation by IGFs, *Mol. Endocrinol.* 6:2038.

McCusker, R. H., and Clemmons, D. R., 1988, Insulin-like Growth Factor binding protein secretion by muscle cells: Effects of cellular differentiation and proliferation, *J. Cell. Physiol.* 137:505.

McCusker, R. H., Camacho-Hubner, C., and Clemmons, D. R., 1989, Identification of the types of Insulin-like Growth Factor-binding proteins that are secreted by muscle cells in vitro, *J. Biol. Chem.* 264:7795.

Olson, E. N., 1990, MyoD family: A paradigm for development?, *Genes. &. Develop.* 4:1454.

Ott, M.-O., Bober, E., Lyons, G., Arnold, H. H., and Buckingham, M., 1991, Early expression of the myogenic regulatory gene, myf-5, in precursor cells of skeletal muscle in the mouse embryo, *Development* 111:1097.

Rhodes, S. J., and Konieczny, S. F., 1989, Identification of MRF4: A new member of the muscle regulatory factor gene family, *Genes. &. Develop.* 3:2050.

Rosenthal, S. M., Brown, E. J., Brunetti, A., and Goldfine, I. D., 1991, Fibroblast Growth Factor inhibits Insulin-like Growth Factor (IGF)-II gene expression and increases IGF-I receptor abundance in BC3H1 muscle cells, *Mol. Endocrinol.* 5:678.

Tollefsen, S. E., Lajara, R., McCusker, R. H., Clemmons, D. R., and Rotwein, P., 1989, Insulin-like Growth Factors (IGF) in muscle development. Expression of IGF-I, the IGF-I receptor, and an IGF binding protein during myoblast differentiation, *J. Biol. Chem.* 264:13810.

Tollefsen, S. E., Sadow, J. L., and Rotwein, P., 1989, Coordinate expression of Insulin-like Growth Factor II and its receptor during muscle differentiation, *Proc. Natl. Acad. Sci. USA.* 86:1543.

Weintraub, H., Davis, R., Tapscott, S., Thayer, M., Krause, M., Benezra, R., Blackwell, T. K., Turner, D., Rupp, R., Zhuang, Y., and Lassar, A., 1991, The myoD gene family: notal point during specification of the muscle cell lineage. *Science.* 251:761.

Wright, W. E., Sassoon, D. A., and Lin, V. K., 1989, Myogenin, a factor regulating myogenesis, has a domain homologous to MyoD, *Cell* 56:607.

IGF-II IN THE PATHOGENESIS OF RHABDOMYOSARCOMA:
A PROTOTYPE OF IGFs INVOLVEMENT IN HUMAN TUMORIGENESIS

Caterina P. Minniti and Lee J. Helman

[1]Pediatric Department, Georgetown University Medical Center
Washington, DC 20007
[2]Pediatric Branch, National Cancer Institute
Bethesda, MD 20892

INTRODUCTION

In recent years there has been a growing interest in the role of peptide growth factors in the regulation of the biologic behavior of several human malignancies. Among those peptides that might have importance in neoplastic pathogenesis are the insulin-like growth factors I and II (IGF-I and IGF-II). These proteins have been shown to stimulate myoblast proliferation and differentiation (1), to promote nutrient uptake and to inhibit proteolysis (2, 3). IGF-II mRNA is highly expressed in human skeletal fetal muscle tissue, but it is not detectable in normal adult skeletal muscle by standard Northern analysis and *in situ* hybridization (4). In mouse myoblasts, IGF-I and IGF-II mRNA levels increase transiently (IGF-II> >IGF-I) within 48-72 hours of the beginning of the myogenic differentiation process. The expression of mRNA is accompanied by secretion of the peptides in the culture medium that peaks at hour 92. IGF-I receptors increase transiently, doubling by 48 hours after the onset of differentiation, while IGF-II receptors increase and remain at a higher number throughout the differentiated state (5, 6). The production of IGF-binding proteins of Mr 29,000 to 32,000 is also induced throughout differentiation (7).

Rhabdomyosarcomas, the most common soft tissue sarcomas of childhood, are embryonal tumors that are thought to arise from primitive skeletal muscle-forming cells as a developmental disturbance during muscle formation. The relationship between the IGFs, and in particular IGF-II, and the pathogenesis of these tumors will be discussed in this chapter.

EXPRESSION OF IGF-II

Analysis of mRNA from human rhabdomyosarcoma cell cultures (RD and Rh30) and fresh human tumor tissues has reveled abundant levels of the 6.0, 4.8, and 1.8 kb mRNA species recognized by a human exon 9 cDNA IGF-II probe, whereas no detectable mRNA was present in normal adult muscle (4) (Fig 1).

Current Directions in Insulin-Like Growth Factor Research,
Edited by D. LeRoith and M.K. Raizada, Plenum Press, New York, 1994

327

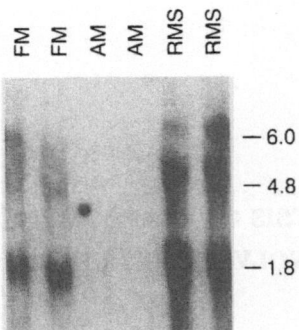

Figure 1. Northern blot analysis of IGF-II expression in normal fetal muscle (FM), rhabdomyosarcoma tumors (RMS), and normal adult muscle (AM). Total RNA (30 μg/lane) was isolated from normal tissues and cell lines, size fractionated on 1% agarose-2.2 m formaldehyde gels, transferred to Nytran membranes (Schleicher & Schuell), and hybridized to 1 x 10⁸ cpm/ml of ³²P-labeled exon 7 human IGF-II cDNA probe (bases 2932-4060). The final wash in 0.1x saline sodium citrate 1 sodium dodecyl sulfate at 65° C for 1 hour. (Reproduced with permission [4]).

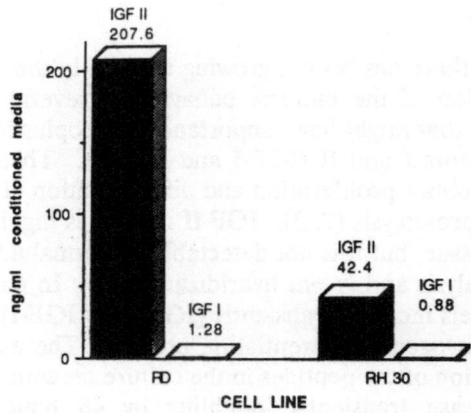

Figure 2. Human rhabdomyosarcoma cells secrete IGF-II peptide. Panel A: Radioimmunoassay determination of IGF-I and -II in conditioned media from RD cells (left) and Rh 30 cells (right). 10 ml of conditioned media were prepared and analyzed for IGF-I and -II using a specific RIA. Panel B: Bio-Gel P-60 chromatography on conditioned media from RD cells. Gel chromatography V total column volume. Fraction number is shown in the x-axis and IGF-II concentration, as determined by RIA for each fraction, is shown on the y-axis. (Reproduced with permission [4]).

When we assayed conditioned media from both cell lines for the presence of IGF-II using a radioimmunoassay, significant levels of IGF-II protein, 207 ng/ml for RD and 42 ng/ml for RH30, (normal human fibroblasts <20 ng/ml) were detected (Fig 2). IGF-II produced by rhabdomyosarcoma tumor cells is mostly in a high molecular form (big IGF-II), with only 25% corresponding to the size of normal serum IGF-II. No IGF-I peptide is secreted into the conditioned media.

An essential requirement for an autocrine growth factor is its production *in vivo* by the tumor cells themselves, and not by the surrounding stroma. This has been

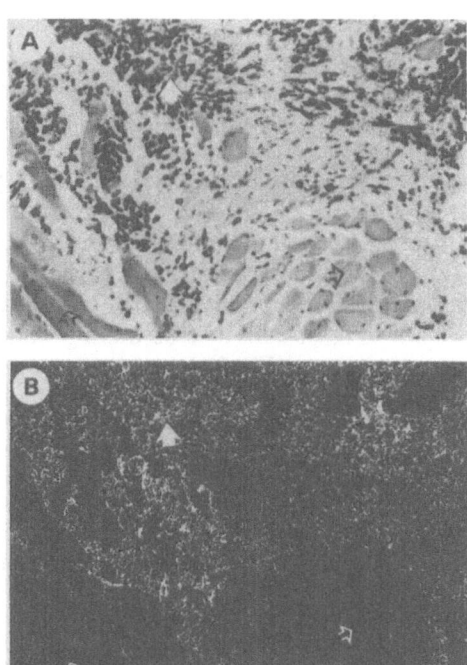

Figure 3. In situ analysis of IGF-II in RNA expression in human rhabdomyosarcomas. Panel A and B are microphotographs (625x) of the same section of an embryonal rhabdomyosarcoma specimen, taken in bright field and in dark field, respectively. The solid arrows point to areas rich in tumor cells, intensely covered by silver grains. The open arrows point to normal muscle entrapped in the tumor, which is negative for IGF-II mRNA expression (silver grains are the same as background).

demonstrated by *in situ* analysis of 26 fresh rhabdomyosarcoma tumor specimens (8) (Fig 3).

All tumors examined expressed IGF-II mRNA, albeit with a wide range of intra and interspecimen signal intensity. This variability cannot be correlated to differences in the histology of the tumors, since both embryonal and alveolar subtypes were positive with different levels of intensity in different specimens. On the other hand, specimens that contained no tumor, or a different type of "small round blue cell tumor" (i.e., Ewing's sarcoma or Peripheral NeuroEctodermal Tumor [PNET]) were consistently negative for IGF-II expression.

EXPRESSION OF IGF RECEPTORS

The biological actions of IGF-II are mediated by specific cell membrane receptors (9). Like insulin, and IGF-I, with which IGF-II shares structural homology, IGF-II is capable of binding the IGF-I and the insulin receptors. IGF-I receptors are heterodimers of α and β polypeptide chains. The α subunits (Mr 135,000) contain the ligand binding site, and the β subunits (Mr 90,000) contain a transmembrane domain, an ATP-binding site, and a tyrosine kinase domain. Both chains are glycosylated and linked together by disulfide bonds. IGF-II also binds with high affinity to a distinct receptor, the mannose 6-P/IGF-II receptor, a single chain polypeptide of a Mr of 270,000 (10, 11).

A typical type I IGF receptor is present on rhabdomyosarcoma tumor cells, as shown by competitive binding experiments using ^{125}I-IGF-I (Fig 4).
Binding of IGF-I to the type I receptor was inhibited in a dose-dependent manner by α-IR3, a blocking mouse monoclonal antibody specific for the type I IGF receptor, but not by an irrelevant mouse IgG, MOPC-21 (12, 13). The same rhabdomyosarcoma tumors and cell lines also expressed the 10 and 6.3 kb mRNA species recognized by a cDNA probe for the type I IGF receptor (4, 14).

In addition to the type I IGF receptor we also demonstrated the presence of the Mannose 6-P/IGF-II receptor in these cells by crosslinking, under reducing conditions, ^{125}I-IGF-II to microsomal membranes (Fig 5).

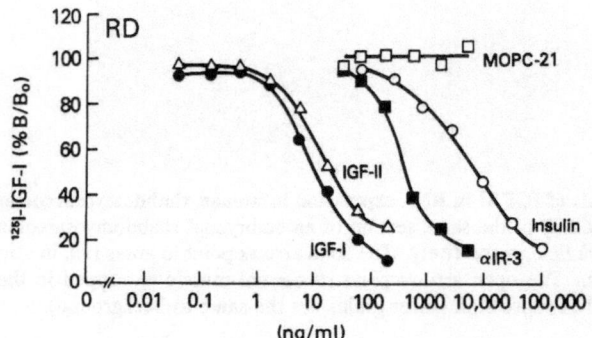

Figure 4. Radioreceptor assays of I[125]-IGF-I on RD and RH-30 cells. Determination of surface IGF type I cells was performed on RD cell suspensions. Cells were incubated overnight at 4° C in the presence of 20,000 cpm of I[125]-IGF-I and the indicated concentration of cold IGF-I (●), IGF-II (▲), insulin (○), α IR-3 (■), and MOPC-21 (□). (Reproduced with permission [4]).

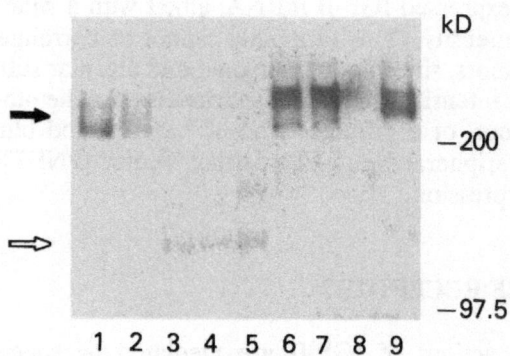

Figure 5. Cross-linking of the IGF-II receptor. RD cell membranes were incubated under reducing conditions with 30,000 cpm of I[125]-IGF-II (lane 9) and the following cold ligands: goat non-immune serum at 1:100 and 1:50 dilution (lanes 1 & 2), anti-type II receptor Ab at 1:200, 1:100, and 1:50 dilution (lanes 3-5), 1 μg/ml insulin (lane 6), 200 ng/ml IGF-I (lane 7), and 200 ng/ml cold IGF-II (lane 8). Lane 9 indicates the labeling of both receptors by I[125]-IGF-II. The upper solid arrow indicates the type II IGF receptor at 270 kDa, the lower clear arrow indicates the α subunit of the type I IGF receptor at 130 kDa. (Reproduced with permission [29]).

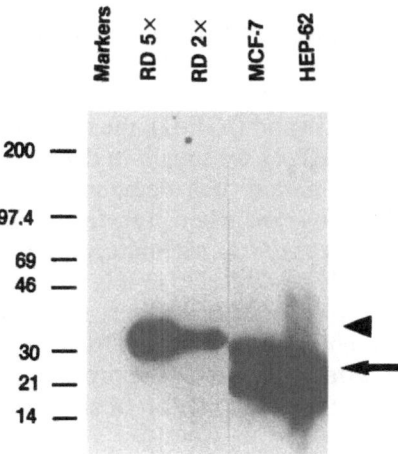

Figure 6. Western ligand blotting of conditioned media from RD tumor cells. Conditioned media samples from confluent RD cell cultures, in serum free medium, were concentrated (5x or 2x) and then electrophoresed on 10% SDS-polyacrylamide minigels (NOVEX) according to the methods of Laemmli (40). Pre-stained protein standards were electrophoresed in a parallel line. The size fractionated proteins were electroblotted onto nitrocellulose. Membrane-immobilized proteins were incubated with ^{125}I-IGF-II and visualized by autoradiography. Lane 1 -- markers, lane 2 -- RD conditioned media 5x, lane 3 -- RD conditioned media 2x, lane 4 -- MCF-7 conditioned media, and lane 5 -- HEP-G2. Full arrow = BP-4, arrowhead = BP-2.

The specificity of the 270 kDa band for the type II receptor is demonstrated by its disappearance in the presence of cold IGF-II or anti-type II receptor Ab, but not in the presence of excess insulin or IGF-I (lanes 3-5 anti-type II receptor Ab; lane 6, cold insulin; lane 7 cold IGF-I; lane 8 cold IGF-II).

EXPRESSION OF IGF BINDING PROTEINS

Nearly all IGFs in blood, extracellular fluids, and cell culture supernatants are noncovalently bound to several distinct forms of specific carrier proteins, called insulin-like growth factor-binding proteins (IGF-BPs) (15). They are postulated to influence local IGF's bioactivity *in vivo* and to extend greatly IGF's plasma half-life (16). A human rhabdomyosarcoma cell line, A673, has previously been reported to synthesize large amounts of a 34 kDa and smaller amounts of a 24 kDa IGF-BP (17). However, a reexamination of the tumor from which the A673 cell line was originated (18) led to its classification as a PNET (M. Tsokos, personal communication). We therefore sought to determine whether RD rhabdomyosarcoma tumor cells could secrete IG-BPs in the conditioned media, after 24 hours of incubation in serum-free medium. Ligand blot analysis with radiolabeled IGF-II showed the presence of the same 34-36 and 24 kDa IGF-BP bands that likely correspond to BP-2 and BP-4, respectively (BP-2> >BP-4) (Fig 6).

No BP-3 or BP-1 was found, even when the total mRNA was hybridized to specific synthetic oligonucleotide DNA probes. These findings are in agreement with data previously published on the IGF-BP's production in normal rodent muscle cells (20).

ESTABLISHMENT OF IGF-II AS AN AUTOCRINE GROWTH FACTOR

Once we showed the presence of all of the components of the IGF system in human rhabdomyosarcoma, the ligand (IGF-II), the receptors (type I and type II), and the binding proteins (BP-2 and BP-4), we sought to determine its biological relevance. Autonomous growth of these tumors was demonstrated by the ability of several rhabdomyosarcoma cell lines to grow when transferred in serum and mitogen-free medium (N_2E), which contains transferrin, selenium, putrescine, and progesterone. We have propagated the cell lines RD and RH30 in this media for over 36 and 24 months respectively. Both cell lines grow in N_2E medium at a rate that is identical to their growth rate in 10% FBS (doubling time 24 hours for RD and 72 hours for RH30, as measured by cell number) after a lag time of approximately 48 hours, thought to be due to the time necessary for the accumulation of sufficient IGF-II peptide amounts into the media.

To date, most of the mitogenic and anabolic effects of IGF-II have been shown to be mediated through the type I receptor (21). Therefore, in order to investigate whether this receptor was mediating the IGF-II mitogenic signal, we treated RD tumor cells with α-IR3, a blocking monoclonal antibody for the type I receptor. This led to a dose-dependent inhibition of [³H]thymidine incorporation, up to 70% inhibition compared to controls with the maximal effect seen at an α-IR3 concentration of 0.8 μg/ml (Fig 7).

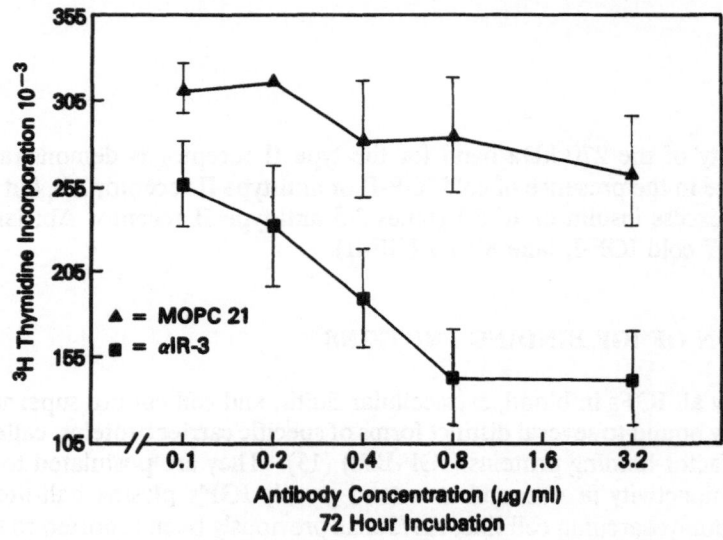

αIR-3 Inhibition of RD Cells in Serum-Free Media

Figure 7. A monoclonal antibody to the type I IGF receptor inhibits the growth of rhabdomyosarcoma cell lines. Dose response of RD cells to α IR-3. Cells were plated in triplicate into 96-well plates with the indicated concentrations of α IR-3. After 72 hours incubation, [³H]thymidine incorporation was determined. Each data point represents the mean of triplicate determinations. (Reproduced with permission [4]).

The finding that not all the mitogenic effect of IGF-II is abolished by pretreatment of RD tumor cells with α-IR3 correlates well with the radioreceptor assay data which shows a maximum of 70% displacement of labeled IGF-II at this antibody concentration (see Fig 4). On the other hand, the presence of additional mechanisms

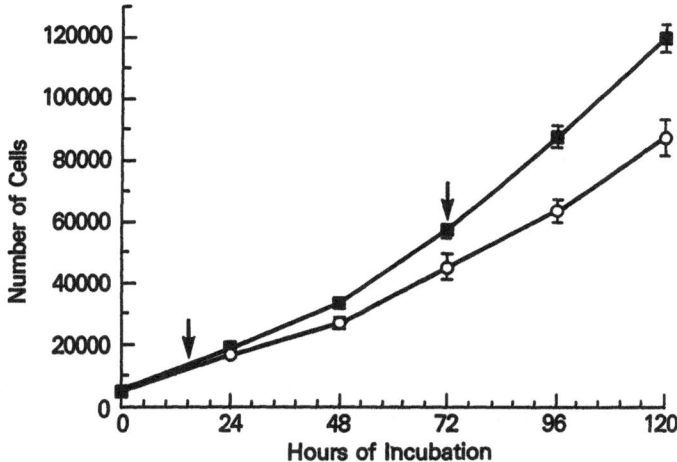

Figure 8. The anti-type II receptor Ab does not inhibit RD cell proliferation. RD cells were plated in triplicate into 96-well plates in the presence of 1:100 dilution of the anti-type II receptor Ab (■) or goat non-immune serum (○) as a control. The arrows identify the times at which the antibody or the control serum were added to the cell cultures. Each data point represents the mean of triplicate determinations. (Reproduced with permission [29]).

of growth, not mediated by the IGF-II loop, cannot be excluded. The possibility that part of the mitogenic activity of the IGF-II is mediated by the type II receptor is very unlikely since, in cell growth experiments, pretreatment of RD tumor cells with an anti-type II receptor antibody (22) does not cause any inhibition of cell proliferation (Fig 8).

IGF-II AS AN AUTOCRINE MOTILITY FACTOR

Rhabdomyosarcomas are highly metastatic tumors, as demonstrated by the fact that most patients with localized disease will eventually succumb to malignant dissemination if only a local treatment approach is utilized (i.e., surgery or radiation therapy)(23). Cell locomotion is a crucial step in the three-stage metastatic process (24, 25) (Fig 9) and IGFs have been shown to stimulate cell motility in various malignancies (26), including melanoma, breast, bladder, and ovarian carcinoma. We therefore sought to determine whether IGF-II might function as an autocrine motility factor in human rhabdomyosarcoma. IGF-II elicited a dose-dependent migration response of rhabdomyosarcoma cells (Fig 10A). To ascertain whether IGF-II-stimulated motility was random (chemokinetic) or directed toward a concentration gradient (chemotactic), a checkerboard analysis was performed (27) in which RD cells were exposed to both uniform and gradient concentrations of IGF-II. Cell motility was predominantly chemokinetic, as demonstrated by optimal motility when the cells were exposed to uniform concentrations of IGF-II above and below the filter (diagonal, Fig 10B). This is consistent with the fact that IGF-II is produced by the tumor cells themselves and might increase the ability of the tumor cells to detach from the tumor mass and move into the circulation.

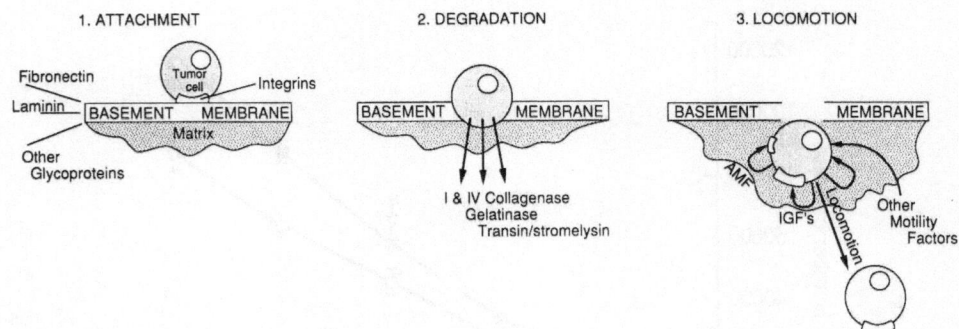

Figure 9. Three-step theory of metastasis. The three steps include: 1. attachment of cells to extracellular matrix; 2. local degradation of the extracellular matrix; and 3. locomotion through the locally degraded matrix. (Reproduced with permission [*Pediatric Clinics of North America* -- Vol. 38, No. 2, April 1991]).

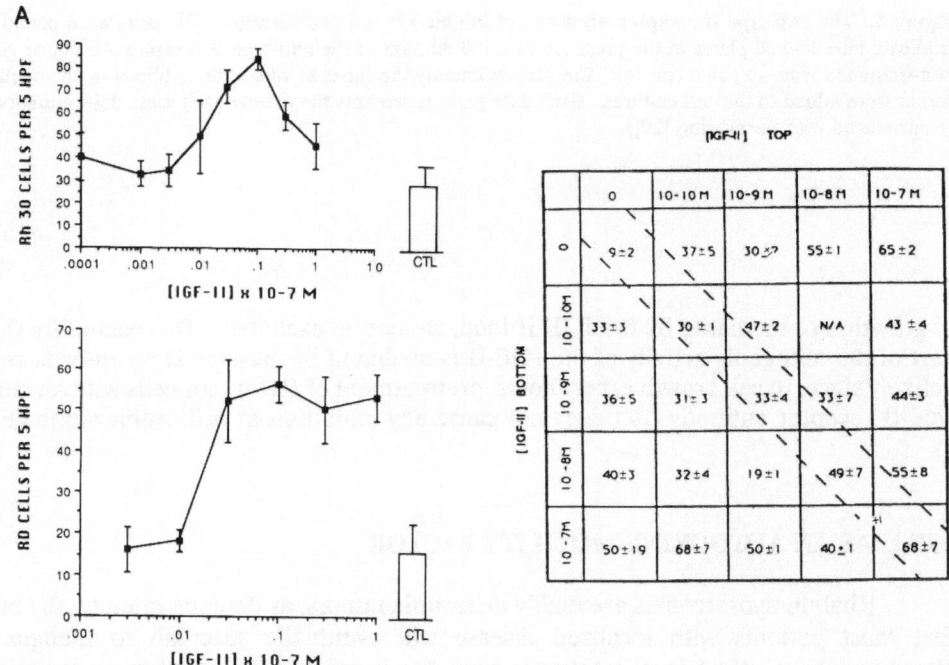

[IGF-II] TOP	0	10-10 M	10-9 M	10-8 M	10-7 M
0	9±2	37±5	30±?	55±1	65±2
10^-10 M	33±3	30±1	47±2	N/A	43±4
10^-9 M	36±5	31±3	33±4	33±7	44±3
10^-8 M	40±3	32±4	19±1	49±7	55±8
10^-7 M	50±19	68±7	50±1	40±1	68±7

Figure 10A. IGF-II stimulates motility of rhabdomyosarcoma cells. Rh30 (upper panel) and RD (lower panel) cells were assayed for their motility response to IGF-II. The motility assay was performed using a modified Boyden chamber. The concentration of IGF-II is plotted against the number of cells migrating through the filter per five high power fields (HPF). Each point represents a triplicate experiment. The bar graph to the right represents the number of cells per five high power fields when no IGF-II was added to the bottom chamber (CTL). (Reproduced with permission [29]).

Figure 10B. Checkerboard analysis of IGF-II induced motility. Motility of RD cells to increasing uniform concentrations of IGF-II. The diagonal indicates wells in which the IGF-II concentration above and below the filter barrier was equal. Lower triangle, numbers of cells migrating toward a positive concentration gradient of IGF-II. The values shown are the number of cells per five high power fields (500x), and the standard error is indicated. N/A, data not available for this point (n=3). (Reproduced with permission [29]).

Since IGF-II binds to different receptors on the cell surface (the type I, the Mannose 6P/IGF-II receptor, or type II receptor and the insulin receptor), we tested the ability of [leu27]IGF-II, an analog with an affinity 100-fold greater for the Mannose 6P/type II receptor than to the type I receptor to stimulate motility (28). This analog was able to elicit RD cell motility at a concentration as low as 10^{-10}M (Fig 11). At this concentration, [leu27]IGF-II does not bind to the type I receptor as shown by radioreceptor binding assays (Fig 12).

To further characterize the role of the type II receptor in mediating this response, we analyzed the ability of antibodies to the type I and II receptors to alter IGF-II mediated motility. As shown in Figure 13, panel B, a goat Ab specific for the type II receptor (22) completely inhibited IGF-II induced motility while the control goat serum had no effect ($p = 0.00095$ and $p = 0.35$, respectively) (29). Figure 13, panel A, shows that 1 μg/ml of α-IR3 did not inhibit the motility response elicited by either IGF-II or [leu27] IGF-II. The treatment of RD cells with MOPC-21 control, a matched mouse IgG (1 μg/ml), did not inhibit migration. These findings strongly imply that rhabdomyosarcoma cell motility response is mediated through the type II receptor.

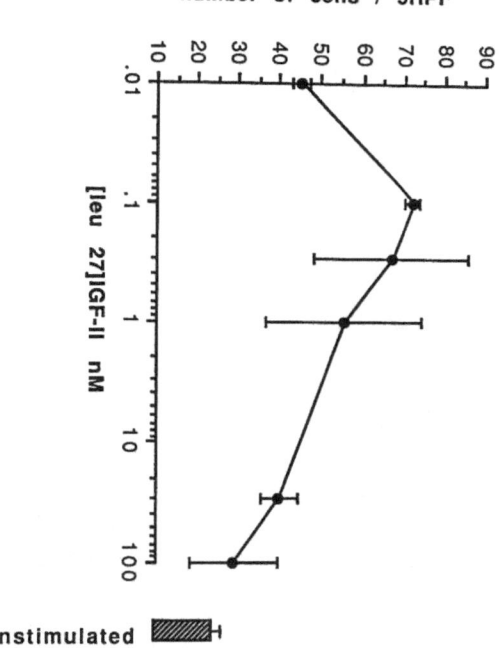

Figure 11. Rhabdomyosarcoma cells migrate to [Leu[27]]IGF-II. RD cells were harvested with 0.25% trypsin and allowed to regenerate the surface receptors for 2 hours at room temperature, and 1.1 x 10[5] cells were placed in the upper wells of the modified Boyden chamber. Increasing concentrations of [Leu[27]]IGF-II ranging from 0.01 to 100 nM, or plain 0.1% BSA/DMEM, used as control or unstimulated motility, were placed in the lower wells as chemoattractant. After 4 hours of incubation at 37° C, the chamber was opened and the number of cells that had migrated in response to the ligand counted. Data are presented as percent of control, obtained by dividing the mean of triplicate determinations of individual data points to the mean of the triplicate determination of the controls. Each data point is a mean ± S.E. of two individual experiments. (Reproduced with permission [29]).

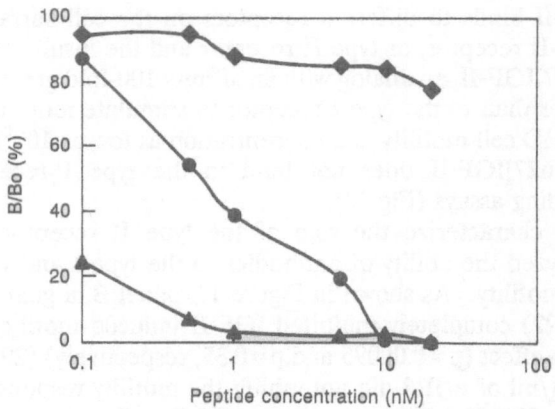

Figure 12. Radioreceptor assay on RD membranes. Rhabdomyosarcoma cell membranes (50 μg of protein) were incubated for 18 hours at 4° C with ^{125}I-IGF-I in the presence or absence of varying concentrations of IGF-I (▲), IGF-II (●), or [Leu^{27}]IGF-II (♦). Results shown are means of duplicate determinations and are representative of two independent experiments. B and Bo are specific binding in the presence and absence, respectively, of excess unlabeled ligand. (Reproduced with permission [29]).

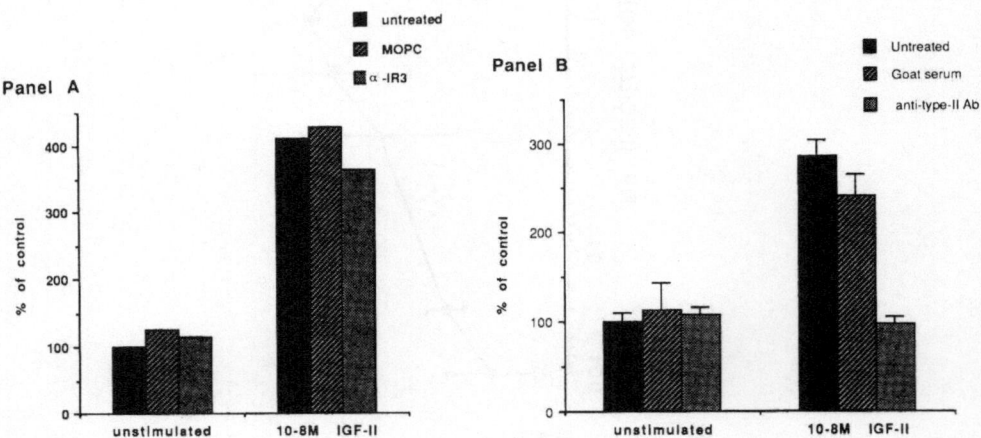

Figure 13. Motility response of rhabdomyosarcoma cells pre-treated with anti-type I or anti-type II receptor Ab. RD cells were harvested, rested, and then incubated in serum free medium with the appropriate Ab. The incubation with α-IR3, shown in panel A, did not inhibit IGF-II- nor [Leu^{27}]IGF-II-induced motility. The data presented in panel A refer to an individual representative experiment. The incubation with the anti-type II receptor antiserum, shown in panel B, totally abolished the motility response of RD cells to 10^{-8}M IGF-II. The difference between treated and untreated cells was statistically significant (p=0.009 on a two-tailed Student's t test), while the slight decrease observed with pre-incubation with goat serum was not significant (p=0.35). The data shown in panel B are the mean ± S.E. of three individual experiments. (Reproduced with permission [29]).

In conclusion, IGF-II functions as an autocrine growth factor through the type I receptor and as an autocrine motility factor through the type II receptor in human rhabdomyosarcoma cell lines.

POTENTIAL CLINICAL IMPLICATIONS

We have defined two independent autocrine loops for IGF-II-mediated growth and motility in human rhabdomyosarcoma. Although the production of IGF-II by rhabdomyosarcomas might represent an expression of the embryonal nature of the tumor, there is little doubt that its presence constitutes a vital part of the biology of this tumor. Therefore, we sought to investigate novel strategies of anti-neoplastic therapy, specifically designed to interfere with the autocrine IGF-II loops.

We chose to study suramin, a polysulphonated naphtylurea (30A, 30B) that, *in vitro*, binds avidly to and neutralizes the biologic effects of a wide range of growth factors. Several reports indicate that suramin interferes with the cellular binding of platelet-derived growth factor (PDGF) (31), basic fibroblast growth factor (bFGF) (32), epidermal growth factor (EGF) (33), transforming growth factor beta (TGF-ß) (34), and IGF-I (35). We first studied the effect of suramin on the growth of human embryonal rhabdomyosarcoma cell line RD and sought to determine the mechanism by which suramin exerted its biological action. Increasing concentrations of suramin ranging from 62.5 to 250 μg/ml added to RD cells in the serum-free medium N2E resulted in a dose- and time-dependent inhibition of [^{3}H]thymidine uptake as depicted in Fig 14. This effect was demonstrable after a minimum of two days of incubation, with an ED50 of ~125 μg/ml (9 x 10^{-5}M), and a parallel reduction in RD cell number. The effect on [^{3}H]thymidine incorporation preceded the reduction in cell number by 24 - 36 hours.

Suramin injected intravenously is known to be 99.7% bound to plasma proteins (36, 37). Therefore, the effects observed in serum-free conditions *in vitro* at a defined suramin concentration may not reflect those observed *in vivo* at a similar serum level. A repeat analysis of the effect of suramin on cell growth and [^{3}H]thymidine uptake in RD cells grown in DMEM supplemented with 10% FBS demonstrated a time and dose dependent decrease in [^{3}H]thymidine uptake and in cell number. These effects required a longer period of exposure to suramin (96 hours) than those observed in serum-free conditions. Inhibition of cell growth was seen at concentrations slightly higher than those needed in serum-free conditions: ED50 of ~225 μg/ml (~1.6 x 10^{-4}M).

A radioreceptor assay was utilized to study the interaction of suramin with IGF binding to the type I receptor. In competitive binding experiments using ^{125}I-labeled IGF-I, binding at 0° C was inhibited in a dose-dependent manner by unlabeled IGF-I (ID50 8.8 ng/ml). IGF-II was five times less potent, insulin was roughly 1000-fold less potent, and suramin 2.5 x 10^4 times less potent than IGF-I in displacing labeled IGF-I (Fig 15).

The curves depicted in Figure 15 are the result of the collective analysis of the data obtained from all experiments, using the computer program ALLFIT (38). In addition, we performed radioreceptor assays on RD cells previously treated with 125 and 250 μg/ml of suramin for 48 hours. The amount of ^{125}I-IGF-I bound (total uptake) and the displacement curve with unlabeled IGF-I was not affected by suramin preincubation, suggesting that the number of receptors was not altered by this treatment. Because growth factor-related events are mediated through binding to specific cell surface receptors, we wanted to determine whether there was a correlation between suramin's ability to inhibit RD cell growth and the inhibition of ligand binding in the radioreceptor assay. We therefore analyzed the curve of growth-inhibition and the curve of IGF-I displacement in the computer program ALLFIT. The two curves could be fit with one equation (Fig 16) (ID50 and ED50 $1.4 \times 10^{-4}M \pm 1.0 \times 10^{-5}M$), showing an excellent correlation between the doses of suramin required for both effects. These findings suggest that the inhibitory action of suramin on RD cell growth is due to an interference with ligand binding to the type I IGF receptor on the cell surface.

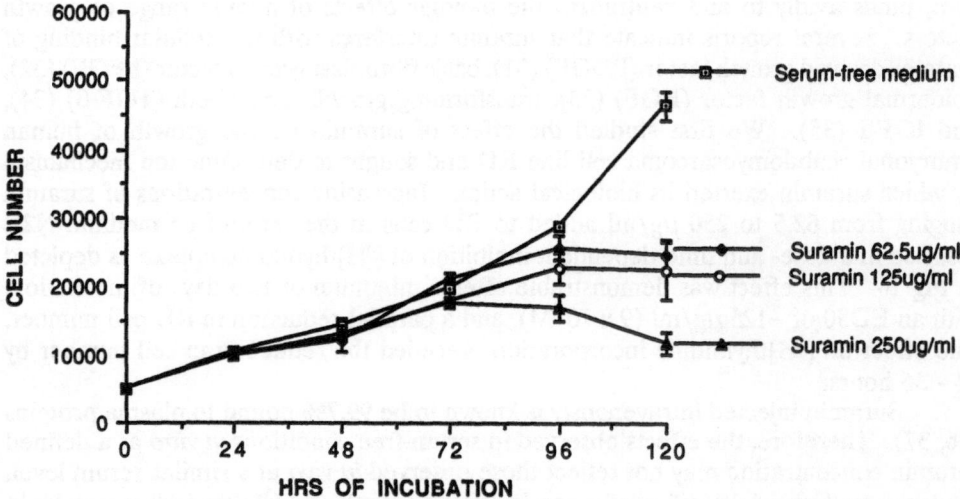

Figure 14. Time and dose curves of suramin inhibition of RD cells grown in serum free medium (N2E). RD cells were harvested with Puck's saline containing 1 mM EDTA and plated in triplicate at 5×10^3/well in 96-well microtiter plates in serum free medium. Suramin was added at 62.5 μg/ml (♦), 125 μg/ml (o), and 250 μg/ml (▲), after allowing the cells to adhere to the wells (12-18 hours). Medium alone (the vehicle in which suramin was diluted) was added to the control wells (□). A: [³H]thymidine uptake was determined after 24, 48, and 72 hours of incubation at 37° C in 6% CO_2. (Reproduced with permission [30B]).

Moreover, addition of pharmacological doses (200 ng/ml) of exogenous IGF-II to suramin-treated RD cells was able to counteract the inhibition of cell growth previously observed (Fig 17). We therefore propose a model of suramin action like a "saturable" system in which, under physiologic conditions, suramin inhibits cell growth by interfering with the binding of IGF-II to the type I receptor. In the presence of excess IGF-II, all suramin is bound and therefore some peptide is able to exert its mitogenic effect on the cells.

338

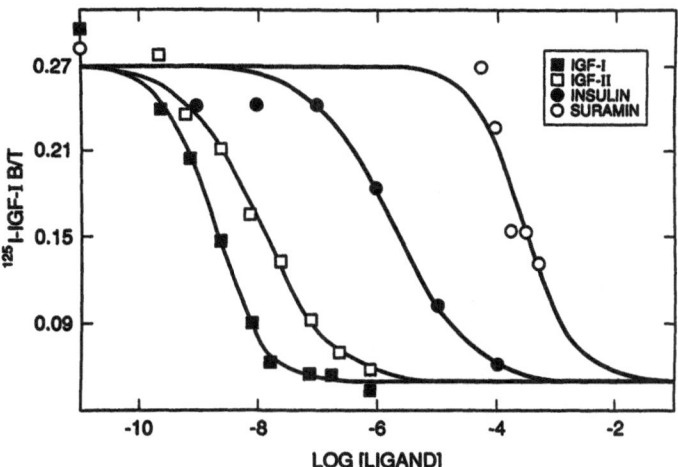

Figure 15. Displacement of specific ^{125}I-IGF-I binding to RD cells by IGF-I (■), IGF-II (□), insulin (●), and suramin (○). RD cells (1 x 10^6) were resuspended in 4 (2 hydroxyethyl)-I-piperazineethanesulfonic acid binding buffer, ph 8.0, with ^{125}I-IGF-I and the indicated concentrations of unlabeled ligands. RD cells were incubated for 18 hours at 0° C, then spun at 4° C, and washed once with cold buffer, and the pellets were counted in duplicate in a Beckman 5500 gamma counter at 10% efficiency. The curves are the result of the cumulative analysis of all experiments performed under the same conditions with the program ALLFIT. (Reproduced with permission [30B]).

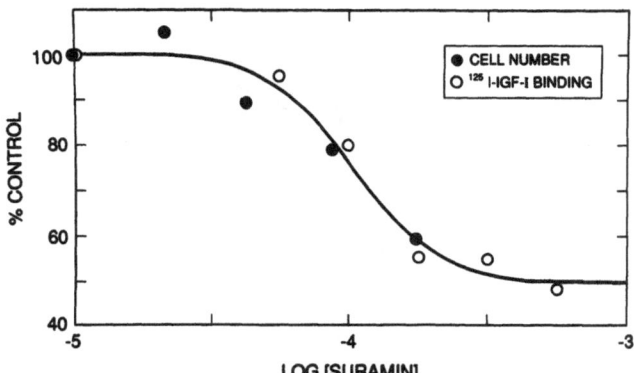

Figure 16. Correlation between the concentrations of suramin that cause inhibition of either cell growth (●) or ^{125}I-IGF-I binding (○). The statistical hypothesis that the experimental data obtained in cell growth or binding studies share common parameters (i.e., the expected maximal response, the slope factor, the ED_{50} or ID_{50} and the minimal response) was tested by first forcing the curves to share these parameters and then verifying that such constraints have minimal effects on several indicators of "goodness of fit." The residual variance test and the runs test were used as indices of the "goodness of fit" (ALLFIT). The F ratio tests were obtained by calculating the ratio of the residual variance for one curve to the overall residual variance for all the other curves. The F ratio test is not significantly different from unity, indicating that the shared parameters are compatible with each other. (Reproduced with permission [30B]).

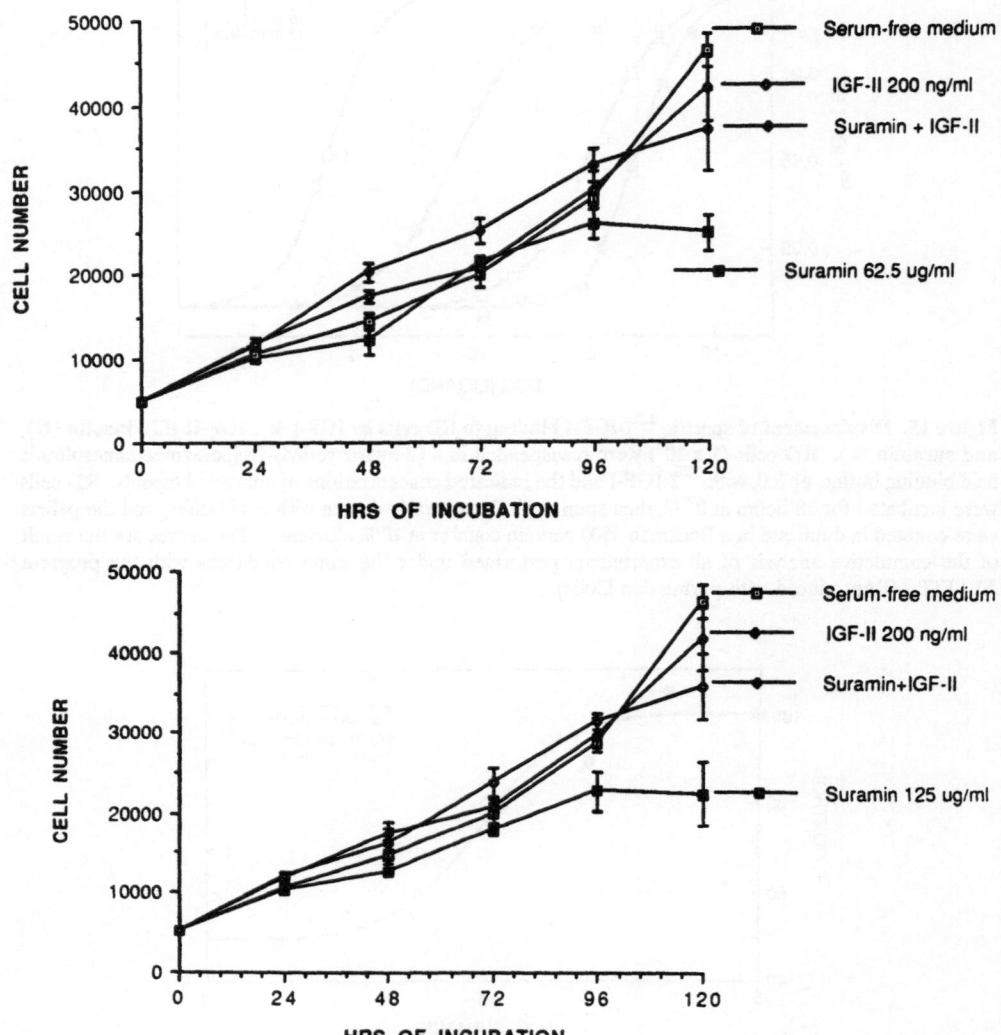

Figure 17. Effect of IGF-II on suramin-induced growth inhibition of RD cells. RD cells (5 x 10³/well) were plated in triplicate in 96-well plates in serum free medium. After allowing the cells to adhere to the wells, we added 62.5 μg/ml (upper panel) or 125 μg/ml (lower panel) of suramin alone (■) or along with 200 ng/ml of human recombinant IGF-II (♦). Two sets of control wells were prepared by adding the vehicle alone (N2E medium) (□) and the growth factor alone (◊). Cell number was assessed after 24, 48, 72, 96, and 120 hours of incubation using the MTT. Points, mean of triplicate determinations; bars, SD. (Reproduced with permission [30B]).

340

CONCLUSIONS

Human rhabdomyosarcoma have been shown to express all the components of the IGFs system: 1) the ligand (IGF-II); 2) the receptors (type I > type II); and 3) the binding proteins (BP-2 >> BP-4). The mechanism leading to high-level IGF-II expression in rhabdomyosarcoma is still under investigation in our laboratory as well as in other's. Because this growth factor is normally present in fetal skeletal muscle cells, it is unclear whether its continued expression in rhabdomyosarcoma is a reflection of the embryonic nature of the tumors or the tumorigenic transformation. Nevertheless, the dysregulated production of IGF-II has a significant effect on rhabdomyosarcoma cell growth and motility, and ultimately on tumor survival. Interfering with the ability of IGF-II to bind its cognate receptor with a polyanionic drug, like suramin, points to a new, very promising field of therapy. Specific anti-tumor therapy requires a deep understanding of growth factors and growth factor receptors as regulators of cell proliferation. The selected target of the future antineoplastic treatment will be the tumorigenic process itself, not the cells produced by the molecular events of the tumorigenic transformation.

REFERENCES

1. Florini JR, Ewton DZ, Falen SL, and Van Wyk JJ: Biphasic concentration dependeancy of stimulation of myoblast differentiation by somatomedins. Am J Physiol 250:C771-8, 1986.
2. Ewton DZ, Falen SL, and Florini JR: The type II insulin-like growth factor (IGF) receptor has low affinity for IGF-I analogs: pleiotypic actions of IGFs on myoblasts are apparently mediated by the type I receptor. Endocrinology 120:115-23, 1987.
3. Beguinot F, Kahn CR, Moses AS, and Smith RJ: Distinct biologically active receptors for insulin, insulin-like growth factor I, and insulin-like growth factor II in cultured skeletal muscle cells. J Biol Chem 260:15892-8, 1985.
4. El-Badry OM, Minniti CP, Kohn EC, Houghton PJ, Daughaday WH, and Helman LJ: Insulin-like growth factor II acts an autocrine growth and motility factor in human rhabdomyosarcoma tumors. Cell Growth Differ 1:325-31, 1990.
5. Tollefsen SE, Lajara R, McCusker RH, Clemmons DR, and Rotwein P: Insulin-like growth factors (IGFs) in muscle development: expression of IGF-I, the IGF-I receptor, and the IGF binding protein during myoblast differentiation. J Biol Chem 264(3):13810-7, 1989.
6. Tollefsen SE, Sadow JL, and Rotwein P: Coordinate expression of insulin-like Growth factor II and its receptor during muscle differentiation. Proc Natl Acad Sci USA 86:1543-7, 1989.
7. McCusker RH and Clemmons DR: Insulin-like growth factor binding protein secretion by muscle cells: Effect of cellular differentiaton and proliferation. J Cell Physiol 137:505-12, 1988.
8. Minniti CP, Tsokos M, Newton, Jr WA, and Helman LJ: Specific expression of insulin-like growth factor-II in rhabdomyosarcoma tumor cells. Am J Clin Pathol, in press.
9. Rechler MM and Nissley SP: The nature and regulation of the receptors for insulin-like growth factors. Annu Rev Physiol 47:425-42, 1985.
10. Lobel P, Dahms NM, and Kornfeld S: Cloning and sequence analysis of the cation-independent mannose 6-phosphate receptor. J Biol Chem 263(5):2563-70, 1988.
11. Kiess W, Blickenstaff GD, Sklar MM, Thomas CL, Nissley SP, and Sahagian GC: Biochemical evidence that the type II insulin-like growth factor receptor is identical to the cation-independent mannose 6-phosphate receptor. J Biol Chem 263(19):9339-44, 1988.
12. Kull FC, Jacobs S, Su Y-F, Svoboda ME, Van Wyk JJ, and Cuatrecasas P: Monoclonal antibodies to receptors for insulin and somatomedin-C. J Biol Chem 258:6561-6, 1983.
13. Rohlik QT, Adams D, Kull FC, and Jacobs S: An antibody to the receptor for insulin-like growth factor inhibits the growth of MCF-7 cells in tissue culture. Biochem Biophys Res Commun 149:276-81, 1987.
14. Ullrich A, Gray A, Tam AW, Yang-Feng T, Tsubokawa M, Collins C, Henzel W, Le Bon T, Dathuria S, Chen E, Jacobs S, Francke U, Ramachandran J, and Fujita-Yamaguchi Y: Insulin-like growth factor I receptor primary structure: comparison with insulin receptor suggests structural determinants that define functional specificity. EMBO J 5:2503-12, 1986.

15. Rechler MM and Nissley SP: Insulin-like growth factors. In: Sporn MB and Roberts AB (eds.) *Handbook of Experimental Pharmacology* 95(I):317-43, Springer-Verlag, Berlin, 1990.

16. Ibid.

17. Romanus JA, Tseng Y-HL, Yang Y W-H, and Rechler MM: The 34 kilodalton insulin-like growth factor binding protein in human cerebrospinal fluid and the A673 rhabdomyosarcoma cell line are human homologous of the rat BRL-3A binding protein. Biochem Biophys Res Comm 163:875-81, 1989.

18. Giard DJ, Aaronson SA, Toolars GJ, Arnstein P, Kersey JH, Dosik H, and Parks WP: *In vitro* cultivation of human tissues: Establishment of cell lines derived from a series of solid tumors. J Natl Cancer Inst 81:1417-23, 1973.

19. Hossenlopt D, Seurin D, Segovia-Guinson B, Hardouin S, and Binaux M: Analysis of serum insulin-like growth factor binding proteins using Western blotting: use of the method for titration of the binding proteins and competitive binding studies. Anal Biochem 154:138, 1986.

20. McCusker RH, Camacho-Hubner C, and Clemmons DR: Identification of the types of insulin-like growth factors binding proteins that are secreted by muscle cells *in vitro*. J Biol Chem 264:7795-800, 1989.

21. Czech MP: Signal transmission by the insulin-like growth factors. Cell 59(2):235-8, 1989.

22. Nolan CM, Creek KE, Grubb JH, and Sly WS: Antibody to the phosphomannosyl receptor inhibits recycling of receptor in fibroblasts. J Cell Biochem 35(2):137-51, 1987.

23. Raney RB, Hays Jr DM, Tefft M, and Triche TJ: Rhabdomyosarcoma and the undifferentiated sarcomas. In: Pizzo PA and Poplack D (eds.) *Principles and Practice of Pediatric Oncolgy*. pp. 635-58, Philadelphia: JB Lippincott Co., 1989.

24. Stauli P and Weiss L: Cell locomotion and tumor penetration. Report on a workshop of the EOR TC cell surface project group. Eur J Cancer 13:1-12, 1977.

25. Russo RG, Foltz CM, and Liotta LA: New invasion assay using endothelial cels grown on native human basement membrane. Clin Exp Metastasis 1:115-27, 1983.

26. Kohn EC, Francis EA, Liotta LA, and Schiffmann E: Heterogeneity of the motility response in malignant tumor cells: a biological basis for the diversity and homing of metastatic cells. Int J Cancer 46:287-92, 1990.

27. Zigmond SH and Hirsch JG: Leukocyte locomotion and chemotaxis. New methods for evaluation and demonstration of a cell-derived chemotactic factor. J Exp Med 137:387-410, 1973.

28. Fischer HD, Gonzalez-Noriega A, and Sly WS: Beta-glucuronidase binding to human fibroblast membrane receptors. J Biol Chem 255(11):5069-74, 1980.

29. Minniti CP, Kohn EC, Grubb JH, Sly WS, Youngman O, Müller HL, Rosenfeld RG, and Helman LJ: The insulin-like growth factor II (IGF-II)/mannose 6-phosphate receptor mediates IGF-II-induced motility in human rhabdomyosarcoma cells. J Biol Chem 267(13):9000-4, 1992.

30A. Hawking F: Suramin: with special reference to onchocerciasis. Adv. Pharmacol Chemother 15:289-322, 1978.

30B. Minniti CP, Maggi M, and Helman LJ: Suramin inhibits the growth of human rhabdomyosarcoma by interrupting the insulin-like growth factor II aitocrine growth loop. Cancer Res 52:1830-5, 1992.

31. Hosang M: Suramin binds to platelet-derived growth factor and inhibits its biological activity. J Cell Biochem 29:265-73, 1985.

32. Sato Y and Rifkin DB: Autocrine activities of basic fibroblastic growth factor: regulation of endothelial cell movement, plasminigen activator synthesis, and DNA synthesis. J Cell Biol 107:1199-205, 1988.

33. Betsholtz C, Johnsson A, Heldin C, and Westermark B: Efficient reversion of simian sarcoma virus-transformation and inhibition of growth factor-induced mitogenesis by suramin. Proc Natl Acad Sci USA 83:6440-4, 1986.

34. Coffey RJ, Leof EB, Shipley G, and Moses HL: Suramin inhibition of growth factor receptor binding and mitogenity in AKR-2B cells. J Cell Physiol 132:143-8, 1987.

35. Pollack M and Richard M: Suramin blockade of insulin-like growth factor I-stimulated proliferation of human osteosarcoma cells. J Natl Cancer Inst 82:1349-52, 1990.

36. Stein CA, La Rocca RV, Thomas R, McAtee N, and Myers CE: Suramin: an anticancer with a unique mechanism of action. J Clin Oncol 7:499-508, 1989.

37. Collins JM, Klecker Jr. RW, Yarchoan R, Lane HC, Fauci AS, Redfiels RR, Broder S, and Myers CE: Clinical pharmacokinetics of suramin in patients with HTLV-III/LAV infection. J Clin Pharmacol 26:22-6, 1986.

38. DeLean A, Munson PJ, and Rodbard D: Simultaneous analysis of families of sigmoidal curves: applications to bioassay and physiological dose-response curves. Am J Physiol 235:E97-102, 1978.

39. Denziot F and Lang R: Rapid colorimetric assay for cell growth and survival: modifications to the tetrazolium dye procedure giving improved sensitivity and reliability. J Immunol Methods 89:271-7, 1986.
40. Laemmli UK: Cleavage of structural proteins during the assembly of the head of bacteriophage T4. Nature 227:680-5, 1970.

343

THE PHYSIOLOGY AND PATHOPHYSIOLOGY OF IGF-I IN THE KIDNEY

Raimund Hirschberg

Division of Nephrology and Hypertension
Harbor-UCLA Medical Center and UCLA School of Medicine
Torrance, CA 90509

INTRODUCTION

Insulin-like growth factor I (IGF-I) is synthesized and released in the nephron[1-10] and released into the systemic circulation by the kidneys[11,12]. Under normal conditions the glomerulus expresses IGF-I mRNA and releases IGF-I peptide[1,2]. Cell culture studies have shown that glomerular mesangial cells, which are mainly smooth muscle-like cells, express IGF-I[13,14]. Furthermore, cultured glomerular epithelial cells also release IGF-I (own unpublished observation). Within the nephron IGF-I is expressed in distal tubules and cortical collecting ducts but not in proximal tubules[1-5]. Scattered proximal tubule cells are positive for IGF-I mRNA by in-situ hybridization, suggesting that very small amounts of IGF-I may be synthesized under normal conditions[1]. In many tissues, growth hormone is the strongest secretagogue for IGF-I. In distal and collecting tubules the synthesis of IGF-I is growth hormone dependent[15,16].

Specific IGF-I receptors are found in the glomerulus, particularly on mesangial cells[14,17]. We have recently demonstrated that IGF-I binds also to cultured human glomerular epithelial cells (unpublished observation). IGF-I receptors are present on proximal tubule epithelial cells, mainly on the baso-lateral membrane but at lesser density receptors are also expressed on the apical (luminal) membrane[18,19]. In contrast, IGF-I receptors could not be demonstrated in distal tubules or collecting ducts[19].

The kidney appears to contribute significantly to the total body synthesis and the plasma levels of IGF-I[8,9,11,12]. In the circulation, most of the IGF-I is bound to specific binding proteins and $\leq 10\%$ of the serum IGF-I may be present in the unbound form[20-23]. The kidneys are among the organs with the highest blood perfusion rates and it is feasible that circulating IGF-I induces endocrinologic effects on the nephron. Indeed, endocrine actions of IGF-I are suggested by the effects of systemically administered IGF-I which induces several responses in a variety of tissues, such as the kidneys (see below). Furthermore, circulating IGF-I is distributed outside the vascular space into peripheral tissues by a selective and specific process[24].

Current Directions in Insulin-Like Growth Factor Research,
Edited by D. LeRoith and M.K. Raizada, Plenum Press, New York, 1994

345

Autocrine and paracrine mechanisms have been suggested as the major mechanisms of action of IGF-I in several tissues[9,25] and are also likely modes of action for IGF-I in the kidney. Conti et al[13] and Doi et al[14] have demonstrated that IGF-I may exert paracrine/autocrine actions in the glomerulus. The lack of IGF-I synthesis in proximal tubules under physiologic conditions but the expression of specific receptors in this part of the nephron appears to conflict with the hypothesis of paracrine action of IGF-I. Similarly, the lack of specific receptors in distal tubules and cortical collecting ducts, which are major sites of IGF-I release, also appears to contradict the hypothesis of autocrine/paracrine action of IGF-I in the nephron. However, the anatomy of the nephron may solve this apparent conflict: The distal tubule neighbors the proximal tubule and the vascular pole of the glomerulus of the same nephron. This anatomic relationship may allow IGF-I that is released from the distal tubule to act on the proximal tubule as well as on the glomerular microvasculature of that nephron and may support a paracrine mode of action of IGF-I on glomerular dynamics and proximal tubule function.

In the glomerulus IGF-I affects the regulation of glomerular perfusion and ultrafiltration. The peptide has also been shown to affect proximal tubule transport. IGF-I can induce nephron hypertrophy and increases the synthesis of certain extracellular matrix proteins. IGF-I may contribute to the compensatory nephron growth that is observed in a number of disease states. Recent studies also suggest an important role of IGF-I in tissue repair after acute renal injury.

EFFECTS OF IGF-I ON GLOMERULAR HEMODYNAMICS AND THE REGULATION OF GLOMERULAR ULTRAFILTRATION

The Acute Effects of IGF-I on GFR and Single Nephron GFR in the Rat: For three decades it has been recognized that excess serum concentrations of growth hormone result in increased levels of glomerular filtration rate (GFR) and renal plasma and blood flow rates (RPF, RBF)[26-32]. In acromegaly, GFR and RPF as well as the kidney mass are increased. In patients with growth hormone deficiency the GFR and RPF are commonly below normal levels[32-35]. However, in these latter patients, the kidney still responds normally to acute stimuli, such as an acute amino acid infusion which quickly raises GFR and RPF to a similar extent in growth hormone deficient patients as in normal subjects[33,36]. This response is believed to be mediated by various vasoactive compounds such as glucagon and vasodilatory prostaglandins[37-39]. Hence, in growth hormone deficiency the GFR is reduced secondary to changes in the functional determinants of glomerular ultrafiltration rather than only due to a reduced glomerular size, and the response to vasoregulators is not impaired.

Parving et al observed in normal subjects that acute infusions of growth hormone do not increase GFR and RPF[40]. However, Christiansen and coworkers treated normal subjects with daily injections of growth hormone for one week and noted a significant rise in GFR without a concomitant increase in kidney volume[41].

After surgical ablation of the pituitary adenoma in acromegalics the serum growth hormone levels fall and the GFR and kidney size decrease to normal values[26-28]. However, the GFR decreases more rapidly than the kidney volume[26-28]. More recently several studies have confirmed that growth hormone induces a rise in GFR[42-44].

Several years ago the author and his associates begun to examine the hypothesis that IGF-I may mediate the rise in renal function that occurs after administration of growth hormone[32,42]. Seven normal adults underwent a single intramuscular injection of human recombinant growth hormone and renal hemodynamics, plasma growth hormone and IGF-I levels were measured during the baseline period, for 5.5 hours after the injection as well as at 24 and 48 hours[42]. The results (Figure 1) confirm the previous findings by Parving et al[40] that growth hormone even at pharmacologically raised plasma levels does not acutely reduce renal vascular resistance or increase RPF and GFR. However, at 24 and 48 hours after the growth hormone injection, when the plasma growth hormone concentrations had returned to baseline levels, GFR and RPF were significantly increased and the renal vascular resistance was reduced (Figure 1). The rise in renal function had occurred concomitant with a rise in plasma IGF-I levels[42] (Figure 1). As a result of this observation the hypothesis was raised that IGF-I increases glomerular hemodynamics and mediates the rise in RPF and GFR that is observed with growth hormone treatment.

To further test this hypothesis short-term studies were conducted in fasted rats undergoing an acute injection and short-term infusion (20 min) of recombinant human IGF-I (rhIGF-I)[45]. RPF (para-aminohippurate clearance) and GFR (inulin clearance) tended to rise already during the 20 min period during which rhIGF-I was infused (Figure 2). Renal function remained elevated for about two hours after cessation of the IGF-I infusion and values returned to baseline thereafter[45]. These latter studies, for the first time, demonstrated that an exogenous infusion of rhIGF-I increases glomerular perfusion and filtration rates,

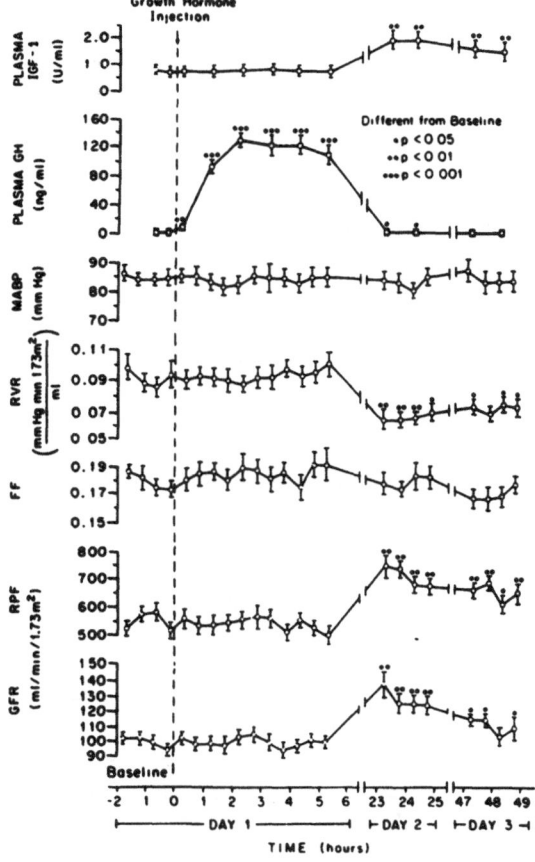

Figure 1. Effect of a single intramuscular injection of human recombinant growth hormone on mean arterial pressure (MABP), renal vascular resistance (RVR), filtration fraction (FF), renal plasma flow (RPF) and glomerular filtration rate in seven normal subject.

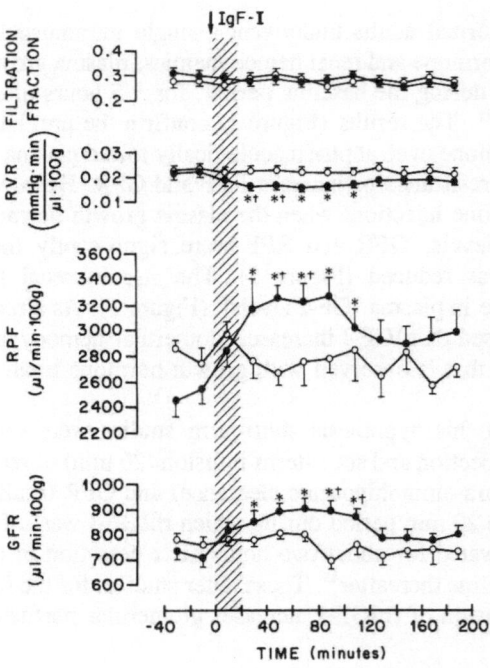

Figure 2. Effects of an injection (arrow) and short-term infusion (striped vertical bar) of recombinant human IGF-I (black circles) or vehicle (open circles) on renal hemodynamics in fasted normal rats. *p<0.05 vs. vehicle controls; +p<0.05 vs. baseline (-40 to 0 min).

and, hence, supported the hypothesis that IGF-I mediates the growth hormone-induced rise in RPF and GFR.

In subsequent renal micropuncture studies we have delineated the mechanisms by which rhIGF-I administration affects the physiologic determinants of single nephron glomerular ultrafiltration rate (SNGFR) in fasted and normal anesthetized rats[46]. Rats received a bolus injection followed by an infusion of rhIGF-I similar to the previous studies. In both the fasted and non-fasted animals, the IGF-I infusion decreased the efferent arteriolar resistance and tended to decrease the afferent arteriolar resistance[46]. Since the systemic mean arterial blood pressure did not change in the animals receiving IGF-I, the decrease in renal arteriolar resistance resulted in an increase in single nephron plasma flow rate (SNPF), which rose by an average of 22% and 16% in the fasted and non-fasted rats, respectively[46] (Figure 3). The rise in SNPF contributed to the increase in SNGFR that was observed in the animals receiving the IGF-I compared to the respective vehicle controls (Figure 3). However, the rise in SNGFR was of excess magnitude compared to the IGF-I-induced rise in SNPF, suggesting that further physiologic regulators of SNGFR must have been involved. The net glomerular ultrafiltration pressure, ΔP, was not increased by IGF-I (Figure 3). In contrast, IGF-I doubled the glomerular ultrafiltration coefficient, LpA (Figure 3). Since all these changes in the determinants of nephron filtration occurred quickly, it is unlikely that the ability of IGF-I to induce nephron hypertrophy and renal growth (see below) played any role in these acute studies. In summary, IGF-I acutely

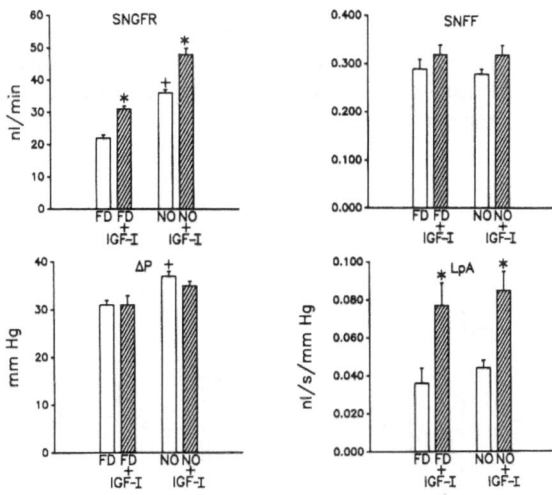

Figure 3. Effects of an infusion of IGF-I (striped bars) or vehicle (open bars) on the determinants of glomerular ultrafiltration in normal (NO) and food deprived (FD) rats. *p<0.05 vs. respective control; +p<0.05 vs. FD-control.

lowers nephron arteriolar resistance and raises GFR through an increase in RPF and the glomerular ultrafiltration coefficient.

IGF-I Increases GFR in Normal Human Subjects: Whereas the above experimental studies in rats help to determine the physiologic mechanisms by which IGF-I affects glomerular function, studies in man are necessary to confirm that the experimental findings also apply to humans. To answer this question we have studied normal subjects that were admitted to a clinical research center for five and one-half day and underwent continuous clearance measurements to determine RPF, GFR and renal vascular resistance[47]. During the first day of study baseline measurements were obtained. For the subsequent three days subjects received subcutaneous injections of rhIGF-I, 60 μg/kg three times daily. Within a few hours after the onset of the treatment with rhIGF-I, RPF and GFR increased and remained elevated for the period of IGF-I treatment (Figure 4). Thereafter, renal function decreased stepwise to baseline values (Figure 4). These data confirm our previous findings in rats, that the IGF-I-induced rise in renal function occurs quickly and does not require treatment with growth hormone.

These studies also confirm observations by Guler et al[21,48], who infused rhIGF-I for three or five days into two normal volunteers and measured the 24 hour creatinine clearance[21] or iothalamate and iodohippurate clearances[48]. In both subjects, the infusion of rhIGF-I induced a rise in creatinine clearance or iothalamate and iodohippurate clearances. These studies clearly indicate that IGF-I has similar effects on glomerular filtration in man as previously demonstrated in the rat. Most likely, although not proven, the physiologic mechanisms by which IGF-I affects the determinants of GFR in man are similar to the rat. The main mechanisms of action of IGF-I on the glomerular vasculature involves arteriolar dilatation and a rise in LpA, possibly secondary to mesangial relaxation.

Possible Hormonal Mediators of the IGF-I-induced rise in Glomerular Ultrafiltration: It is yet unknown whether the effect of IGF-I to induce renal arteriolar dilation and to raise

LpA results directly form vascular smooth muscle cell relaxation upon binding of IGF-I to its specific receptor and is the result of subsequent signal transduction processes in the cell. In contrast, some experimental studies have suggested that other autacoids mediate the effects of IGF-I on the glomerulus.

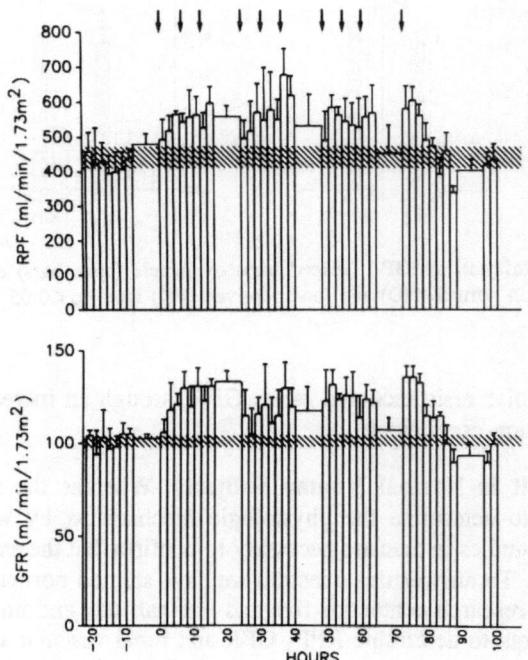

Figure 4. Effects of subcutaneous injections of IGF-I (arrows) in four normal men. Each vertical bar represents one 2- or 8-hour clearance period. The striped horizontal bar represents the weighed grand mean ± 2 SEM of values before onset of IGF-I treatment (baseline).

The author investigated in fasted rats whether blockade of the cyclooxigenase activity would prevent the IGF-I-induced rise in GFR[45]. Animals received a constant infusion of indomethacin before, during and after an injection and short-term infusion (20 min) of rhIGF-I. As shown in Figure 5, administration of IGF-I concomitant with the cyclooxigenase inhibitor indomethacin failed to increase GFR and RPF. These results may suggest that vasodilatory prostanoids, such as PGI_2, mediate the IGF-I-induced vasodilation and rise in RPF and GFR. On the other hand, there is an alternative explanation for these results as well: The vasodilating prostanoids may play a permissive role allowing for the IGF-I-induced arteriolar vasodilation to occur. For example, blockade of vasodilating prostaglandins by indomethacin may induce an abnormal balance between vasodilatory (PGI_2) and vasoconstrictory (angiotensin II) compounds. In a milieu of unbalanced vasoconstrictors, IGF-I may not be able to induce vasodilation.

Haylor and his associates[49] also investigated in anesthetized rats whether the administration of a cyclooxigenase antagonist would prevent the IGF-I induced acute decrease in renal vascular resistance and increase in RPF. These investigators demonstrated some effect of the prostaglandin inhibitor to reduce but not prevent the acute effect of IGF-I on the renal vasculature.

Additional studies in our laboratory were aimed at investigating a possible mediator role for glucagon. Due to the insulin-like effects acute administration of IGF-I could result in hypoglycemia which would acutely induce elevations in glucagon. This hormone is known to reduce renal vascular resistance and to raise RPF and GFR[37]. To examine this question somatostatin was administered concomitant with IGF-I[45]. In control rats somatostatin tended to reduce RPF and GFR, although not significantly, but it did not prevent the significant rise in renal function during concomitant administration of IGF-I (Figure 5).

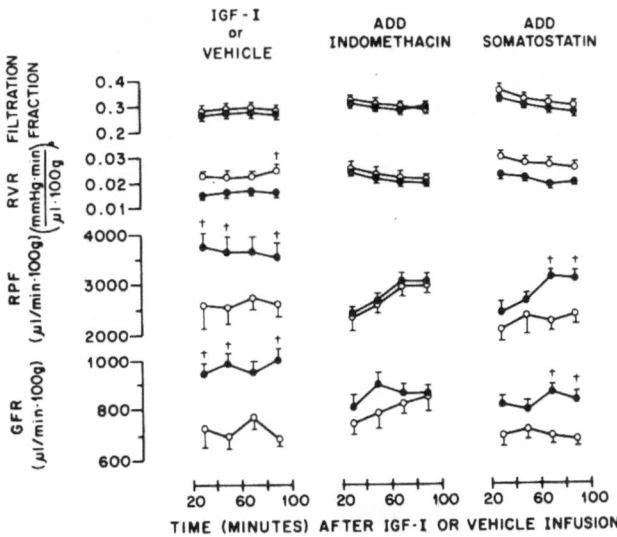

Figure 5. Renal hemodynamics in normal rats receiving either vehicle, indomethacin or somatostatin concomitant with an infusion of IGF-I (black circles), or receiving these agents without IGF-I (control, open circles); +p<0.05 vs. control.

There is further evidence indicating that the IGF-I - induced acute rise in renal function is independent of the insulin-like actions of IGF-I. Baumann and associates[50] infused IGF-I into anesthetized rats during euglycemic clamp. IGF-I increased the GFR by up to 35%. The increase in GFR was dose-dependent with half-maximal stimulation at a total IGF-I plasma concentration of about 24 nM. These studies indicate that IGF-I affects renal function through specific receptors independent of acute effects on glucose metabolism[50.]

Angiotensin II which is synthesized and released in the nephron and acts as an autacoid on the resistance vasculature, contracts efferent and to a lesser extent also afferent arterioles and raises SNGFR by augmenting the glomerular ultrafiltration pressure[51,52]. However, as shown above, the ultrafiltration pressure is not affected by IGF-I. Thus, angiotensin II is an unlikely mediator for the IGF-I (or growth hormone) induced effects on renal function. Haffner and associates[43] studied this question indirectly in eight normal subjects who received growth hormone injections, 4.5 units twice daily for 3 days, with or without concomitant administration of an angiotensin converting enzyme inhibitor. The effects of growth hormone to induce a rise in GFR was not obliterated by the ACE inhibitor, suggesting that angiotensin II does not mediate the effects of growth hormone (and IGF-I) on renal function.

Endothelium-derived relaxation factor (EDRF) appears to be a possible mediator for some (or all) effects of IGF-I on glomerular function, although direct proof is lacking and would be very difficult if not impossible to obtain. EDRF, possibly identical with nitric oxide (NO)[53], is a powerful, short-lived vasodilator that is released from vascular endothelial cells upon stimulation with several compounds, such as acetylcholine, bradykinin, histamine, thrombin, substance P, ATP, and hydralazine[54-58]. It has been hypothesized that EDRF is released from the renal arterial endothelium and acts on resistance vessels[59]. NO may also be released from microvascular endothelial cells and may diffuse to adjacent vascular smooth muscle cells of that microvessel to reduce vascular tone in these resistance vessels in the nephron[59]. Haylor and associates[49] studied the effects of an acute infusion of rhIGF-I on renal blood flow and renal vascular resistance in rats using a magnetic flow probe. These authors confirmed our previous findings that IGF-I acutely reduces renal vascular resistance and raises RBF. Even very low infusion rates of rhIGF-I (2.5 ng/100 g BW/min) induced this effect. These investigators also observed that L-NAME (N^G-nitro-L-arginine-methylester), a powerful inhibitor of NO-synthase activity, completely abolished the IGF-I-induced increase in RBF. This finding supports the possibility that NO (EDRF) mediates the IGF-I-induced effects on glomerular dynamics[49].

There is also indirect support for the hypothesis that NO may mediate the rise in LpA that contributes to the IGF-I-induced rise in GFR. EDRF relaxes mesangial cells in vitro[60,61]. Possibly, IGF-I increases the synthesis and release of EDRF from mesangial cells in vivo and thus relaxes the mesangium which would lead to a rise in LpA.

The effects of EDRF/NO on vascular smooth muscle cells use cGMP as second messenger[62]. In previous studies the author has measured the urinary cGMP excretion in rats that were infused with rhIGF-I. Indeed, urinary cGMP increased significantly during and for about 1.5 hours after the short-term infusion of rhIGF-I in normal fasted rats[63]. Although, the above studies provide only pieces of experimental evidence for a role of EDRF/NO to mediate the IGF-I-induced effects on glomerular function, final proof for this attractive hypothesis remains to be obtained.

Evidence for a Role of Endogenous IGF-I to Contribute to the Regulation and Maintenance of GFR: Both circulating and/or renal IGF-I levels and GFR (or SNGFR) co-vary in some disease states. Both are elevated in a number of physiologic and pathophysiologic conditions, such as acromegaly[30,64,65], pregnancy (particularly during the third trimester)[66], feeding high protein diets[67-69], and in certain animal models of chronic renal failure as well as in early diabetic nephropathy[70]. In contrast, IGF-I levels as well as GFR are both reduced in growth hormone deficiency or panhypopituitarism[26-28,33], feeding low protein diets[33,68,71-73], in starvation and in states of malnutrition[74-79]. These observations suggest that endogenous IGF-I may contribute to the maintenance of an adequate level of GFR.

We have performed a series of experimental studies in rats to provide further evidence for a role of endogenous IGF-I to contribute to the regulation of GFR. Since changes in serum or renal IGF-I levels do not occur quickly (within minutes) rather than within many hours or several days, endogenous IGF-I may participate in the long-term rather than the acute regulation of GFR.

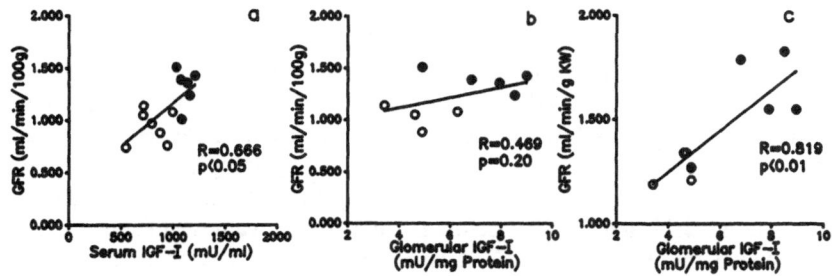

Figure 6. Correlations between (a) GFR and serum IGF-I levels, (b) GFR and glomerular IGF-I levels, and (c) GFR, corrected for kidney weight and glomerular IGF-I levels, in normal rats, fed a high (black circles) or a low protein diet (open circles).

First, we have investigated whether feeding a high protein diet to normal rats for 10-14 days, which is known to raise GFR, would also increase renal glomerular and/or serum IGF-I compared to animals pair-fed a low protein diet[69].

Feeding the high protein diet resulted in an increase in serum IGF-I levels as well as the IGF-I tissue concentrations in the liver and in isolated glomeruli. Dardevet et al[80] confirmed that total plasma IGF-I levels in rats increase linearly as a function of dietary protein intake. In our study, as shown in Figure 6, the serum levels of IGF-I correlated reasonably well with the GFR, but the glomerular IGF-I levels correlated with the GFR only if the latter was corrected for renal mass. Although these data do not proof that IGF-I regulates GFR, they suggest a contributory role of IGF-I to up-regulate GFR through actions on the renal microvasculature and/or through the induction of renal and/or glomerular hypertrophy.

Second, GFR and SNGFR are increased in normal rats at the end of a six to seven day subcutaneous infusion of rhIGF-I, 125 μg/day, into normal rats[81]. GFR and SNGFR are

Table 1. Effects of treatment of normal rats with IGF-I or a peptidic antagonist to the action of growth hormone releasing hormone on glomerular function and volume.

	Serum IGF-I (U/ml)	GFR μl/min/100g	SNGFR nl/min	LKW/BW g/100 g	Glom Tuft Volume (x10$^4 \mu$m^3)
Control	1.14±0.14	0.86±0.03	34±2	0.388±0.010	27±3
IGF-I	2.07±0.43*	1.01±0.04*	41±2*	0.419±0.011*	35±2*
GHRH-Ant	0.38±0.06*	0.70±0.03*	27±2*	0.383±0.012	27±3

*p<0.05 vs control; Data are mean±SEM.

reduced by administration of a synthetic growth hormone releasing hormone antagonist which lowers both serum growth hormone and IGF-I levels[81-84]. As shown in Table 1 the low-dose administration of IGF-I increased renal mass, presumably by inducing nephron hypertrophy. Treatment with IGF-I increased GFR and SNGFR significantly. This effect on GFR was achieved with a low infusion dose of IGF-I that increased the serum levels moderately within a range that can be observed in physiologic and pathophysiologic conditions. In contrast, the treatment with the growth hormone releasing hormone antagonist lowered GFR and SNGFR below normal control values[81] (Table 1). This latter treatment reduced the endogenous IGF-I levels most likely secondary to a decrease in growth hormone activity. It is noteworthy that the reduction in GFR and SNGFR in the latter rats occurred without reducing renal mass or glomerular tuft volume (Table 1) suggesting a functional role of IGF-I on the maintenance of GFR. These studies indicate that moderate chronic elevations in circulating IGF-I increase the GFR and SNGFR, whereas the reduction in endogenous IGF-I reduces glomerular function.

Table 2. Determinants of glomerular ultrafiltration in growth hormone deficient vehicle treated controls and in growth hormone deficient rats receiving des (1-3) IGF-I or growth hormone for one week.

	SNGFR nl/min	SNPF nl/min	R_A	R_E	P_G mmHg	LpA nl/s/mmHg
			10^9 dyn s cm^{-5}			
vehicle (n=7)	21±1	85±6	45.3±4.4	20.6±2.1	47±1	0.046±0.004
des (1-3) IGF-I (n=7)	36±3*	149±11*	23.1*±2.5	9.7*±0.8	40±2*	0.127*±0.036
Growth hormone (n=7)	39±4	180±10*	20.6*±1.0	8.2*±0.7	44±2	0.125*±0.019

*$p < 0.05$ vs vehicle control; Data are mean±SEM.
SNGFR, SNPF, SNFF = single nephron GFR, single nephron plasma flow, single nephron filtration fraction; $R_{A,E}$ = afferent and efferent arteriolar resistance; P_G = glomerular hydrostatic pressure; LpA = glomerular ultrafiltration coefficient.

In a third set of studies we measured the determinants of glomerular ultrafiltration in growth hormone deficient rats and in growth hormone deficient rats that were treated either with growth hormone or des(1-3)IGF-I for about one week[85]. The results indicate (Table 2) that in growth hormone deficiency the reduced GFR is not only caused by a low glomerular tuft volume ("glomerular hypotrophy") but also by an increase in afferent and efferent arteriolar resistance and a low LpA compared to values commonly found in normal rats. Both treatment with growth hormone or truncated IGF-I raised glomerular function into the normal range in these growth hormone deficient animals. These findings indicate that the deficiency of growth hormone and IGF-I reduces GFR. The finding that only the administration of IGF-I in growth hormone deficiency is required to restore GFR to normal indicates, that the reduced glomerular filtration in growth hormone deficient rats results from a lack of IGF-I rather than of growth hormone per-se. This latter conclusion suggests a role of IGF-I to contribute to the maintenance of normal GFR.

These three experimental studies support the hypothesis that endogenous IGF-I participates in the long-term regulation of glomerular ultrafiltration.

Effects of IGF-I on the Glomerular Ultrafiltration of Plasma Proteins and on Microalbuminuria: As described above the administration of IGF-I increases the glomerular ultrafiltration coefficient, LpA. This term is the product of Lp, the pressure-dependent permeability of the ultrafiltration barrier for water and small solutes and A, the surface area that is available for filtration in a given glomerulus. IGF-I may affect both of these parameters which cannot be measured individually with current experimental techniques. There is some evidence that IGF-I may raise the permeability for larger molecules, such as plasma proteins. Dullaart and coworkers measured the microalbuminuria in acromegalics before and during treatment with octreotide, a somatostatin analogue that lowers growth hormone and IGF-I levels[64]. This treatment lowered the microalbuminuria significantly in this group of patients. We have previously demonstrated that exogenous administration of IGF-I increases the fractional clearance of albumin as well as IgG in normal human subjects[47]. In these latter studies the microalbuminuria does not exceed normal limits.

Effects of IGF-I on Glomerular Filtration in Chronic Renal Failure: The striking effect of IGF-I as well as of growth hormone (most likely through elevating IGF-I activity) on GFR in normal rats and healthy human subjects suggests that this peptide may be useful for the treatment of patients with chronic renal failure. Miller and coworkers[86] treated normal and subtotally nephrectomized rats with IGF-I and demonstrated a rise in GFR in the normal but not in the rats with chronic renal failure. Haffner et al[87] measured the GFR during growth hormone administration in normal subjects and in patients with chronic renal failure. These authors found that the growth hormone-induced rise in GFR cannot be induced in patients with chronic renal failure. Similar results were obtained in a groups of children with chronic renal failure as well as in further experimental studies in rats[88,89]. However, chronic renal failure is a pathophysiologic state that results from a variety of renal diseases and the GFR may be decreased by a variety of structural and functional abnormalities in the nephron. For example, in the model of the subtotally nephrectomized rat, the remnant nephrons display hyperfiltration mainly due to an elevated glomerular ultrafiltration pressure. Possibly, in this model the SNGFR in remnant glomeruli is already maximally elevated and cannot be further increased by any manipulation including the administration of IGF-I. However, this widely used experimental model may not accurately describe the changes in the determinants of glomerular ultrafiltration that occur in many renal diseases in humans. O'Shea et al[90] recently described four patients with moderate chronic renal failure in whom the administration of rhIGF-I induced a marked increase in GFR. Further studies are clearly needed to investigate whether treatment with rhIGF-I will benefit some patients with chronic renal failure.

EFFECTS OF IGF-I ON RENAL TUBULAR FUNCTION

Tubular Sodium and Water Absorption: As outlined above the proximal tubule displays specific IGF-I receptors on the basolateral membrane and at lesser density also on the apical membrane[19]. Since the major physiologic activity of the proximal tubule is devoted to the absorption of water and solutes from the tubular fluid it is feasible to hypothesize that IGF-I may play a role in tubular transport.

Several earlier studies have investigated the role of growth hormone on tubular absorption. Two of these studies indicated that growth hormone administration to humans induces water and sodium retention and edema formation[91,92]. However, these studies were performed with somewhat crude growth hormone extracts from human or monkey pituitaries and some of the effects may have resulted from impurities of the preparations. In a single growth hormone deficient patient who underwent treatment with recombinant human growth hormone we observed a fall in the fractional excretion of sodium but the patient did not develop edema[32]. In seven normal subjects who received each a single injection of growth hormone we did not find consistent evidence for increased tubular sodium absorption[42]. In a group of normal subjects who were treated with rhIGF-I for three days the fractional excretion of sodium was not consistently changed[47].

Quigley and Baum[93] directly perfused rabbit proximal tubules with rhIGF-I. These authors found no significant effect of IGF-I on the tubular absorption of sodium. Thus, the overwhelming evidence indicates that IGF-I does not affect the proximal tubule transport of sodium under normal conditions.

The proximal tubule also displays specific IGF-II receptors which are about evenly distributed between the apical and basolateral membranes[19]. Yanagawa and associates[94] demonstrated a powerful effect of IGF-II to increase the sodium transport in isolated rabbit proximal tubule apical membrane vesicles. This latter study demonstrated for the first time a possible purpose for the luminal IGF-II receptors, namely to participate in the regulation of proximal tubule sodium absorption.

Tubular Phosphate Absorption: Administration of growth hormone to hypophysectomized rats increases the proximal tubular phosphate absorption and raises phosphate retention[95]. The effect of growth hormone to raise phosphate absorption in normal subjects was already noted several years ago by Corvilain and coworkers[29]. Recently, Bonjour and his associates[96] showed that the stimulatory effect of growth hormone on tubular phosphate transport in rats is mediated through IGF-I.

We also studied the fractional excretion of phosphate ($FEPO_4$) in normal adults before and during subcutaneous administration of rhIGF-I, 60 μg/kg three times daily, for three days[47]. The $FEPO_4$ decreased quickly after onset of the rhIGF-I treatment and was reduced by about 50%, on average, during the period of rhIGF-I administration (Figure 7). The physiological circadian rhythm of phosphate excretion was maintained but the undulations occurred around a reduced average (Figure 7). In these subjects, serum phosphate levels did not increase, suggesting that the absorbed phosphate was stored in tissues, possibly in bone. These findings confirm previous data from human and experimental studies and further suggest a direct stimulatory effect of IGF-I on tubular phosphate transport.

Quigley and Baum[93] perfused rabbit renal proximal tubules with either growth hormone or IGF-I. Growth hormone had no effect on the absorption of phosphate, bicarbonate or water. In contrast, IGF-I stimulated the uptake of phosphate from the perfusate by up to 46%, but did not affect sodium, bicarbonate or water absorption. The stimulation of phosphate transport was greater when the peptide was presented to the apical membrane as compared to the basolateral membrane[93]. This latter finding suggests an important function for the apical tubular IGF-I receptors to participate in the regulation of tubular phosphate transport.

These findings raise important questions as to the physiologic or pathophysiologic role of the luminal IGF-I receptors in the proximal tubule. The proximal tubule does not release IGF-I under physiologic conditions or releases only extremely small amounts. Distal tubular IGF-I appears to be released on the basolateral side of the cells, rather than through the luminal membrane. Even if release of IGF-I into the distal tubular lumen would occur, the peptide is unlikely to travel upstream to reach the proximal tubule apical IGF-I receptors. Thus, a paracrine/autocrine mode of action of IGF-I on the apical proximal tubule receptor is unlikely. IGF-I could reach the proximal tubule lumen by means of glomerular ultrafiltration. However, due to the binding to IGFBPs in serum most of the circulating IGF-I is excluded from glomerular ultrafiltration under physiologic conditions. Ongoing research in the author's laboratory indicates that small amounts of IGF-I are present in glomerular ultrafiltrate in normal rats but estimated tubular fluid concentrations may be below effective levels. In contrast, in rats with the nephrotic syndrome, we have demonstrated that glomerular ultrafiltration of IGF-I into the proximal tubule occurs at physiologically significant rates and the proximal tubule fluid contains meaningful concentrations of IGF-I, that could induce cellular effects through luminal IGF-I receptors (own unpublished observation).

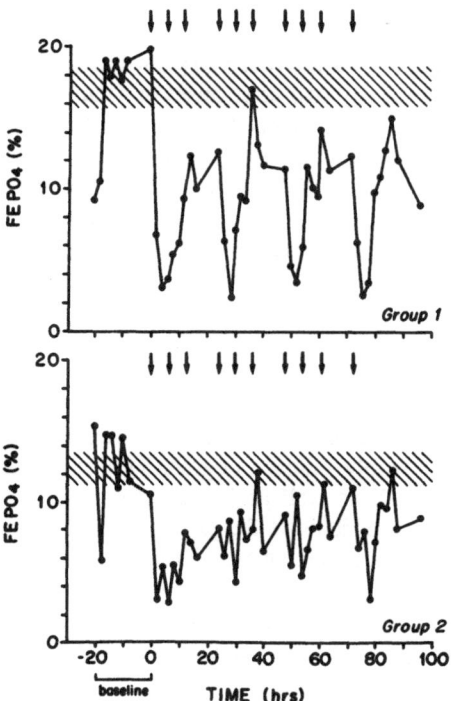

Figure 7. Fractional excretion of PO₄ in normal subjects (Group 1, upper panel) and saline loaded normal subjects (Group 2, lower panel) treated with subcutaneous injections of IGF-I (arrows). Each black circle represents the mean value of one clearance period. The horizontal bar indicates the weighed grand mean ± 2 SEM during baseline before onset of IGF-I administration.

IGF-I AND NEPHRON HYPERTROPHY AND GLOMERULAR SCLEROSIS

As indicated above the administration of IGF-I to rats increases the kidney wet weight and the glomerular tuft volume[81,85,97]. Guler and associates[98] demonstrated that the kidney is among the most sensitive organs to undergo growth in-vivo upon treatment with IGF-I.

El Nahas and coworkers investigated the role of IGF-I in renal compensatory hypertrophy[99]. In these studies subtotally nephrectomized rats were fed either a high-, medium- or low-protein diet. The tissue IGF-I concentration in the hypertrophied remnant kidney in the animals fed the high or the medium levels of dietary protein were greater than in those fed the low protein diet.

Fagin and Melmed[10] were the first to describe a rise in tissue IGF-I and IGF-I mRNA concentrations in remnant rat kidneys after subtotal nephrectomy. Mulroney and her associates[100,101] have recently described that IGF-I and IGF-I gene expression increase

357

in the contralateral growing kidney after unilateral nephrectomy in immature rats. These studies suggest that IGF-I contributes to the renal compensatory hypertrophy that occurs after substantial loss of nephrons.

Flyvbjerg and associates[70,102] investigated the relationship of renal growth and renal IGF-I concentrations in four rat models of renal hypertrophy. After unilateral nephrectomy in rats the renal IGF-I levels in the contralateral kidney increased by about 50% within one day after surgery[102] as was also found by Mulroney et al in immature, but not in adult rats[100,101]. In hypophysectomized rats that received growth hormone injections, kidney IGF-I levels rose by almost 300% on day two after onset of treatment[102] and predated the onset of renal growth. Potassium depletion which was induced by feeding potassium-free diets to normal rats induced doubling of the renal IGF-I content[102]. In experimental streptozotocin diabetes in the rat renal IGF-I concentrations rose by $\approx 50\%$ within two days after disease onset[70,102]. In all these experimental models, the rise in renal IGF-I levels preceded the onset of compensatory renal growth. These data suggest that IGF-I induces or helps to induce the hypertrophy of nephrons in all of these forms of experimental chronic renal failure.

Nephron hypertrophy has been associated with progressive renal failure by several investigators[103-106], but may not be the major cause[99,107]. In models of chronic progressive renal failure in the rat, worsening glomerular sclerosis develops which results from increased production (and possibly reduced degradation) of extracellular matrix proteins, such as collagens, in glomeruli as well as in the interstitium. These findings raise the question whether IGF-I may contribute to the progressive glomerular sclerosis which is a hallmark of most experimental and human glomerular diseases that result in chronic renal failure. Recent data by Peng et al[108] indicate that the collagen synthesis in glomeruli of subtotally nephrectomized rats increases prior to the onset of hypertrophy. Hence, glomerular hypertrophy may not be a necessary pre-requisite for the development of glomerular sclerosis.

We have recently re-examined the relationship of glomerular sclerosis and nephron hypertrophy in subtotally nephrectomized rats[97]. In these latter studies, animals were either pair-fed a low protein diet which tends to reduce the degree of glomerular hypertrophy and the rate of progression of glomerular sclerosis and renal failure. Other groups of rats were pair-fed a high protein diet which augments both, the hypertrophy and the glomerular sclerosis. Treatment of rats fed the low protein diet with IGF-I, 50 μg three times daily s.c., increased glomerular hypertrophy to a similar level as the high protein diet. This treatment, however, induced a lesser degree of expression of collagen types I and IV and of procollagen α_1(IV) mRNA in isolated glomeruli[97]. In animals that received the high protein diet, the additional treatment with the somatostatin analogue octreotide prevented the hypertrophy of glomeruli but did not prevent the accumulation of collagens or the procollagen α_1(IV) mRNA in glomeruli and did not reduce the histologic degree of glomerular sclerosis. These data suggest that the glomerular hypertrophy does not per-se determine glomerular sclerosis. Furthermore, IGF-I may modestly contribute to the accumulation of extracellular matrix proteins in glomeruli. Indeed, we recently demonstrated that IGF-I has a modest effect on cultured rat mesangial cells to increase the procollagen mRNA levels[109].

Mathews and coworkers and Quaife and associates described the successful genetic transformation of mice to over-express chimeric genes composed of the coding regions for bovine growth hormone, human growth hormone releasing hormone or human IGF-I fused with the mouse metallothionein I promoter[110-112]. Doi and coworkers and Pesce et al.

investigated these mice with regard to the degree of glomerular hypertrophy and sclerosis[113-116]. To summarize the results, for a comparable degree of glomerular hypertrophy growth hormone transgenic mice developed more severe glomerular sclerosis than the IGF-I transgenic counterparts.

EFFECTS OF IGF-I IN EXPERIMENTAL ACUTE ISCHEMIC RENAL FAILURE

Recently, several groups of investigators examined the effects of IGF-I on the recovery from ischemic acute renal failure (ARF) in rats[122-124]. In this classic animal model, ARF is induced by clamping of both renal arteries for certain periods of time (i.e., 60 min). This period of renal ischemia is followed by severe renal failure which may lead to the animal's death or may slowly recover during the subsequent several days or few weeks. Ischemic ARF induces only minor injury to the glomerulus but results in severe necrosis of proximal tubules, particularly the S_2 and S_3 segments[122-129]. Most of the renal failure (reduced GFR) is attributed to the incompetence of the proximal tubule rather than primary glomerular filtration failure[125].

In normal rats IGF-I is undetectable in proximal tubule[1-5,130], but becomes expressed during the recovery from acute ischemic renal injury[5,130] suggesting an important role of IGF-I in the natural mechanism of repair after induction of acute tubular necrosis. Recent evidence indicates that IGF-I is expressed in new tubule cells that grow and multiply in the injured tubule[5,130].

We recently examined the effects of treatment with exogenous rhIGF-I in rats with severe acute ischemic renal failure on the functional and structural recovery of the nephron[122]. In this study acute ischemic renal failure was induced in rats using the classic model described above. Control animals underwent sham surgery. Beginning at 5 hours after the surgery, when the renal damage was established, one group of renal failure and control rats, respectively, received s.c. injections of rhIGF-I, 50 μg/100 g BW, three times per day for three days. A second group of acute renal failure and control rats received vehicle. Before surgery (baseline), immediately before the onset of IGF-I treatment (5 hours post surgery) and at 24, 48 and 72 hours, serum creatinine and urea nitrogen were measured. At 72 hours, GFR and RPF were measured with clearance techniques. The kidneys were then removed to perform histological examinations and to estimate the expression of the proliferating cell nuclear antigen (PCNA) by in-situ immunohistochemistry. PCNA is a protein of the DNA-polymerase-δ complex. In similar but separate studies the ^{3}H-thymidine incorporation into DNA in glomeruli, tubules and renal cortex were measured. For these latter studies rats were given intraperitoneal injections of ^{3}H-thymidine at 71 hours after surgery and the kidneys were obtained exactly one hour later. The thymidine incorporation was measured in DNA extracts from renal tissues and by histoautoradiography[122].

As shown in Figure 8, urea nitrogen and creatinine serum levels rose less steeply and decreased more rapidly in ARF rats that received IGF-I. After three days of treatment with the peptide, GFR was three times greater in treated as compared to untreated animals with acute renal failure[122]. Treatment with rhIGF-I increased the ^{3}H-thymidine incorporation into DNA in-vivo several fold in glomeruli, tubules and renal cortical tissues, respectively, suggesting that IGF-I increased DNA synthesis and possibly the rate of mitosis in the kidney leading to the more rapid recovery. IGF-I also raised the expression of PCNA, particularly in S_2/S_3 segments of proximal tubules.

The improvement in azotemia and the greater GFR in ARF rats that were treated with IGF-I as compared to the untreated animals may have resulted, in part, from the effects of the peptide on glomerular dynamics. However, the effect of IGF-I on proximal tubule cell mitosis and restoration of proximal tubule integrity is likely to be of major importance in regaining renal function after acute ischemic renal injury.

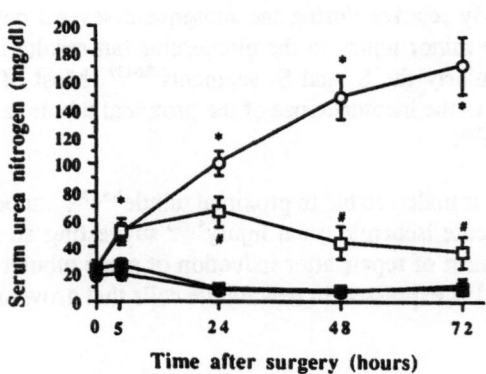

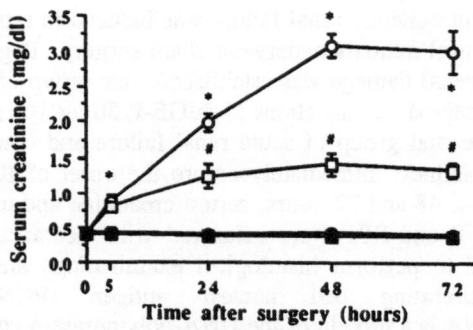

Figure 8. Serum urea nitrogen and creatinine values in four groups of rats after induction of ischemic acute renal failure (ARF) at time 0. Open circles: ARF rats receiving vehicle; open squares: ARF rats receiving IGF-I; closed circles: Sham rats receiving vehicle; closed squares: Sham rats receiving IGF-I. Vehicle or IGF-I was administered every eight hours after the values at 5 hours were obtained. *p < 0.05 vs. ARF + vehicle; #p < 0.05 vs. Sham + vehicle; [From Ref. 122].

Miller and associates[123] also examined the effects of rhIGF-I in acute ischemic renal failure in the rat. In this latter study animals were treated with continuous subcutaneous infusions of rhIGF-I, 100 μg/day, for seven days. GFR, measured by inulin-clearance on day 2 after ARF, was two-fold greater in the rats receiving IGF-I as compared to vehicle treated ARF rats.

Rabkin and associates[124] treated rats with ischemic ARF with des(1-3)IGF-I, 250 μg/kg/day, by subcutaneous infusion. These authors noted a lesser rise in serum creatinine and earlier recovery in IGF-I treated as compared to untreated animals.

The experimental studies reviewed above suggest, that the hemodynamic and mitogenic effects of IGF-I may render it useful for the medical treatment of patients with acute renal failure. Furthermore, the in-vivo findings indicate that IGF-I may play an important role in tissue repair after acute injury.

In summary, IGF-I plays important physiologic roles in the long-term regulation of glomerular ultrafiltration and in tubular phosphate absorption. Furthermore, IGF-I induces nephron growth and appears to contribute to the compensatory renal hypertrophy that is observed during the course of several experimental chronic renal diseases. IGF-I may contribute modestly to the synthesis of extracellular matrix protein by mesangial cells in vitro and the nephron in vivo. The mitogenic effects of IGF-I contribute to the tubule repair after acute ischemic nephron injury. These experimental observations may lead to the pharmacological use of IGF-I in patients with acute and chronic renal disease.

REFERENCES

1.   G. Matejka, P. Eriksson, B. Carlsson, and E. Jennische, Distribution of IGF-I mRNA and IGF-I binding sites in the rat kidney, Histochem 97:173-180 (1992).
2.   J. Bortz, P. Rotwein, D. DeVol, P. Bechtel, V. Hansen, and M. Hammerman, Focal expression of insulin-like growth factor I in rat kidney collecting duct, J Cell Biol 107:811-819 (1988).
3.   G. Andersson, A. Skottner, and E. Jennische, Immunocytochemical and biochemical localization of insulin-like growth factor I in the kidney of rats before and after uninephrectomy. Acta Endocrinol (Copenhagen), 119:555-560 (1988).
4.   G. Andersson, L. Ericson, and E. Jennische, Ultrastructural localization of IGF-I in the rat kidney; an immuinocytochemical study, Histochem 94:263-267 (1990).
5.   G. Matejka, and E. Jennische, IGF-I binding and IGF-I mRNA expression in the postischemic regenerating rat kidney. Kidney Int, 42:1113-1123 (1992).
6.   W. Lowe, M. Adamo, H. Werner, C. Roberts, and D. LeRoith, Regulation by fasting of rat insulin-like growth factor I and its receptor; Effects on gene expression and binding, J Clin Invest, 84:619-626 (1989).
7.   P. McConahey, and J. Dehnel, Preliminary studies of 'sulfation factor' production by rat kidney. J Endocrinol, 52:587-588 (1972).
8.   A. D'Ercole, and L. Underwood, Estimation of tissue concentrations of somatomedin C/insulin-like growth factor I, Meth Enzymol, 146:227-233 (1987).
9.   A. D'Ercole, D. Stiles, and L. Underwood, Tissue concentrations of somatomedin C: Further evidence for multiple sites of synthesis and paracrine and autocrine mechanism of action. Proc Natl Acad Sci USA, 81:935-939 (1984).
10.   J. Fagin and S. Melmed, Relative increase in insulin-like growth factor I messanger ribonucleic acid levels in compensatory renal hypertrophy. Endocrinology, 120:718-724 (1987).
11.   R. Schimpff, M. Donnadieu, and M. Duval, Serum somatomedin activity measured as sulphation factor in peripheral, hepatic and renal veins of mongrel dogs: Basal levels, Actoa Endocrinol (Copenhagen), 93:67-72 (1980).
12.   R. Schimpff, M. Donnadieu, and M. Duval, Serum somatomedin activity measured as sulphation factor in peripheral, hepatic and renal veins of mongrel dogs: Early effects of intravenous injection of growth hormone, Acta Endocrinol (Copenhagen), 93:155-161 (1980).
13.   F. Conti, L. Striker, S. Elliot, D. Andreani, and G. Striker, Synthesis and release of insulinlike growth factor I by mesangial cells in culture, Am J Physiol, 255:F1214-F1219 (1988).
14.   T. Doi, L. Striker, S. Elliot, F. Conti, and G. Striker, Insulinlike growth factor-1 is a progression factor for human mesangial cells, Am J Pathol, 134:395-404 (1989).

15.   S. Rogers, S. Miller, and M. Hammerman, Growth hormone stimulates IGF-I gene expression in isolated rat renal collecting duct, Am J Physiol, 259:F474-479 (1990).

16.   S. Miller, P. Rotwein, J. Bortz, P. Bechtel, V. Hansen, S. Rogers, and M. Hammerman, Renal expression of IGF I in hypersomatotropic states, Am J Physiol, 259:F251-F257 (1990).

17.   F. Conti, L. Striker, M. Lesniak, K. MacKay, J. Roth, and G. Striker, Studies on binding and mitogenic effect of insulin and insulin-like growth factor I in glomerular mesangial cells, Endocrinology, 122:2788-2795 (1988).

18.   M. Hammerman, and J. Gavin, Binding of IGF-I and IGF-I-stimulated phosphorylation in canine renal basolateral membranes, Am J Physiol, 251:E32-E41 (1986).

19.   M. Hammerman, and S. Rogers, Distribution of IGF receptors in the plasma membrane of proximal tubular cells, Am J Physiol, 253:F841-F847 (1987).

20.   H. Guler, J. Zapf, and E. Froesch, Short-term metabolic effects of recombinant human insulin-like growth factor I in healthy adults, N Engl J Med 317:137-140 (1987).

21.   H. Guler, C. Schmid, J. Zapf, and E. Froesch, Effects of recombinant insulin-like growth factor I on insulin secretion and renal function in normal human subjects, Proc Natl Acad Sci USA 86:2868-2872 (1989).

22.   M. Baxter, and J. Martin, Structure of the M 140,000 growth hormone - dependent insulin-like growth factor binding protein complex: determination by reconstitution and affinity labeling. Proc Natl Acad Sci USA 86:6898-6902 (1989).

23.   J. Zapf, and E. Froesch, Extrapancreatic tumor hypoglycemia: The role of insulin-like growth factors and insulin-like growth factor binding proteins,in: "Growth hormone and insulin-like growth factor I in human and experimental diabetes," A. Flyvbjerg, H. Ørskov, and K. Alberti, eds., John Wiley & Sons Ltd, Chichester, 1993.

24.   S. Hodgkinson, G. Spencer, J. Bass, S. Davis, and P. Gluckman, Distribution of ciruclating insulin-like growth factor-I (IGF-I) into tissues, Endocrinology, 129:2085-2093 (1991).

25.   Z. Dai, A. Stiles, B. Moats-Staats, J. Van Wyk, and A. D'Ercole, Interaction of secreted insulin-like growth factor-I (IGF-I) with cell surface receptors is the dominant mechanism of IGF-I's autocrine actions, J Biol Chem, 267:19565-19571 (1992).

26.   D. Ikkos, H. Ljunggren, and R. Luft, Glomerular filtration rate and renal plasma flow in acromegaly, Acta Endocrinol (Copenhagen), 21:226-236 (1956).

27.   T. Falkheden, Renal function following hypophysectomy in man, Acta Endocrinol (Copenhagen), 42:571-590 (1963).

28.   T. Falkheden, and I. Wickbom, Renal function and kidney size following hypophysectomy in man, Acta Endocrinol (Copenhagen), 48:348-354 (1965).

29.   J. Corvilain, M. Abramow, and A. Bergans, Some effects of human growth hormone on renal hemodynamics and on tubular phosphate transport in man, J Clin Invest, 41:1230-1235 (1962).

30.   H. Gershberg, H. Heinemann, and H. Stumpf, Renal function studies and autopsy report in a patient with gigantism and acromegaly, J Clin Endocrinol Metab, 17:377-385 (1957).

31.   H. Gershberg, Metabolic and renotropic effects of human growth hormone in disease, J Clin Endocrinol Metab, 20:1107-1119 (1960).

32.   R. Hirschberg, and J. Kopple, Increase in renal plasma flow and glomerular filtration rate during growth hormone treatment may be mediated by insulin-like growth factor I, Am J Nephrol, 8:249-253 (1988).

33.   R. Hirschberg, and J. Kopple, Role of growth hormone in the amino acid-induced acute rise in renal function in man, Kidney Int, 32:382-387 (1987).

34.   R. Hirschberg, and J. Kopple, Effects of growth hormone on GFR and renal plasma flow in man, Kidney Int 32 (Suppl 22):S21-S24 (1987).

35.   K. Kleinman, and R. Glassock, Glomerular filtration rate fails to increase following protein ingestion in hypothalamo-hypophyseal-deficient adults, Am J Nephrol, 6:169-174 (1986).

36.   L. Ruilope, J. Rodico, B. Miranda, R. Robles, J. Sancho-Rof, and C. Romero, Renal effects of amino acid infusion in patients with panhypopituitarism, Hypertension, 11:557-559 (1988).

37.   R. Hirschberg, R. Zipser, L. Slomowitz, and J. Kopple, Glucagon and prostaglandins are mediators of amino acid-induced rise in renal hemodynamics, Kidney Int, 33:1147-1155 (1988).

38.   P. TerWee, J. Rosman, S. van der Geest, W. Sluiter, and A. Donker, Renal hemodynamics during separate and combined infusion of amino acids and dopamine, Kidney Int 29:870-874 (1986).

39.   P. Castellino, B. Coda, and R. DeFronzo, Effect of amino acid infusion on renal hemodynamics in humans, Am J Physiol, 251:F132-F140 (1986).

40.   H. Parving, I. Noer, C. Mogensen, and P. Svendsen, Kidney function in normal man during short-term growth hormone infusion, Acta Endocrinol (Copenhagen), 89:796-800 (1978).

41.   J. Christiansen, J. Gammelgaard, H. Ørskov, A. Anderson, S. Temler, and H. Parving, Kidney function and size in normal subjects before and during growth hormone administration for one week, Eur J Clin Invest, 11:487-490 (1981).

42.   R. Hirschberg, H. Rabb, R. Bergamo, and J. Kopple, The delayed effect of growth hormone on renal function in humans, Kidney Int, 35:865-870 (1989).

43.   D. Haffner, E. Ritz, O. Mehls, J. Rosman, W. Blum, U. Heinrich, and A. Hübinger, Growth hormone induced rise in glomerular filtration rate is not obliterated by angiotensin-converting enzyme inhibitors, Nephron, 55:63-68 (1990).

44.   E. Ritz, B. Tönshoff, S. Worgall, G. Kovacs, and O. Mehls, Influence of growth hormone and insulin-like growth factor I on kidney function and kidney growth, Pediatr Nephrol 5:509-512 (1991).

45.   R. Hirschberg, and J. Kopple, Evidence that insulin-like growth factor I increases renal plasma flow and glomerular filtration rate in fasted rats, J Clin Invest, 83:326-330 (1989).

46.   R. Hirschberg, J. Kopple, R. Blantz, and B. Tucker, Effects of recombinant human insulin-like growth factor I on glomerular dynamics in the rat, J Clin Invest 87:1200-1206 (1991).

47.   R. Hirschberg, G. Brunori, J. Kopple, and H. Guler, Effects of insulin-like growth factor I on renal function in normal men, Kidney Int, 43:387-397 (1993).

48.   H. Guler, K. Eckardt, J. Zapf, C. Bauer, and E. Froesch, Insulin-like growth factor I increases glomerular filtration rate and renal plasma flow in man, Acta Endocrinol (Copenhagen), 121:101-106 (1989).

49.   J. Haylor, I. Singh, and A. El Nahas, Nitric oxide synthesis inhibitor prevents vasodilation by insulin-like growth factor I., Kidney Int 39:333-335 (1991).

50.   U. Baumann, T. Eisenhauer, and H. Hartmann, Increase of glomerular filtration rate and renal plasma flow by insulin-like growth factor I during euglycemic clamping in anaesthetized rats, Eur J Clin Invest, 22:204-209 (1992).

51.   B. Myers, W. Deen, and B. Brenner, Effects of norepinephrine and angiotensin II on the determinants of glomerular ultrafiltration and proximal tubule fluid reabsorption in the rat, Circ Res, 37:101-110 (1975).

52.   C. Baylis, and B. Brenner, The physiologic determinants of glomerular ultrafiltration, Rev Physiol Biochem Pharmacol 80:1-45 (1978).

53.   R. Palmer, A. Ferrige, and S. Moncada, Nitric oxide release accounts for the biological activity of endothelium-derived relaxing factor, Nature, 327:524-525 (1989).

54.   R. Furchgott, P. Cherry, J. Zawadzki, and D. Jothianandan, Endothelial cells as mediators of vasodilation of arteries, J Cardiovasc Pharmacol 6:S336-S343 (1984).

55.   M. Young, and S. Vatner, Regulation of large coronary arteries, Circ Res, 59:579-596 (1986).

56.   P. Vanhoutte, and V. Miller, Heterogeneity of endothelium-dependent responses in mamallian blood vessels, J Cardiovasc Pharmacol 7:S12-S23 (1985).

57.   R. Furchgott, Role of endothelium in responses of vascular smooth muscle, Circ Res 53:557-573 (1983).

58.   R. Furchgott, and J. Zawadzki, The obligatory role of endothelial cells in the relaxation of arterial smooth muscle by acetylcholine, Nature, 288:373-376 (1980).

59.   V. Kon, R. Harris, and I. Ichikawa, A regulatory role for large vessels in organ circulation, J Clin Invest, 85:1728-1733 (1990).

60.   P. Shultz, A. Shorer, and L. Raij, Effects of endothelium-derived relaxation factor and nitric oxide on rat mesangial cells, Am J Physiol, 258:F162-F168 (1990).

61.   J. Tolins, R. Palmer, S. Moncada, and L. Raij, Role of endothelium-derived relaxing factor in regulation of renal hemodynamic responses, Am J Physiol, 258:H655-H662 (1990).

62.   G. Burton, S. MacNeil, A. De Jonge, and J. Haylor, Cyclic GMP release and vasodilatation induced by EDRF and atrial natriuretic factor in the isolated perfused kidney of the rat, Br J Pharmacol, 99:364-368 (1990).

63.   R. Hirschberg, Die aminosäure und hormoninduzierte Modulation der Nierenfunktion und ihre mögliche Bedeutung für die Progression der chronischen Niereninsuffizienz, Habilitationsschrift, Free University of Berlin, School of Medicine, 1989.

64.   R. Dullaart, S, Meijer, P. Marbach, and W. Sluiter, Effect of a somatostatin analogue, octreotide, on renal haemodynamics and albuminuria in acromegalic patients, Eur J Clin Invest 22:494-502 (1992).

65.   N. Hizuka, K. Takano, K. Asakawa, I. Sukegawa, I. Fukuda, H. Demura, M. Iwashita, T. Adachi, and K. Shizume, Measurement of free form of insulin-like growth factor I in human plasma, Growth Regul 1:51-55 (1991).

66. S. Gargosky, K. Moyse, P. Walton, J. Owens, J. Wallace, J. Robinson, and P. Owens, Circulating levels of insulin-like growth factors increase and molecular forms of their serum binding proteins change with human pregnancy, Biochem Biophys Res Comm 170:1157-1163 (1990).

67. W. Isley, L. Underwood, and D. Clemmons, Dietary components that regulate serum somatomedin-C concentrations in humans, J Clin Invest, 71:175-182 (1983).

68. T. Prewitt, A. D'Ercole, B. Switzer, and J. Van Wyk, Relationship of serum immunoreactive somatomedin-C to dietary protein and energy in growing rats, J Nutr 112:144-150 (1982).

69. R. Hirschberg, and J. Kopple, Response of insulin-like growth factor I and renal hemodynamics to a high and low protein diet in the rat, J Am Soc Nephrol, 1:1034-1040 (1991).

70. A. Flyvbjerg, O. Thorlacius-Ussing, R. Næraa, J. Ingerslev, and H. Ørskov, Kidney tissue somatomedin C and initial renal growth in diabetic and uninephrectomized rats, Diabetologia, 31:310-314 (1988).

71. M. Bolze, R. Reeves, F. Lindbeck, and M. Elders, Influence of selected amino acid deficiencies on somatomedin, growth and glycoaminoglycan metabolism in weanling rats, J Nutr 115:782-787 (1985).

72. M. Maes, Y. Amand, L. Underwood, D. Maiter, and J. Ketelslegers, Decreased serum insulin-like growth factor I response to growth hormone in hypophysectomized rats fed a low protein diet: Evidence for a postreceptor defect, Acta Endocrinol (Copenhagen), 117:320-326 (1988).

73. D. Maiter, M. Maes, L. Underwood, T. Fliesen, G. Gerard, and J. Ketelslegers, Early changes in serum concentrations of somatomedin-C induced by dietary protein deprivation in rats: Contributions of growth hormone receptor and post-receptor defects, J Endocrinol 118:113-120 (1988).

74. H. Payne-Robinson, I. Smith, and M. Golden, Plasma somatomedin-C in jamaican children recovering from severe malnutrition, Clin Res 34:866A (1986).

75. D. Clemmons, L. Underwood, R. Dickerson, R. Brown, L. Hak, R. MacPhee, M. Heizer, and W. Heizer, Use of plasma somatomedin-C/insulin-like growth factor I measurements to monitor the response to nutritional repletion in malnourished patients, Am J Clin Nutr, 41:191-198 (1985).

76. C. Maase, and H. Zondek, Über eigenartige Ödeme, Dtsch Med Wochenschr 43:484-485 (1917).

77. G. Alleyne, The effect of severe protein calorie malnutrition on the renal function of jamaican children, Pediatr 39:400-411 (1967).

78. S. Klahr, and G. Alleyne, Effects of chronic protein-calorie malnutrition on the kidney, Kidney Int, 3:129-141 (1973).

79. I. Ichikawa, M. Purkerson, S. Klahr, J. Troy, M. Martinez-Maldonado, and B. Brenner, Mechanism of reduced glomerular filtration rate in chronic malnutrition, J Clin Invest 65:982-988 (1980).

80. D. Dardevet, M. Manin, M. Balage, C. Sornet, and J. Grizard, Influence of low- and high-protein diets on insulin and insulin-like growth factor-I binding to skeletal muscle and liver in the growing rat, Br J Nutr 65:47-60 (1991).

81. R. Hirschberg, and J. Kopple, The growth hormone-insulin-like growth factor I axis and renal glomerular function, J Am Soc Nephrol 2:1417-1422 (1992).

82. M. Lumpkin, S. Mulroney, and A. Haramati, Inhibition of pulsatile growth hormone secretion and somatic growth in immature rats with a synthetic growth hormone releasing factor antagonist, Endocrinology 124:1154-1159 (1989).

83. M. Lumpkin, and J. McDonald, Blockade of growth hormone-releasing factor (GRF) activity in the pituitary and hypothalamus of conscious rats with a peptidic GRF antagonist, Endocrinology 124:1522-1531 (1989).

84. S. Mulroney, M. Lumpkin, and A. Haramati, Antagonist to GH-releasing factor inhibits growth and renal Pi absorption in immature rats, Am J Physiol 257:F29-F34 (1989).

85. R. Hirschberg, Effects of growth hormone and IGF-I on glomerular ultrafiltration in growth hormone-deficient rats, Regul Peptides, in press (1993).

86. S. Miller, V. Hansen, and M. Hammerman, Effects of growth hormone and IGF-I on renal function in rats with normal and reduced renal mass, Am J Physiol, 259:F747-F751 (1990).

87. D. Haffner, S. Zacharewics, O. Mehls, U. Heinrich, and E. Ritz, The acute effect of growth hormone on GFR is obliterated in chronic renal failure, Clin Nephrol 32:266-269 (1989).

88. B. Tönshoff, C. Tönshoff, J. Pinkowski, W. Blum, U. Heinrich, and O. Mehls, Effects of recombinant human growth hormone on growth and renal function in children with chronic renal failure, Acta Paediatr Scand 370 (Suppl E):193 (1990).

89. O. Mehls, E. Ritz, G. Kovacs, R. Fine, S. Worgal, and R. Mak, Effects of human recombinant growth hormone and IGF-I on growth and GFR in uremic rats, Kidney Int 37:513Abstr (1990).

90. M. O'Shea, S. Miller, and M. Hammerman, Effects of IGF-I on renal function in patients with chronic renal failure, Am J Physiol 264:F917-F922 (1993).

91. D. Bergenstal, and M. Lipsett, Metabolic effects of human growth hormone and growth hormone of other species in man, J Endocrinol Metab 20:1424-1436 (1960).

92. J. Beck, E. McGarry, I. Dyrenfurth, and E. Venning, The metabolic effects of human and monkey growth hormone in man, Ann Intern Med 49:1090-1094 (1958).

93. R. Quigley, and M. Baum, Effects of growth hormone and insulin-like growth factor I on rabbit proximal convoluted tubule transport, J Clin Invest 88:368-374 (1991).

94. N. Yanagawa, D. Sheikh-Hamad, and O. Jo, Insulin-like growth factor II directly increases renal brush border membrane sodium transport, J Am Soc Nephrol 2:447 (Abstr) (1991).

95. J. Caverzasio, R. Faundez, H. Fleisch, and J. Bonjour, Tubular adaptation to Pi restriction in hypophysectomized rats, Pflüger's Arch 392:17-21 (1981).

96. J. Bonjour, and J. Caverzasio, IGF-I a key controlling element in phosphate homeostasis during growth, in: "Modern concepts of insulin-like growth factors," E. Spencer, ed., Elsevier, New York (1991).

97. R. Hirschberg, and C. Nast, Glomerular hypertrophy and segmental glomerular sclerosis are not linked in subtotally nephrectomized rats, J Am Soc Nephrol 3:739 (Abstr) (1992).

98. H. Guler, J. Zapf, E. Scheiwiller, and E. Froesch, Recombinant human insulin-like growth factor I stimulates growth and has distinct effects on organ size in hypophysectomized rats, Proc Natl Acad Sci USA 85:4889-4893 (1988).

99. A. El Nahas, J. Le Carpentier, A. Bassett, and D. Hill, Dietary protein and insulin-like growth factor I content following unilateral nephrectomy, Kidney Int 36 (Suppl 27):S15-S19 (1989).

100. S. Mulroney, A. Haramati, H. Werner, C. Bondy, C. Roberts, and D. LeRoith, Altered expression of insulin-like growth factor I (IGF-I) and IGF receptor genes after unilateral nephrectomy in immature rats, Endocrinology 130:249-256 (1992).

101. S. Mulroney, A. Haramati, C. Roberts, and D. LeRoith, Renal IGF-I mRNA levels are enhanced following unilateral nephrectomy in immature but not adult rats, Endocrinology, 128:2660-2662 (1991).

102. A. Flyvbjerg, H. Ørskov, K. Nyborg, J. Frystyk, S. Marshall, K. Bornfeldt, H. Arnqvist, and K. Jorgensen, Kidney IGF-I accumulation occurs in four different conditions with rapid initial kidney growth in rats, in: "Modern concepts of insulin-like growth factors," E. Spencer, ed., Elsevier, New York (1991).

103. A. Fogo, and I. Ichikawa, Evidence for a pathogenic linkage between glomerular hypertrophy and sclerosis, Am J Kidney Dis, 17:666-669 (1991).

104. A. Fogo, B. Hawkins, P. Berry, A. Glick, M. Chiang, R. MacDonell, and I. Ichikawa, Glomerular hypertrophy in minimal change disease predicts subsequent progression to focal glomerular sclerosis, Kidney int, 38:115-123 (1990).

105. A. Fogo, and I. Ichikawa, Evidence for a central role of glomerular growth promoters in the development of sclerosis, Sem Nephrol, 9:329-342 (1989).

106. Y. Yoshida, A. Fogo, and I. Ichikawa, Glomerular hemodynamics vs. hypertrophy in experimental glomerular sclerosis, Kidney Int, 35:6540660 (1989).

107. H. Lafferty, and B. Brenner, Are glomerular hypertension and hypertrophy independent risk factors for progression of renal disease? Sem Nephrol, 10:294-304 (1990).

108. S. Peng, G. Lee, C. Nast, R. Guillermo, P. Levin, C. Ihm, R. Glassock, and S. Adler, Increments in procollagen α_1(IV) mRNA levels are not required for glomerular hypertrophy, J Am Soc Nephrol 2:687 (Abstr) (1991).

109. S. Feld, R. Hirschberg, A. Artishevsky, R. Glassock, and S. Adler, IGF-I increases procollagen α^1(IV) mRNA levels in cultured mesangial cells, J Am Soc Nephrol 2:573 (Abstr) (1991).

110. L. Mathews, R. Hammer, R. Brinster, and R. Palmiter, Expression of insulin-like growth factor I in transgenic mice with elevated levels of growth hormone is correlated with growth, Endocrinology, 123:433-437 (1988).

111. L. Mathews, R. Hammer, R. Behringer, A. D'Ercole, G. Bell, R. Brinster, and R. Palmiter, Growth enhancement of transgenic mice expressing human insulin-like growth factor I, Endocrinology, 123:2827-2833 (1988).

112. C. Quaife, L. Mathews, C. Pinkert, R. Hammer, R. Brinster, and R. Palmiter, Histopathology associated with elevated levels of growth hormone and insulin-like growth factor I in transgenic mice, Endocrinology, 124:40-48 (1989).

113. T. Doi, L. Striker, C. Quaife, F. Conti, R. Palmiter, R. Behringer, R. Brinster, and G. Striker, Progressive glomerulosclerosis develops in transgenic mice chronically expressing growth hormone and growth hormone releasing factor but not in those expressing insulinlike growth factor-I, Am J Pathol, 131:398-403 (1988).

114. T. Doi, L. Striker, C. Gibson, L. Agodoa, R. Brinster, and G. Striker, Glomerular lesions in mice transgenic for growth hormone and insulinlike growth factor I, Am J Pathol, 137:541-552 (1990).

365

115. T. Doi, L. Striker, K. Kimata, E. Peten, Y. Yamada, and G. Striker, Glomerulosclerosis in mice transgenic for growth hormone: increased mesangial extracellular matrix is correlated with kidney mRNA levels, J Exp Med 173:1287-1290 (1991).

116. C. Pesce, L. Striker, E. Peten, S. Elliot, and G. Striker, Glomerulosclerosis at both early and late stages is associated with increased cell turnover in mice transgenic for growth hormone, Lab Invest, 65:601-605 (1991).

117. H. Humes, D. Cieslinski, T. Coimbra, J. Messana, and C. Galvao, Epidermal growth factor enhances renal tubule cell regeneration and repair and accelerates the recovery of renal function in postischemic acute renal failure, J Clin Invest, 84:1757-1761 (1989).

118. J. Norman, Y. Tsau, A. Bacay, and L. Fine, Epidermal growth factor accelerates functional recovery from ischaemic acute tubular necrosis in the rat: Role of the epidermal growth factor receptor, Clin Sci 78:445-450 (1990).

119. R. Harris, R. Hoover, H. Jacobson, and K. Badr, Evidence for glomerular actions of epidermal growth factor in the rat, J Clin Invest, 82:1028-1039 (1988).

120. S. Kanda, K. Nomata, P. Saha, N. Nishimura, J. Yamada, H. Kanatake, and Y. Saito, Growth factor regulation of the renal cortical tubular cell by epidermal growth factor, IGF-I, acidic and basic fibroblast growth factor, and transforming growth factor-ß in serum free culture, Cell Biol Int Rep 13:687-699 (1989).

121. S. Kupfer, L. Underwood, R. Baxter, and D. Clemmons, Enhancement of the anabolic effects of growth hormone and IGF-I by use of both agents simultaneously, J Clin Invest 91:391-396 (1993).

122. H. Ding, J. Kopple, A. Cohen, and R. Hirschberg, Recombinant human insulin-like growth factor-I accelerates recovery and reduces catabolism in rats with ischemic acute renal failure, J Clin Invest, 91:2281-2287 (1993).

123. S. Miller, D. Martin, J. Kissane, and M. Hammerman, IGF-I accelerates recovery from ischemic acute tubular necrosis in the rat, Proc Natl Acad Sci USA 89:11876-11880 (1992).

124. R. Rabkin, A. Sorensen, D. Mortensen, and R. Clark, Insulin-like growth factor I enhances recovery from acute renal failure induced by ischemia, J Am Soc Nephrol 3:713 (Abstr) (1992).

125. Y. Yagil, B. Myers, and R. Jamison, Course and pathogenesis of postischemic acute renal failure in the rat, Am J Physiol, 255:F257-F264 (1988).

126. R. Williams, C. Thomas, L. Navar, and A. Evan, Hemodynamic and single nephron function during maintenance phase of ischemic acute renal failure in the dog, Kidney Int 19:503-515 (1981).

127. J. Barnes, R. Osgood, H. Reineck, and J. Stein, Glomerular alterations in an ischemic model of acute renal failure, Lab Invest, 45:378-386 (1981).

128. F. Toback, Regeneration after acute tubular necrosis, Kidney Int 41:226-246 (1992).

129. B. Myers, and S. Moran, Hemodynamically mediated acute renal failure, N Engl J Med 314:97-105 (1986).

130. G. Andersson, and E. Jennische, IGF-I immunoreactivity is expressed by regenerating renal tubular cells after ischaemic injury in the rat, Acta Physiol Scand, 132:13-23 (1991).

REGULATION OF IGFBP-4 AND -5 EXPRESSION IN RAT GRANULOSA CELLS

Xin-Jun Liu and Nicholas Ling

Department of Molecular Endocrinology
The Whittier Institute for Diabetes and Endocrinology
9894 Genesee Avenue
La Jolla, CA 92037

INTRODUCTION

One of the central questions in reproductive physiology concerns how follicles in the ovary are selected during folliculogenesis. Although it has been well established that folliculogenesis is initiated and maintained by the gonadotropin, follicle-stimulating hormone (FSH), how follicle selection takes place in the ovary is still an enigma.[1] Based on the notion that selection of the dominant follicle might be due to the action of a locally produced inhibitor in the follicular fluid which could block the effects of FSH on the non-selected follicles, our laboratory initiated a project in 1988 to isolate and identify the FSH inhibitor present in porcine ovarian follicular fluid. Using a well-defined rat granulosa cell culture assay to monitor the purification, a polypeptide was isolated which exhibited a potent inhibition of FSH-stimulated estradiol production in those cells.[2] Chemical characterization of the inhibitor, however, revealed that its structure corresponded to insulin-like growth factor binding protein-3 (IGFBP-3).[3] Subsequent studies carried out by our laboratory identified that the rat ovary produced five IGFBPs in a tissue-specific manner, with the mRNA for IGFBP-2 mainly expressed by theca interstitial cells, IGFBP-3 by luteinized granulosa cells, IGFBP-4 by the granulosa of atretic antral follicles, IGFBP-5 by the granulosa of atretic preantral follicles and IGFBP-6 by the theca externa, stroma and smooth muscle cells[4-6]. IGFBP-1 was not detected in the rat ovary. These findings indicate that the transcription of each IGFBP gene is regulated differently in the rat ovary and that each IGFBP may have a different physiological function. This hypothesis is strengthened by our finding that exogenously added IGFBP-2, -3, -4 and -5 were able to attenuate the FSH-stimulated production of estradiol and progesterone in cultured rat granulosa cells[7,8] and ovarian intrabursal injection of IGFBP-3 was able to decrease the ovulation rate of pregnant mare's serum gonadotropin/human choriogonadotropin-primed animals.[9] Thus the actions of the IGFBPs in the ovary appear to counter the effects of FSH and the expression of IGFBP-4 and -5 in atretic follicles is consistent with the hypothesis that follicle selection might be associated with the ability of the selected follicles to inhibit IGFBP-4 and -5 production.

To further explore the mechanisms underlying the regulation of IGFBP-4 and -5 expression in rat granulosa cells, we have employed an *in vitro* culture system in serum-free medium to study the effects of pituitary hormones and IGF-I.

Current Directions in Insulin-Like Growth Factor Research,
Edited by D. LeRoith and M.K. Raizada, Plenum Press, New York, 1994

367

MATERIALS AND METHODS

Animals

Animals were handled in accord with the NIH Guide for the Care and Use of Laboratory Animals, and the procedures used were approved by the Whittier Institute Animal Care and Use Committee. Twenty-one days old intact female rats (Wistar, Harlan Sprague Dawley, Indianapolis, IN) were implanted subcutaneously with a 10 mm silastic capsule containing approximately 10 mg diethylstilbestrol. Four days later, the animals were killed using a carbon dioxide chamber and the ovaries collected.

Hormones and Reagents

The rat IGFBPs were purified from adult rat serum as previously described.[10] Androstenedione, diethylstilbestrol and bovine serum albumin (BSA) were from Sigma Chemical Co. (St. Louis, MO). McCoy's 5a Medium (serum-free modified RPMI1629), L-glutamine and antibiotics were from Gibco (Santa Clara, CA). Ovine FSH, luteinizing hormone (LH) and prolactin (PRL) were obtained from the National Hormone and Pituitary Program, NIDDK (Baltimore, MD). Human [Thr^{59}]IGF-I was synthesized in our laboratory based on a published procedure.[10]

Granulosa Cell Culture

Granulosa cell culture was performed according to a previously described method.[7] The cells were plated in 6-well tissue culture plates (Nunc, Roskilde, Denmark) at a density of 2 x 10^6 viable cells per 2 ml of culture medium (McCoy's 5a medium supplemented with penicillin (100 U/ml), streptomycin sulfate (100 µg/ml), L-glutamine (2 mM) and 10^{-7} M androstenedione). Cells were cultured at 37° C under 95% air and 5% CO_2 with various concentrations of hormones and reagents.

Electrophoresis and Western Blotting Analysis

Sodium dodecylsulfate-polyacrylamide gel electrophoresis (SDS/PAGE) was performed by using the buffer system of Laemmli[11] in the absence of sulfhydryl reducing agents with the Xcell mini-cell electrophoresis system (Novex, Encinitas, CA). Conditioned media were concentrated 10 times by using Immersible-CX10 ultrafilters (10,000 molecular weight cut-off, Millipore, Bedford, MA). Samples were loaded onto either 12% or 16% polyacrylamide gels (Tris/glycine percentage gel, Novex), electrophoresed at a constant voltage of 100 V for 1.5 - 2.0 h and then transferred to 0.45 µm nitrocellulose membranes (Bio-Rad, Richmond, CA) for 2 h at 150 mA using the Mini Blot module of the Xcell mini-cell system (Novex). The molecular weight of the binding proteins was estimated by comparison with the prestained molecular weight standards (Bio-Rad).

Immunoblotting was performed by first treating the nitrocellulose membrane with casein/TBST buffer (1% casein/Tris buffered saline with 0.1% Tween 20) for 1 h at room temperature. This was followed by incubation with diluted antibodies in casein/TBST buffer for overnight at 4° C. The IGFBP-4 antiserum was used at a 1:400 dilution and the IGFBP-5 antiserum at a 1:200 dilution.[8] The membranes were washed three times with TBST buffer and then incubated for 2 h at 23° C with peroxidase-conjugated goat anti-rabbit IgG (Calbiochem, San Diego, CA) diluted 1:5,000 in casein/TBST buffer. The membranes were washed three times with TBST buffer, after which they were incubated with ECL Western blotting reagents (Amersham, Arlington Heights, IL) for 1 min and then covered with Saran Wrap for exposure to Hyperfilm-ECL (Amersham) for 30 sec to 30 min.

Northern Blotting Analysis

After removing the culture medium for Western blotting, the cells were washed in phosphate-buffered saline. Total RNA was prepared by using the guanidine isothiocyanate extraction method[12] and the extracted RNAs electrophoresed on a 0.66 M formaldehyde-agarose gel. The fractionated RNAs were transferred onto nylon membrane filters and hybridization was performed at 68° C for 15 h with a [^{32}P]labeled antisense cRNA probe prepared from the SmaI-HindIII 444 basepair restriction fragment of the rat IGFBP-4 clone RBP4-501,[5] or from the SacII-HindIII 300 basepair restriction fragment of the rat IGFBP-5 clone RBP5-501.[6] The hybridization buffer consisted of 50% formamide, 6x SSPE [60 mM Na_2HPO_4 (pH 7.0), 1.08 M NaCl, and 6 mM EDTA], 0.5% sodium dodecylsulfate (SDS), 5x Denhardt's solution [0.1% Ficoll, 0.1% polyvinylpyrrolidone and 0.1% BSA], 0.2 mg/ml yeast tRNA and 0.2 mg/ml denatured salmon sperm DNA. After hybridization, the filters were washed with vigorous agitation in 200 mM NaCl/30 mM sodium citrate/0.1% SDS for 15 min at room temperature and then incubated in 15 mM NaCl/15 mM sodium citrate/0.1% SDS for 20 min at 68° C. Autoradiography was performed on an X-ray film (Kodak, Rochester, NY) with an intensifying screen at -80° C.

Digestion of IGFBP with granulosa-derived protease and characterization of the digested fragments

Five micrograms of IGFBP in 10 μl 0.1 M HEPES buffer, pH 7.4, were incubated for 8 h at 37° with 40 μl of 10x concentrated granulosa cell culture medium harvested from cells which had been preconditioned with 100 ng/ml of FSH for 72 h. The incubated mixture was fractionated on a 12% SDS/PAGE gel. After electrophoresis, the separated protein fragments were blotted onto a PVDF membrane (Millipore) and the membrane stained lightly with Coomassie blue to reveal the protein bands. The stained bands were cut out with a razor blade and subjected to amino acid sequence analysis on a model 470A gas-phase protein sequenator coupled to an on-line mode 120A PTH analyzer (Applied Biosystems Inc., Foster City, CA).

RESULTS

Dose-response and time-course effects of FSH on IGFBP-4 and -5 expression

In a prior study,[8] we have determined that serum-free culture of diethylstilbestrol-primed rat granulosa cells spontaneously expressed IGFBP-4 and -5 mRNAs and accumulated their proteins in the culture medium. However, production of the mRNAs and proteins could be inhibited by incubation with 100 ng/ml FSH for 72 h. To further assess the potency of FSH on the expression of IGFBP-4 and -5, an increasing dosage of FSH from 1 to 100 ng/ml was added to the medium and the granulosa cells cultured for 48 h. As shown in Fig. 1A and 1B, at low doses of FSH (≤3 ng/ml), there was a slight stimulation of IGFBP-4 and -5 mRNAs accumulation in comparison with control which received no FSH treatment. Higher concentrations of FSH, however, were able to inhibit the expression of both IGFBP-4 and -5 mRNAs in a dose-dependent manner, with inhibition of IGFBP-5 mRNA noticeable at 10 ng/ml, while IGFBP-4 mRNA required 30 ng/ml to show inhibition. Concentration of the two proteins in the culture media, harvested from the same cells used for the mRNA analyses, paralleled the mRNA responses with slight increase of the 24 and 28 kDa IGFBP-4 protein accumulation at doses of FSH up to 10 ng/ml, but marked decrease of the intact IGFBP-4 and -5 proteins at 30 ng/ml or high FSH concentrations (Fig. 1C and 1D). Moreover, induction of a protease which degraded IGFBP-4 and -5 to yield fragmentation bands at 17.5 and 21 kDa, respectively, was also noted at 30 ng/ml and higher concentrations of FSH.[8]

Since the response of the IGFBP-4 and -5 mRNAs and proteins to FSH was dose-dependent, it was necessary to determine the time-course effects of a low-dose (3 ng/ml) and a high-dose (30 ng/ml) of FSH. Treatment of the cells with both the low and high doses of FSH caused an initial increased accumulation of IGFBP-4 mRNA in comparison with control, which received no FSH, up to 24 h (Fig. 2A). Thereafter, the levels of IGFBP-4 mRNA started to decrease and reached barely detectable levels after 72 h at the high dose, whereas decrease of IGFBP-4 mRNA was similar to control at the low dose after 24 h. By contrast, accumulation of IGFBP-5 mRNA (6.0 kb) was slightly stimulated above control in both high and low doses of FSH up to 24 h (Fig. 2B). The high-dose of FSH was able to continuously maintain the mRNA levels for up to 72 h at the same concentration as it was at 24 h, but the low-dose of FSH was not able to suppress the mRNA levels resulting in a steady increase of the mRNA levels similar to control. Intact 24 kDa IGFBP-4 protein levels in the culture medium, harvested from the same cells employed for the mRNA analyses, were stimulated progressively as in control by the low-dose of FSH up to 48 h and maintained at the high level at 72 h (Fig. 3A), whereas the high-dose of FSH resulted in degradation of the

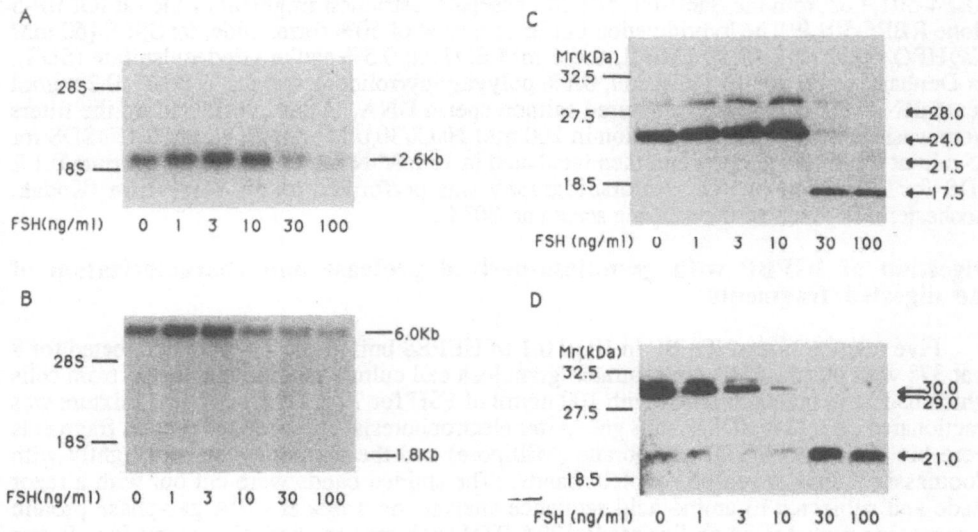

Figure 1. Dose-response effects of FSH on IGFBP-4 mRNA levels (panel A), IGFBP-5 mRNA levels (panel B), IGFBP-4 protein accumulation (panel C), and IGFBP-5 protein accumulation (panel D) in 48 h culture of rat granulosa cells.

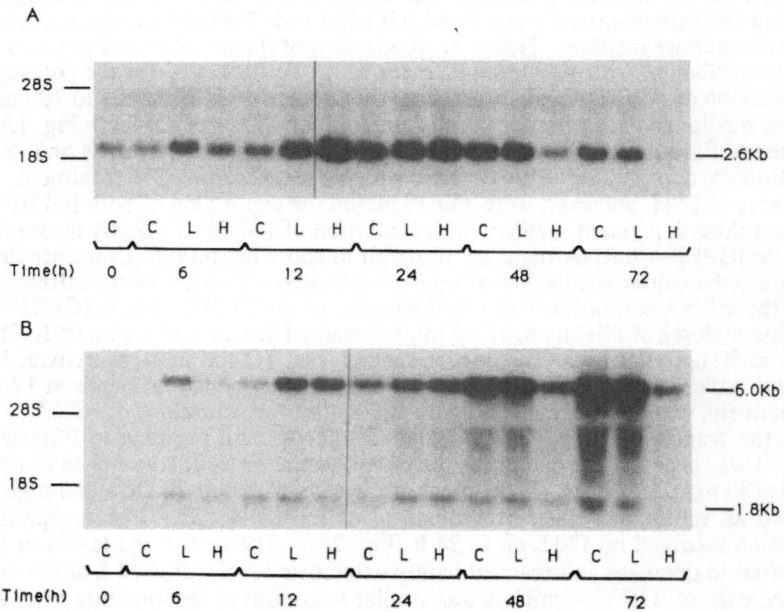

Figure 2. Time-course effects of FSH on IGFBP-4 mRNA levels (panel A) and IGFBP-5 mRNA levels (panel B) in cultured rat granulosa cells. C = control without FSH, L = low-dose with 3 ng/ml FSH, H = high-dose with 30 ng/ml FSH.

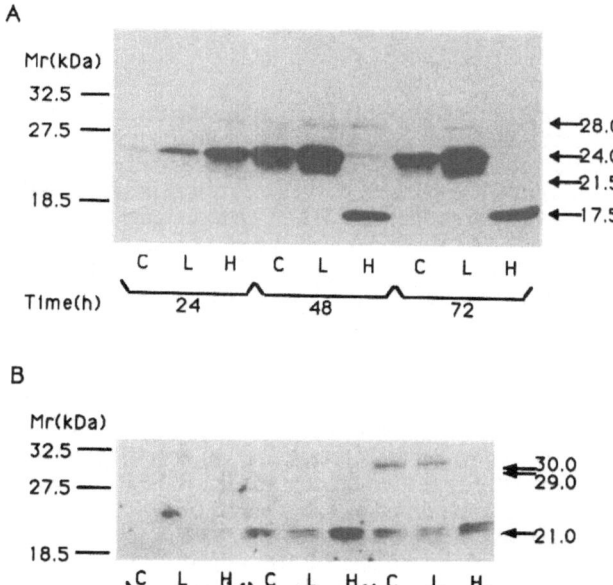

Figure 3. Time-course effects of FSH on IGFBP-4 protein accumulation (panel A) and IGFBP-5 protein accumulation (panel B) in culture media of rat granulosa cells. C = control without FSH, L = low-dose with 3 ng/ml FSH, H = high-dose with 30 ng/ml FSH.

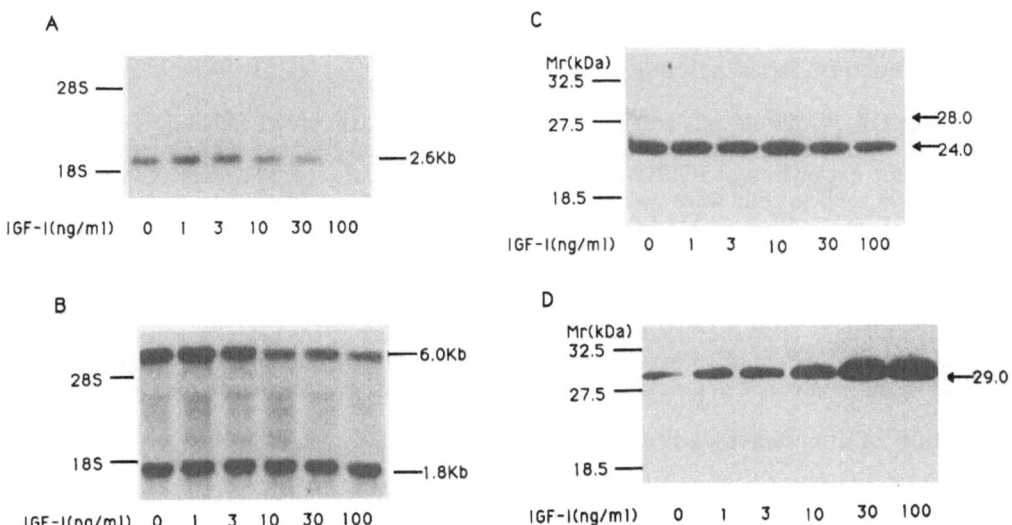

Figure 4. Dose-response effects of IGF-I on IGFBP-4 mRNA levels (panel A), IGFBP-5 mRNA levels (panel B), IGFBP-4 protein accumulation (panel C), and IGFBP-5 protein accumulation (panel D) in 48 h culture of rat granulosa cells.

intact protein and accumulation of the degraded fragment at 17.5 kDa after 48 h or longer incubation (Fig. 3A). Similar effects were found for the IGFBP-5 protein (Fig. 3B).

Dose-response and time-course effect of IGF-I on IGFBP-4 and -5 expression

As noted above, the effects of FSH suppression of IGFBP-4 and -5 mRNA accumulation was not manifested until after 48 h of incubation. This phenomenon suggests that the suppressive action could be mediated by a locally produced factor which was stimulated by FSH. Since IGF-I is one of the factors that is stimulated by FSH in granulosa cells and synergized with the actions of FSH,[13] we tested the dose-response effects of IGF-I alone on the accumulation of IGFBP-4 and -5 mRNAs. As shown in Fig. 4A, increasing concentrations of IGF-I were able to suppress IGFBP-4 mRNA accumulation in a dose-dependent manner. In the case of IGFBP-5, only the high molecular weight mRNA species (6.0 kb) was inhibited by the high doses of IGF-I, whereas the low molecular weight species (1.8 kb) exhibited no change in concentration (Fig. 4B). At the protein level, increasing doses of IGF-I did not inhibit the IGFBP-4 protein concentrations in the medium, harvested from the same IGF-I-treated cells, in comparison with control (Fig. 4C), except at the highest dose of IGF-I (100 ng/ml) tested where there was a slight inhibition. IGFBP-5 protein concentrations, by contrast, were increased in a dose-dependent manner (Fig. 4D). Moreover, no protease activity that could degrade the IGFBP-4 and IGFBP-5 proteins were noted in the culture media in all doses of IGF-I tested (Fig. 4C and 4D), indicating that IGF-I does not induce the protease in granulosa cells.

Time-course studies of the effects of 100 ng/ml IGF-I on IGFBP-4 and -5 mRNA accumulation showed that the levels of IGFBP-4 mRNA started to decrease in comparison with control at 24 h, and more decrease of the mRNA was noted at longer incubation times (Fig. 5A). The effect of IGF-I on IGFBP-5 mRNA by contrast was increased at 6 to 12 h, compared to control (Fig. 5B). But at 24 h and longer time periods, the mRNA levels were decreased in comparison with control. IGFBP-4 protein levels in the medium harvested from the same cell cultures showed a slight increase up to 24 h, but then started to decrease at 48 and 72 h in comparison with control (Fig. 6A), and no degradation of the intact IGFBP-4 protein was detected in all incubation times. By contrast, IGFBP-5 protein levels in the culture medium were increased at every time point in comparison with control (Fig. 6B), and also no degradation of the protein was observed throughout the incubations.

Effects of LH and PRL on IGFBP-4 and -5 protein levels in rat granulosa cell culture medium after priming with FSH for 48 h

When rat granulosa cells are primed with FSH, expression of LH and PRL receptors are induced.[14] To determine whether the newly synthesized LH and PRL receptors could regulate the expression of IGFBP-4 and -5, granulosa cells were pretreated with 30 ng/ml FSH for 48 h. The cells were washed and then treated with an increasing concentration of LH or PRL for another 48 h. As shown in Fig. 7A and 7B, increasing doses of LH were able to block the increase of IGFBP-4 and -5 protein levels in the culture medium, respectively, in comparison with control (lane 3), which received no LH treatment. At doses of LH higher than 3 ng/ml, induction of proteolytic activity which degraded IGFBP-4 and -5 was also noted. Thus the effects of FSH can be mimicked by LH after priming of the granulosa cells by FSH. By contrast, increasing dosage of PRL had no effect on the levels of IGFBP-4 and -5 proteins in the culture medium in comparison with control (data not shown).

Location of the cleavage site in rat IGFBP-5

Since the FSH-induced protease in granulosa cells may play an important role on the regulation of IGFBP-4 and -5, an experiment was carried out in an attempt to locate the cleavage sites in these two proteins treated by the granulosa-derived protease. The recovered 21 kDa fragment obtained from the digestion of IGFBP-5 possessed a sequence, Gly-Ser-Phe-Val-Xaa-Xaa-Glu-Pro-Xaa-Asp-Glu-Lys-Ala-Leu, which is identical to the NH$_2$-terminal amino acid sequence of IGFBP-5,[10] except for the deletion of the first amino acid. Xaa denotes a residue which could not be identified in the sequence analysis. The 15 kDa fragment yielded a sequence, Lys-Phe-Val-Gly-Gly-Ala-Glu-Asn-Thr-Ala, which corresponds to the internal sequence starting at residue-144 of IGFBP-5, indicating that the

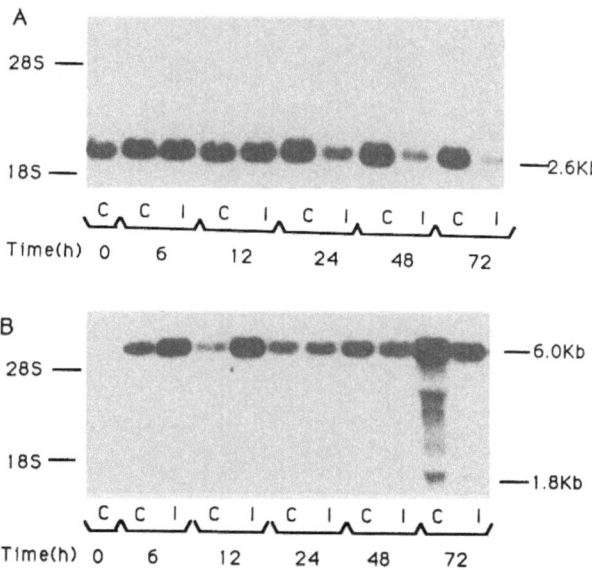

Figure 5. Time-course effects of IGF-I on IGFBP-4 mRNA levels (panel A) and IGFBP-5 mRNA levels (panel B) in cultured rat granulosa cells. C = control without IGF-I, I = 100 ng/ml IGF-I.

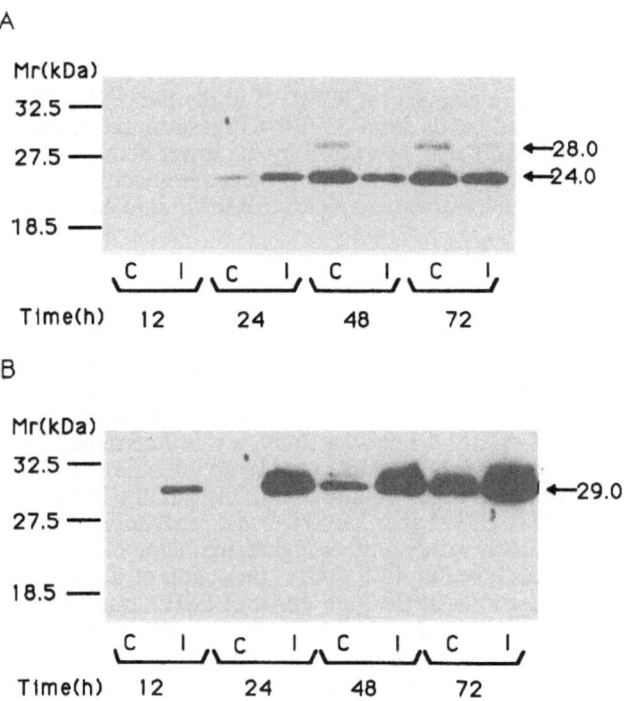

Figure 6. Time-course effects of IGF-I on IGFBP-4 protein accumulation (panel A) and IGFBP-5 protein accumulation (panel B) in culture media of rat granulosa cells. C = control without IGF-I, I = 100 ng/ml IGF-I.

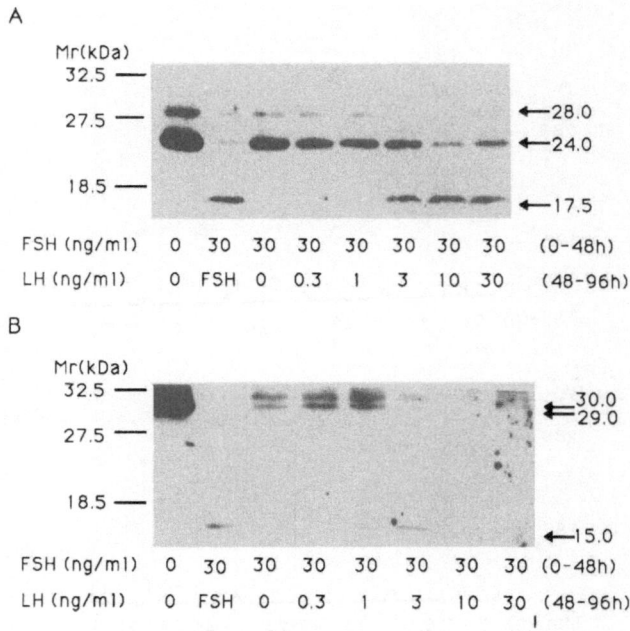

A

Mr(kDa)
32.5 —
27.5 —   ←—28.0
                                                                    ←—24.0
18.5 —
                                                                    ←—17.5

FSH (ng/ml)   0   30   30   30   30   30   30   30   (0-48h)
LH (ng/ml)   0   FSH   0   0.3   1   3   10   30   (48-96h)

B

Mr(kDa)
32.5 —   ⇐—30.0
                                                                    ⇐—29.0
27.5 —

18.5 —

                                                                    ←—15.0

FSH (ng/ml)   0   30   30   30   30   30   30   30   (0-48h)
LH (ng/ml)   0   FSH   0   0.3   1   3   10   30   (48-96h)

Figure 7. Dose-response effects of LH on IGFBP-4 protein accumulation (panel A) and IGFBP-5 protein accumulation (panel B) in 48 h culture media of rat granulosa cells which had been pre-treated with 30 ng/ml FSH for 48 h. FSH denotes the cells were treated with another 30 ng/ml FSH after the initial FSH treatment.

granulosa-derived protease cleaved rat IGFBP-5 at the Ser^{143}-Lys^{144} bond of the protein. Sequencing of the excised bands from IGFBP-4 digestion unfortunately did not yield any meaningful sequences. This may be caused by the lower activity of the protease to cleave IGFBP-4. However, the granulosa-derived protease is specific towards IGFBP-4 and -5, because we have determined that it did not cleave IGFBP-2, -3 and -6 (data not shown).

DISCUSSION

Previous studies carried out in our laboratory have shown that rat granulosa cells express the IGFBP-4 and -5 mRNAs and produce the two proteins spontaneously under serum-free culture conditions.[8] FSH at 100 ng/ml was able to inhibit this process after 72 h. In addition, the high concentration of FSH also induced the production of a protease which degraded IGFBP-4 and -5.[8] In the present study, it was determined that the FSH effect was dose-dependent. At low doses of FSH (≤3 ng/ml), no inhibition of IGFBP-4 and -5 mRNAs as well as protein levels was observed after 48 h. Instead, the low doses of FSH stimulated the accumulation of the IGFBP-4 and -5 mRNAs and production of the IGFBP-4 protein. At higher doses of FSH, namely at 30 ng/ml or higher, inhibition of the mRNAs and proteins for IGFBP-4 and -5 was observed at 48 h. Also, induction of a protease that could degrade IGFBP-4 and -5 was apparent at the high doses of FSH. Interestingly, even at 30 ng/ml FSH, stimulation of the IGFBP-4 and -5 mRNAs was observed at the first 12 h. Only at longer incubation times was inhibition of the mRNAs observed. This biphasic phenomenon, which had also been reported by Adashi, et al.,[15] suggests that the inhibitory effects of FSH on IGFBP-4 and -5 mRNA expression may be caused by the induction of a locally produced factor, such as IGF-I, which is stimulated by FSH.[13] Indeed, incubation with 100 ng/ml IGF-I was able to inhibit IGFBP-4 mRNA production within 24 h, but not IGFBP-5 mRNA production. These results demonstrated that even though rat granulosa cells produce both IGFBP-4 and -5, the two proteins and their mRNAs are regulated differently. Furthermore, IGF-I did not induce the granulosa-derived protease to degrade the IGFBP-4 and -5 proteins. The effects of FSH on granulosa cell IGFBP-4 and -5 expression could be mimicked by LH once the cells have been primed with FSH for 48 h to induce LH receptors. Since the intracellular signalling pathway coupled to the FSH and LH receptors is through activation of adenylate cyclase to increase cAMP,[16,17] cAMP may be the common intracellular mediator for the regulation of IGFBP-4 and -5 by FSH and LH. This hypothesis is reinforced by the lack of effect of PRL after priming by FSH, since the intracellular mediator of PRL action does not involve cAMP.[18]

Overall, the regulation of IGFBP-4 and -5 expression in rat granulosa cells is a complex phenomenon involving pituitary hormones as well as locally derived factors, and further experiments are required to define the physiological role of the IGFs and their binding proteins in ovarian function.

ACKNOWLEDGMENTS

We thank Drs. G.F. Erickson, L. DePaolo and S. Shimasaki for helpful discussion, M. Regno-Lagman and R. Schroeder for excellent technical assistance, and E. Exum for typing the manuscript. The work presented in this paper was supported by NICHD Program Project Grant HD-09690 and NICHD contract NO1-HD-0-2902.

REFERENCES

1. J.S. Richards, Maturation of ovarian follicles: Actions and interactions of pituitary and ovarian hormones on follicular cell differentiation, *Physiol.Rev.* 60:51 (1980).
2. M. Ui, M. Shimonaka, S. Shimasaki, and N. Ling, An insulin-like growth factor-binding protein in ovarian follicular fluid blocks follicle-stimulating hormone-stimulated steroid production by ovarian granulosa cells, *Endocrinology* 125:912 (1989).
3. S. Shimasaki, M. Shimonaka, M. Ui, S. Inouye, F. Shibata, and N. Ling, Structural characterization of a follicle-stimulating hormone action inhibitor in porcine ovarian follicular fluid. Its identification as the insulin-like growth factor-binding protein, *J.Biol.Chem.* 265:2198 (1990).
4. A. Nakatani, S. Shimasaki, G.F. Erickson, and N. Ling, Tissue specific expression of four insulin-like growth factor binding proteins (1, 2, 3 and 4) in the rat ovary, *Endocrinology* 129:1521 (1991).
5. G.F. Erickson, A. Nakatani, N. Ling, and S. Shimasaki, Cyclic changes in insulin-like growth factor binding protein-4 (IGFBP-4) messenger ribonucleic acid in the rat ovary, *Endocrinology* 130:625 (1992).
6. G.F. Erickson, A. Nakatani, N. Ling, and S. Shimasaki, Localization of insulin-like growth factor-binding protein-5 messenger ribonucleic acid in rat ovaries during the estrous cycle, *Endocrinology* 130:1867 (1992).
7. T.A. Bicsak, M. Shimonaka, M. Malkowski, and N. Ling, Insulin-like growth factor-binding protein (IGF-BP) inhibition of granulosa cell function: Effect on cyclic adenosine 3',5'-monophosphate, deoxyribonucleic acid synthesis, and comparison with the effect of an IGF-I antibody, *Endocrinology* 126:2184 (1990).
8. X.-J. Liu, M. Malkowski, Y.-L. Guo, G.F. Erickson, S. Shimasaki, and N. Ling, Development of specific antibodies to rat insulin-like growth factor-binding proteins (IGFBP-2 to -6): Analysis of IGFBP production by rat granulosa cells, *Endocrinology* 132:1176 (1993).
9. T.A. Bicsak, N. Ling, and L.V. DePaolo, Ovarian intrabursal administration of insulin-like growth factor-binding protein inhibits follicle rupture in gonadotropin-treated immature female rats, *Biol.Reprod.* 44:599 (1991).
10. S. Shimasaki, M. Shimonaka, H.P. Zhang, and N. Ling, Identification of five different insulin-like growth factor binding proteins (IGFBPs) from adult rat serum and molecular cloning of a novel IGFBP-5 in rat and human, *J.Biol.Chem.* 266:10646 (1991).
11. U.K. Laemmli, Cleavage of structural proteins during the assembly of the head of bacteriophage T4, *Nature* 227:680 (1970).
12. J.M. Chirgwin, A.E. Przybyla, R.J. MacDonald, and W.J. Rutter, Isolation of biologically active ribonucleic acid from sources enriched in ribonuclease, *Biochemistry* 18:5294 (1979).
13. E.Y. Adashi, C.E. Resnick, A.J. D'Ercole, M.E. Svoboda, and J.J. Van Wyk, Insulin-like growth factors as intraovarian regulators of granulosa cell growth and function, *Endocr.Rev.* 6:400 (1985).
14. A.J. Hsueh, E.Y. Adashi, P.B. Jones, and T.H. Welsh,Jr., Hormonal regulation of the differentiation of cultured ovarian granulosa cells, *Endocr.Rev.* 5:76 (1984).

15. E.Y. Adashi, C.E. Resnick, A. Hurwitz, E. Ricciarelli, E.R. Hernandez, and R.G. Rosenfeld, Ovarian granulosa cell-derived insulin-like growth factor binding proteins: Modulatory role of follicle-stimulating hormone, *Endocrinology* 128:754 (1991).

16. R. Sprengel, T. Braun, K. Nikolics, D.L. Segaloff, and P.H. Seeburg, The testicular receptor for follicle stimulating hormone: Structure and functional expression of cloned cDNA, *Mol.Endocrinol.* 4:525 (1990).

17. K.C. McFarland, R. Sprengel, H.S. Phillips, M. Köhler, N. Rosemblit, K. Nikolics, D.L. Segaloff, and P.H. Seeburg, Lutropin-choriogonadotropin receptor: An unusual member of the G protein-coupled receptor family, *Science* 245:494 (1989).

18. J.M. Boutin, C. Jolicoeur, H. Okamura, J. Gagnon, M. Edery, M. Shirota, D. Banville, I. Dusanter-Fourt, J. Djiane, and P.A. Kelly, Cloning and expression of the rat prolactin receptor, a member of the growth hormone/prolactin receptor gene family, *Cell* 53:69 (1988).

INSULIN-LIKE GROWTH FACTOR (IGF) BINDING PROTEIN-1 IS AN
ANTIGONADOTROPIN: EVIDENCE THAT OPTIMAL FOLLICLE-STIMULATING
HORMONE ACTION IN OVARIAN GRANULOSA CELLS IS CONTINGENT UPON
AMPLIFICATION BY ENDOGENOUSLY-DERIVED IGFs[*]

E.Y. Adashi[1], C.E. Resnick[1], R.G. Rosenfeld[2], D.R.
Powell[3], R. Koistinen[4], E.M. Rutanen[4,5], and
M. Seppala[4]

[1]Division of Reproductive Endocrinology, Departments
of Obstetrics, Gynecology, and Physiology,
University of Maryland School of Medicine,
Baltimore, Maryland, 21201
[2]Department of Pediatrics, Stanford University
Medical Center, Stanford, California, 94305
[3]Department of Pediatrics, Baylor College of
Medicine, Houston, Texas, 77054
[4]Department of Obstetrics/Gynecology, Helsinki
University Central Hospital, Helsinki, Finland
[5]Minerva Institute for Medical Research, Helsinki,
Finland

Implicit in the antigonadotropic property of IGFBPs is the
presumption that they sequester endogenously-generated IGFs
thereby diminishing their access to cognate cell surface
receptors and hence their cellular hormonal action. This
communication examines this concept and the relevance of
endogenous IGF to FSH action.

Key Words: IGF, Ovary, Binding Proteins,
Gonadotropins

Current Directions in Insulin-Like Growth Factor Research,
Edited by D. LeRoith and M.K. Raizada, Plenum Press, New York, 1994

377

RESULTS

To examine the functional impact of IGFBP-1 on gonadotropin-promoted differentiation, granulosa cells were cultured for 72h in the absence (C) or presence of FSH (100ng/ml), with or without increasing concentrations (10-1,000ng/ml) of IGFBP-1.

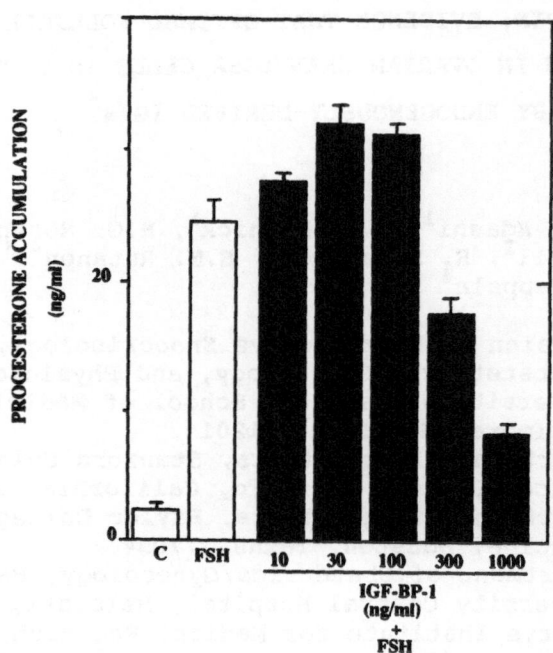

Figure 1. *IGFBP-1 as an Antigonadotropin: Dose-Dependence* – Granulosa cells were cultured under serum-free conditions for 72h in the absence (C) or presence of FSH (100ng/ml), with or wihtout increasing concentrations (10-1,000ng/ml) of IGFBP-1. Medium progesterone content was determined by RIA as described under Methods. Results represent the mean ± SE of three separate experiments, each of which represents three independent determinations.

Although the latter was without significant effect on progesterone accumulation when applied by itself (not shown), its concurrent provision produce dose-dependent (EC_{50}=380±21ng/ml) attenuation of FSH hormonal action and a maximal inhibitory effect (P<0.05) of 65% (Fig. 1). The addition of excess amounts of FSH (up to 1,000ng/ml) failed to reverse the antigonadotropic action of IGFBP-1 (not shown). Importantly, this inhibitory effect of IGFBP-1 proved relatively specific in that treatment with comparable concentrations (up to 1,000ng/ml) of alpha fetoprotein, an unrelated amniotic fluid-derived protein, was without effect (not shown). Denatured (i.e. DTT-pretreated or boiled) IGFBP-1 (1,000ng/ml) proved equally ineffective (not shown). Although lower doses (10-100ng/ml) of IGFBP-1 appeared to modestly enhance FSH hormonal action, the differences observed did not reach statistical significance (P>0.05).

To evaluate the possibility that the antigonadotropic potential of IGFBP-1 is associated with its IGF-binding property, use was made of a unique set of IGFBP-1-directed monoclonal antibodies known(1) to bind to an epitope at or near the IGF binding site (Mab 6302) or to an non-IGF-binding epitope (Mab F5-3F4). To revalidate the unique properties of these two antibodies, use was made of a standardized PEG precipitation assay wherein IGFBP-1 (100ng/tube) was incubated for 1h at room temperature with 2×10^5 cpm of $[^{125}I]$IGF-I in the absence or presence of decreasing dilutions ($1:10^{-4} - 1:10^{-2}$) of either Mab 6302 or Mab F5-3F4. As shown (Fig. 2), decreasing dilutions of Mab 6302 (but not Mab F5-3F4) progressively diminished the ability of IGFBP-1 to specifically sequester $[^{125}I]$IGF-I. Further validation studies made use of a

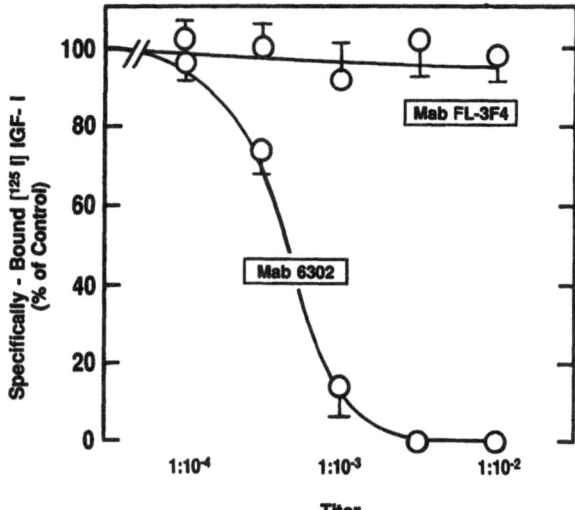

Titer
IGF BP-1-Directed Monoclonal Antibody

Figure 2. *Functional Validation of Mab 6302 and Mab FL-3F4: Inhibition of [^{125}]IGF-I Binding to IGFBP-1* - IGFBP-1 (100ng/tube) was incubated for 1h at room temperature with 2×10^5 cpm of $[^{125}I]$IGF-I in the absence or presence of the indicated decreasing dilutions of either Mab 6302 or Mab FL-3F4. Specifically-bound $[^{125}I]$IGF-I was determined as described under Methods for the determination of IGF binding activity. Results represent the mean ± SE of three separate experiments, each of which represents three independent determinations.

conventional radioligand receptor assay wherein rat granulosa cell membranes were incubated with $[^{125}I]$IGF-I (10^5cpm/tube) in the absence or presence of IGFBP-1 (250ng/ml), with or without decreasing titres ($1:10^{-4}-1:10^{-2}$) of Mab 6302. As shown (Fig. 3), decreasing titres of Mab 6302 (but not Mab FL-3F4; now shown) progressively and completely reversed the ability of IGFBP-1 to compete for binding to $[^{125}I]$IGF-I.

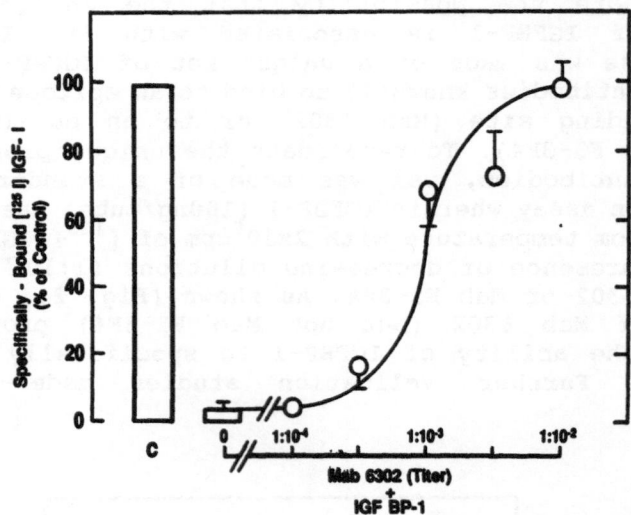

Figure 3. *Functional Validation of Mab 6302: Reversal of the Ability of IGFBP-1 (250ng/ml) to Compete for Binding to $[^{125}]$IGF-I in a Conventional Radioligand Receptor Assay* – Rat granulosa cell membranes were incubated with $[^{125}I]$IGF-I (10^5cpm/tube) in the absence or presence of IGFBP-1 (250ng/ml), with or without the indicated decreasing titers of Mab 6302. C=specifically-bound $[^{125}I]$IGF-I in the absence of IGFBP-1 (arbitrarily designated 100%). Specifically-bound $[^{125}I]$IGF-I was determined as described under Methods for radioligand receptor assays. Results represent the mean ± SE of three separate experiments, each of which represents three independent determinations.

Using the above validated reagents, FSH (100ng/ml)-primed/IGFBP-1 (500ng/ml)-treated granulosa cells were cultured for 72h in the absence or presence of 1:3,000 dilution of the IGFBP-1-directed Mab 6302. As shown (Fig. 4), the concurrent provision of Mab 6302, the ability of which to block the IGF binding site was previously(38) and presently (Figs. 2 and 3) demonstrated, all but eliminated the antigonadotropic action of IGFBP-1 thereby strongly suggesting that its antigonadotropic potential is contingent upon an intact IGF binding site. Although not shown, Mab 6302 (1:3,000 dilution) proved without effect on either the basal or FSH (100ng/ml)-supported accumulation of progesterone. Importantly, IGFBP-1 action remained unaffected in the face of 1:3,000 dilution of Mab FL-3F4 (Fig. 4), an IGFBP-1-directed monoclonal antibody known(1) [and herein confirmed] to bind to a non-IGF-binding epitope.

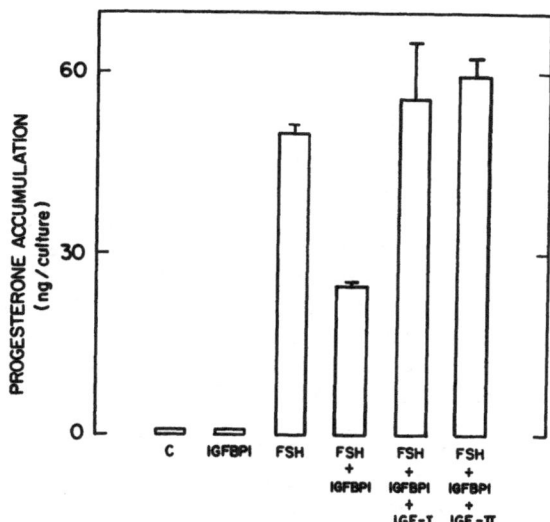

Figure 4. *IGFBP-1 as an Antigonadotropin: Dependence on an Intact IGF Binding Site* - FSH (100ng/ml)-primed/IGFBP-1 (500ng/ml)-treated granulosa cells were cultured for 72h in the absence or presence of 1:3,000 dilution of IGFBP-1-directed monoclonal antibodies 6302 or FL-3F4. Medium progesterone content was determined by RIA as described under Methods. Results represent the mean ± SE of three independent determinations. Mab 6302 = IGFBP-1-directed monoclonal antibody to IGF binding site epitope. Mab FL-3F4 = IGFBP-1-directed monoclonal antibody to a non-IGF binding epitope.

DISCUSSION

Soluble IGFBPs may subserve an inhibitory function by "trapping" extracellular (granulosa cell-elaborated) IGFs thereby preventing their access to their cognate cell surface receptor. In so doing, IGFBPs may in effect be acting as antigonadotropins in that IGFs have been shown to amplify the gonadotropic signal(2). Our present observations appear to support these earlier predictions in that IGFBP-1 proved a potent inhibitor of FSH hormonal action (Fig. 1) thereby providing the first demonstration of its antigonadotropic potential.

The above notwithstanding, relatively lower concentrations of IGFBP-1 (10-100ng/ml) appeared to enhance rather than inhibit FSH hormonal action (Fig. 1). Although the differences observed did not reach statistical significance (P>0.05), the present observations appear in keeping with those reported by others wherein IGFBP-1 has been shown to enhance IGF-I binding and action under specified experimental circumstances(3-12). Although our current observations do not provide conclusive confirmation of the latter observations, one cannot rule out at this time the possibility that appropriate concentrations of IGFBPs and IGF-I may in fact produce meaningful enhancement of FSH hormonal action at the level of the rat granulosa cell.

Studies are currently under way to define experimental conditions wherein such amplification may in fact be reliably documented.

The present report revealed an apparent need in relatively high concentrations of IGFBP-1 (EC_{50}=15nM) required to disclose its inhibitory potential in vitro. Although the exact reason(s) underlying this apparent high-dose requirement remain uncertain, similar observations were reported by other investigators. In this connection, Frauman et al. employed IGFBP-1 concentrations of up to 1000ng/ml (40nM) to inhibit TSH-stimulated DNA synthesis by cultured FRTL5 cells(8). Note is also made that comparable concentrations were required for the demonstration of the antigonadotropic potential of IGFBP-2 (12-14nM) or IGFBP-3 (5-9nM)(13). That notwithstanding, consideration was also given to the possibility that the actual content of exogenously-added IGFBP-1 in the culture medium is substantially lower than projected due to non-specific adsorption to the substratum of the dish. Although an attractive possibility, preliminary studies wherein the medium content of IGFBP-1 was measured by RIA argue against such possibility (Adashi, E.Y., Powell, D.R., and Lee, P.D.K.; unpublished). Lastly, consideration must be given to the possibility that the high dose requirements of human IGFBP-1 may reflect a decreased affinity of this principle for rat IGFs.

To further evaluate the mechanism(s) underlying the antigonadotropic activity of IGFBP-1, use was made of an IGFBP-1-directed monoclonal antibody (Mab 6302) the ability of which to block the IGF binding site was previously(1) and presently (Figs. 2 and 3) demonstrated. Our present findings reveal that the concurrent provision of Mab 6302, all but eliminated IGFBP-1 action, thereby strongly suggesting that its antigonadotropic potential is contingent upon an intact IGF binding site. In contrast, an IGFBP-1-directed monoclonal antibody known(1) [and herein confirmed] to bind to non-IGF binding epitopes was without effect. These findings strongly suggest that the ability of IGFBP-1 to inhibit FSH hormonal action is most likely due to sequestration of endogenously-generated ligand(s). In so doing, IGFBP-1 effectively reduces the bioavailable fraction of receptor-active IGFs. As such, these findings once again implicate endogenously-derived IGFs in FSH hormonal action for which they may serve as amplifiers(2).

The present findings provide the first demonstration of the antigonadotropic potential of IGFBP-1. As such, IGFBP-1 can be added to a growing list of IGFBPs the ability of which to compromise gonadotropin hormonal action has been demonstrated. Indeed, both IGFBP-2 and IGFBP-3 have been shown to constitute potent antigonadotropic principles(13-15) the mechanism of action of which may be due, if only in part, to mechanism(s) other than the sequestration of endogenously-derived IGFs(13). Regardless of the mechanism of action involved, the present and previous(13-15) observations suggest the existence of several IGFBPs capable of attenuating gonadotropin hormonal action. Although the reasons(s) underlying the multiplicity of intraovarian (antigonadotropic) IGFBPs remains uncertain, both the present and previous findings are in keeping with the notion that optimal FSH hormonal action is contingent upon the bioavailability of granulosa cell-derived IGFs and the consequent amplification of the gonadotropic signal(2). Stated differently, "intrinsic" FSH hormonal action, as assessed in a hypothetical IGF vacuum (i.e. IGFBP-replete circumstances) may well be relatively limited (Fig. 5). In contrast, "augmented"

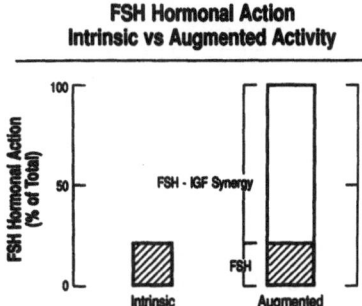

Figure 5. *Schematic Representation of FSH Hormonal Action as a Function of Synergy with Endogenously-Derived IGFs* - An arbitrary scale depicts FSH hormonal action in terms of percent of total.

FSH hormonal action, apparent only under IGF-replete (presumably IGFBP-deplete) conditions, is comprised of a relatively modest "intrinsic" component complemented by a substantial synergistic component representing FSH/IGF interaction. According to this view, FSH is but one determinant of a complex equation dictating its impact at the target tissue level. Preliminary studies (not shown) suggest that similar conclusions apply to LH hormonal action at the level of the ovarian theca-interstitial cell. If so, our current observations suggest a more generally applicable principle wherein putative intraovarian regulators may be viewed as key in situ modulators of gonadotropin hormonal action.

ACKNOWLEDGEMENTS

Supported in part by NIH Research Grant HD-19998 and RCDA HD-0067 (EYA), NIH Research Grants DK28229, DK36054, CA-42106, and RCDA DK-01275 (RGR), DK39773 (DRP), and grants from the Sigrid Juselius Foundation (EMR), the Academy of Finland (RK), and the Finnish Social Insurance Institution (MS).

REFERENCES

1. E.-M. Rutanen, T. Karkkainen, C. Lundqvist, F. Pekonen, O. Ritvos, P. Tanner, M. Welin, and T. Weber, Monoclonal antibodies to the 27-34K insulin-like growth factor binding protein, Biochem. Biophys. Res. Commun. 152:208-215 (1988).

2. E.Y. Adashi, C.E. Resnick, M.E. Svoboda, and J.J. Van Wyk, Somatomedin-C as an amplifier of follicle-stimulating hormone action: Enhanced accumulation of adenosine 3',3'-cyclic monophosphate, Endocrinology 118:149-156, (1986).

3. D.R. Clemmons, R.G. Elgin, V.K.M. Han, S.J. Casella, A.J. D'Ercole, and J.J. Van Wyk, Cultured fibroblast monolayers secrete a protein that alters the cellular binding of somatomedin-C/insulin-like growth factor I, J. Clin. Invest. 77:1548-1556 (1986).

4. M.A. De Vroede, L.Y.-H. Tseng, P.G. Katsoyannis, S.P. Nissley, and M.M. Rechler, Modulation of insulin-like growth factor I binding to human fibroblast monolayer cultures by insulin-like growth factor carrier proteins released to the incubation media, J. Clin Invest. 77:602-613 (1986).

5. R.G. Elgin, W.J. Busby Jr., and D.R. Clemmons, An insulin-like growth factor (IGF) binding protein enhances the biologic response to IGF-I, Proc. Natl. Acad. Sci. U.S.A. 84:3254-3258 (1987).

6. D.R. Clemmons, V.K.M. Han, R.G. Elgin, and A.J. D'Ercole, Alterations in the synthesis of a fibroblast surface associated 35K protein modulates the binding of somatomedin-C/insulin-like growth factor I, Mol. Endocrinol. 1:339-347 (1987).

7. W.H. Busby Jr., D.G. Klapper, and D.R. Clemmons, Purification of a 31,000-Dalton insulin-like growth factor binding protein from human amniotic fluid, J. Biol. Chem. 263:14203-14210 (1988).

8. A.G. Frauman, S. Tsuzaki, and A.C. Moses, The binding characteristics and biologic effects in FRTL5 cells of placental protein-12, an insulin-like growth factor-binding protein purified from human amniotic fluid, Endocrinology 124:2289-2296 (1989).

9. W.H. Busby, P. Hossenlopp, M. Binous, and D.R. Clemmons, Purified preparations of the amniotic fluid-derived insulin-like growth factor-binding protein contain multimeric forms that are biologically active, Endocrinology 125:773-777 (1989).

10. R.S. Bar, B.A. Booth, M. Boes, and B.L. Dake, Insulin-like growth factor-binding proteins from vascular endothelial cells: Purification, characterization, and intrinsic biological activities, Endocrinology 125:1910-1920 (1989).

11. D.R. Clemmons and L.I. Gardner, A factor contained in plasma is required for IGF binding protein-1 to potentiated the effect of IGF-I on smooth muscle cell DNA synthesis, J. Cell. Physiol. 145:129-135 (1990).

12. R.H. McCusker, C. Camacho-Hubner, M.L. Bayne, M.A. Cascieri, and D.R. Clemmons, Insulin-like growth factor (IGF) binding to human fibroblast and glioblastoma cells: The modulatory effect of cell released IGF binding proteins (IGFBPs), J. Cell Physiol. 144:244-253 (1990).

13. T.A. Bicsak, M. Shimonaka, M. Malkowski, and N. Ling, Insulin-like growth factor binding protein (IGF-BP) inhibition of granulosa cell function: Effect on cyclic adenosine 3', 3'-monophosphate, deoxyribonucleic acid synthesis, and comparison with the effect of an IGF-I antibody, Endocrinology 126:2184-2189 (1990).

14. M. Iu, M. Shimonaka, S. Shimasaki, and N. Ling, An insulin-like growth factor-binding protein in ovarian follicular fluid blocks follicle-stimulating hormone-stimulated steroid production by ovarian granulosa cells, Endocrinology 125:912-916 (1989).

15. S. Shimasaki, M. Shimonaka, M. Ui, S. Inouye, F. Shibata, and N. Ling, Structural characterization of a follicle-stimulating hormone action inhibitor in porcine ovarian follicular fluid, J. Biol. Chem. 265:2198-2202 (1990).

*Address all correspondence and reprint requests to:
  Dr. Eli Y. Adashi, Division of Reproductive Endocrinology,
  Department of Obstetrics/Gynecology and Physiology,
  University of Maryland, 405 W. Redwood Street, Third Floor,
  Baltimore, Maryland 21201.
  Phone: #(410)328-2304   FAX: #(410)328-8389

INSULIN-LIKE GROWTH FACTOR-I AND INSULIN-LIKE GROWTH FACTOR BINDING PROTEINS IN THE ZUCKER FATTY RAT: A CASE FOR DIFFERENTIAL TISSUE REGULATION

Marie C. Gelato and Michael Berelowitz

Division of Endocrinology and Metabolism, Department
of Medicine, SUNY, Stony Brook, NY 11794

INTRODUCTION

Obesity is one of the most common health care problems in the Western world.[1] It is associated with hypertension, hyperlipidemia and a predisposition to Type II Diabetes Mellitus, which probably accounts, in part, for the high risk these patients have of accelerated atherosclerosis. Given the numbers of people affected and the predisposition these patients have to multiple medical problems, an understanding of the pathophysiology involved in the causes for, and perpetuation of, obesity would aid our ability to possibly prevent and more effectively treat this disease.

Diminished growth hormone (GH) secretion is a well known association of obesity.[2] It is not the cause of the obese state but may in part be responsible for its perpetuation. It is known from both human and animal studies, that the decreased GH in obesity may be related to neuroendocrine defects in the pituitary and hypothalamus.[3,4] Insulin-like growth factor-I (IGF-I) which is an important component of the GH regulatory system has been reported to be abnormal in both human and animal models of obesity.[2,5] It has been proposed that increased levels of IGF-I in peripheral circulation may be in part responsible for the diminished GH secretion in obesity.[3] The IGF-I system has not been extensively studied in obesity. This article will review current literature regarding diminished GH secretion in obesity and the available data on IGF-I and then present studies undertaken in our laboratories to better define the IGF-I system in one experimental model of obesity, the Zucker fatty rat.

BACKGROUND

The Zucker fatty rat is an experimental model of human obesity in which obesity is inherited as a single autosomal recessive gene, termed "fa".[6] Obesity in the

Current Directions in Insulin-Like Growth Factor Research,
Edited by D. LeRoith and M.K. Raizada, Plenum Press, New York, 1994

387

Zucker fatty rat is visibly distinguishable by body shape at about 20 days of age, however, an obesity-associated defect in thermogenesis and an increase in body fat content can be demonstrated as early as the first week of life.[7] These animals have also been noted to develop GH deficiency at a later age.[8] Plasma levels of GH[9], pulsatile GH release[4] and pituitary GH concentration[4] are reduced in adult obese male rats when compared to lean litter mate controls. More recently, Leidy et al[10] have shown that pulsatile release of GH, both peak amplitude and mean GH concentration are reduced in female obese Zucker rats to the same degree as that reported in male rats. These investigators found that the decrease in GH secretion is observed as early as 6 weeks of age in both male and female animals. These data correlate well with the mRNA levels for GH measured in the pituitary which are decreased as early as 5 weeks of age in both Zucker fatty male and female rats.[8] In one study, the changes in GH mRNA were associated with the time when decreased protein deposition is initially seen in the obese animals.[8]

In humans, similar data are available. Both in obese children and adults, serum GH levels are diminished.[2,11,12] Twenty four hour profiles show both decreased amplitude and mean GH concentrations.[13] In addition, GH responses to a variety of stimuli are blunted or abolished in obesity.[2,3]

The mechanism subserving decreased GH secretion in obesity is not fully understood. Growth hormone is normally under opposing regulation of two hypothalamic peptides, GH-releasing hormone (GHRH), which causes stimulation of both GH release and synthesis, and somatostatin (SRIF), which inhibits GH release from the anterior pituitary gland[14], as well as negative feedback regulation by IGF-I.[15] Thus, either diminished GHRH secretion or increased production of SRIF or IGF-I could result in decreased GH. In obese animal models, both SRIF and GHRH have been demonstrated to be abnormal. Leidy and colleagues reported that hypothalamic content of SRIF is increased in both male and female obese rats in comparison to their lean litter mate controls.[10] GHRH production is also low.[4] However, there is also decreased pituitary GH responsiveness to GHRH which occurs in both obese humans[16] and rats.[17] In rats, this disturbance persists in vitro suggesting the pituitary cells of obese animals produce an agent inhibitory to GHRH-induced GH release.[17] These data suggest a combined pituitary-hypothalamic defect occurring as a result of obesity.

The IGF-I system has not been extensively studied in this model. An early paper by Gahagan et al[18] reported that somatomedin activity as measured by bioassay was significantly lower in obese Zucker rats compared to lean litter mate controls. Later work by Renier et al.[17] demonstrated by specific RIA that IGF-I levels were significantly lower in Zucker fatty male rats at 6, 8 and 12 weeks age as compared to lean animals. However, these data have not been confirmed in more recent studies. Leidy et al.[10] showed in both male and female Zucker fatty rats that there were no differences in IGF-I levels between obese rats and their lean litter mate controls. More recently, two groups of investigators have reported that IGF-I levels are increased in the sera of obese Zucker rats, particularly in female rats.[5,19] In addition, these investigators demonstrated that obese Zucker rats are more sensitive to the effects of GH in stimulating IGF-I production, i.e., for each dose of GH administered to obese rats there was a greater rise in circulating IGF-I levels than seen at the same dose in lean controls.[5] To correlate, obese humans, both adults and children, have been reported to have either normal or elevated serum levels of IGF-I[2], and the blunted GH responses observed in these subjects to GHRH and other secretogogues has been proposed to be secondary to abnormal IGF-I levels.[19] Since IGF-I appears to be an important factor in the abnormal GH dynamics observed in obesity, we undertook preliminary studies to characterize the IGF-I system in the Zucker fatty rat. Our initial studies focused on female and male obese Zucker rats and their lean litter mate controls at 10-12 weeks of age. This age was selected because the GH deficit is maximal at this stage of development.

MATERIALS AND METHODS

The initial studies were done to determine peripheral levels and tissue concentrations of IGF-I and its binding proteins as well as in some tissues the levels of prepro IGF-I mRNA.

Serum and Tissue IGF-I Concentrations

Zucker fatty rats were obtained from Charles River (Wilmington, MA) and were housed for 3-4 days before being studied. They were allowed free access to food and water. On the day the animals were sacrificed, blood was obtained from the trunk and all tissues were removed immediately and placed on dry ice. All samples were stored at -70 C until assayed.

Serum samples were acid-ethanol extracted.[20] Tissue samples were extracted for measurement of IGF-I peptide content using 0.5 N HCL (1g/5 ml) as previously described.[21] IGF-I was separated from its binding proteins by passing aliquots of tissue extracts supernatant through C18 Sep-pak cartridges.[21] IGF-I was measured by RIA.[22]

Prepro IGF-I mRNA

Additional tissue samples were extracted for total RNA using methods developed previously.[21] RNA concentrations were estimated based on absorbance at 260 nm, and identical concentrations of RNA from each sample were used for nuclease protection assay. Aliquots (3-4 mcg) of total RNA were electrophoresed in 1% agarose gel, stained using ethidium bromide, then examined visually to confirm quality and integrity and provide an estimate of quantity of RNA extracts.

A RNA probe specific to the 5' region of rat IGF-I gene provided by Drs. C. Roberts and D. LeRoith (NIH, NIDDK, Bethesda, MD) was used to detect and quantitate IGF-I transcripts.[21] The riboprobe was labeled in vitro with [32P] UTP (Amersham, Arlington Heights, IL) by methods previously published.[23] Solution hybridization/nuclease protection assays were performed as developed by our laboratories.[24]

IGF-Binding Protein Measurements

In order to identify the IGF-BP species present in liver and pituitary, two types of studies were done:

a) ligand blotting to determine the sizes of the IGF-BPs[25] and
b) affinity crosslinking and immunoprecipitation to identify the individual BPs in these tissues.[26] Two polyclonal antibodies to human (h) IGF-BP-1 and bovine (b) IGF-BP-2 were used, kindly supplied by Dr. David Clemmons (Chapel Hill, NC).

RESULTS

IGF-I concentration in serum and tissues of male and female obese Zucker rats and lean littermate controls are presented in Table 1. In male obese Zucker rats, IGF-I content is increased in pituitary ($p < 0.05$), hypothalamus (not significant) and cerebral cortex ($p < 0.05$) compared to lean controls. In contrast, serum IGF-I concentration and IGF-I content in all peripheral tissues was decreased. In female obese rats, the pattern of IGF-I content was similar to male obese rats in neural tissues. Thus, IGF-I content was increased in pituitary, cerebral cortex and hypothalamus (significant only in pituitary and cortex) (Table 1). However, IGF-I concentrations in peripheral tissues of female obese rats varied. IGF-I content was increased in muscle, heart, ovaries but decreased in liver and kidney tissue and serum IGF-I levels were increased. Overall, the levels of IGF-I were higher in female rats compared to male animals.

Prepro IGF-I mRNA expression in liver and kidney of male and female obese rats was decreased compared to lean controls in keeping with the peptide concentrations (data not shown).

IGF-binding proteins (IGF-BPs) were determined by ligand blotting in liver and pituitary of male and female obese and lean rats. In liver, IGF-BPs are present as multiple bands ranging in size from 45-25 K in both male (Figure 1) and female (Figure 2) obese animals. The smaller bands, 30 and 25 K, appear less dense in the obese animals (males and females). In pituitary of lean and obese male and female rats, there is a predominant band at approximately 30 K (only male data shown, Figure 3). Lesser bands are seen at 47-43 K and 25 K. In the obese animals, the 30 and 25 K bands are diminished.

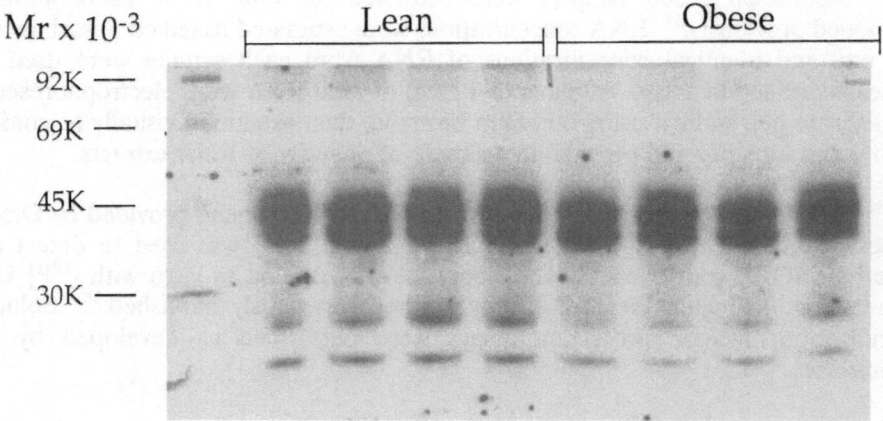

Fig. 1   Ligand blot of liver tissue from lean and obese male rats. Tissue was homogenized in a TRIS-TRITON buffer. The equivalent of 200 μg wet weight of liver tissue was subjected to SDS-PAGE. The samples were then transferred to nitrocellulose paper and developed with radiolabeled IGF-I. Specificity was tested by running companion blots with excess unlabeled (1μg) IGF-I. Radiolabeled molecular weight (Mr) standards are lysozyme (14K), Carbonic anhydrase (30K), ovalbumin (45K), BSA (69K), phosphorylase-B (92.5K) and myosin (200K).

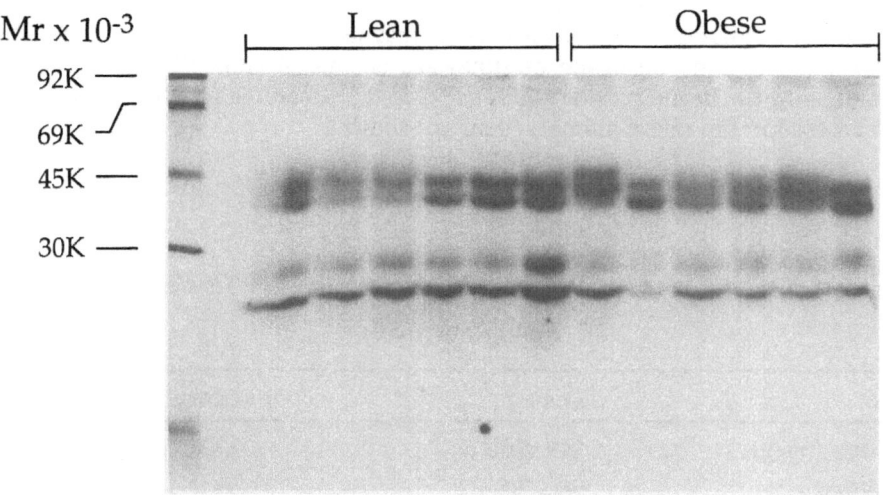

Fig. 2   Ligand blot of liver tissue from lean and obese female rats. Tissue was prepared as noted in legend to Figure 1. The radiolabeled molecular weight standards are the same noted in the legend to Figure 1.

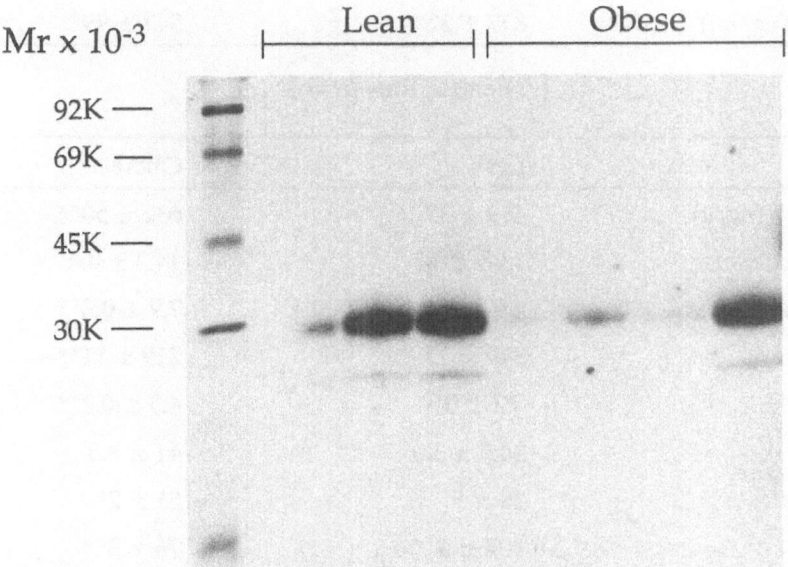

Fig. 3   Ligand blot of pituitary tissue from lean and obese male rats. Tissue was prepared as noted in legend to Figure 1. The molecular weight markers are the same as those noted in legend to Figure 1.

391

To further identify the specific IGF-BPs seen on ligand blotting, affinity cross-linking and immunoprecipitation studies were done using specific antisera to (h) IGF-BP-1 and (b) IGF-BP-2. In the liver of male and female obese rats, both IGF-BP-1 and -BP-2 (30 K band on ligand blotting) are decreased (IGF-BP-1 > IGF-BP-2) (data not shown). In the pituitary, IGF-BP-2 is the predominant BP measured and it is decreased in the obese animals (data not shown).

Table 1. Insulin-Like Growth Factor-I Concentrations in Tissues and Serum

Male Rats (n=9)

	Lean	Obese
Pituitary (ng/g)	149 ± 10	230 ± 40*
Hypothalamus	20 ± 3	27 ± 3
Cerebral Cortex	1.6 ± 0.2	2.5 ± 0.1*
Kidney	129 ± 18	188 ± 25
Muscle	4.1 ± 0.8	2.0 ± 0.2*
Heart	57 ± 5	37 ± 4*
Testes	54 ± 5	25 ± 2*
Liver	65 ± 5	43 ± 3*
Serum (ng/ml)	877 ± 22	547 ± 99*

Female Rats (n=6)

	Lean	Obese
Pituitary (ng/g)	479 ± 35	632 ± 50**
Hypothalamus	13.8 ± 0.2	14.2 ± 0.6
Cerebral Cortex	4.9 ± 0.4	7.9 ± 0.2**
Kidney	319 ± 17	219 ± 11**
Muscle	3.2 ± 0.3	4.3 ± 0.2**
Heart	34.5 ± 3.0	41 ± 3.0
Ovaries	29 ± 5	51 ± 2*
Liver	104 ± 6	76 ± 3**
Serum (ng/ml)	994 ± 41	1524 ± 84**

Data expressed as Mean ± S.E.

* $P < 0.05$
**$P < 0.01$

DISCUSSION

These data confirm the changes in peripheral levels of IGF-I reported by other investigators, i.e., decreased serum levels in male obese rats [18,17] and increased levels in female rats.[19] Our data further extend these observations by demonstrating abnormalities in IGF-I tissue content in obese animals. In both male and female obese rats, IGF-I content is significantly increased in pituitary and cerebral cortex. In male obese rats, the peripheral tissues had diminished IGF-I content, whereas, in female obese rats, the peripheral tissues had variable IGF-I content (decreased in liver and kidney, and increased in ovary and muscle). Some of these changes were paralleled by decreases in prepro IGF-I mRNA (liver and kidney). In addition to the changes in peptide content, there are also changes in some of the IGF-BPs, decreased levels of BP-1 and BP-2.

These data suggest differential regulation of IGF-I between neural and peripheral tissues in the Zucker fatty rat. The increased levels of IGF-I in the pituitary of the obese rats could in part explain the diminished GH secretion observed in this model. It would also potentially explain the persistent in vitro defect observed in this model, i.e., the diminished responsiveness to GHRH of pituitary cells in culture from obese animals.[17]

It is not clear why there is this dysregulation of IGF-I in this model. The obese rat seems to be "hyper" responsive to GH in that the same dose of GH causes a more pronounced increase in IGF-I in the obese animal as compared to its lean litter mate control.[5] So that even though GH levels are low, there is greater stimulation of IGF-I. Other factors may be involved in the increased production of IGF-I in this model as well. It has been shown that insulin can stimulate the production of IGF-I in some systems.[27] The Zucker fatty rat is hyperinsulinemic as early as 3-5 weeks of age, just shortly before the defect in GH secretion is noted.[28] Thus, it is possible that increased levels of insulin may stimulate production of IGF-I in certain tissues. It is also known that estrogen can influence IGF-I levels. Both increases and decreases in IGF-I production have been reported with estrogen use.[29,30] A recent report suggests that estrogen administration to castrated male and female rats increases levels of prepro IGF-I mRNA in the anterior pituitaries of these animals.[31] Thus, in the obese rat, due to the increase in body fat, there may be more peripheral conversion of androgens to estrogens.[2,3] However, there appears to be a difference in the responsiveness of the male and female obese rats, since overall the female rats show more consistent increases in IGF-I levels than do the male rats. This may be due to the amount of substrate available for peripheral conversion of androgens to estrogens in the female obese animals, i.e., from human studies obese women tend to have high levels of androstenedione which can be peripherally converted to estrogens.[2] In men, testosterone levels fall proportionate to the level of adiposity.[2,3] However, the free levels of testosterone tend to remain normal until the individual reaches greater then 200% of ideal body weight. It is known that testosterone levels are low in the male Zucker fatty rat.[32] So perhaps in the male rat there is not as much substrate for peripheral conversion of androgens to estrogens and the estrogen levels are not as high in the males as they may be in the female Zucker fatty rats. It is also possible that the female rats are more sensitive to the effects of estrogens. These hypotheses remain to be confirmed.

In addition to the changes in IGF-I peptide concentrations which can alter GH dynamics, the changes in IGF-BP-1 and BP-2 seen in pituitary and liver of obese rats may actually enhance the actions of IGF-I in these tissues as well. It has been shown that the IGF-BPs can modulate the action of IGF-I in some systems, i.e., fibroblasts and decidual cells.[33] Several mechanisms may explain how these BPs interact with the peptide to alter its action.[33] They can be present in such high concentrations that they bind all available IGF-I and keep it from interacting with the receptor. A change in their physical state, such as phosphorylation, can alter the affinity of the BP for IGF-I. Phosphorylated BPs have a higher affinity for the peptide and, thus, can limit the interaction of IGF-I with its receptor. So that in the case of the Zucker fatty rat, less BP in the tissues may enhance the amount of peptide that is free and able to attach to the receptor, and consequently enhance its action at these sites.

We have shown in rats made diabetic by streptozotocin that IGF-BP-1 and BP-2 are regulated by insulin.[34] Thus, levels of BP-1 and BP-2 in sera and tissues are markedly increased in diabetic rats and giving insulin restores the levels of IGF-BP-1 and -2 to normal. Thus, in the Zucker fatty rat, the changes in the BPs may reflect the changes in the circulating levels of insulin.

In summary, IGF-I and some of the IGF-BPs are abnormal in the Zucker fatty rat. These changes may be a reflection of the abnormal metabolic and hormonal milieu reported in these animals. There is clearly a differential tissue regulation which is apparent in this model. The increases of IGF-I in the pituitary of these obese animals may contribute to the decreased GH secretion noted in this model. The diminished GH is probably not a causative agent in obesity but rather a result and probably enhances its perpetuation.

ACKNOWLEDGMENTS

This work was presented in part at the 74th Annual Meeting of the Endocrine Society, San Antonio, Texas, 1992. The work was supported by grants to Marie C. Gelato from the Diabetes Research and Education Foundation and the New York Obesity Research Center.

REFERENCES

1. J. Hirsch, L.B. Salans. Obesity, in: "Principles and Practice of Endocrinology and Metabolism", K.L. Becker, J.P. Bilezikian, W.J. Bremmer, et.al., eds., J.B. Lippincott Company, Philadelphia (1990).
2. A.R. Glass. Endocrine aspects of obesity, Medical Clinics of North America 73:139 (1989).
3. P.G. Kopelman, Neuroendocrine function in obesity, Clin Endocrinol 28:675 (1988).
4. G.S. Tannenbaum, M. Lapointe, W. Gurd, J.A. Finkelstein, Mechanisms of impaired growth hormone secretion in genetically obese Zucker rats: roles of growth hormone-releasing factor and somatostatin, Endocrinology 127:3087 (1990).
5. L. Nguyen-Yamamoto, C.L. Deal, J.A. Finkelstein, G. Van Vliet, Increased response of IGF-I to GH in obese Zucker rats: a post-receptor phenomenon, The 74th Annual Meeting of the Endocrine Society, San Antonio, Texas, Abstract No. 686 p.223 (1992).
6. L.M. Zucker, T.F. Zucker, Fatty, a new mutation in the rat, J. Hered 52:275 (1961).
7. E. Planche, M. Joliff, P. de Gasquet, X. Leliepvre, Evidence of a defect in energy expenditure in 7 day old Zucker rat (fa/fa), Am J Physiol 245:E107 (1983).

8. I. Ahmad, A.W. Steggles, A.J. Carrillo, J.A. Finkelstein, Developmental changes in levels of growth hormone mRNA in Zucker rats, J Cellular Biochem 43:59 (1990).

9. J.A. Finkelstein, P. Jervais, M. Menadue, J.O. Willoughby, Growth hormone and prolactin secretion in genetically obese Zucker rats, Endocrinology 118:1233 (1986).

10. J.W. Leidy, Jr., T.M. Romano, W. J. Millard, Developmental changes of the growth hormone axis in lean and obese Zucker male and female rats, The 74th Annual Meeting of the Endocrine Society, San Antonio, Texas, Abstract No. 463, 167 (1992).

11. J.L. Chaussain, E. Binet, A. Schlumberger, Serum somatomedin activity in obese children, Pediatr Res 9:667 (1975).

12. G. Giordano, F. Minuto, E. Foppiano, The behavior of somatomedin in obese subjects, Recent Adv Obesity Res 1:143 (1975).

13. J.D. Veldhuis, A. Iranmanesh, K.K. Ho, et al., Dual defects in pulsatile growth hormone secretion and clearance subserve the hyposomatropism of obesity in man, J Clin Endocrinol Metab 72:51 (1991).

14. G.S. Tannenbaum, N. Ling, The interrelationship of growth hormone (GH) - releasing factor and somatostatin in generation of the ultradian rhythm of GH secretion. Endocrinology 115:1952 (1984).

15. M. Berelowitz, M. Szabo, L.H. Frohman, S. Firestone, L. Chu, Somatomedin-C mediates growth hormone negative feedback by effects on both the hypothalamus and pituitary, Science 212:1279 (1981).

16. T. Williams, M. Berelowitz, S.W. Joffe, et al., Impaired growth-hormone responses to growth-hormone-releasing factor in obesity: a pituitary defect reversed with weight reduction, N Engl J Med 311:1403 (1984).

17. G. Renier, P. Gaudreau, N. Deslauriers, P. Brazeau, In vitro and in vivo growth hormone responsiveness to growth hormone-releasing factor in male and female Zucker rats, Neuroendocrinology 50:454 (1989).

18. J.H. Gahagan, R.J. Martin, R.M. Leach, Serum somatomedin activity in two animal models as measured using chick epiphyseal plate cartilage bioassay, Proc Soc Exp Biol Med 163:455 (1980).

19. B.B. Bercu, S-W. Yang, R. Masuda, C-S. Hu, R.F. Walker, Effects of coadministered growth hormone (GH) - releasing hormone and GH-releasing hexapeptide on maladaptive aspects of obesity in Zucker rats, Endocrinology 131:2800 (1992).

20. W.H. Daughaday, K.A. Parker, S. Borowsky, B. Trivedi, M. Kapadia, Measurement of somatomedin-related peptides in fetal, neonatal, and maternal rat serum by insulin-like growth factor (IGF) I radioimmunoassay, IGF-II radioreceptor assay (RRA) and multiplication-stimulating activity RRA after acid-ethanol extraction, Endocrinology 110:575 (1982).

21. D. Olchovsky, J.F. Bruno, T.L. Wood, M.C. Gelato, J. M. Gilbert, M. Berelowitz, Altered pituitary growth hormone regulation in streptozotocin-diabetic rats: a combined defect of hypothalamic somatostatin and growth hormone releasing factor. Endocrinology 126:53 (1991).

22. E. Spatola, O.H. Pescovitz, K. Marsh, N.B. Johnson, S.A. Berry, M.C. Gelato, Interaction of growth hormone-releasing hormone with the insulin-like growth factors during prenatal development in the rat, Endocrinology 129:1193 (1991).

23. W. L Lowe, Jr., C.T. Roberts, Jr., S.R. Lasky, D. LeRoith, Differential expression of alternative 5'-untranslated regions in mRNAs encoding rat insulin-like growth factor-I, Proc Natl Acad Sci USA 84:8946 (1987).

24. J. D. White, K.D. Stewart, J.F. McKelvy, Measurement of neuroendocrine peptide mRNA in discrete brain regions, in: "Methods in Enzymology", P. M. Conn, ed., Academic Press, Orlando, Vol 124:548 (1986).

25. P. Hossenlopp, D. Seurin, B. Segonia-Quinson, S. Hardouin, M. Binoux, Analysis of serum insulin-like growth factor binding proteins using Western blotting: use of the method for titration of the binding proteins and competitive binding sites, Anal Biochem 154:138 (1986).

26. T.G. Unterman, D.T. Oehler, R.E. Becker, Identification of a Type I insulin-like growth factor binding protein (IGF-BP) in serum from rats with diabetes mellitus, Biochem Biophys Res Commun 163:882 (1989).

27. M. Labib, D. Teale, V. Marks, Insulin-like growth factor I in patients with hypoglycemia, Ann Clin Biochem 27:107 (1990).

28. J.M. Fletcher, P. Haggarty, K.W.J. Wahle, P.J. Reeds, Hormonal studies of young lean and obese Zucker rats, Horm Metabol Res 18:290 (1986).

29. G. Norstedt, A. Levinovitz, H. Eriksson, Regulation of uterine insulin-like growth factor I mRNA and insulin-like growth factor II mRNA by estrogen in the rat, Acta Endocrinol 120:466 (1989).

30. L.J. Murphy, H.G. Friesen, Differential effects of estrogen and growth hormone on uterine and hepatic insulin-like growth factor I gene expression in the ovariectomized hypophysectomized rat, Endocrinology 122:325 (1988).

31. K.M. Michels, W-H. Lee, A. Seltzer, J.M. Saavedra, C.A. Bondy, Up-regulation of pituitary [125I] insulin-like growth factor-I (IGF-I) binding and IGF binding protein -2 and IGF-I gene expression by estrogen, Endocrinology 132:23 (1993).

32. R.A. Young, R. Frink, C. Longcope, Serum testosterone and gonadotropins in the genetically obese male Zucker rat, Endocrinology 111:977 (1982).

33. D.R. Clemmons, C. Camacho-Hubner, J.I. Jones, R.H. McCusker, W.H. Busby Jr., Insulin-like growth factor binding proteins: mechanism of action at the cellular level, in: Modern Concepts of Insulin-Like Growth Factors, E. Martin Spencer, ed., Elsevier, New York (1991).

34. M.C. Gelato, D. Alexander, K. Marsh, Differential tissue regulation of insulin-like growth factor binding proteins in experimental diabetes mellitus in the rat, Diabetes 41:1511 (1992).

CHARACTERIZATION OF THE IGF REGULATORY SYSTEM IN BONE

Subburaman Mohan and David J. Baylink

Departments of Medicine, Biochemistry and Physiology
Loma Linda University and
Pettis VA Medical Center
Loma LInda, CA 92357

INTRODUCTION

There are two major dynamic processes in bone: bone formation and bone resorption which determine the volume of bone and ultimately its strength. Because bone is a tissue which is continuously remodeling, bone formation is an essential part of skeletal maintenance even after the growth period where bone formation is obviously essential. Moreover, bone has a remarkable regenerative potential which enables it to remodel itself in response to changing physical demands, and to repair itself after injury. In the majority of the adult skeleton, bone formation is confined to specific remodeling sites. Whether bone is lost or gained will depend on the extent of bone formation in each one of these sites. With regard to the mediators of bone formation, growth factors have received much attention. Of the several growth factors present in bone (IGF-I, IGF-II, TGF-β1, TGF-β2, PDGF, basic FGF, acidic FGF and BMPs), there is now sufficient evidence to document an important role for the IGFs in the local regulation of bone formation. In this paper we describe the components of the IGF regulatory system and discuss evidence supporting a role for IGFs in the local regulation of bone formation.

IGF SYSTEM AND BONE

Studies in our and other laboratories provide strong support to the concept that IGFs function in an autocrine/paracrine manner to regulate the proliferative and differentive functions of bone forming osteoblasts. The findings that support this concept include: 1) IGFs are the most abundant growth factors produced by bone cells and stored in bone[1,2], 2) The production of IGFs by bone cells is regulated by both local and systemic effecters of bone metabolism[1,2], 3) IGFs stimulate both bone cell proliferation and collagen synthesis, two essential components of bone formation in vitro[1,2], 4) IGFs increase bone formation in vivo in both rats and humans[3,4], 5) 40-50% of total basal bone cell proliferation was blocked by inhibiting the actions of endogenous IGFs in serum free culture[5], 6) The amount of IGFs stored in bone is not invariant[6,7]. Accordingly, the levels of both IGF-I and IGF-II are increased in bones from osteoarthritic subjects who have elevated bone density[6]. Also the concentration of IGF-I in bone decreases with age in both males and females[7] thus raising the possibility that the decreased growth factor

Current Directions in Insulin-Like Growth Factor Research,
Edited by D. LeRoith and M.K. Raizada, Plenum Press, New York, 1994

397

level may in part may be responsible for the impaired bone coupling seen in osteoporosis.

COMPONENTS OF THE IGF SYSTEM

The IGF system in bone is made up of several components including, IGF-I, IGF-II, IGF variants, type I IGF receptor, type II IGF receptor, IGF binding proteins (IGFBPs 1 through 6) and IGFBP proteases. There are a number of potential explanations why two IGF receptors, six IGFBPs, and unknown number of IGFBP proteases could be involved in regulating the actions of 2 IGFs. First, we now have accumulating evidence to strongly suggest that IGFs are important autocrine and paracrine regulator of cell proliferation and differentiation in a number of tissues including bone[1,2]. We therefore predict that the functions of IGFs in a given tissue viz. thyroid gland might be some what different from those in bone and thus the actions of IGFs should be regulated in a tissue specific manner. Second, many hormones and growth factors appear to recruit the IGF system to modulate their effects on cell proliferation and differentiation in the target tissues. For example, we have evidence that a number of systemic hormones (PTH, 1,25-dihydroxy vitamin D_3, progesterone) and local growth factors (BMP-7) may modulate their effects on bone cell proliferation in part by regulating the actions of IGFs[8]. Similarly, there is some evidence that TSH may regulate thyroid cell functions by modulating the actions of IGFs[9]. Third, there is evidence that the actions of IGFs are regulated differently depending on the developmental stage of the tissue[10]. Based on the complexity of IGF actions described above, it can be predicted that several components are required to serve multiple inputs. Thus, in order for us to understand how the IGF system functions as a whole in bone, we need to understand the regulation of the different components of the IGF system which include IGF-I, IGF-II, IGF receptors, IGFBPs and IGFBP proteases.

IGF-I & IGF-II SYNTHESIS AND BONE

IGF-II appears to be the most abundant mitogen produced by serum free cultures of human bone cells. IGF-I is produced by human bone cells at 50-100 fold lower rate than IGF-II[1,2]. In contrast to human bone cells, adult mouse and rat bone cells in culture produce and adult rat and mouse bone extracts contain several fold more IGF-I than IGF-II. Thus the relative distribution of IGF-I and IGF-II appears to be different between rodents and humans[1,2]. With regard to the regulation of IGF production, there is now ample evidence which suggests that both systemic and local agents regulate the synthesis of IGF-I and IGF-II in bone cells. For example, PTH, a systemic hormone, has been shown to increase the level of IGF-I both at the protein and mRNA level in cultures of fetal rat bone cells[11]. The synthesis of IGF-I in rat bone cells appears to be mediated by cAMP. We obtained similar results with intact new born mouse calvaria in organ culture[1]. The effects of PTH on IGF regulatory system appears to be different between murine and human bone cells since PTH had no effect on the synthesis of either IGF-I or IGF-II in human bone cells[12]. In contrast to PTH, 1,25-dihydroxy vitamin D_3 inhibited production of IGF-I in new born mouse calvaria organ culture and in serum-free cultures of mouse osteoblastic cell line, MC3T3-E1[13]. Recent studies also demonstrate that the IGF production is regulated by sex steroid hormones. Estradiol has been shown to increase the release of IGF-I and TGF-β in UMR 106 rat osteosarcoma cells. In addition, progesterone has been shown to stimulate human bone cell proliferation and production of IGF-II both at the protein and mRNA level.

In addition to the systemic regulation, there is also evidence for the regulation of IGF-I and IGF-II production by local factors. For example, low amplitude frequency-specific electric fields increase the release of IGF-II in TE85 human osteoblast-like osteosarcoma cells[14]. We consider electric fields to be a local factor because it is our working assumption that exercise which is a local effecter of bone mediates its actions on bone through creating electric fields. In addition, IGF-II and TGF-β1 have been shown to modulate production of IGF-I in mouse osteoblast cell line[15]. We have also recently found that osteogenic protein-1 (BMP-7) stimulates human bone cell proliferation and increases IGF-II production in human bone cells[16]. Because IGFs are anabolic for bone cells in culture, it seems probable that systemic and local stimulators of bone formation may mediate their effects on bone by local elaboration of IGFs.

IGF RECEPTORS AND BONE

Bone cells have been shown to contain both type I and type II IGF/mannose 6 phosphate receptors[1]. The question of which IGF receptor(s) mediate the actions of IGF-II is still controversial. In this regard, we have found that IGF-II induced bone cell proliferation is inhibited by type II receptor blocking antibodies suggesting that type II IGF receptor may in part mediate the proliferative actions of IGF-II in bone cells. Consistent with this concept, we have found that testosterone treatment increases type II IGF receptor number in human bone cells and that pretreatment of human bone cells with testosterone potentiated the action of IGF-II. In addition, we have recently found that treatment of human bone cells with progesterone increased mRNA level for type II IGF receptor and bone cell proliferation. Thus, our findings are consistent with the idea that IGF receptor regulation is one of the components of the IGF regulatory system that can be modulated.

IGF BINDING PROTEINS AND BONE

Studies in our and in other laboratories provide strong support to the concept that IGFBPs function to regulate the local actions of IGFs in several tissues including bone: 1) IGFs occur in most tissues including bone as complexes with IGFBPs (there is very little free IGFs in biological fluids), 2) IGFBPs have been shown to modulate the actions of IGFs both in a positive and negative manner, 3) The production of IGFBPs appears to be regulated in a tissue specific manner and 4) The production of IGFBPs appears to be much more regulated than the production of IGFs themselves[8,17]. Thus IGFBPs are important components of the IGF regulatory system.

IGFBPs Produced By Bone Cells and Stored in Bone

We have shown that human bone cells in serum-free culture produce in addition to IGF-II, a number of IGFBPs (25, 29, 34, 38.5 and 41.5 kDa forms). Similarly, osteoblast-like cells derived from fetal mouse and rat calvaria have been shown to secrete multiple IGFBPs[2,18]. Human bone cells in culture (except MG63 human osteosarcoma cells) did not produce immunoreactive IGFBP-1 but have been shown to produce IGFBPs 2-6[18]. Comparison of the amounts and types of IGFBPs secreted by untransformed normal human bone cells derived from various skeletal sites as well as by different human osteosarcoma cell lines (G2, MG63, TE85, TE89, SaOS and U2) revealed that both types and amounts of IGFBPs secreted by different cell lines are different. Furthermore, the 25 kDa IGFBP-4 and the 38.5-41.5 kDa IGFBP-3 appear to be the two most abundant IGFBPs produced by bone cells under serum free conditions. It is possible that these differences in IGFBP production could be attributed to 1) the differences in autocrine/paracrine factors produced by various bone cell types under serum-free conditions, 2) the genetic differences between the various osteosarcoma cell lines studied and 3) the different stages of osteoblast differentiation these osteosarcoma cell lines may represent.

We have also shown that human bone extract contains 43-68, 29 and 24 kDa IGFBPs of which the 29 kDa IGFBP was the major form. Subsequent studies showed that the 29 kDa IGFBP was very similar to the IGFBP-5 purified from serum and U2 osteosarcoma cell conditioned medium and that the 24 kDa IGFBP was a breakdown product of the 29 kDa IGFBP-5[19]. Furthermore, it was found that the 29 kDa IGFBP-5 purified from human bone extract exhibited several fold higher affinity for IGF-II than IGF-I and in addition bound hydroxyapatite with very high affinity. These data may explain why IGF-II is several fold more abundant than IGF-I in human bone.

IGFBP Functions and Bone

The traditional role suggested for IGFBPs is that they provide a storage and transportation function for the IGFs. However, recent work indicates besides storage and transportation, several other functions could be attributable to the IGFBPs[17]. In bone, there is evidence that IGFBPs modulate local IGF actions. Campbell and Novak have reported that purified IGFBP-1 inhibited IGF-I induced cell proliferation in human MG63 osteosarcoma cells[8]. In contrast to these results, we found that IGFBP-1 up to 30 ng/ml had no effect on chick calvaria cell proliferation. Purified recombinant IGFBP-2 inhibited IGF-I induced bone cell proliferation at an apparent dose ratio of 10:1. Ernst & Rodan have shown that IGFBP-3 augmented the effects of IGF-I in rat osteoblast cultures[8]. We have found that the biological actions of IGFBP-3 on bone cells vary depending on culture conditions. Under acute conditions, IGFBP-3 inhibited basal and IGF induced bone cell proliferation. When IGFBP-3 was added before IGF-I, this IGFBP had a biphasic effect on IGF induced cell proliferation i.e., it potentiated IGF effect at low concentration but inhibited IGF effect at high concentration. Similar findings have been reported for IGFBP-3 in fibroblasts[20].

With regard to IGFBP-4, our previous studies have shown that IGFBP-4 inhibits IGF-I and IGF-II stimulated cell proliferation in embryonic chick calvaria cells and MC3T3-E1 mouse osteoblasts under all culture conditions tested. In addition, we have recently found that exogenous addition of antisense oligonucleotide to IGFBP-4 mRNA stimulated proliferation of human bone cells in vitro[21]. The inhibitory effect of IGFBP-4 on IGF induced cell proliferation is not specific to bone cells, since Cheung et al.[22] have also shown that IGFBP-4 inhibited IGF induced cell proliferation in a neuroblastoma cell line. In addition, Culouscou and Shoyab[23] have shown that the colon cancer cell growth inhibitor purified from CM of the HT29 human colon adenocarcinoma cell line was in fact IGFBP-4. Thus, the above findings are consistent with the interpretation that IGFBP-4 is a potent inhibitor of cell proliferation in bone cells as well as in other cell types.

In contrast to IGFBP-4, IGFBP-5 purified from human bone matrix extract potentiated IGF-II action in mouse osteoblasts[19]. Similar potentiating effects of IGFBP-5 purified from U2 human osteosarcoma cell conditioned medium has been reported[24]. Based on the findings that IGFBPs can modulate IGF actions either positively or negatively and based on the findings that the production of IGFBPs are regulated, it is obvious that the balance between the stimulatory and inhibitory classes of IGFBP will determine the degree and extent of IGF induced cellular response in target tissues. The functions of IGFBPs to modulate local IGF actions would be analogous to the functions of soluble cytokine receptors, viz., interleukin 1, which have been shown to modulate cytokine signals both in a positive and negative manner.

Regulation of IGFBP Production By Bone Cells In Vitro

Recent studies suggest that the concentration of IGFBPs in bone cell microenvironment can be regulated at multiple levels including transcriptional, post-transcriptional, translational and post-translational. In this regard, we have shown that the production of IGFBPs by bone cells is regulated by both local and systemic effecters of bone metabolism. PTH, a systemic calcium regulating hormone, increased production of IGFBP-4 both at the protein and mRNA levels in normal human bone cells and in UMR 106-01 rat osteosarcoma cells[25,26]. Consistent with these in vitro findings, we have shown that elderly women with hip fractures had an increase in serum PTH and an increase in serum IGFBP-4 levels[27]. Since cyclic AMP has been shown to be a potential second messenger for PTH in bone cells, we tested the effects of agents which increased cyclic AMP production (forskolin, prostaglandin E_2 and isobutylmethyl xanthine) and dibutyryl cyclic AMP on IGFBP-4 production. We found that dibutyryl cyclic AMP and agents which increased cyclic AMP inhibited cell proliferation, decreased IGFBP-3 and increased IGFBP-4, both at the protein and mRNA levels in TE85 and SaOS-2 human osteosarcoma cells[8]. Consistent with these data, dibutyryl cyclic AMP increased IGFBP-4 gene transcription rates 3-5 fold in TE85 and SaOS-2 cells suggesting that IGFBP-4 gene may possess a cyclic AMP response element[28].

1,25 dihydroxy vitamin D_3, another major calcium regulating hormone, has been shown to stimulate IGFBP-4 production and decrease cell proliferation in MC3T3-E1 mouse osteoblasts. 1,25 dihydroxy vitamin D_3 also stimulated IGFBP-4 production both at the protein and mRNA levels in normal bone cells and in human osteosaroma cells[29]. Consistent with these in vitro findings, treatment of human subjects with oral 1,25 dihydroxy vitamin D_3 for psoriasis resulted in a significant increase in serum IGFBP-4 concentration compared to pretreatment levels[29]. These observations suggest that 1,25 dihyroxy vitamin D_3 plays an important role in the regulation of IGFBP-4 secretion in vitro and in vivo.

In contrast to the stimulatory effects of PTH and 1,25 dihyroxy vitamin D_3 on IGFBP-4 production in human bone cells, progesterone, which increased human bone cell proliferation and IGF-II production, had opposite effects on IGFBP-4 production. Treatment of human bone cells with progesterone increased IGFBP-5 mRNA level by more than five fold and decreased IGFBP-4 mRNA level by about 50%. Progesterone, however, had no effect on IGFBP-3 or IGFBP-6 mRNA level[30]. In contrast to progesterone, dexamethasone decreased cell proliferation and inhibited production of IGFBP-5 in human bone cells.

The production of IGFBPs in bone cells is regulated not only by systemic agents but also by local effecters of bone metabolism (Table 1). In this regard, IGF-I and IGF-II which stimulate human bone cell proliferation decreased the level of IGFBP-4 in the conditioned medium of TE89 human osteosarcoma cells and untransformed normal human bone cells derived from rib[31]. In contrast, both IGF-I and IGF-II stimulated production of IGFBP-5 both at the protein and mRNA level in U2 human osteosarcoma cells and in normal human bone cells derived from rib[32]. Thus our data has shown that both IGF-I and IGF-II 1) stimulate human bone cell proliferation, 2) inhibit production of IGFBP-4 (which inhibits human bone cell proliferation) and 3) stimulate production of IGFBP-5 (which stimulates human bone cell proliferation). However, it is not known at this time whether all hormones which stimulate IGF production also produce this entire cascade of responses or whether some hormones influence the production of IGFBPs and not the IGFs or vice versa.

Table 1. Regulation of Cell Proliferation and IGFBP Production in Human Bone Cells.

Agents	Cell prol.	IGFBP-3	IGFBP-4	IGFBP-5
IGF-I	↑	↑	↓	↑
IGF-II	↑	↑	↓	↑
BMP-7	↑	↑	↓	↑
Progesterone	↑	-	↓	↑
Dibutyryl cyclic AMP	↓	↓	↑	ND
1,25-$(OH)_2$ vitamin D_3	↓	↑	↑	ND
Dexamethasone	↓	↓	ND	↓

↑ = increase; ↓ = decrease; - = No change; ND = Not determined

IGFBP PROTEASES AND BONE

The newest component of the IGF regulatory system is the IGFBP proteases. The degradation of IGFBP by specific proteases may be as important as synthesis in determining IGFBP abundance in the local extracellular fluid. Consistent with the recent discovery of IGFBP-3 protease in the serum of pregnant women[33,34], we found evidence for the presence of IGFBP-4 and IGFBP-5 fragments during the purification of these IGFBPs, thus suggesting that human bone cells in culture do in fact produce IGFBP-4 and IGFBP-5 proteases. Subsequently, we have found evidence that human bone cells in culture produce proteases which are relatively specific to IGFBP-4 and IGFBP-5. In addition, we have found evidence that IGFs can promote or inhibit proteolytic degradation of IGFBP-4 and IGFBP-5 respectively in human bone cells[35]. These data indicate that IGFBP proteases may provide a mechanism by which the IGFs may increase their own availability and/or activity in biological fluids.

There is only one other published report with regard to the presence of IGFBP protease in bone cell conditioned medium. Campbell et al.[36] have recently proposed that the serine protease plasmin selectively destroys IGFBP moiety of IGF-IGFBP complexes in bone. In these studies, plasmin was shown to hydrolyze multiple IGFBPs. In addition, plasmin has also been shown to hydrolyze latent TGF-β complex and release active TGF thus suggesting that the protease activity of plasmin may not be specific to IGFBPs. In contrast to plasmin, IGFBP-4 and IGFBP-5 proteases produced by human bone cells appear to be relatively specific. Since IGFBPs appear to modulate the functions of IGFBPs, we believe that studies on identity, regulation and action of IGFBP-4 and IGFBP-5 proteases are essential to our understanding of how the IGF system functions in bone.

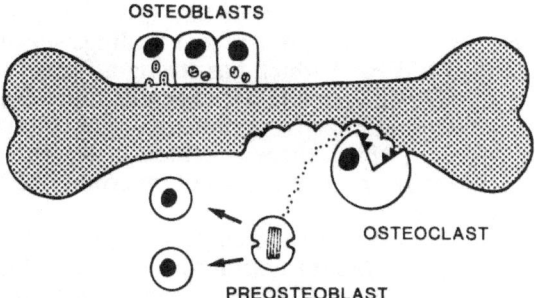

OSTEOBLASTS

OSTEOCLAST

PREOSTEOBLAST

Figure 1. Model for the role of IGFs in the coupling of bone formation to resorption. IGFs deposited in bone are postulated to be released during osteoclastic bone resorption to act on precursor osteoblasts and mature osteoblasts to insure site-specific bone replacement (reproduced from Ref. 37, with permission from the authors).

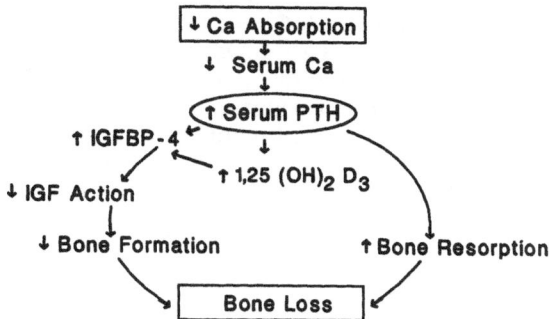

Figure 2. Model illustrating how IGF system could be involved in mediating the PTH and 1,25 dihyroxy vitamin D_3 induced uncoupling of bone formation to resorption in women with calcium deficiency (adapted from Ref. 2).

PHYSIOLOGICAL SIGNIFICANCE OF IGF SYSTEM IN BONE

Bone volume is maintained under normal conditions by a coupling mechanism in which bone formation increases in proportion to resorption. In post- menopausal and senile osteoporosis, however, there appears to be an impairment in the coupling mechanism which results in bone loss. Recent studies in several laboratories suggest that coupling of bone formation to bone resorption is mediated in part by growth factors, including IGFs, stored in bone (Fig. 1). Several findings support this concept. First, there is evidence that growth factors such as IGF-II become fixed in bone. In this regard, our recent findings suggest that IGF-II is fixed in bone by means of IGFBP-5 which has strong affinity for hydroxyapatite and selective affinity for IGF-II over IGF-I[19]. Second, there is evidence that the amount of mitogenic activity released from bone in organ culture is proportional to the extent of resorption[37]. Third, there is evidence that the concentration of growth factors such as IGFs stored in bone is not invariant but changes with disease states and hormonal status[6,7].

Recent in vivo and in vitro evidence also suggest that the IGF-II system may be involved in the uncoupling of bone formation to resorption (Fig. 2). Elderly hip fracture patients had elevated serum PTH levels and elevated inhibitory IGFBP-4 levels as compared to normal age matched subjects[27]. In addition, serum IGFBP-4 levels were elevated after oral 1,25-dihydroxy vitamin D_3 treatment compared to pretreatment levels. These in vivo findings are consistent with our in vitro findings on the stimulatory effects of PTH and 1,25-dihyroxy vitamin D_3 on IGFBP-4 production in human bone cells[2]. Based on the in vitro and in vivo findings, we have proposed a model involving the IGF system to explain the mechanism of bone loss in elderly subjects with secondary hyperparathyroidism or in young individuals who take very little calcium. These subjects have decreased serum calcium, which leads to an increase in serum PTH which in turn increases serum 1,25-dihydroxy vitamin D_3 level. In calcium deficiency state, there are two means by which an increase in the amount of net calcium delivered from bone to blood can be accomplished: 1) increased bone resorption and 2) decreased bone formation. The effect of PTH and 1,25-dihydroxy vitamin D_3 to increase bone resorption is well known. In addition, the increase in serum PTH and 1,25-dihydroxy vitamin D_3 may inhibit bone formation by increasing the production of IGFBP-4 by bone cells. Further elucidation of the relationship between IGF system and calcium regulating hormones may have major implications for the diagnosis and treatment of age-related osteoporosis.

REFERENCES

1. S. Mohan, D.J. Baylink. Bone growth factors. *Clin Orthoped Rel Res* 263:30 (1991).
2. S. Mohan, D.J. Baylink. The role of insulin-like growth factor-II in the coupling of bone formation to resorption. *in*: "Modern Concepts of Insulin-Like Growth Factors," E.M. Spencer, ed., Elsevier, New York (1991).
3. J. Zapf, E.R. Froesch. Insulin-like growth factors/somatomedins: structure, secretion, biological actions and physiological role. *Horm Res* 24:121 (1986).
4. A.G. Johansson, E. Lindh, S. Ljunghall. Insulin-like growth factor I stimulates bone turnover in osteoporosis. *Lancet* 339:1619 (1992).
5. S. Mohan, C.M. Bautista, J.E. Wergedal, D.J. Baylink. Isolation of an inhibitory insulin-like growth factor (IGF) binding protein from bone cell-conditioned medium: A potential local regulator of IGF action. *Proc Natl Acad Sci USA* 86:8338 (1989).

6. S. Mohan, J. Dequeker, R. Van den Eyned, J. Peeters, J. Aerssens, D.J. Baylink. Increased insulin-like growth factor (IGF)-I and IGF-II in bone from patients with osteoarthritis. *J Bone and Mine Res* 6:S131 (1991).

7. V. Nicolas, A. Prewett, P. Bettica, et.al. Evidence for a progressive decline of IGF-I in human bone in both males and females as a function of age. *J Bone Min Res* 7:S255 (1992).

8. S. Mohan. Insulin-like growth factor binding proteins in bone cell regulation. *Growth Regulation* 3:65 (1993).

9. R.M. Maciel, A.C. Moses, G. Villone, D. Tramontano, S.H. Ingbar. Demonstration of the production and physiological role of insulin-like growth factor II in rat thyroid follicular cells in culture. *J Clin Invest* 82:1846 (1988).

10. A. Gray, A.W. Tam, T.J. Dull, J. Hayflick, J. Pintar, W.K. Cavenee, A. Koufos, A. Ullrich. Tissue-specific and developmentally regulated transcription of the insulin-like growth factor 2 gene. *DNA* 6:283 (1987).

11. T.L. McCarthy, M. Centrella, E. Canalis. Parathyroid hormone enhances the transcript and polypeptide levels of insulin-like growth factor I in osteoblast-enriched cultures from fetal rat bone. *Endocrinology* 124:1247 (1989).

12. R.D. Finkelman S. Mohan, T.A. Linkhart, S.M. Abraham, J.P. Boussy, D.J. Baylink. PTH stimulates the proliferation of TE-85 human osteosarcoma cells by a mechanism not involving either increased cAMP or increased secretion of IGF-I, IGF-II or TGF beta. *Bone Mineral* 16:89 (1992).

13. S.H. Scharla, D.D. Strong, S. Mohan, D.J. Baylink, T.A. Linkhart. 1,25 dihydroxyvitamin D_3 differentially regulates the production of IGF-I and IGFBP-4 in mouse osteoblasts. *Endocrinology* 129:3139 (1991).

14. R.J. Fitzsimmons, D.D. Strong, S. Mohan, D.J. Baylink. Low-amplitude, low-frequency electric field-stimulated bone cell proliferation may in part be mediated by increase IGF-II release. *J Cell Physiol* 150:84, 1992.

15. F.T. Tremollieres, D.D. Strong, D.J.Baylink, S. Mohan. Progesterone and promegestone stimulate human bone cell proliferation and insulin-like growth factor-II production. *Acta Endocrinol (Copenh)* 126:329 (1992).

16. R. Knutsen, K. Sampath, D.J. Baylink, S. Mohan. Evidence that osteogenic protein-1 (OP-1) may modulate its effects on human bone cell proliferation (HBC) by regulating the local production of insulin-like growth factors. *J Bone and Min Res* 7:S104 (1992).

17. S. Shimasaki, N. Ling. Identification and molecular characterization of insulin-like growth factor binding proteins (IGFBP-1, -2, -3, -4, -5 and -6). *Prog in Growth Factor Res* 1:243 (1991).

18. S. Mohan, D.J. Baylink. Isolation and characterization of insulin-like growth factor binding proteins produced by human bone cells in vitro. *J Bone Min Res* 6:S141 (1991).

19. C.M. Bautista, D.J. Baylink, S. Mohan. Isolation of a novel insulin-like growth factor (IGF) binding protein from human bone: A potential candidate for fixing IGF-II in human bone. *Biochem Biophys Res Commun* 176:756 (1991).

20. C. Conover. Potentiation of insulin-like growth factor (IGF) action by IGF binding protein-3: Studies of underlying mechanism. *Endocrinology* 130:3191 (1992).

21. M. Rashmi, D.D. Strong, D.J. Baylink, S. Mohan. Stimulation of human bone cell proliferation by an antisense oligodeoxyribonucleotide to IGFBP-4 mRNA. *J Bone Min Res* 7:S124 (1992).

22. P.T. Cheung, E.P. Smith, S. Shimasaki, N. Ling, S.D. Chernausek.

Characterization of an insulin-like growth factor binding protein (IGFBP-4) produced by B104 rat neuronal cell line: Chemical and biological properties and differential synthesis by sublines. *Endocrinology* 129:1006 (1991).

23. J.M. Culouscou, M. Shoyab. Purification of a colon cancer cell growth inhibitor and its identification as an insulin-like growth factor binding protein. *Cancer Res* 51:2813 (1991).

24. D.L. Andress, R.S. Birnbaum. A novel insulin-like growth factor binding protein secreted by osteoblast-like cells. *Biochem Biophys Res Commun* 176:213 (1991).

25. D. LaTour, S. Mohan, T.A. Linkhart, D.J. Baylink, D.D. Strong. Inhibitory insulin-like growth factor binding protein: Cloning, complete sequence and physiological regulation. *Molec Endocrinol* 4:1806 (1990).

26. O. Torring, A.F. Firek, H. Heath III, C.A. Conover. Parathyroid hormone and parathyroid hormone-related peptide stimulate insulin-like growth factor-binding protein secretion by rat osteoblast-like cells through a adenosine 3',5'-monophosphate-dependent mechanism. *Endocrinology* 128:1006 (1991).

27. C. Rosen, L.R. Donahue, S. Hunter, M. Holick, H. Kavookjian, A. Kirschenbaum, S. Mohan, D.J. Baylink. The 24/25 kDa serum insulin-like growth factor-binding protein is increased in elderly women with hip and spine fractures. *J Clin Endocinol Metab* 74:24 (1992).

28. K. Lee, S. Mohan, D.J. Baylink, D.D. Strong. Evidence that transcriptional regulation of insulin-like growth factor binding protein-4 (IGFBP-4) gene in human bone cells. *J Bone Min Res* 7:S232 (1992).

29. S.H. Scharla, C. Rosen, D.D. Strong, S. Mohan, D.J. Baylink, T.A. Linkhart. Stimulatory effect of 1,25(OH)$_2$D$_3$ on IGFBP-4 secretion in bone cells. *2nd International Workshop on IGFBPs*, Opio, France (1992).

30. U.G. Lempert, D.D. Strong, D.J. Baylink, S. Mohan. Effects of progesterone on the mRNA levels of IGFBP-3, IGFBP-4, IGFBP-5 and IGFBP-6 in human osteoblastic cells. *2nd International Workshop on IGFBPs*, Opio, France (1992).

31. S. Mohan, D.D. Strong, U.G. Lempert, J.E. Wergedal, D.J. Baylink. Studies on regulation of insulin-like growth factor binding protein (IGFBP)-3 and IGFBP-4 production in human bone cells. *Acta Endocinology* 127:555 (1992).

32. U.G. Lempert, C. Bautista, D.D. Strong, D.J. Baylink, S. Mohan. Purification of a novel human bone derived inulin-like growth factor binding protein (hBD-IGFBP): A potential candidate for fixing IGF-II in bone. *J Bone Min Res* 6:S250 (1991).

33. L.C. Guidice, E.M. Farrel, H. Pham, G. Lamson, R.G. Rosenfeld. Insulin-like growth factor bidning proteins in maternal serum throughout gestation and in the puerperium: Effects of a pregnancy-associated serum protease activity. *J Clin Endocrinol Metab* 71:806 (1990).

34. P. Hossenlopp, B. Segovia, C. Lassarre, M. Roghani, M.Bredon, M. Binoux. Evidence of enzymzatic degradation of insulin-like growth factor binding proteins in the 150K complex during pregnancy. *J Clin Endocrinol Metab* 71:797 (1990).

35. S. Kanzaki, Malpe R. Baylink D.J., S. Mohan. Evidence that human bone cells (HBC) in culture produce insulin-like growth factor binding protein (IGFBP)-4 and -5 proteases. *75th Annual Meeting, The Endocrine Society* In Press (1993).

36. P.G. Campbell, J.F. Novak, T.B. Yanosick, J.H. McMaster. Involvement of the plasmin system in dissociation of the insulin-like growth factor binding protein complex. *Endocrinology* 130:1401 (1992).

37. J.R. Farley, N. Tarbaux, L.A. Murphy, T. Masuda, D.J. Baylink. In vitro evidence that bone formation may be coupled to bone resorptioon by release of mitogen(s) from resorbing bone. *Metabolism* 36:314 (1987).

REGULATION OF IGF ACTIVITY IN BONE

Thomas L. McCarthy and Michael Centrella

Yale University School of Medicine
Section of Plastic and Reconstructive Surgery
333 Cedar Street, P.O. Box 3333
New Haven, Connecticut 06510

BONE PHYSIOLOGY

Although the skeleton's principal function is most often associated with the determination of size and shape of vertebrates, it also serves as a reservoir of calcium (and phosphate), from which the body maintains this important divalent cation within a critical range required for membrane polarization, muscle function, and other key physiological processes. Parathyroid hormone (PTH), the most important hormone for the minute to minute regulation of blood calcium levels, regulates the catabolic process of bone resorption. Interestingly, although osteoclasts are the cells responsible for dissolving bone collagen matrix and calcium release, receptors for PTH in bone reside chiefly on osteoblasts, the anabolically active bone cell that produces collagen and the other matrix associated macromolecules that constitute the organic scaffolding of bone and upon which mineral deposition occurs. In order to maintain a structurally intact skeleton, the amount of bone lost through resorption must be matched by an equal amount of bone produced by formation. For more than a decade this concept has been referred to as coupling, and one or more growth factors have been hypothesized to function in this capacity as coupling factors in bone remodelling[1].

METABOLIC EFFECTS OF IGFs FOR BONE

Both IGF-I and IGF-II increase collagen synthesis in primary fetal rat osteoblasts and calvarial explant cultures[2]. This enhanced level of collagen synthesis exceeds the increased rate of non-collagen protein synthesis, resulting in an effective increase in the percent of total protein synthesis that is collagen. Using histomorphometry and calvarial explant cultures, IGF-I has been shown to increase collagen synthesis, matrix apposition and cell replication[3]. Cell replication is primarily increased in the osteoprogenitor region, where osteoblast differentiation occurs, but also was detected in the zone of mature osteoblasts and

Current Directions in Insulin-Like Growth Factor Research,
Edited by D. LeRoith and M.K. Raizada, Plenum Press, New York, 1994

407

periosteum, a more fibroblastic region. Recently, IGF-I was found to support the formation and activation of osteoclasts, the cells that carry out bone resorption, suggesting that IGF-I may also function as a bone resorption regulator that mediates interactions between osteoblasts and osteoclasts[4]. These, and other studies demonstrate the importance of IGFs, and their influence on bone metabolism. Together with the finding that both IGF-I and IGF-II are made by bone cells[5-7], it is likely that IGFs are autocrine and paracrine growth and differentiation factors for this tissue.

REGULATION OF IGF PRODUCTION BY BONE CELLS

While the importance of IGF-I as the mediator of growth hormone's stimulatory effect on longitudinal bone growth had been known for some time[8], the anabolic potential of intermittent PTH treatment was originally described in the early 1980s[9]. Several years later, in 1989 we reported the first demonstration that PTH augments IGF-I expression in primary fetal rat cell cultures enriched with osteoblasts[10]. Subsequently, we established a reliable *in vitro* fetal rat calvarial explant organ culture protocol that mimicked the anabolic effects of PTH observed *in vivo*. Using this tool, along with the application of neutralizing antibodies to IGF-I, we provided initial evidence for the participation of IGF-I in the coupling of bone matrix production (collagen synthesis) to bone resorption[11]. Since then we have further characterized the mechanism of PTH action on IGF-I synthesis in bone, and have determined that cAMP serves as the intracellular second signal for the stmulatory effect of PTH on IGF-I. Furthermore, forskolin, dibutyryl cAMP and isobutylmethyl xanthine, each of which increases cAMP levels by alternate mechanisms, effectively stimulate IGF-I production[12].

Multiple second signals are generated in osteoblasts in response to PTH, and these include: cAMP with a subsequent activation of protein kinase A (PKA)[13], elevated calcium and inositol trisphosphate with a subsequent activation of protein kinase C (PKC)[14-16], and the generation of prostaglandin E_2 (PGE_2)[17]. While cAMP regulates IGF-I production by fetal rat osteoblasts, it does not induce immediate changes in IGF-II transcript or polypeptide levels. Selective modulators of intracellular calcium and PKC, such as ionomycin and phorbol myristate acetate (PMA), do not stimulate either IGF-I or IGF-II synthesis in the fetal rat osteoblast culture model. On the other hand, PGE_2 potently elevates IGF-I transcript and polypeptide levels with a time course similar to PTH[18]. Because PGE_2 causes a rapid and pronounced increase in cAMP, as does PTH, it is likely that cAMP also mediates the stimulatory effect of PGE_2 on IGF-I expression. Inhibition of prostaglandin synthesis with indomethacin does not modify the stimulatory effect of PTH treatment, indicating that the PTH effect does not function through a prostaglandin intermediary[18]. Furthermore, PGE_2 does not influence IGF-II levels in primary fetal rat osteoblasts. Interestingly, IGF-II synthesis in this culture model appears to be constitutive, and treatment with a variety of hormones that influence bone metabolism, including cortisol, estrogen, testosterone, triiodothyronine, $1,25(OH)_2$vitaminD3 ($1,25(OH)_2D_3$), growth hormone, and insulin do not produce immediate or large changes in its expression[19].

Another calcium regulating hormone, $1,25(OH)_2D_3$ was found to alter the amounts of IGF-I released in the culture medium in primary human bone cells and mouse calvarial explant cultures, but opposite effects were noted in these two models. In human bone cell cultures, $1,25(OH)_2D_3$ increased IGF-I synthesis, while in mouse calvariae the hormone suppressed IGF-I levels[20,21]. Further studies with mouse calvariae demonstrated an increase in IGF-II release by both PTH and $1,25(OH)_2D_3$. Additional differences among various culture models include the observation that $1,25(OH)_2D_3$ has no effect on IGF-I levels being produced in the clonal mouse osteoblastic cell line MC3T3-E1[22]. Consequently, these differences may relate to variations in experimental protocols, in specific detection methods, or in the ability of cells at specific stages of differentiation to produce or release IGF-I in response to selected hormones.

The higher incidence of osteoporosis in postmenopausal females and the subsequent bone sparing effects of estrogen treatment has motivated the search for how these effects occur on the molecular level, and for optimal ways to use estrogen or estrogen analogues in the clinical setting. 17β-estradiol treatment produces modest increases in rat osteoblast proliferation and collagen synthesis *in vitro*, both of which are inhibited by neutralizing antibodies to IGF-I[23]. Furthermore, Northern analysis indicates an increase in IGF-I steady-state transcript levels in 17β-estradiol treated cultures independent of new protein synthesis. These findings suggest one possible mechanism for the anabolic potential of estrogen on bone, independent of the current interest in the inhibitory effects of the hormone on synthesis of molecules that enhance osteoclast development.

Cortisol, the principal naturally occuring glucocorticoid, has a dose-dependent inhibitory effect on IGF-I synthesis. More importantly, cortisol decreases the stimulatory effect of PTH on IGF-I synthesis, and directly inhibits bone cell collagen synthesis[24]. A peculiar wrinkle in this scheme is the observation that cortisol enhances the anabolic effect of exogenous IGF-I on collagen expression in intact bone cultures of fetal rat calvariae[25]. While these data at first appear difficult to accommodate, when examined in the context of the inhibitory effect of cortisol on IGF-I synthesis, the increase in sensitivity in the intact bone cultures may result from a deprivation of endogenously produced IGF-I. Furthermore, changes in specific IGFBP expression, and alterations in IGF receptor binding parameters (as described below) may contribute to the overall influence of cortisol on bone cell responsiveness to exogenous IGF-I. These results will necessarily be dependent on changes in cortisol concentration as well as duration of exposure, both of which are critical elements in the way that tissues respond to this steroid.

Although many of the details are still lacking, an additional route for regulation of IGF-I expression and responsiveness may be through class I histocompatibility molecules[26]. These findings may implicate a partial restriction by way of tissue or genetic specificity, but the physiological or pathological circumstances in which these effects becomes important are presently unresolved.

POTENTIAL ALTERNATE MECHANISMS TO REGULATE ENDOGENOUS IGF ACTIVITY IN BONE

As in other tissue and cell culture systems, IGF activity in bone can be regulated at multiple levels. As already discussed, IGF synthesis and/or release by skeletal tissue may be altered by a variety of other growth regulators. In addition, local synthesis and differential expression of specific IGF binding proteins (IGFBPs) may control the amount of IGFs available for binding to specific IGF receptors. Alterations in type 1 and type 2 IGF receptors (increases or decreases in receptor number or affinity) that are independent of local or circulating IGF concentrations may be influenced by other regulatory agents. Finally, the actual biochemical effects of the IGFs can be increased or inhibited by other growth regulators, whose own mechanisms of regulation may enhance or interfere with the intracellular signals induced by binding to IGF receptors. The remainder of this chapter will address the regulation of IGFBP expression and modulation of IGF receptor binding kinetics. While the ultimate step that regulates IGF activity is the generation of the appropriate intracellular signals once IGF-I or IGF-II binds to cell surface receptors, little information in bone is presently available in this regard, and will likely remain a focus of future research.

IGFBP EXPRESSION BY BONE CELLS

Fetal rat osteoblasts express a variety of IGFBPs, but we and others have not detected IGFBP-1 transcript or polypeptide expression and have been unable to determine culture conditions that support the synthesis of IGFBP-1[27]. Although the full complement of IGFBPs that bone cells synthesize is not yet known, a range of results have accumulated from our studies and those of others regarding the various IGFBPs. It is important to note, however, that the array of IGFBPs produced by bone cells likely will vary among the osteoblast-like cell lines studied, as well as among the different species used to isolate primary bone cell cultures[28].

Primary fetal rat bone cells express transcripts for IGFBP-2, -3, -4, -5, and -6. IGFBP-2 polypeptide is expressed in high amounts in fetal rat bone cell cultures, and its mRNA levels are not altered by treatment with cAMP inducing agents, by growth hormone, or by IGF-I or IGF-II[27] (and our unpublished observations). The synthetic glucocorticoid dexamethasone has been reported to decrease the appearance of IGFBP-2, while $1,25(OH)_2D_3$ modestly enhances its appearance in the culture medium of first passage fetal rat osteoblasts[29]. We have observed similar results in cortisol treated primary fetal rat bone cell cultures, but this effect was not reflected by a comparable decrease in IGFBP-2 mRNA transcripts. Therefore, the mechanism by which this occurs is currently unclear, and may reflect post-transcriptional events. 17β-estradiol is reported to have complex, biphasic effects on IGFBP expression by first passage fetal rat osteoblast cultures, inhibiting IGFBP-2 levels at low doses, and stimulating at high doses[30]. This observation has not been further examined or confirmed by other investigators.

In contrast to the high amounts of constitutive IGFBP-2 expression by bone cells, IGFBP-3, IGFBP-4, and IGFBP-5 mRNA and polypeptide are found in low to moderate amounts in untreated primary cultures of fetal rat osteoblasts. Agents that induce cAMP potently enhance the mRNA and polypeptide levels of IGFBP-3[31] and IGFBP-5 (which may enhance IGF activity), and moderately increase IGFBP-4[32] (which is inhibitory for IGF actions)[27], (and our unpublished observation). Treatment with growth hormone also increases IGFBP-3 expression in bone cells, as seen in other tissue systems, but this effect is small by comparison to the effect of cAMP inducing agents[33]. $1,25(OH)_2D_3$ has been reported to increase IGFBP-3 expression in MG-63 human osteosarcoma cell cultures, and to increase IGFBP-4 in clonal mouse osteoblastic MC3T3-E1 cultures[34,35].

Of the six IGFBPs examined, only IGFBP-5 mRNA levels rise after treatment with IGF-I or IGF-II. However, an apparent increase in IGFBP-2, IGFBP-3, IGFBP-4 and IGFBP-5 polypeptide, as determined by Western ligand blot analysis with ^{125}I-IGF-I, was observed in the same IGF treated cultures[27] (and our unpublished observation). Consequently, both the IGFs and the IGFBPs may be stabilized by their association, limiting their sensitivity to proteolytic degradation. Finally, while primary fetal rat osteoblasts express transcripts for IGFBP-6, little is known about the level of protein expression or its biological function. Nonetheless, no changes in IGFBP-6 expression have been found yet in our bone cell culture model (our unpublished observations).

IGF RECEPTORS ON BONE CELLS

In serum-free cultures, fetal rat osteoblasts express abundant type 1 IGF receptors that are similar in structure to insulin receptors and preferentially bind IGF-I[36]. These cells also express approximately 2-3 fold more high affinity type 2 binding sites that preferentially bind IGF-II. Type 2 sites possess an amino acid domain that binds mannose-6-phosphate (M-6-P) or glycoproteins that contain M-6-P residues, in addition to the domain that binds IGF-II. Cross-receptor ligand binding is observed in many species, including the rat. However, in some organisms type 2 sites fail to bind IGF-II, in agreement with the widely held belief that biologically functional binding for both IGF-I and IGF-II occurs at type 1 IGF receptors[37].

Bone growth regulators that induce cAMP synthesis, such as PTH and PGE_2, increase the number of type 2 IGF binding sites that are present on primary fetal rat osteoblasts, as determined by Scatchard analysis and ligand-receptor cross linking studies. Little effect is observed on type 1 IGF receptor binding. The resulting increase in IGF-II binding occurs over a wide IGF-II concentration range. IGF-I binding to the type 2 receptor is also increased, however, this effect is only observed at high input concentrations of IGF-I[38]. In contrast, PMA, an agent that activates protein kinase C, rapidly increases binding and the apparent receptor number of type 1 IGF receptors in primary fetal rat osteoblasts[39].

Cortisol treatment, which reduces bone cell activity and IGF-I expression, also decreases type 2 binding site number after short term treatment, but has no effect on type 1 receptors[40]. In contrast, an earlier study using prolonged glucocorticoid exposure showed an increase in total IGF-I binding[41]. This later effect may result from a decrease in basal IGF-I synthesis, and a subsequent failure in autologous IGF binding and receptor down-regulation. Furthermore, $1,25(OH)_2D_3$ increased type 1 IGF receptor number while having little effect on IGF-I expression in the murine clonal osteoblastic cell line MC3T3-E1[22].

PHYSIOLOGICAL IMPORTANCE OF PTH, PGE$_2$ AND IGFs TO BONE

Recently, using *in situ* hybridization, Edwall *et al.* have shown an increased expression of IGF-I transcripts in regenerating bone following a traumatic fracture. They further demonstrated that the prostaglandin synthesis inhibitor indomethacin reduced IGF-I transcript expression in the healing bone[42]. In another study, Wronski *et al.* using ovariectomized rats as a model of postmenopausal osteoporosis, showed that PTH was more effective than estrogen, the traditional agent of choice, at restoring bone mass, and was also more effective than resorption inhibitors of the bisphosphonate class[43]. These and other studies, along with our data (summarized in Table 1), strongly point to the importance of PTH, PGE_2 and IGF-I in maintaining bone mass and integrity.

Table 1

PTH likely affects local skeletal IGF activity via multiple mechanisms.

1. PTH activates protein kanase A via elevated cAMP levels resulting in:
     Increased IGF-I synthesis, while IGF-II levels remain constant
     Increased synthesis of IGFBP-3, IGFBP-4, & IGFBP-5
     Increased IGF-II binding resulting from increased number of type 2 IGF receptors

2. PTH activates protein kinase C via elevated calcium and inositol trisphosphate:
     Increased IGF-I binding resulting from increased number of type 1 IGF receptors

3. PTH elevates PGE_2 synthesis, which elevates cAMP levels with the consequences described above

---

CONCLUSION

In an effort to understand how the activity of locally produced IGF is regulated, we and a number of investigators of bone cell biology have begun to dissect the response of osteoblasts to a variety of hormones and agents with functionally discrete mechanisms of action or that generate alternate second signals. While the information presented here demonstrates potential mechanisms that may influence IGF actions in bone, the ultimate control of endogenous IGF activity appears likely to result from complex interactions. Although at first glance the job is daunting, we must consider the simultaneous effects of these agents on IGF synthesis and release from bone matrix, the complex array of IGFBPs produced by skeletal tissue, changes in IGF receptor binding kinetics, and the coupling of IGF receptors to signal transducing pathways. More work will follow in order to then integrate the interacting effects of multiple hormones on bone metabolism, the alterations that may occur in certain disease states, and the way these observation may be exploited to improve management of bone integrity, fracture repair, and tissue reconstruction in the clinical setting.

REFERENCES

1. G.A. Howard, G.A., B.L. Bottemiller, R.T. Turner, J.I. Rader, and D.J. Baylink, Parathyroid hormone stimulates bone formation and resorption in organ culture: evidence for a coupling mechanism, *Proc. Natl. Acad. Sci. USA* 78:3204-3208 (1981).

2. T.L. McCarthy, M. Centrella, and E. Canalis, Regulatory effects of insulin-like growth factors I and II on bone collagen synthesis in rat calvarial cultures, *Endocrinology* 124:301-309 (1989).

3. J.M. Hock, M. Centrella, and E. Canalis, Insulin-like growth factor I (IGF-I) has independent effects on bone matrix formation and cell replication, *Endocrinology* 122:254-260 (1988).

4. H. Mochizuki, Y. Hakeda, N. Wakatsuki, N. Usui, S. Akashi, T. Sato, K. Tanaka, M. Kumegawa, Insulin-like growth factor-I supports formation and activation of osteoclasts, *Endocrinology* 131:1075-1080 (1992)

5. E. Canalis, T. McCarthy, and M. Centrella, Isolation and characterization of insulin-like growth factor I (Somatomedin-C) from cultures of fetal rat calvariae, *Endocrinology* 122:22-27 (1988).

6. C.A. Frolik, L.F. Ellis, and D.C. Williams, Isolation and characterization of insulin-like growth factor II from human bone, *Biochem. Biophys. Res. Commun.* 151:1011-1018 (1988).

7. S. Mohan, J.C. Jennings, T.A. Linkhart, and D.J. Baylink, Primary structure of human skeletal growth factor: homology with human insulin-like growth factor II, *Biochim. Biophys. Acta* 966:44-55 (1988).

8. Lowe, W.L. Jr., Biological actions of the insulin-like growth factors; in "Insulin-like Growth Factors: Molecular and Cellular Aspects", D. LeRoith, Editor; CRC Press, Inc., Boca Raton, FL (1991).

9. D.M. Slovik, R.M. Neer, and J.T. Potts, Short term effects of synthetic human parathyroid hormone-(1-34) administration on bone mineral metabolism in osteoporotic patients, *J. Clin. Invest.* 68:1261-1271 (1981).

10. T.L. McCarthy, M. Centrella, and E. Canalis, Parathyroid hormone enhances the transcript and polypeptide levels of insulin-like growth factor I in osteoblast-enriched cultures from fetal rat bone, *Endocrinology* 124:1247-1253 (1989).

11. E. Canalis, M. Centrella, W. Burch, and T.L. McCarthy, Insulin-like growth factor I mediates selective anabolic effects of parathyroid hormone in bone cultures, *J. Clin. Invest.* 83:60-65 (1989).

12. T.L. McCarthy, M. Centrella, and E. Canalis, Cyclic AMP induces insulin-like growth factor I synthesis in osteoblast-enriched cultures, *J. Biol. Chem.* 265:15353-15356 (1990).

13. L.R. Chase, and G.D. Aurbach, The effect of parathyroid hormone on the concentration of adenosine 3',5'-monophosphate in skeletal tissue *in vivo*, *J. Biol. Chem.* 245:1520-1526 (1970).

14. D.T. Yamaguchi, T.J. Hahn, A. Iida-Klein, C.R. Kleeman, and S. Muallem, Parathyroid hormone-activated calcium channels in an osteoblast-like clonal osteosarcoma cell line, *J. Biol. Chem.* 262:7711-7718 (1987).

15. R. Civitelli, I.R. Reid, S. Westbrook, L.V. Avioli, and K.A. Hruska, Parathyroid hormone elevates inositol polyphosphates and diacylglycerol in a rat osteoblast-like cell line, *Am. J. Physiol.* 255:E660-E667 (1988).

16. A.B. Abou-Samra, H. Jueppner, D. Westerberg, J.T. Potts Jr., and G.V. Segre, Parathyroid hormone causes translocation of protein kinase-C from cytosol to membranes in rat osteosarcoma cells, *Endocrinology* 124:1107-1113 (1989).

17. J. Klein-Nulend, P.N. Bowers, and L.G. Raisz, Evidence that adenosine 3',5'-monophosphate mediates hormonal stimulation of prostaglandin production in cultured mouse parietal bones, *Endocinology* 126:1070-1075 (1990).

18. T.L. McCarthy, M. Centrella, L.G. Raisz, and E. Canalis, Prostaglandin E_2 stimulates insulin-like growth factor I synthesis in osteoblast-enriched cultures from fetal rat bone, *Endocrinology* 128:2895-2900 (1991).

19. T.L. McCarthy, M. Centrella, and E. Canalis, Constitutive synthesis of insulin-like growth factor-II by primary osteoblast-enriched cultures from fetal rat calvariae, *Endocrinology* 130:1303-1308 (1992).

20. C. Chenu, A. Valentin-Opran, P. Chavassieux, S. Saez, P.J. Meunier, and P.D. Delmas, Insulin like growth factor I hormonal regulation by growth hormone and by 1,25(OH)$_2$D$_3$ and activity on human osteoblast-like cells in short-term cultures, *Bone* 11:81-86 (1990).

21. T.A. Linkhart and M.J. Keffer, Differential regulation of insulin-like growth factor-I (IGF-I) and IGF-II release from cultured neonatal mouse calvaria by parathyroid hormone, transforming growth factor-β, and 1,25-dihydroxyvitamin D$_3$, *Endocrinology* 128:1511-1518 (1991).

22. H. Kurose, K. Yamaoka, S. Okada, S. Nakajima, and Y. Seino, 1,25-Dihydroxyvitamin D$_3$ [1,25(OH)$_2$D$_3$] increases insulin-like growth factor I (IGF-I) receptors in clonal osteoblastic cells. Study on interaction of IGF-I and 1,25(OH)$_2$D$_3$, *Endocrinology* 126:2088-2094 (1990).

23. M. Ernst and G.A. Rodan, Estradiol regulation of insulin-like growth factor-I expression in osteoblastic cells: evidence for transcriptional control, *Mol. Endocrinol.* 5:1081-1089 (1991).

24. T.L. McCarthy, M. Centrella, and E. Canalis, Cortisol inhibits the synthesis of insulin-like growth factor-I in skeletal cells, *Endocrinology* 126:1569-1575 (1990).

25. B.E. Kream, D.N. Petersen, and L.G. Raisz, Cortisol enhances the anabolic effects of insulin-like growth factor I on collagen synthesis and procollagen messenger ribonucleic acid levels in cultured 21-day fetal rat calvariae, *Endocrinology* 126:1576-1583 (1990).

26. M. Centrella, T.L. McCarthy, and E. Canalis, β$_2$-Microglobulin enhances insulin-like growth factor I receptor levels and synthesis in bone cell cultures, *J. Biol. Chem.* 264:18268-18271 (1989).

27. T.L. McCarthy, M. Centrella, and E. Canalis, Regulation of insulin-like growth factor (IGF) binding protein (IGF-BP) expression in primary rat osteoblast (Ob) enriched cultures, *J. Bone Min. Res.* 6:abstract 480 (1991).

28. C. Hassager, L.A. Fitzpatrick, E.M. Spencer, B.L. Riggs, and C.A. Conover, Basal and regulated secretion of insulin-like growth factor binding proteins in osteoblast-like cells is cell line specific, *J. Clin. Endo. Metab.* 75:228-233 (1992).

29. T.L. Chen, L.Y. Chang, R.L. Bates and A.J. Perlman, Dexamethasone and 1,25-dihydroxyvitamin D3 modulation of insulin-like growth factor-binding proteins in rat osteoblast-like cell cultures, *Endocrinology* 128:73-80 (1991).

30. T.L. Chen, F. Liu, R.L. Bates and R.L. Hintz, Further characterization of insulin-like-growth factor binding proteins in rat osteoblast-like cell cultures: modulation by 17β-estradiol and human growth hormone, *Endocrinology* 128:2489-2496 (1991).

31. C. Schmid, I. Schläpfer, M. Waldvogel, J. Zapf, E. R. Froesch, Prostaglandin E$_2$ stimulates synthesis of insulin-like growth factor binding protein-3 in rat bone cells in vitro, *J. Bone Mineral Res.* 7:1157-1163 (1992)

32. D. LaTour, S. Mohan, T. A. Linkhart, D. J. Baylink, D. D. Strong, Inhibitory insulin-like growth factors-binding protein: cloning, complete sequence, and physiological regulation, *Mol. Endocrinol.* 4:1806-1814 (1990)

33. M. Ernst and G.A. Rodan, Increased activity of insulin-like growth factor (IGF) in osteoblastic cells in the presence of growth hormone (GH); positive correlation with the presence of the GH-induced IGF-binding protein BP-3, *Endocrinology* 127:807-814 (1990).

34. T. Moriwake, H. Tanaka, S. Kanzaki, J. Higuchi, and Y. Seino, 1,25-Dihydroxyvitamin D$_3$ stimulates the secretion of insulin-like growth factor binding protein 3 (IGFBP-3) by cultured human osteosarcoma cells, *Endocrinology* 130:1071-1073 (1992).

35. S.H. Scharla, D.D. Strong, S. Mohan, D.J. Baylink, and T.A. Linkhart, 1,25-Dihydroxyvitamin D$_3$ differentially regulates the production of insulin-like growth factor I (IGF-I) and IGF-binding protein-4 in mouse osteoblasts, *Endocrinology* 129:3139-3146 (1991)

36. M. Centrella, T.L. McCarthy, and E. Canalis, Receptors for insulin-like growth factors I and II in osteoblast-enriched cultures from fetal rat bone, *Endocrinology* 126:39-44 (1990).

37. M.P. Czech, Signal transmission by the insulin-like growth factors, *Cell* 59:235-238 (1989).

38. T.L. McCarthy, E. Canalis, and M. Centrella, Prostaglandin E$_2$ (PGE$_2$) enhances IGF binding in primary osteoblast-enriched (Ob) cell cultures, *J. Bone Min. Res.* 6:abstract 242 (1991).

39. T.L. McCarthy, M. Centrella, and E. Canalis, Alternate second signals mediate differential changes in insulin-like growth (IGF) binding in primary cultures of fetal rat bone cells, *J. Bone Min. Res.* 7:abstract 498 (1992).

40. T.L. McCarthy, E. Canalis, and M. Centrella, Cortisol differentially modulates insulin-like growth factor I and II binding on osteoblast-enriched fetal rat bone cells, *J. Bone Min. Res.* 5:abstract 258 (1990).

41. A. Bennett, T. Chen, D. Feldman, R.L. Hintz, and R.G. Rosenfeld, Characterization of insulin-like growth factor I receptors on cultured rat bone cells: regulation of receptor concentration by glucocorticoids, *Endocrinology* 115:1577-1583 (1984).

42. D. Edwall, R.T. Prisell, A. Levinovitz, E. Jennische, and G. Norstedt, Expression of insulin-like growth factor I messenger ribonucleic acid in regenerating bone after fracture: influence of indomethacin, *J. Bone Min. Res.* 7:207-213 (1992).

43. T.J. Wronski, C.-F. Yen, H. Qi, and L.M. Dann, Parathyroid hormone is more effective than estrogen or bisphosphonates for restoration of lost bone mass in ovariectomized rats, *Endocrinology* 132:823-831 (1993).

INDEX